Visual Symmetry

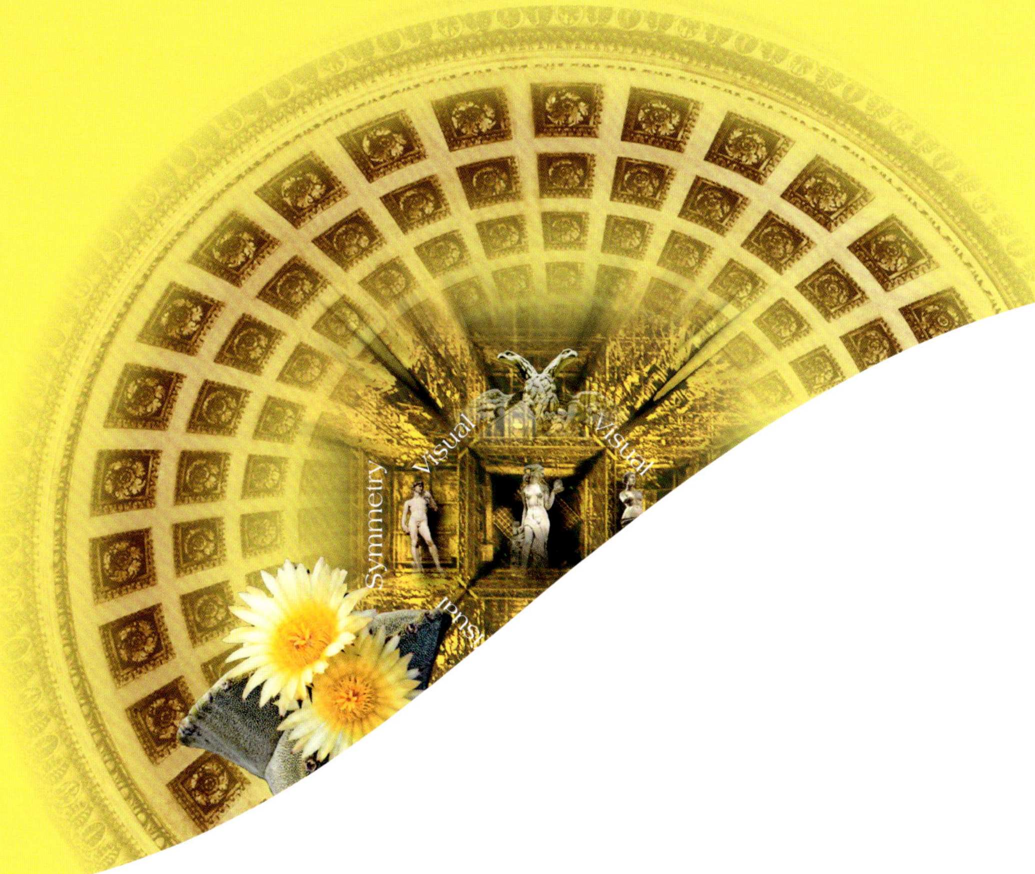

Visual Symmetry

Magdolna Hargittai
István Hargittai

Budapest University of Technology and Economics, Hungary

World Scientific

NEW JERSEY · LONDON · SINGAPORE · BEIJING · SHANGHAI · HONG KONG · TAIPEI · CHENNAI

For our children and grandchildren

Published by
World Scientific Publishing Co. Pte. Ltd.
5 Toh Tuck Link, Singapore 596224
USA office: 27 Warren Street, Suite 401-402, Hackensack, NJ 07601
UK office: 57 Shelton Street, Covent Garden, London WC2H 9HE

Library of Congress Cataloging-in-Publication Data

Hargittai, Magdolna.
Visual symmetry / by Magdolna Hargittai, István Hargittai.
p. cm.
ISBN 978-981-283-531-4 (alk. paper)
1. Symmetry--Pictorial works. 2. Symmetry (Physics)--Pictorial works. I. Hargittai, István. II. Title.
Q172.5.S95H37513 2009
500--dc22

2009006808

Page layout and cover design by Ng Chin Choon.

Printed in Singapore by Mainland Press.

Preface

In creating this book we followed the Chinese adage according to which a picture is worth a thousand words. We believe that our interest in symmetry accompanies our lives; ever since as a baby we discover for the first time that we have two hands even though we may not realize the similarity and difference between our right and left hands right away. We live in a world of symmetries in which the absence of symmetry may be as interesting and often more important than its presence. Our reference point is usually symmetry, and being aware of it makes it easier to recognize its absence and various imperfections. We have been consciously interested in symmetry for about four decades and started collecting images of symmetry and its conspicuous absence from the time our children were born because at that time photography became part of our daily routine. Over the years, we have developed an awareness of symmetry that has helped us notice and recognize many beautiful and instructive patterns that we might have missed otherwise. The awakening of this awareness is what we hope to share with our readers.

Magdolna Hargittai and István Hargittai
Budapest, Summer 2008

Photographs by the authors unless indicated otherwise. For acknowledgements, see, p. 207.

Contents

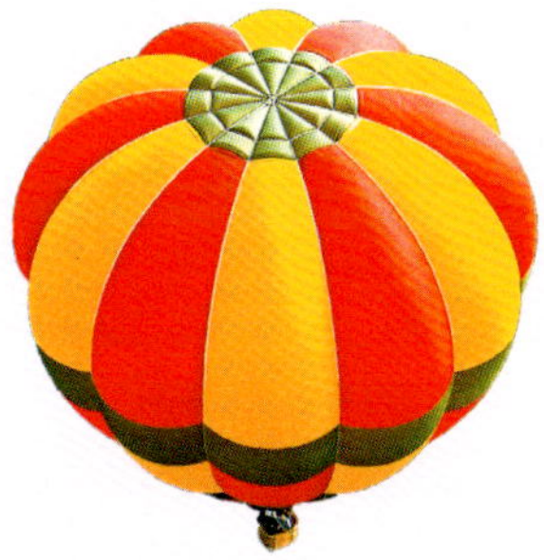

Introduction

When something is symmetrical, there is a one-to-one correspondence among its parts. The word comes from the Greek *symmetria*, meaning "the same measure." In most languages it appears similar to the Greek word.

The symmetry concept may be interpreted as rigorously geometrical, but may also be thought of as expressing harmony and proportion. In this book, we will have this more relaxed approach to symmetry.

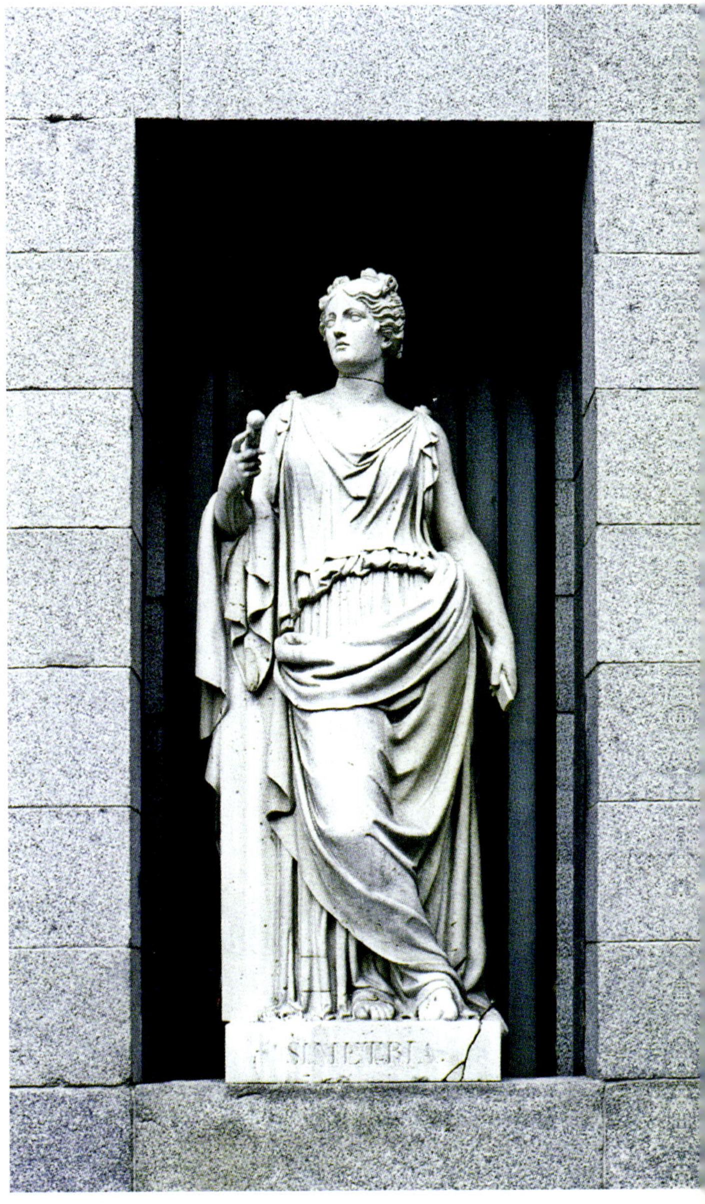

Simétria at the Prado in Madrid

Bilateral symmetry occurs when two halves of a whole are each other's mirror images. When we talk about "mirror image" we usually think of a whole object and its reflection in a mirror. A real mirror creates a perfect correspondence between an object and its image. However, we all recognize mirror images even if the reflection is not perfect, such as it may happen when an object is reflected in a water surface. This is what we see in the image from Seoul, Republic of Korea. There is a pagoda and its reflection. This picture may be taken as a symbolic expression of our approach to symmetry. The surface of the water is not perfectly flat because a mild breeze is rippling the water's surface slightly distorting the reflection of the pagoda, but there is no doubt that it is there. By a nice coincidence, there is yet another reflection of the pagoda in the drawing of the young artist at the lower right corner of the photograph. It is another less-than-perfect reflection, yet it is still a reflection.

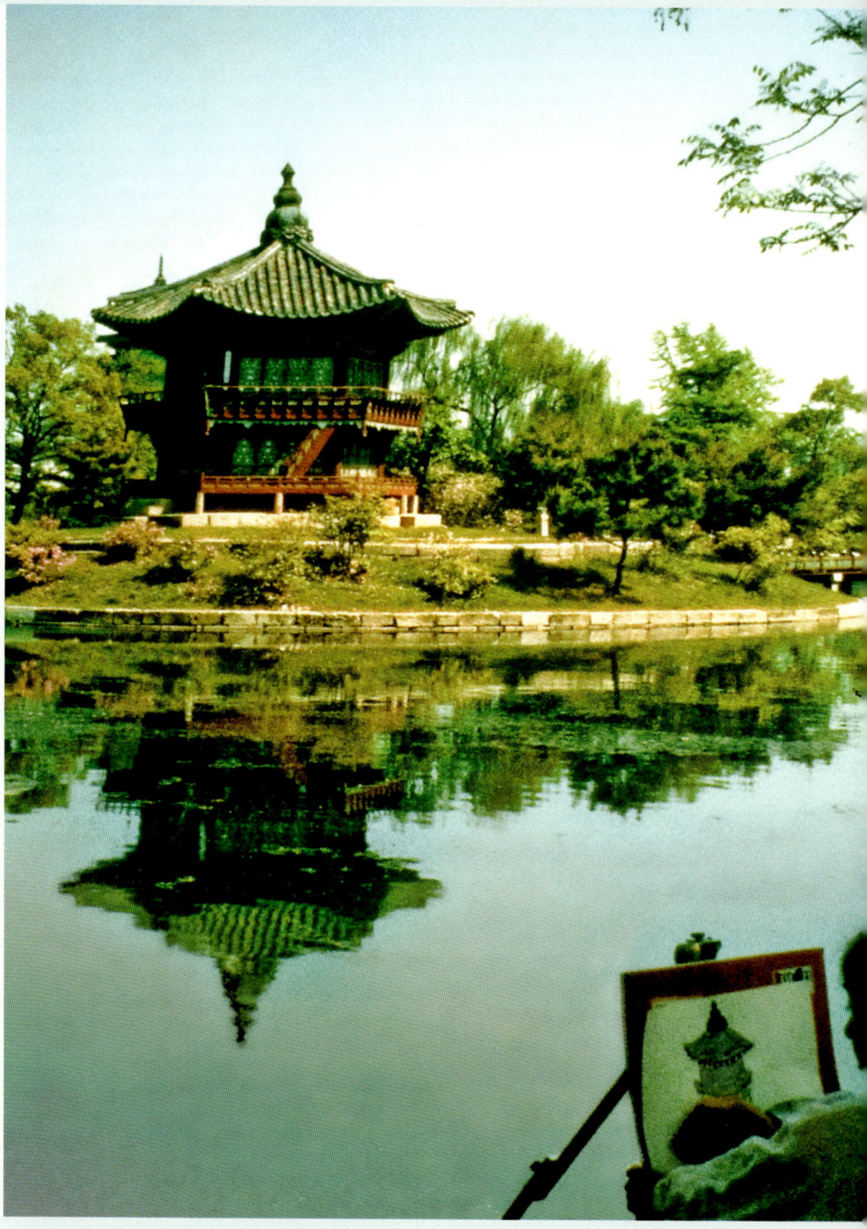

Pagoda in Seoul, Republic of Korea

The symmetry of the decoration above the entrance to a palace in Erice suggests harmony, yet there are slight discrepancies between the left and right sides in their original designs. In addition, the weather has left slightly differing marks on them during the centuries.

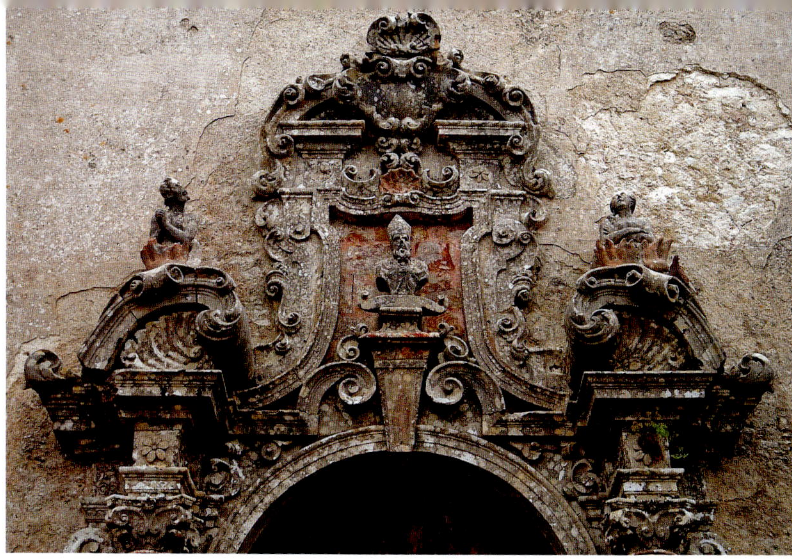

Ornaments, Erice, Sicily, Italy

The entrance to the Italian crypt radiates harmony yet its bilateral symmetry is far from perfect. The way the two angels keep their hands and legs is not the same; one is a male angel and the other is a female angel — still we perceive the picture as "bilaterally symmetrical". This is the attitude to symmetry that we are going to follow throughout this book.

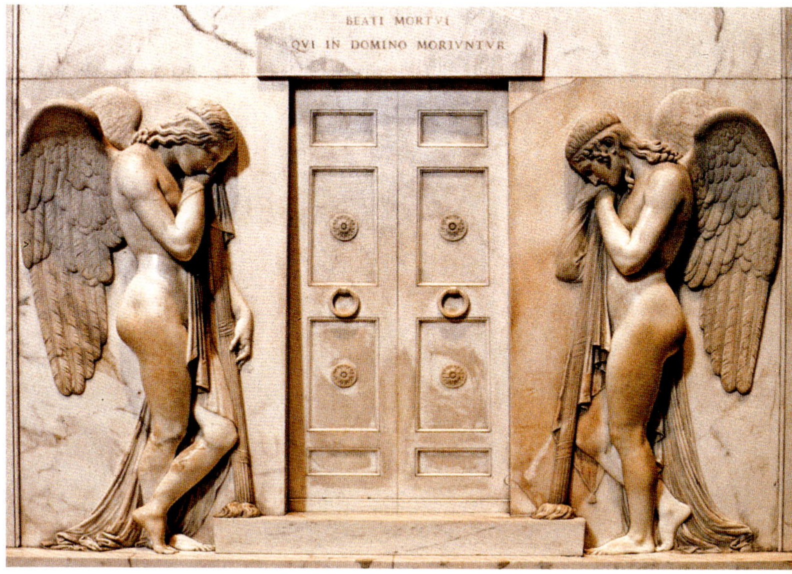

Crypt, Italy

Another example of bilateral symmetry with many deviations from it in the small details.

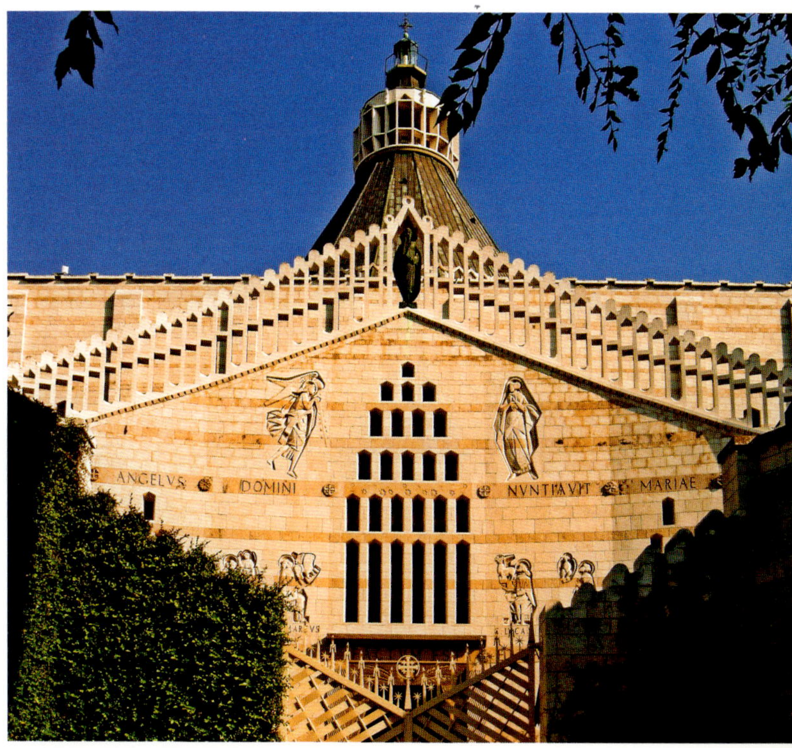

The Church of Annunciation in Nazareth, Israel

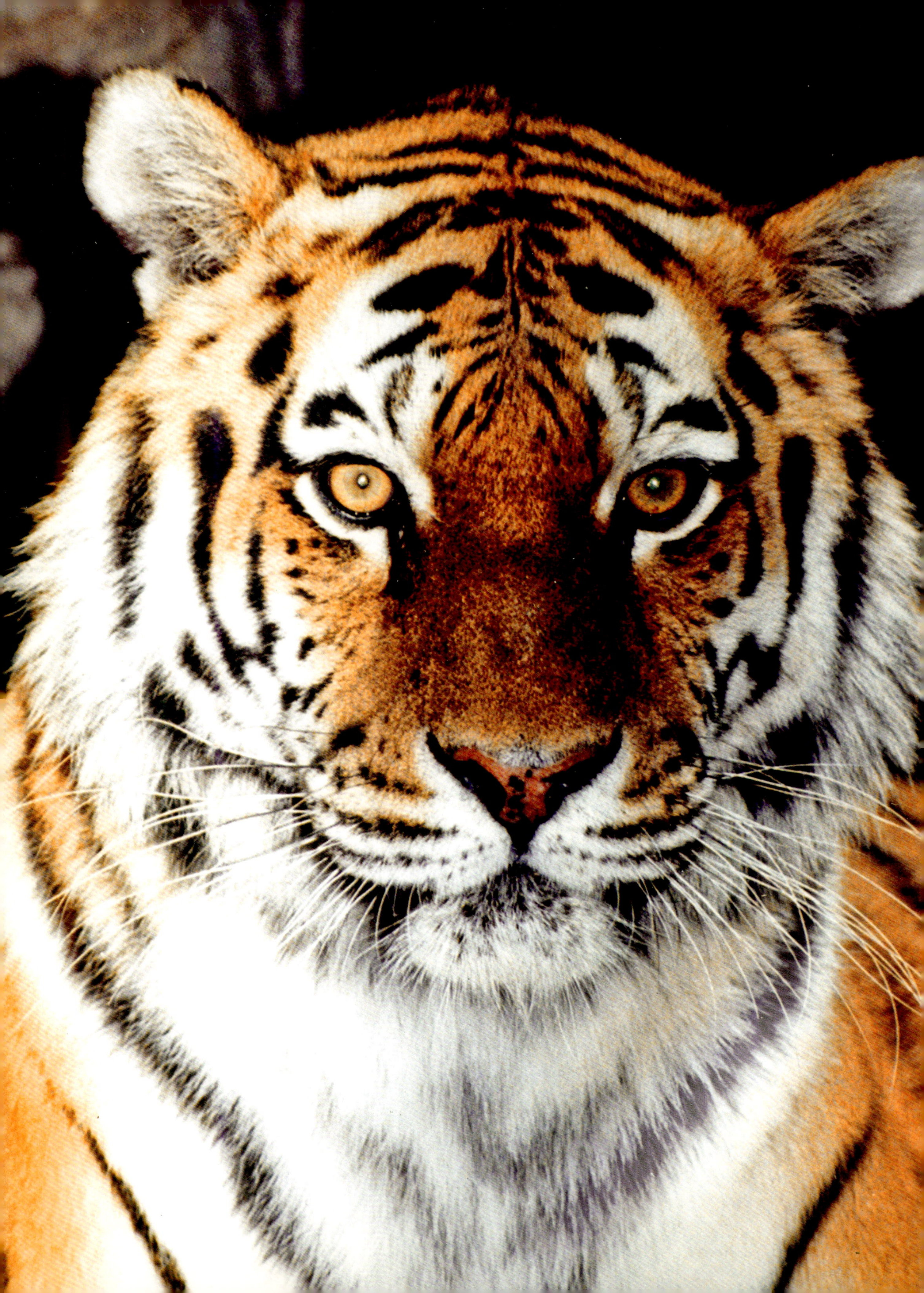

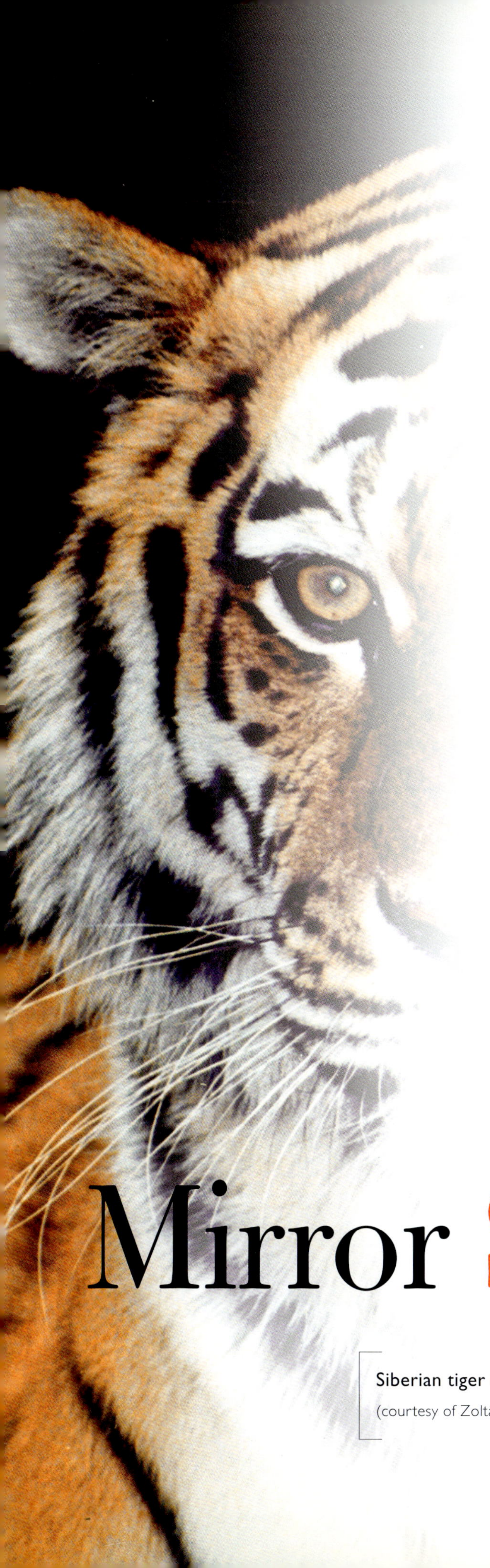

William Blake (1757–1827) produced a powerful expression of symmetry in his poem *The Tyger.*

Tyger! Tyger! burning bright
In the forests of the night,
What immortal hand or eye
Could frame thy fearful symmetry?

"Fearful symmetry" alludes to something in addition to harmony and proportion mentioned above. It may be taken as perfect symmetry being alien from life.

Mirror Symmetry

Siberian tiger in the Budapest Zoo
(courtesy of Zoltán Bagosi, Budapest)

Floral Kingdom

Bilateral symmetry was already mentioned in the Introduction. Often it is simply called *mirror symmetry*. This is the most common occurrence of symmetry — so much so that this is what most of us think of when the word "symmetry" is mentioned. It appears in nature as well as in man-made objects.

The symmetries of plants are diverse. Some flowers, orchids among them, display bilateral symmetry conspicuously. Over 25,000 kinds of orchids have been described and they all exhibit mirror symmetry. Doubling the left-hand side and the right-hand side of an orchid shows its near-perfect bilateral or mirror symmetry. The human face has bilateral symmetry, too, but by far not as perfect as the symmetry of orchids, as we will soon see.

Orchids from Oahu, Hawaii

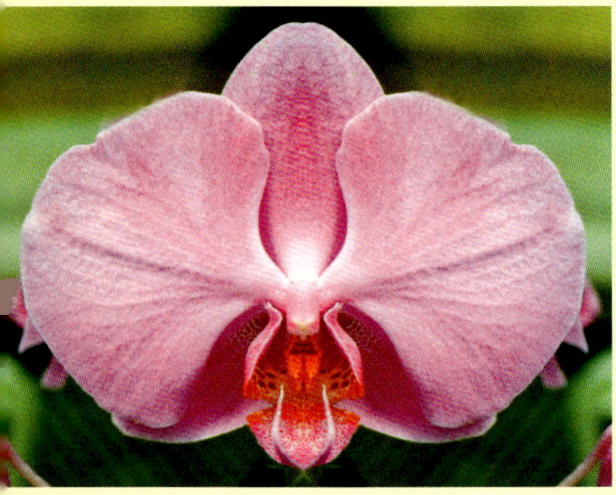

The left half of the orchid and its mirror image

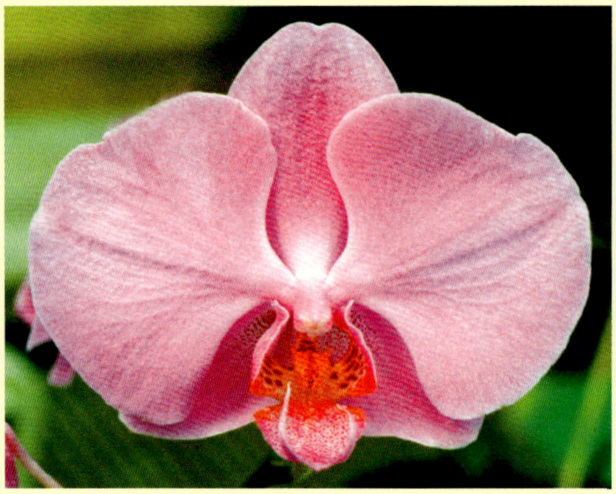

The orchid

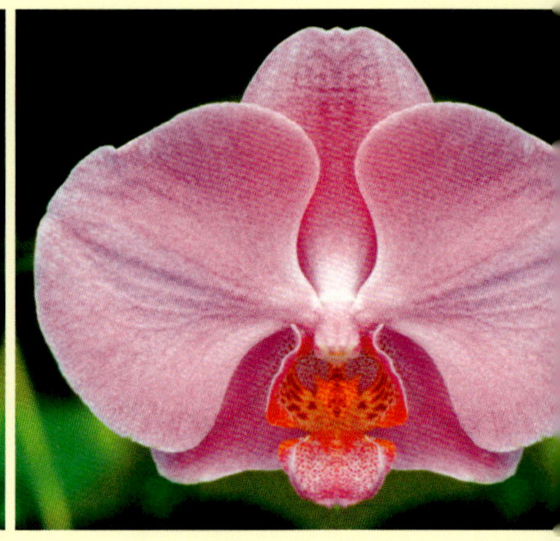

The right half of the orchid and its mirror image

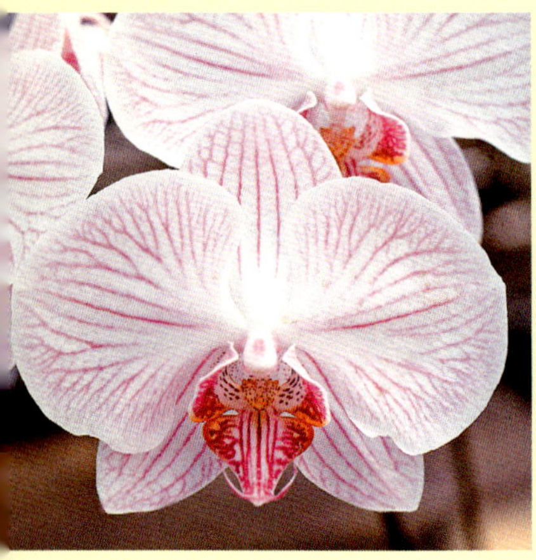

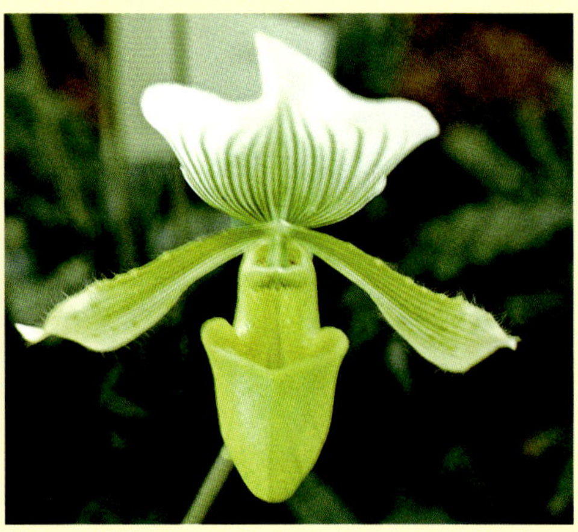

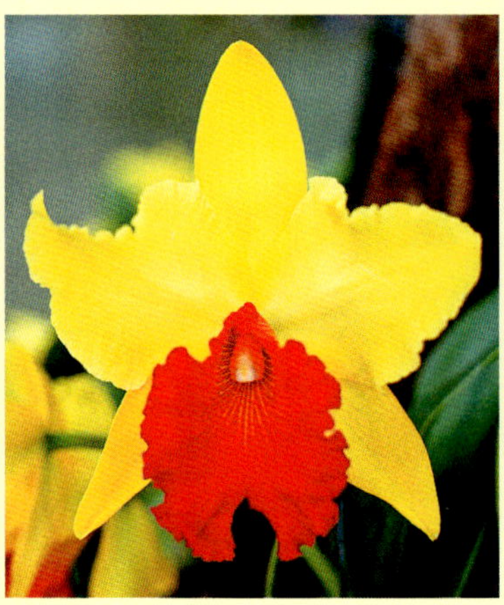

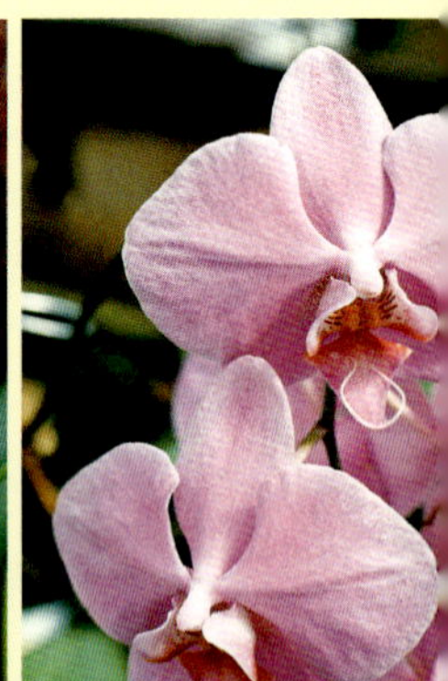

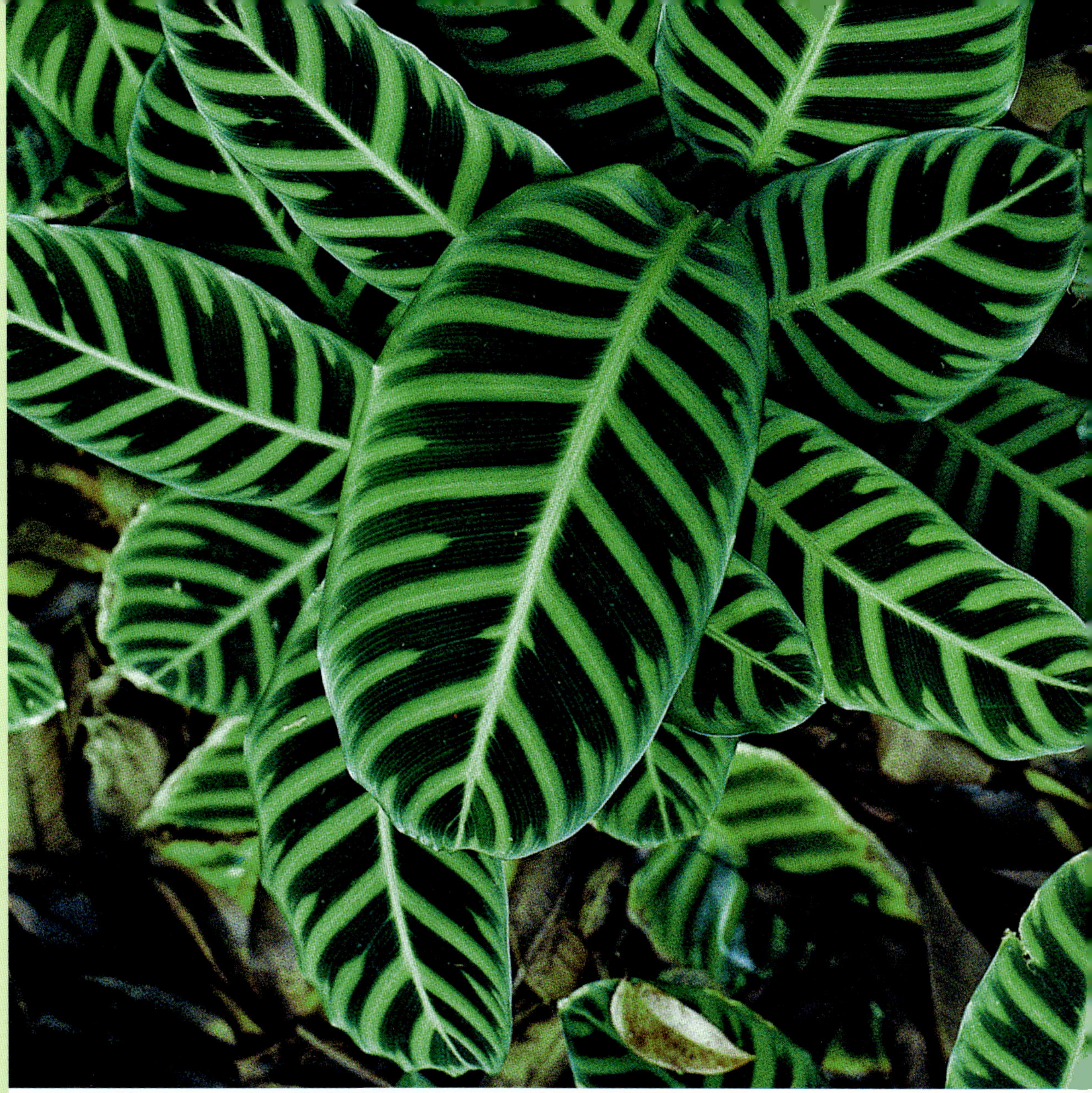

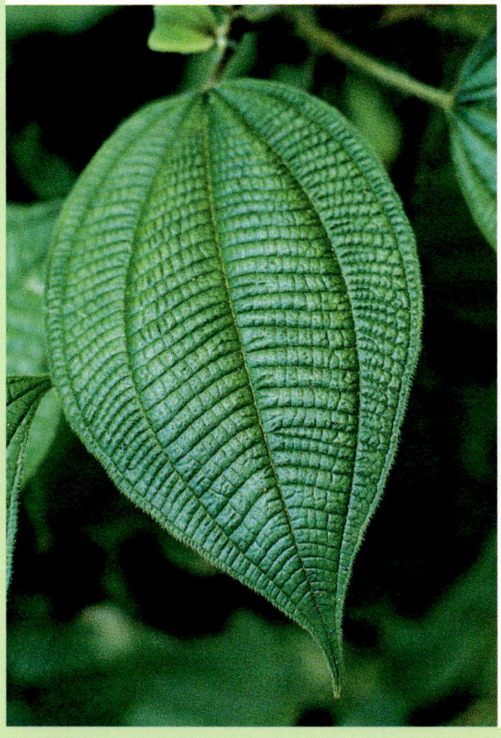

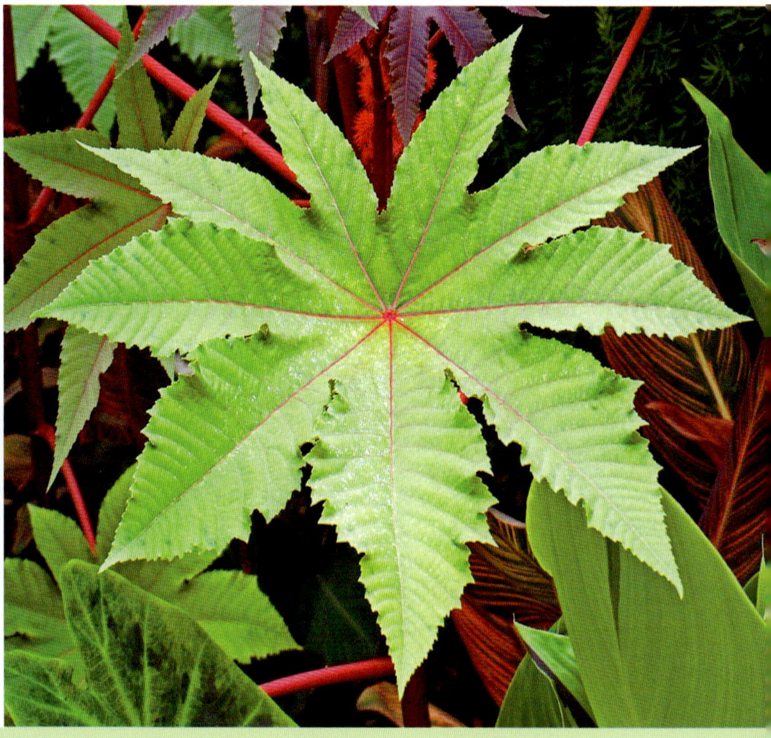

Most leaves have mirror symmetry. Our examples come from different parts of the world.

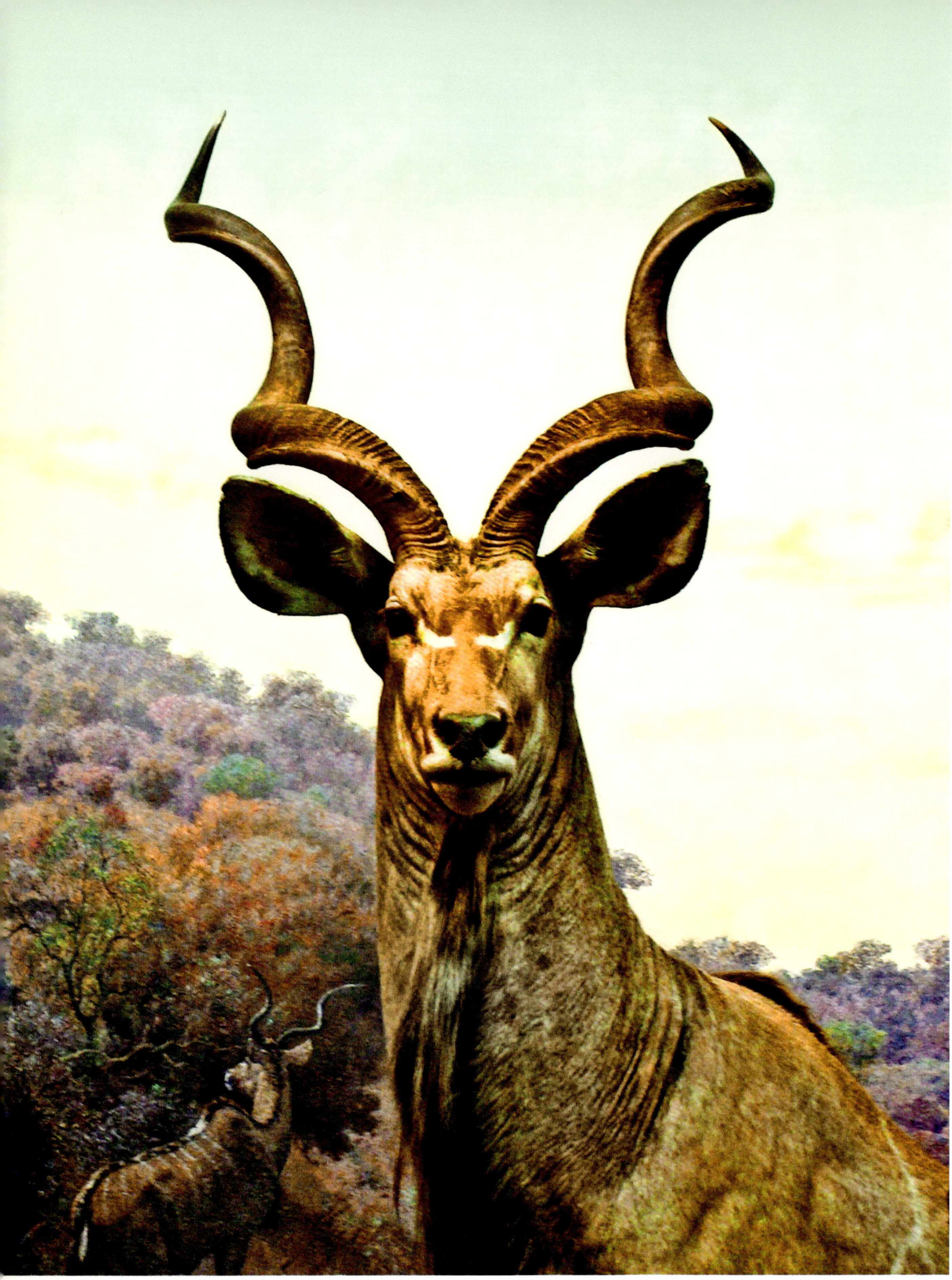

Greater Koodoo (*Tragelaphus strepsiceros*, a native to Tanzania) diorama at the American Museum of Natural History

(photograph by M. Hargittai, 2003; used with permission from the American Museum of Natural History, New York)

Animal Kingdom

Most animals have bilateral symmetry. Some striking examples are shown in these two pages. Three of the images (Cape buffalo (*Syncerus caffer*) — 9-3; Korean stock owl (*Bubo bubo*) — 9-4 and Gibbon (*Hylobates*) — 9-7) depict animals in the Budapest Zoo.

(courtesy of Zoltán Bagosi)

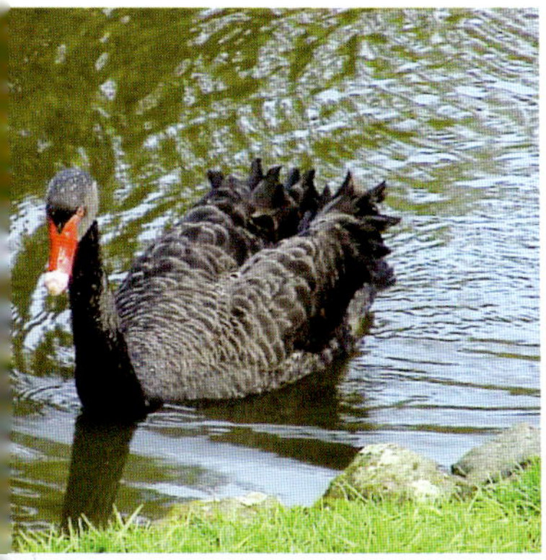

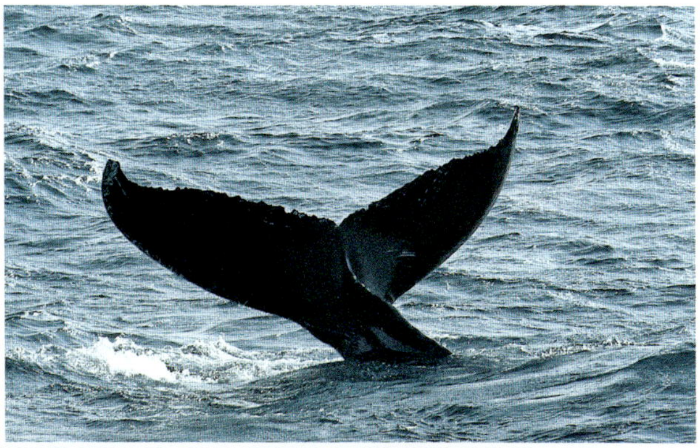

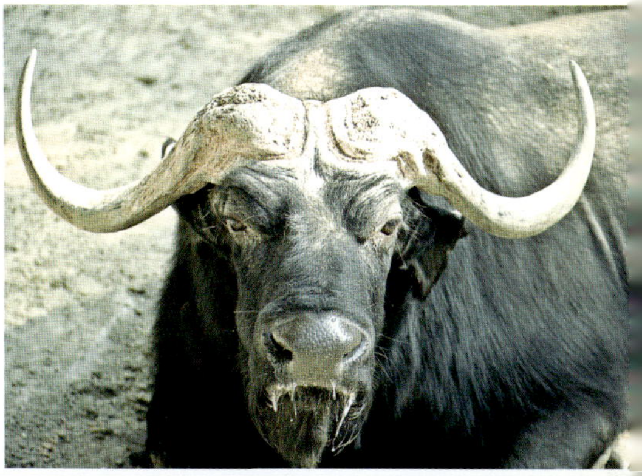

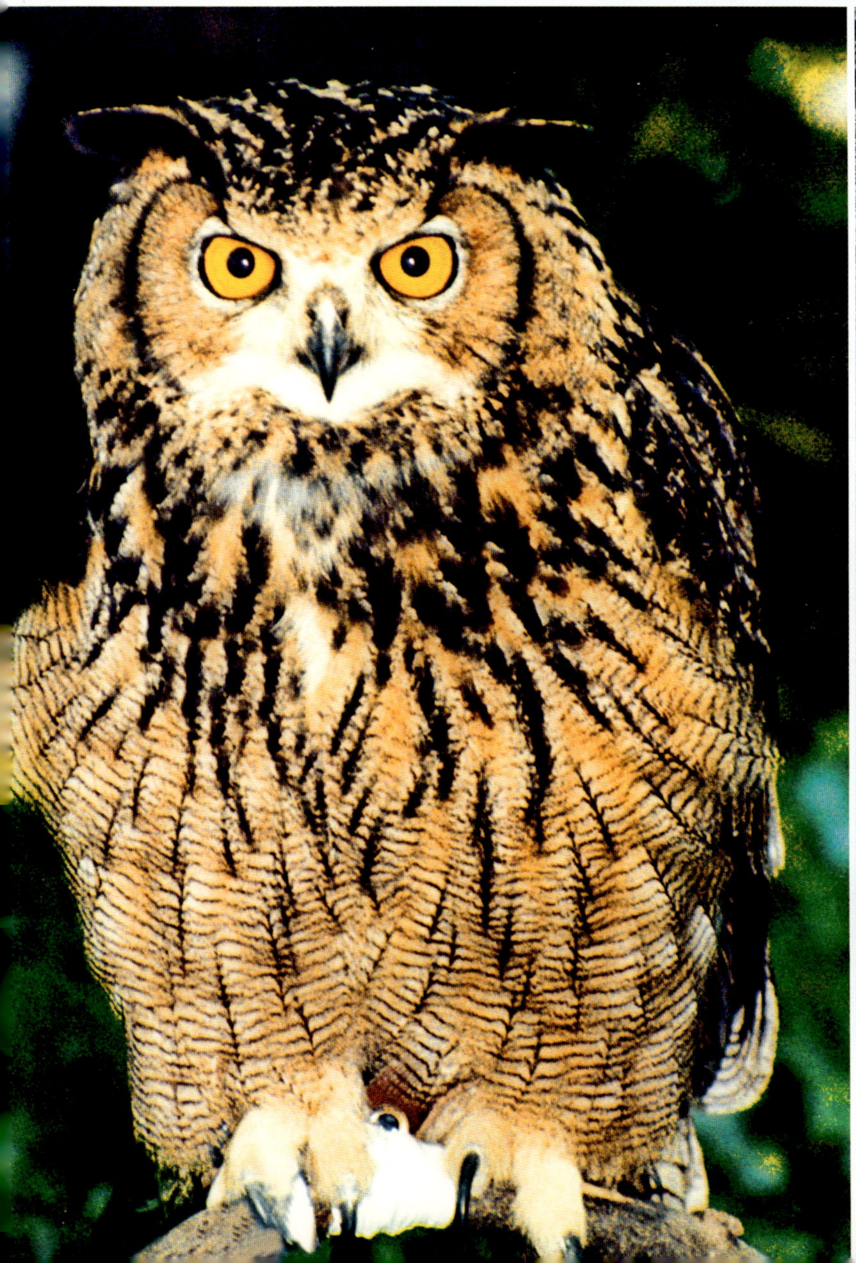

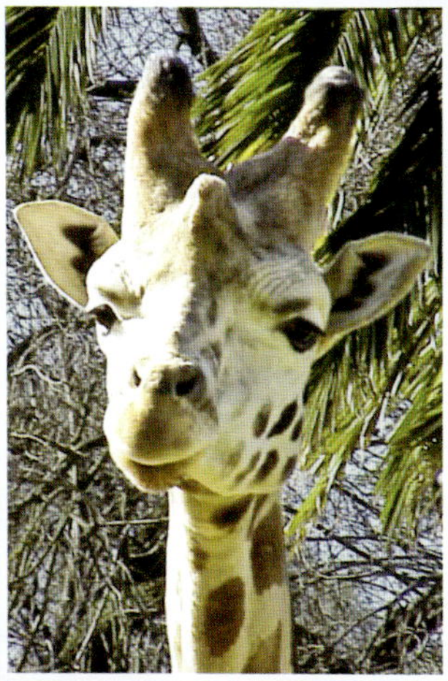

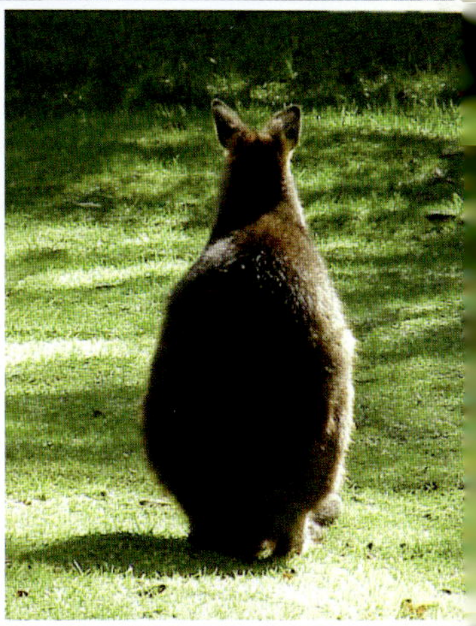

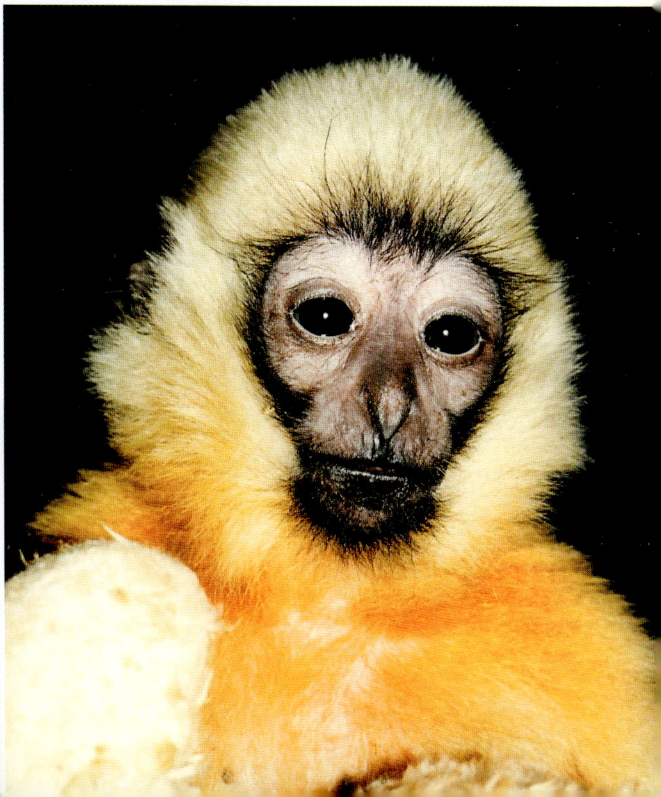

Human Body

The human body has bilateral symmetry. However, it is only in its external appearance and even then it is approximate rather than fulfilled with geometrical rigor. Motion especially diminishes the impression of bilateral symmetry. There are considerable discrepancies from bilateral symmetry in the interior of the human body. The most conspicuous is perhaps that the heart is on the left-hand side in most people.

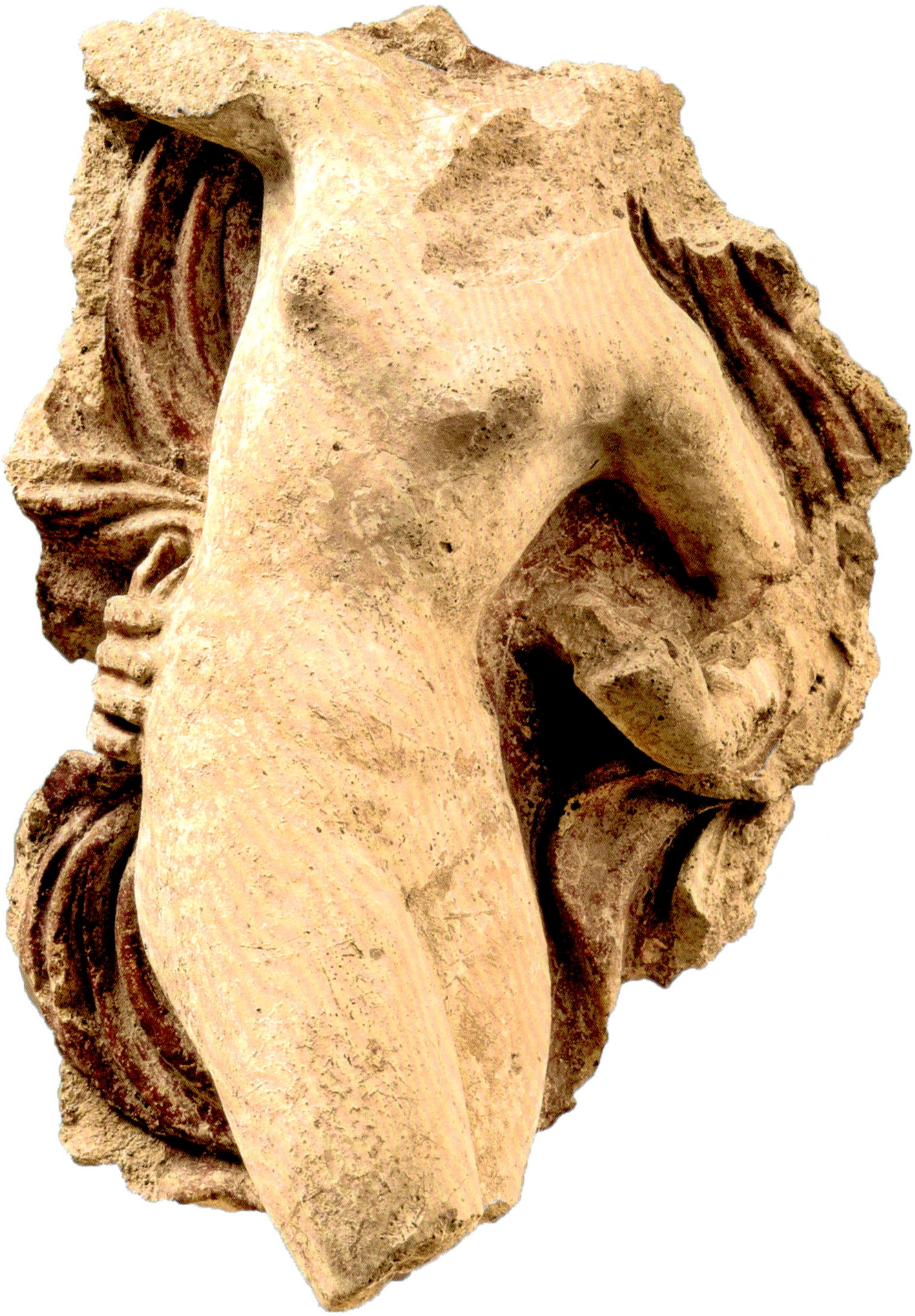

Torso in an Etruscan sculpture from about 350–328 BCE in the Antique Collection of the Museum of Fine Arts in Budapest

(courtesy of the Museum of Fine Arts, Budapest, Hungary)

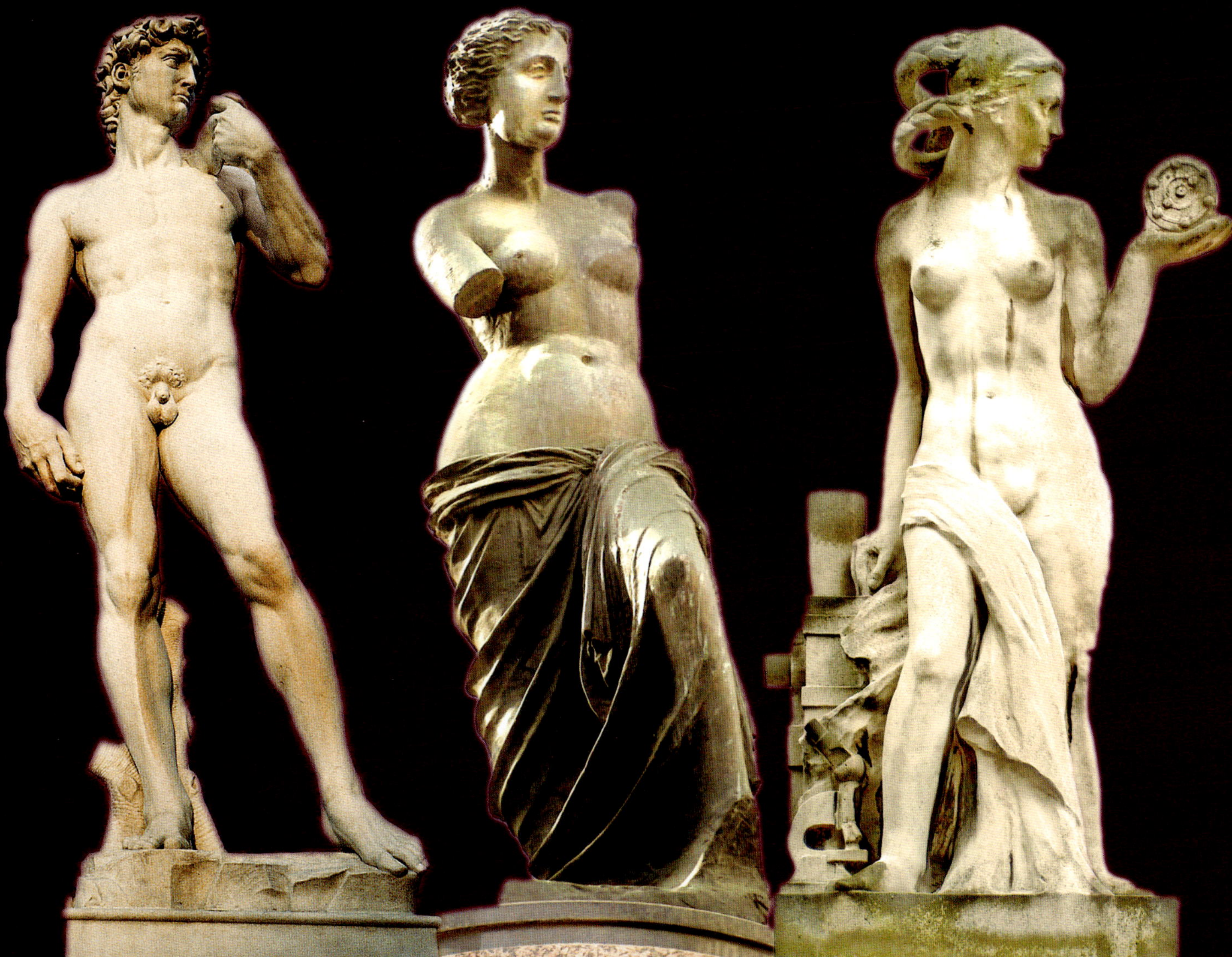

From left to right:

David by Michelangelo in Florence, Italy

Woman sculpture symbolizing an antique-shop chain in Budapest

Woman sculpture in Paris

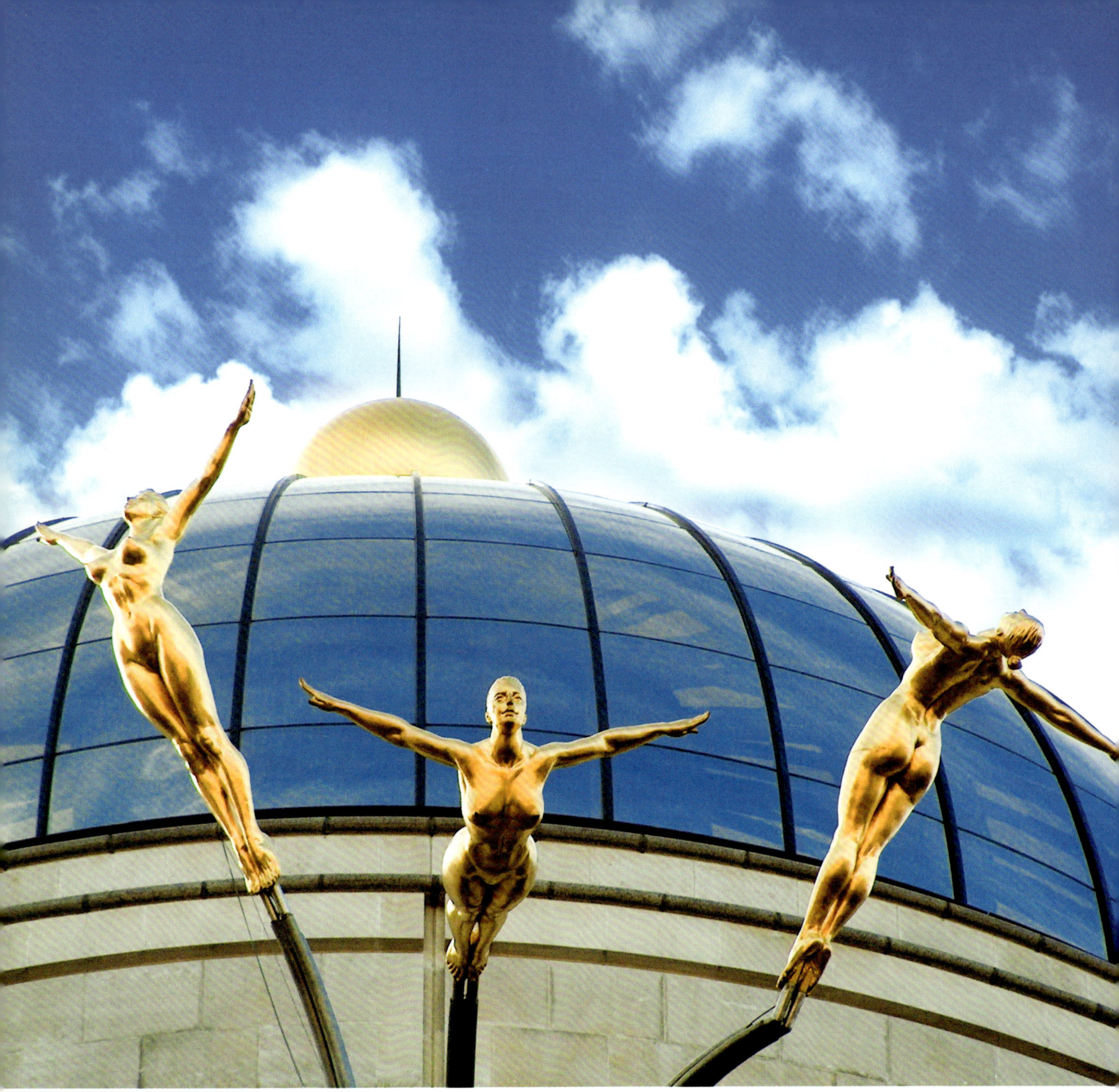

Even though motion may diminish the apparent symmetry of the human body, it also gives it emphasis when it is part of gymnastics or other sports and dancing. This is illustrated here by the sculptures at the top of a building in Piccadilly Circus in London, UK, and by postage stamps of different countries.

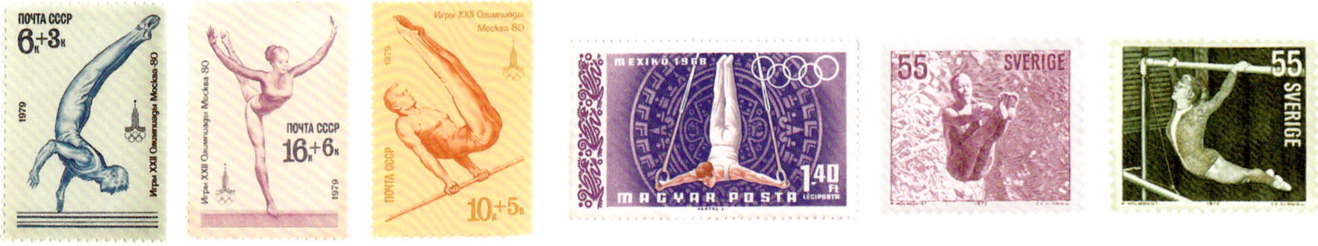

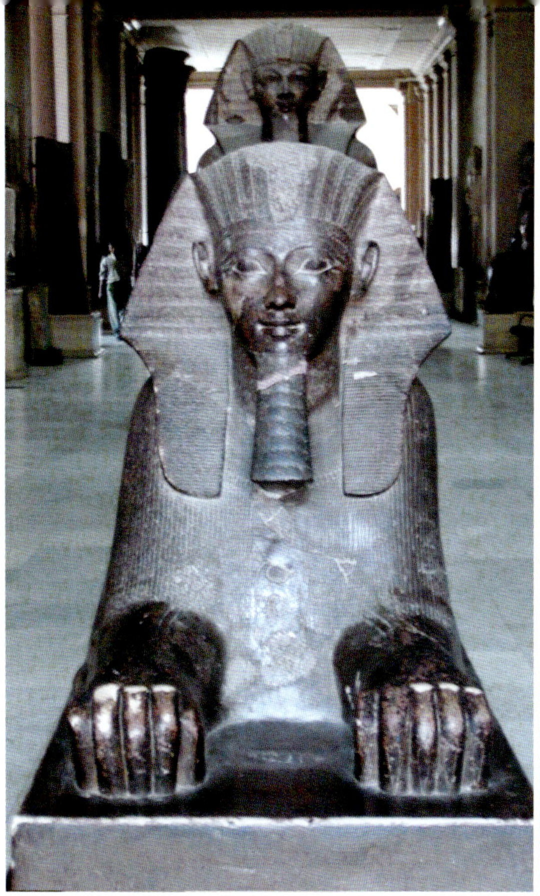

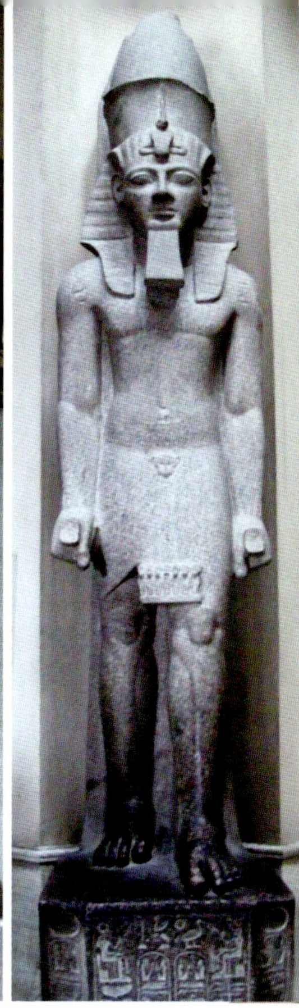

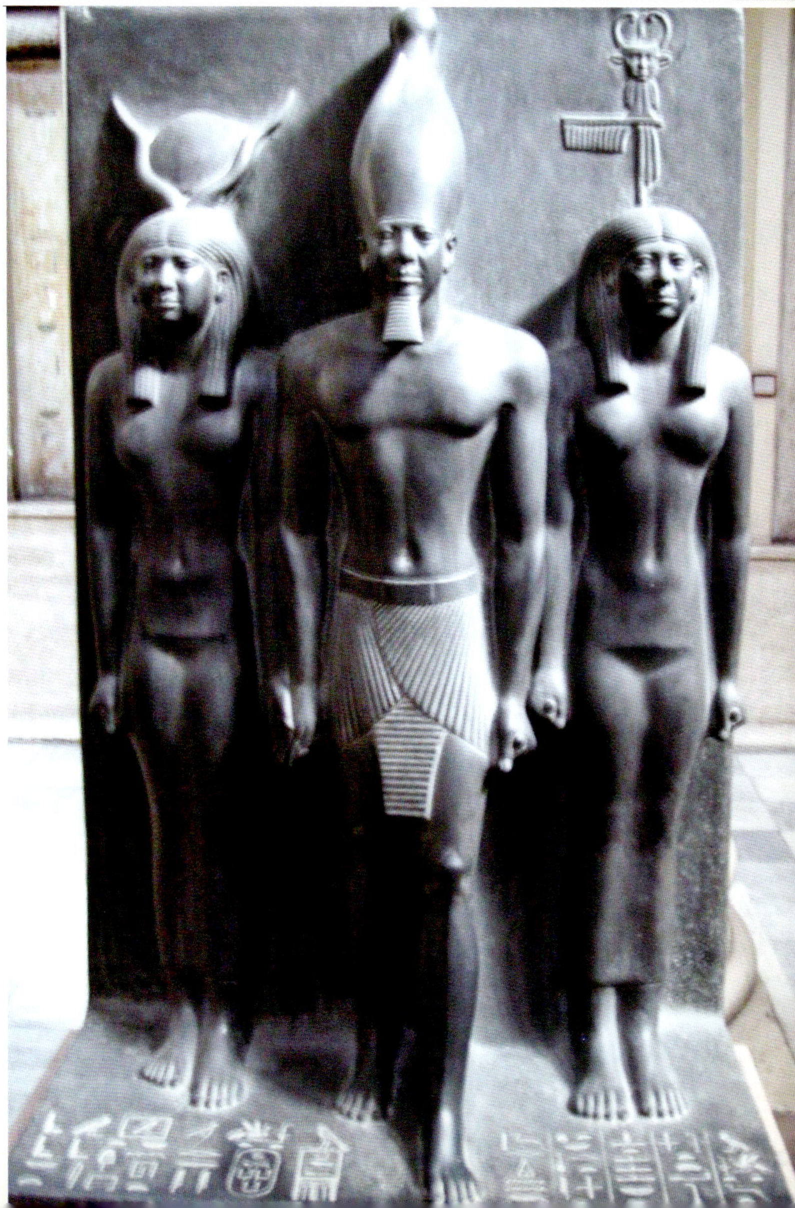

Antique Egyptian sculptures stress the symmetry of the motionless human body. Our examples come from the Egyptian Museum in Cairo, Egypt.

Clockwise:

Sphinx of Hatshepsut

Ramses II, XIX dynasty

King Mykerinos with the goddess Hathor and a local goddess

(courtesy of László Vámhidy, Pécs, Hungary)

Various cultures often emphasize the perfection of bilateral symmetry of the human body.

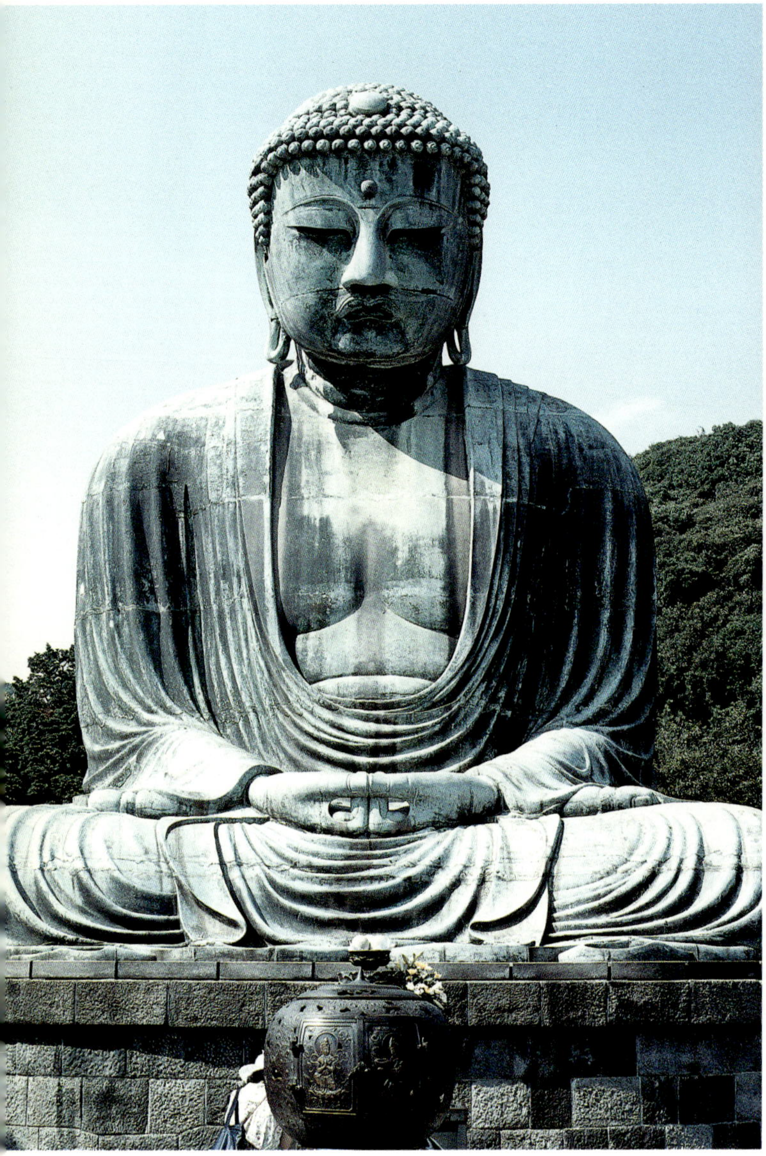

Buddha in Japan

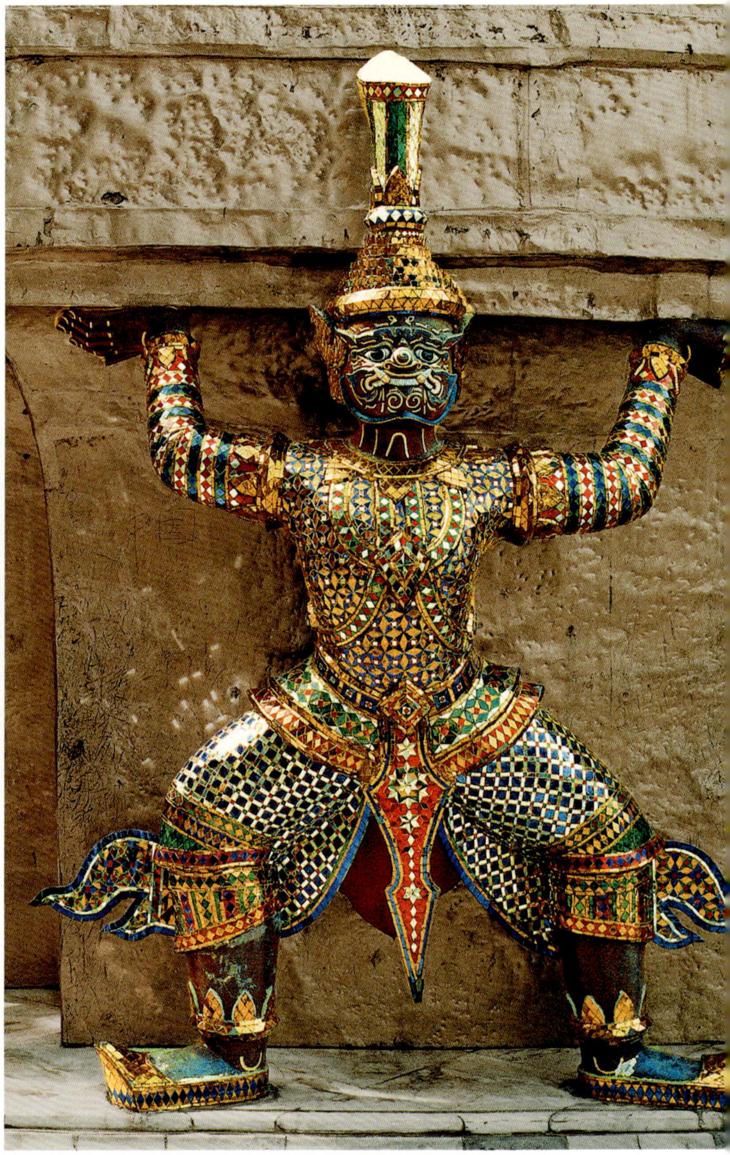

Decoration of the Royal Palace in Bangkok, Thailand

The perfection of bilateral symmetry is also often emphasized in depicting faces.

Guard in a temple in the Republic of Korea

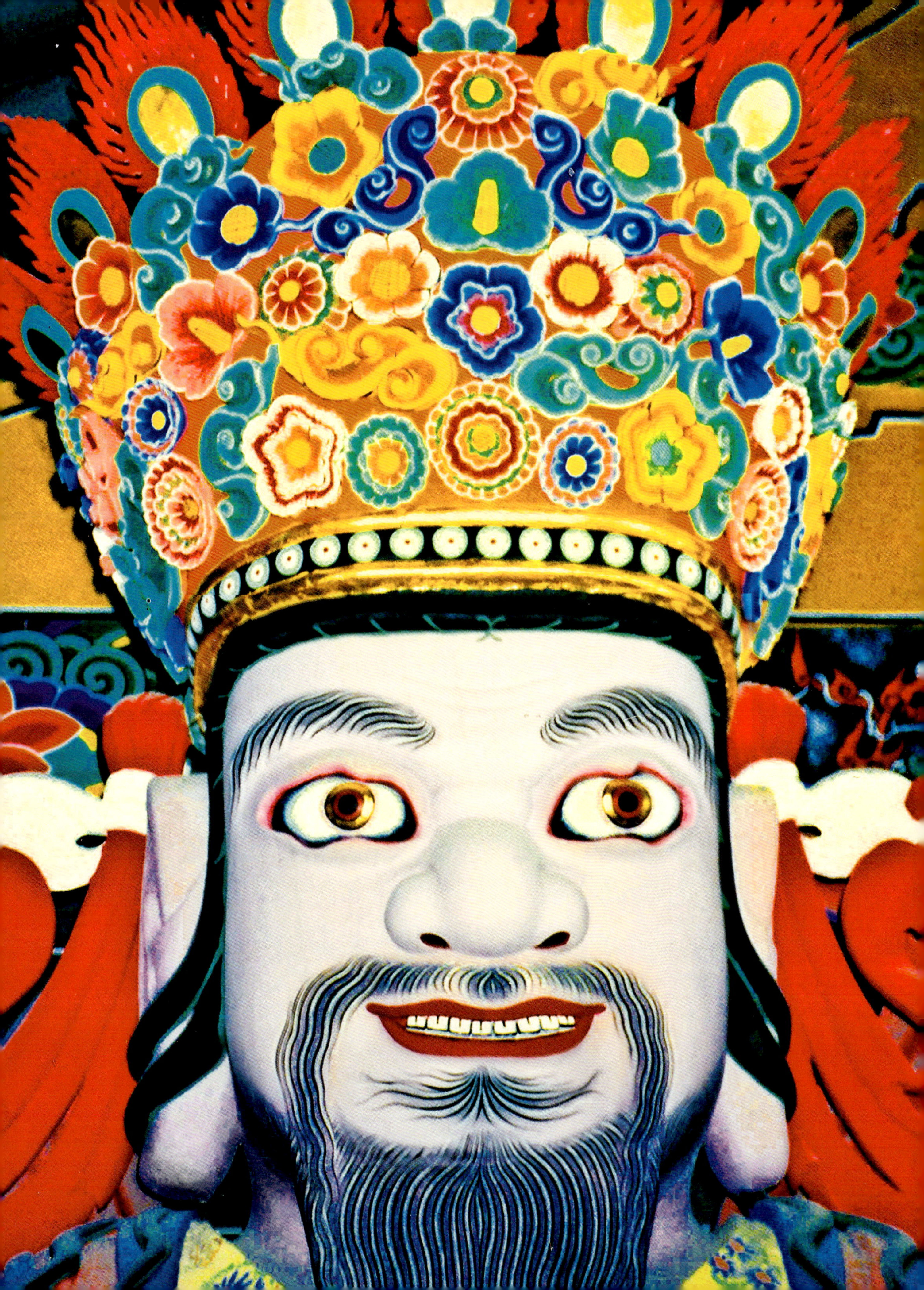

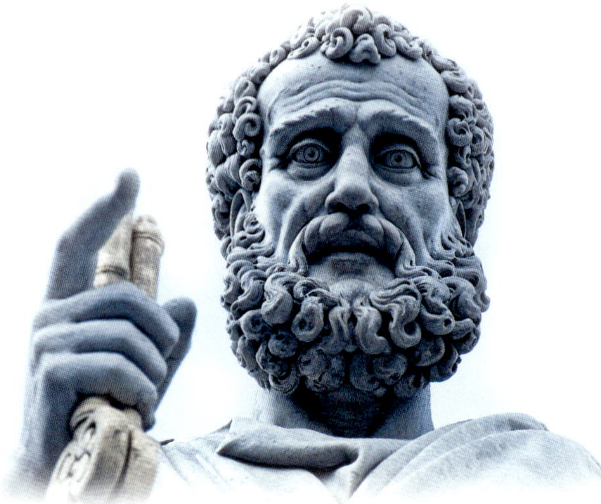

It has often been perceived that the perfect symmetry of the human face expresses "perfection" in general in the person. This is why the sculptures and busts of saints, politicians, or even great scientists show idealized symmetrical faces.

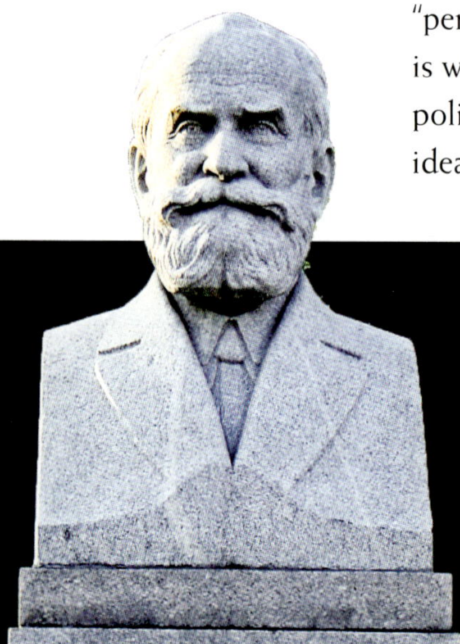

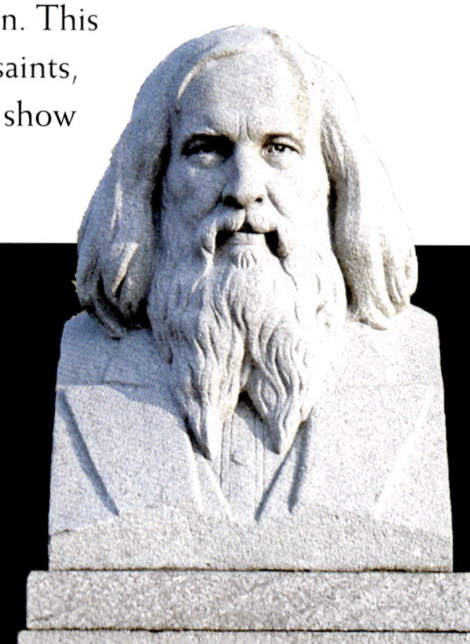

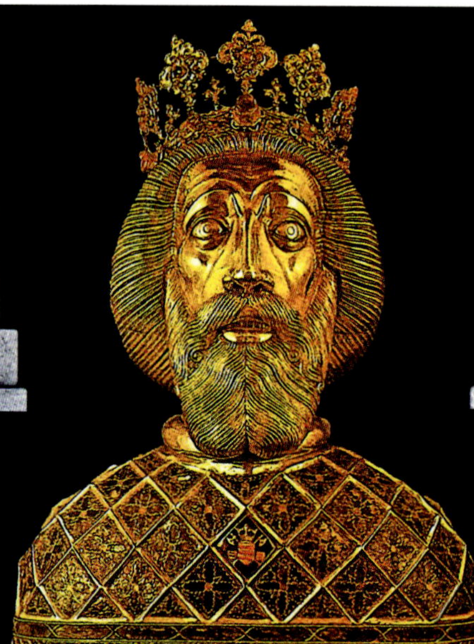

Top: St. Peter in St. Peter's Square, Vatican City

Middle: Russian scientists I. P. Pavlov and D. I. Mendeleev in Moscow

Bottom: St. Ladislaus, King of Hungary in Györ, Hungary

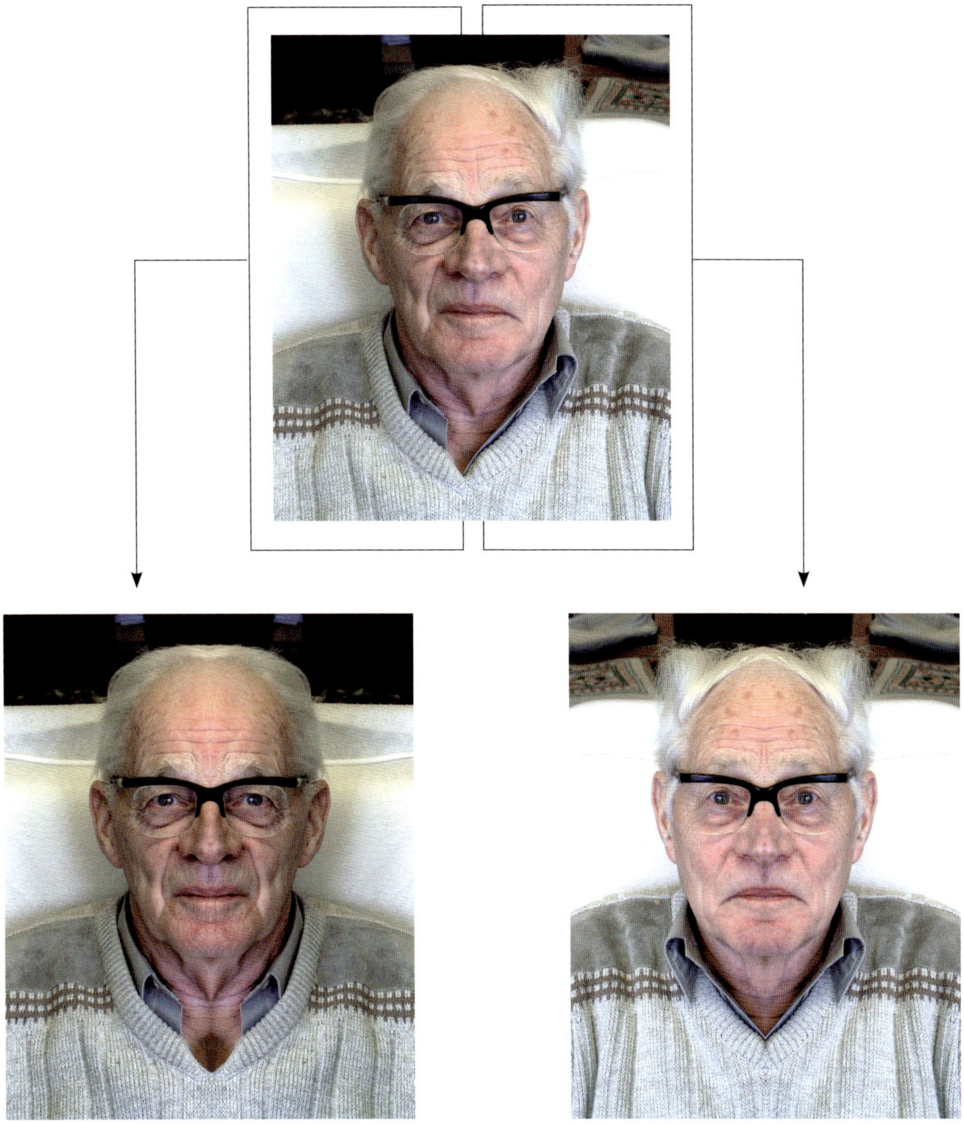

There have been speculations that the right side of the human face is more "public," while the left side is more "private." Others have argued that the right side is more "representative" of the whole face than the left side. There is much research going on about this question. It has been said that one of the American presidents allowed to photograph himself from one side only, fearing that the other side might reveal too much of his feelings or personality.

Top: The real face of the British scientist Alan L. Mackay

Bottom left: The right side of his face with its own reflection

Bottom right: The left side of his face with its own reflection

The two pictures at the bottom are strikingly different, reflecting the differences between the right and left sides of Professor Mackay's face (with kind permission of Alan L. Mackay, London).

Some artists stress the mirror symmetry of the human face by displaying the reflection plane. Others give added emphasis to the differences between the two sides of the face.

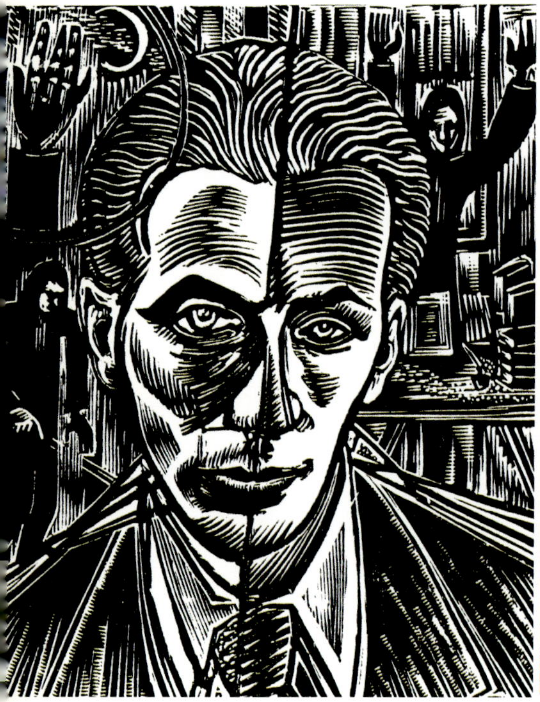

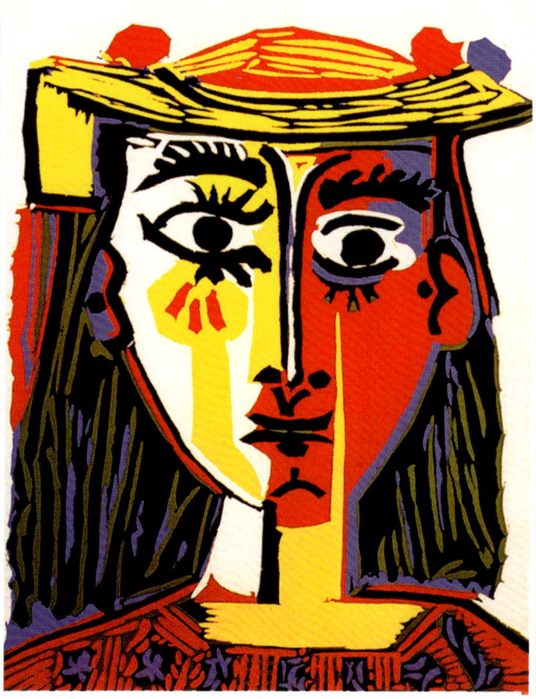

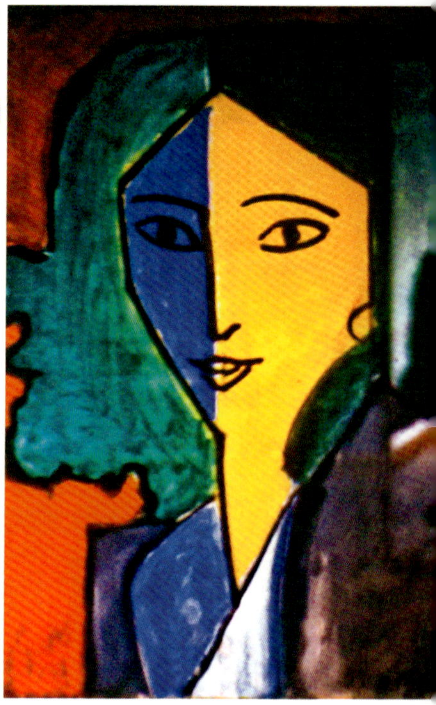

Left: György Buday, *Miklós Radnóti*
(courtesy of the late artist)

Middle: Pablo Picasso, *Woman with a hat*
(courtesy of the Museum of Fine Arts, Budapest)

Right: Henri Matisse, *Portrait of Lydia Delektorskaya*
(courtesy of The State Hermitage Museum, St. Petersburg)

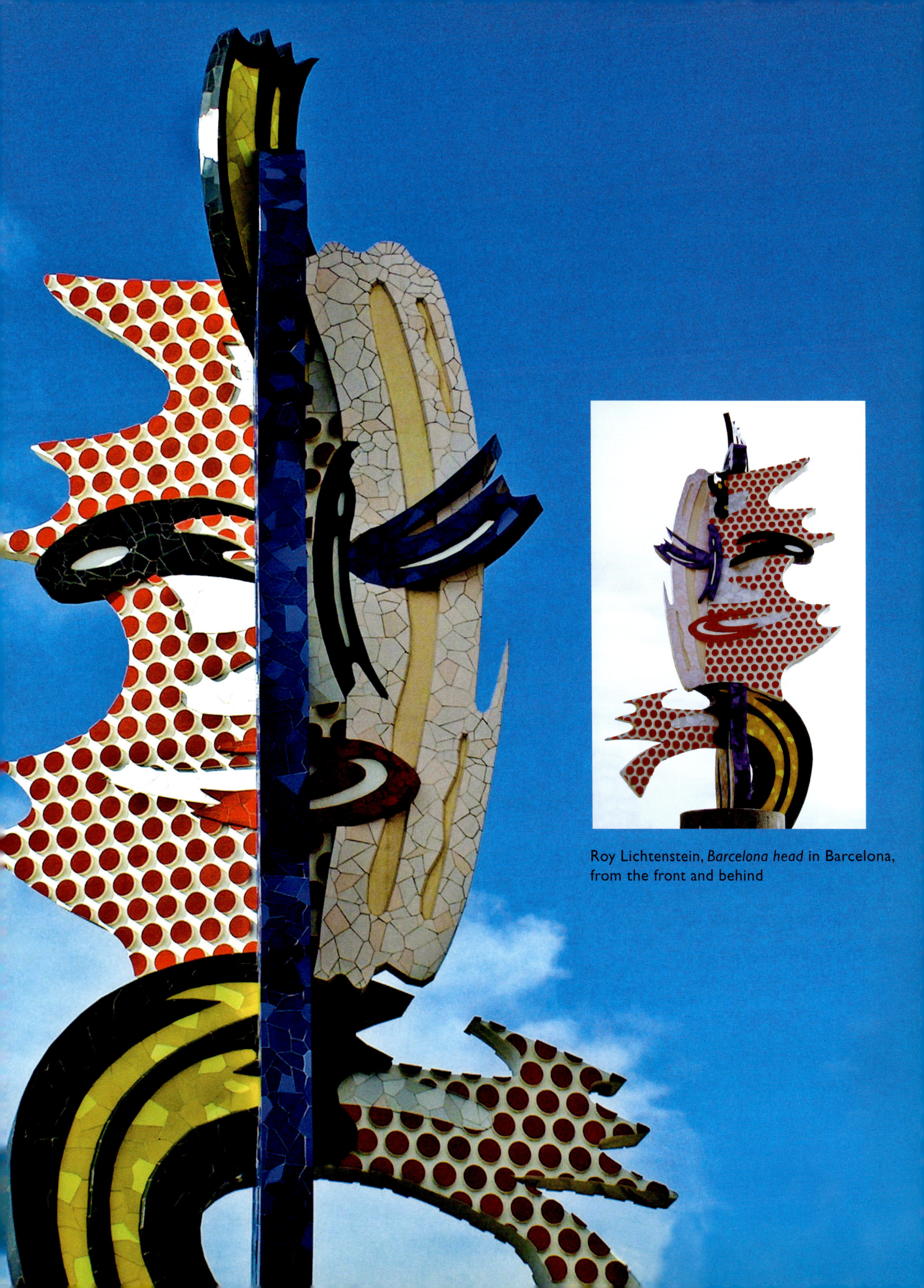

Roy Lichtenstein, *Barcelona head* in Barcelona, from the front and behind

Architecture

Bilateral symmetry often appears in human creations; architecture, design of squares, gardens, sculptures, and more.

Top left: Heroes' Square, Budapest
Top right: Gate of the Bahai Temple in Haifa, Israel
Bottom: Gate to the park next to the Buckingham Palace in London

(page 21, from the top):

Parliament building in Budapest
Moscow State Lomonosov University, Moscow
Schönbrunn Palace in Vienna

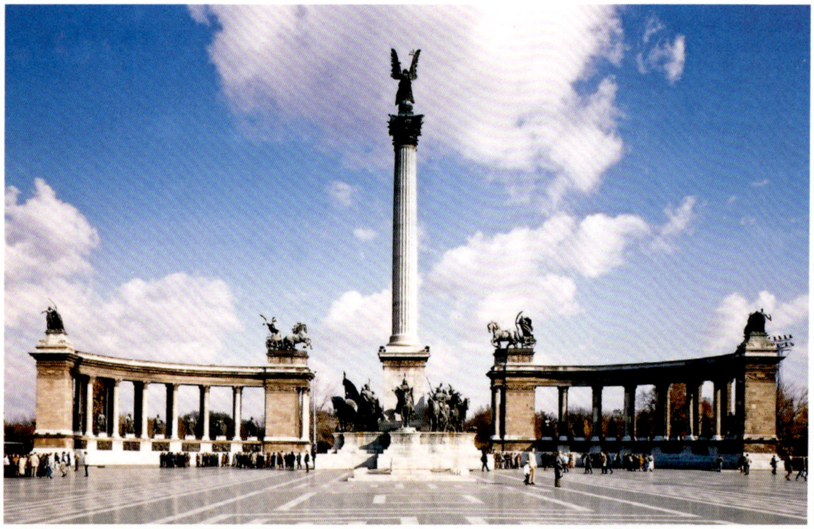

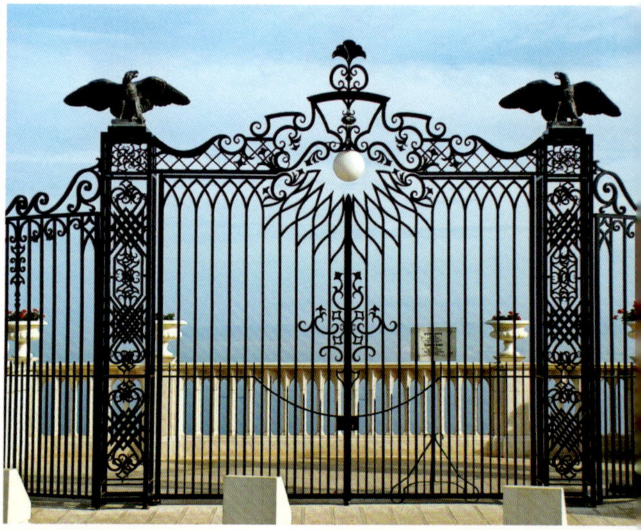

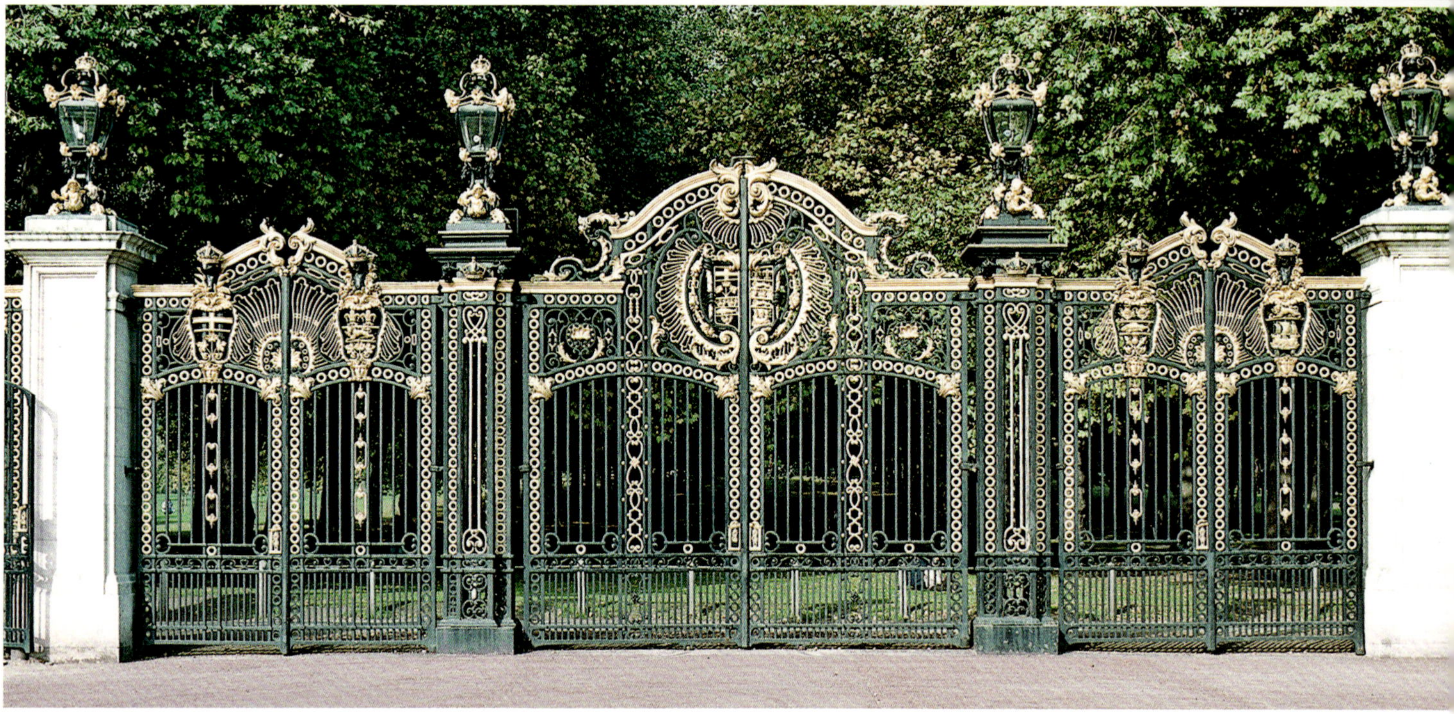

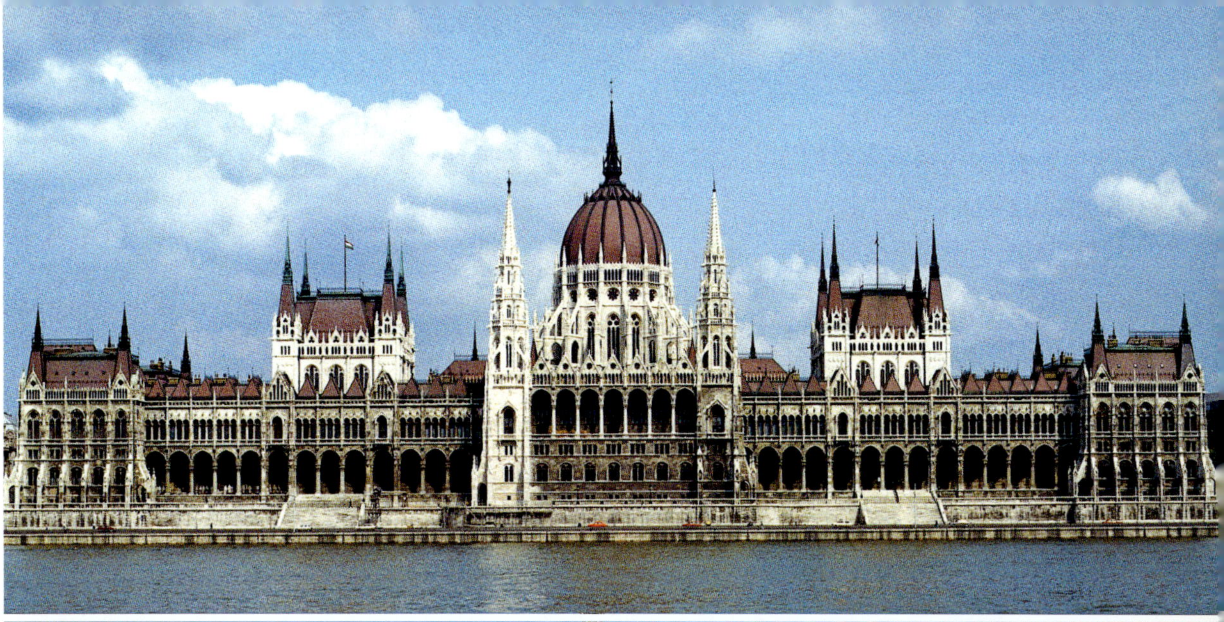

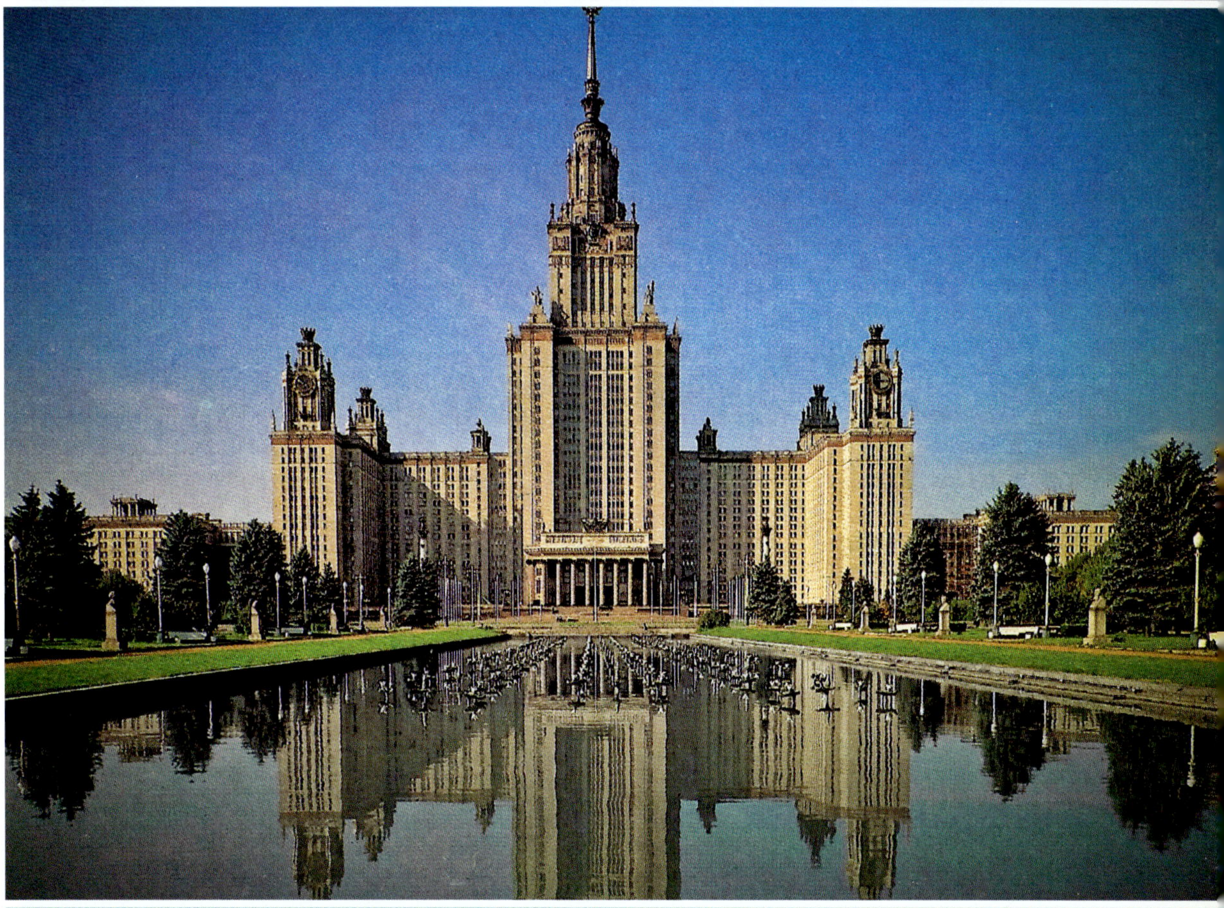

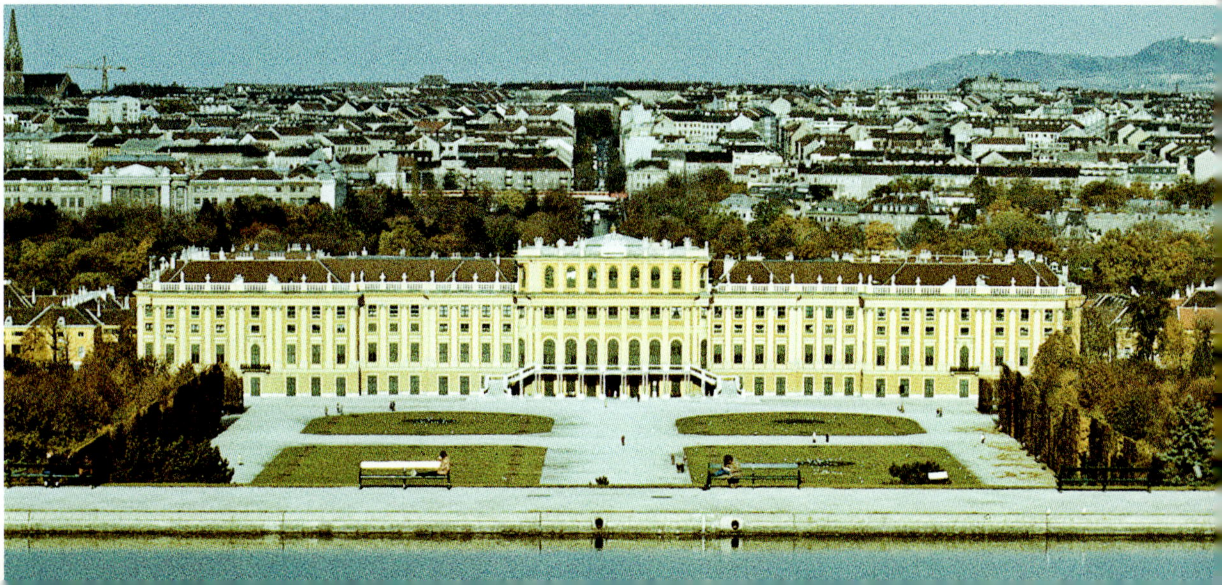

Top left: Notre Dame Cathedral, Paris
Top right: The Cathedral in Milan
Bottom left: Temple in Japan
Bottom right: St. Peter's Square in Vatican City

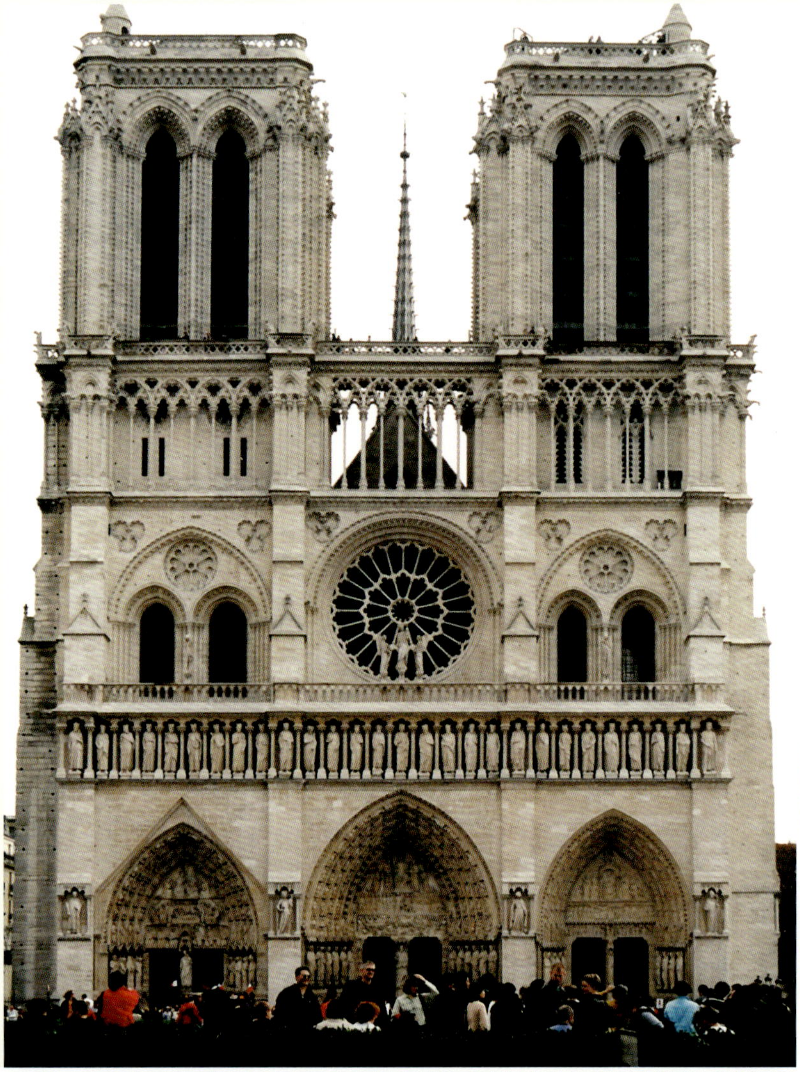

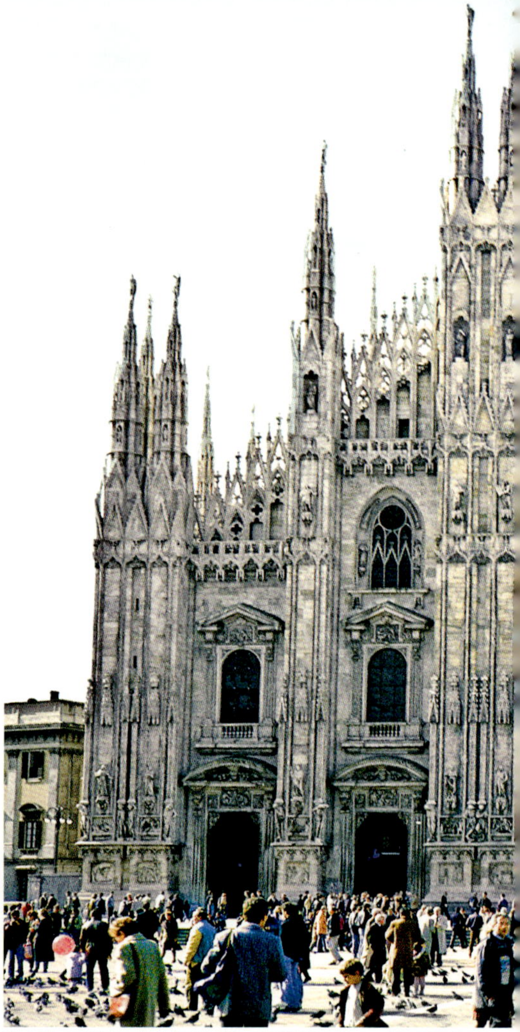

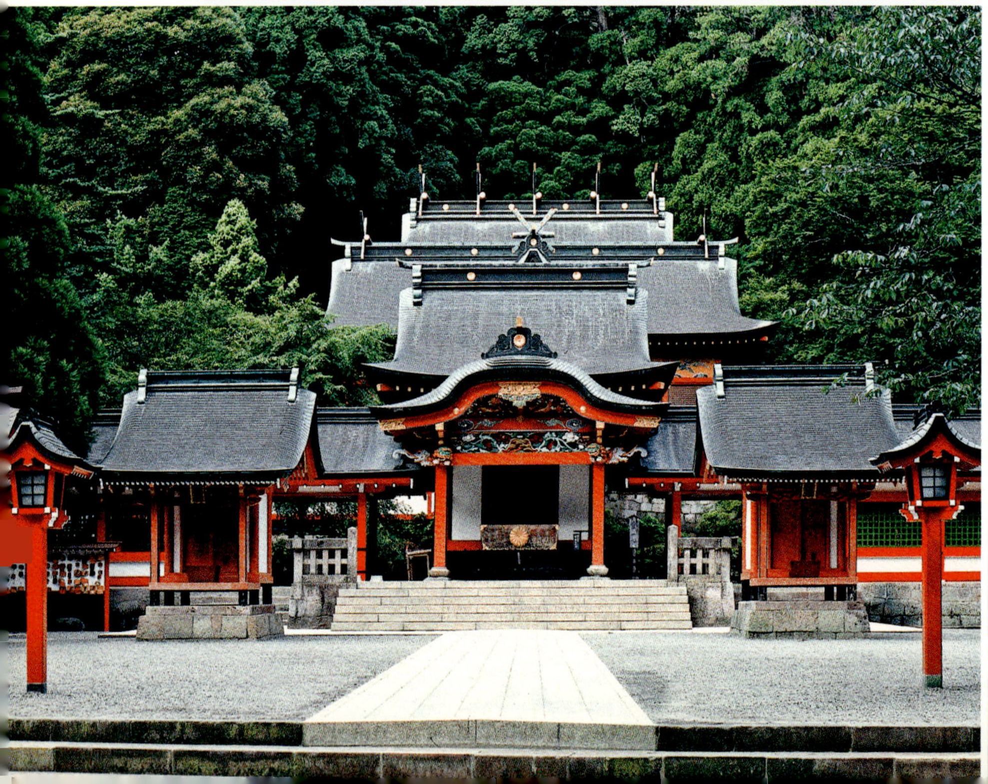

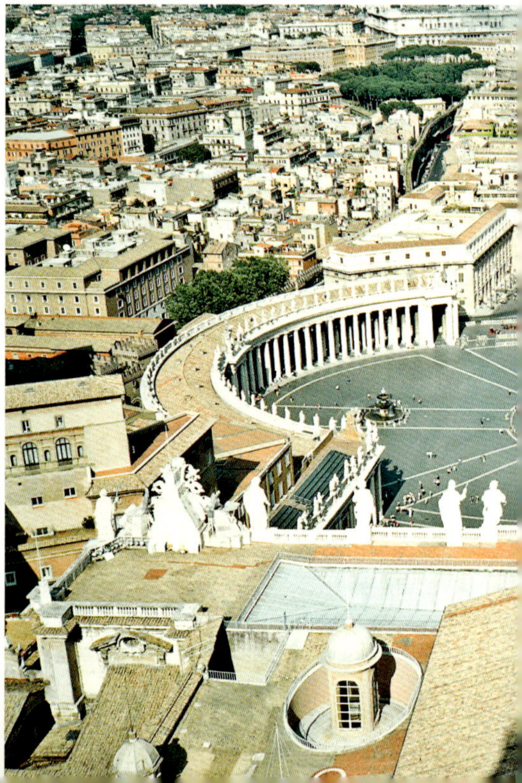

Top right: The Blue Mosque in Istanbul
Bottom right: The staircase leading to the *Capitolium* in Rome

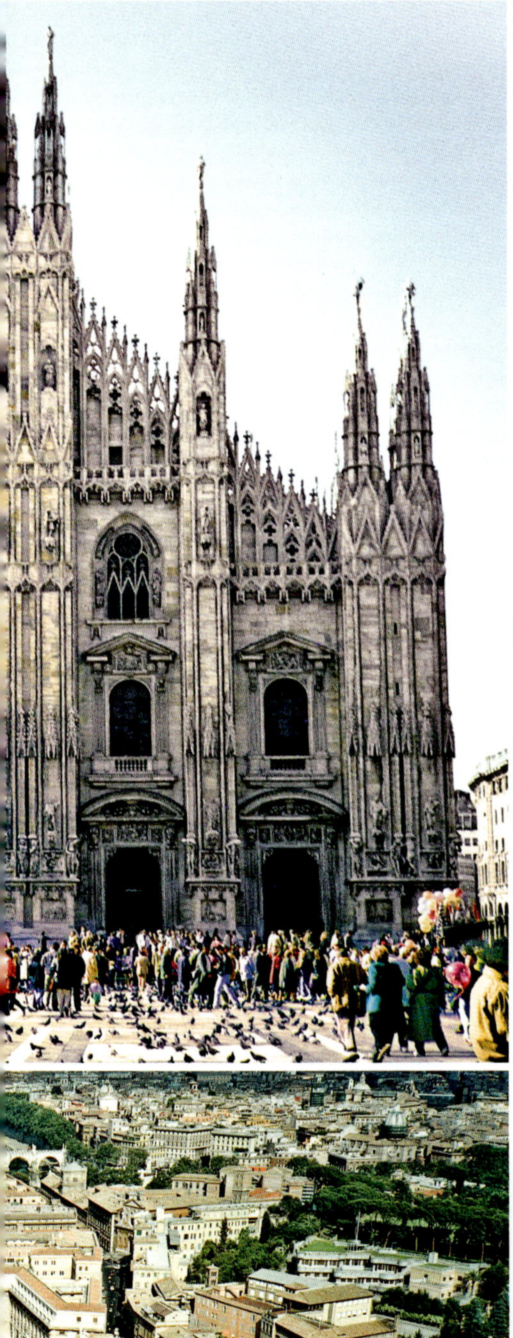

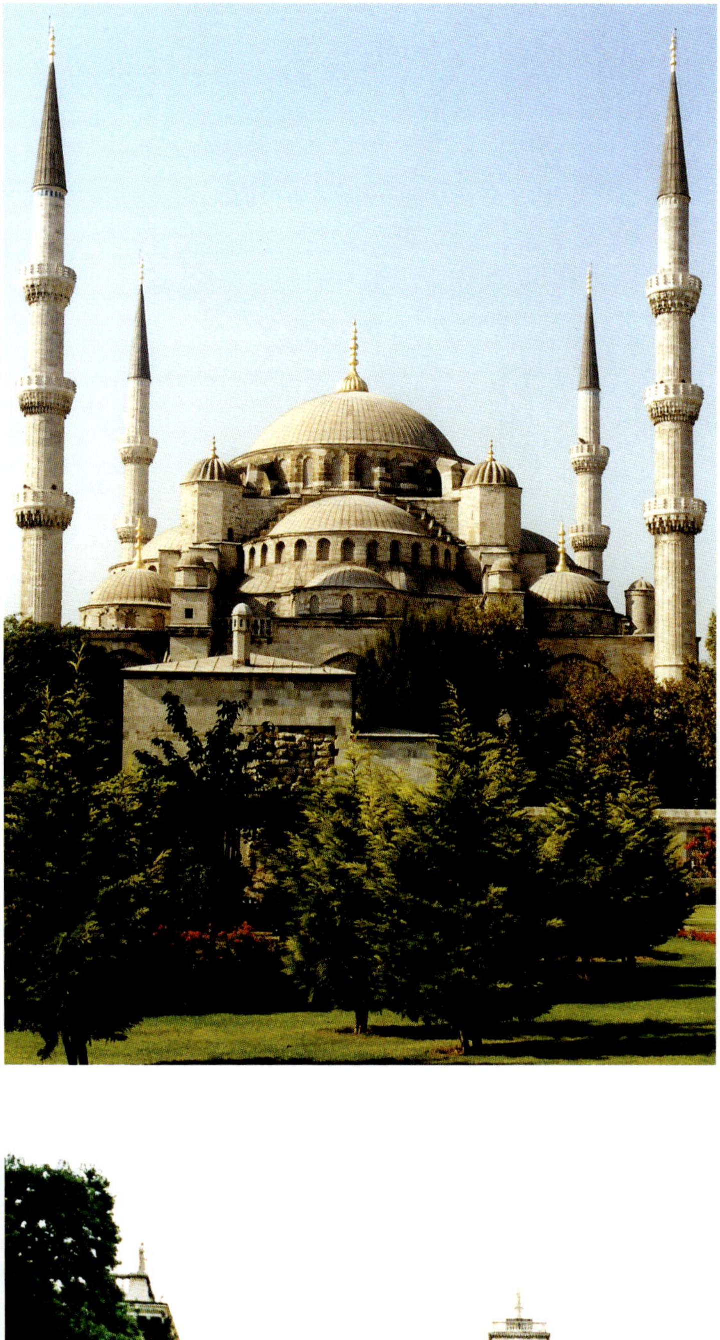

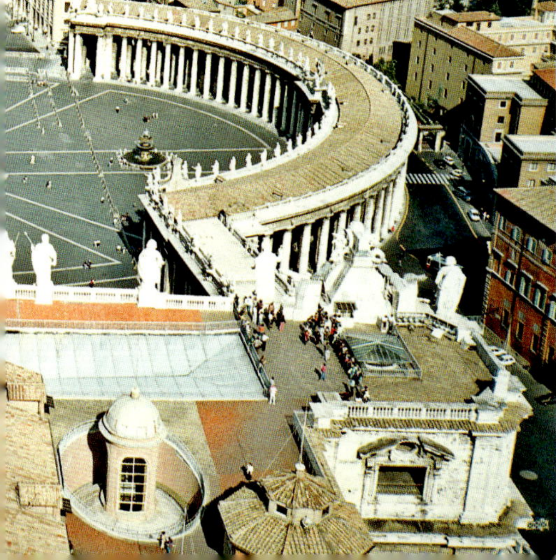

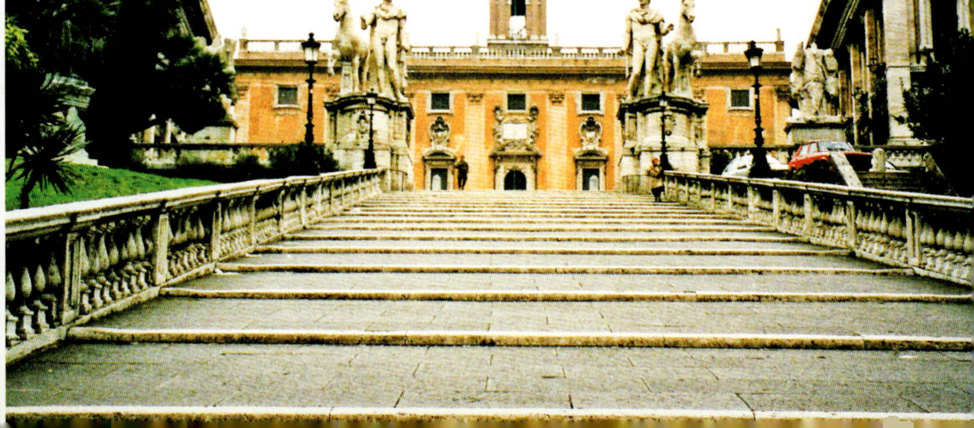

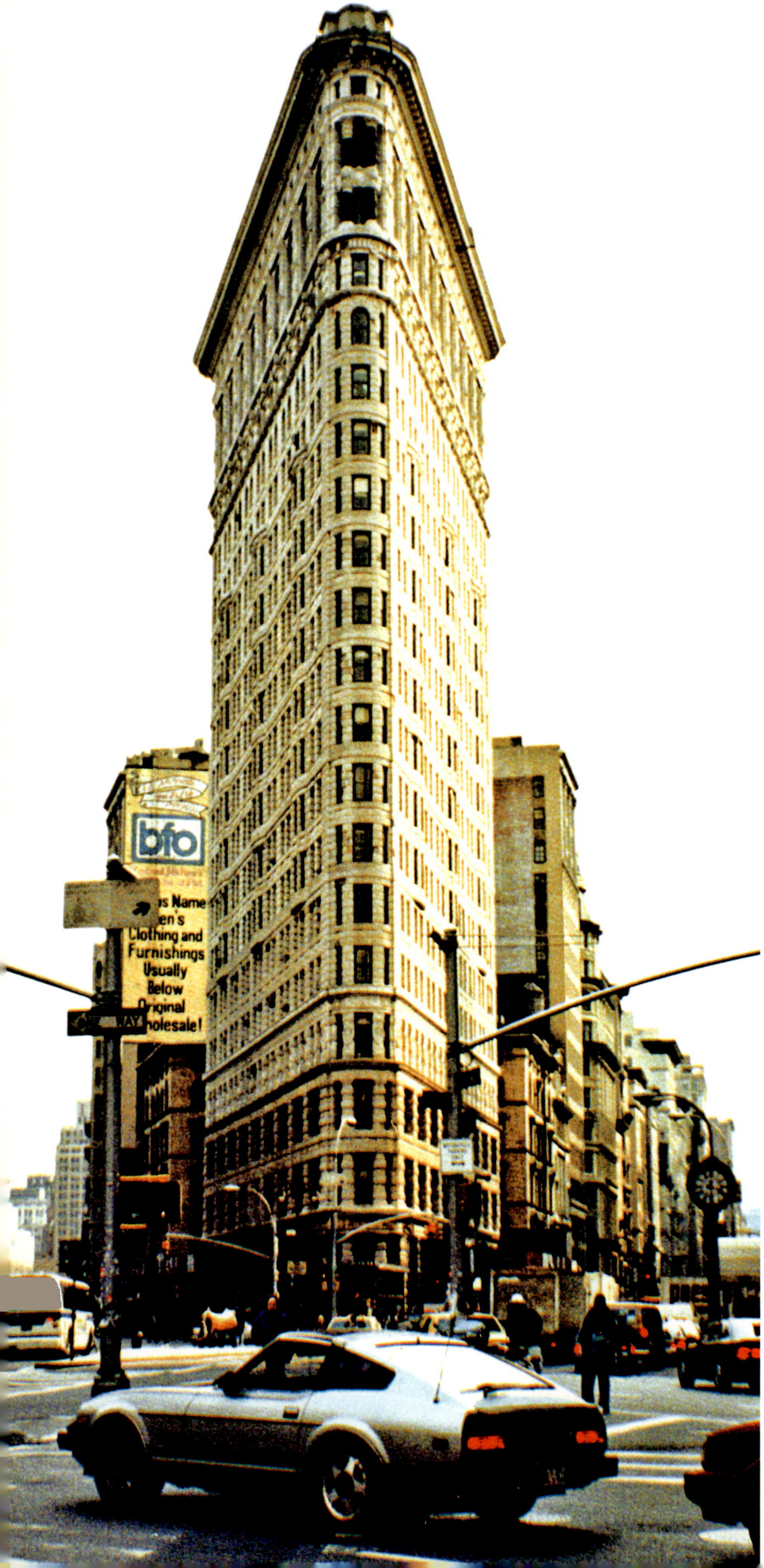

The Flatiron building at Broadway and 23rd Street in New York City

(page 25)

The smaller details of this cathedral, again, do not correspond to perfect bilateral symmetry but the overall image does.

Antoni Gaudi, La Sagrada Familia in Barcelona

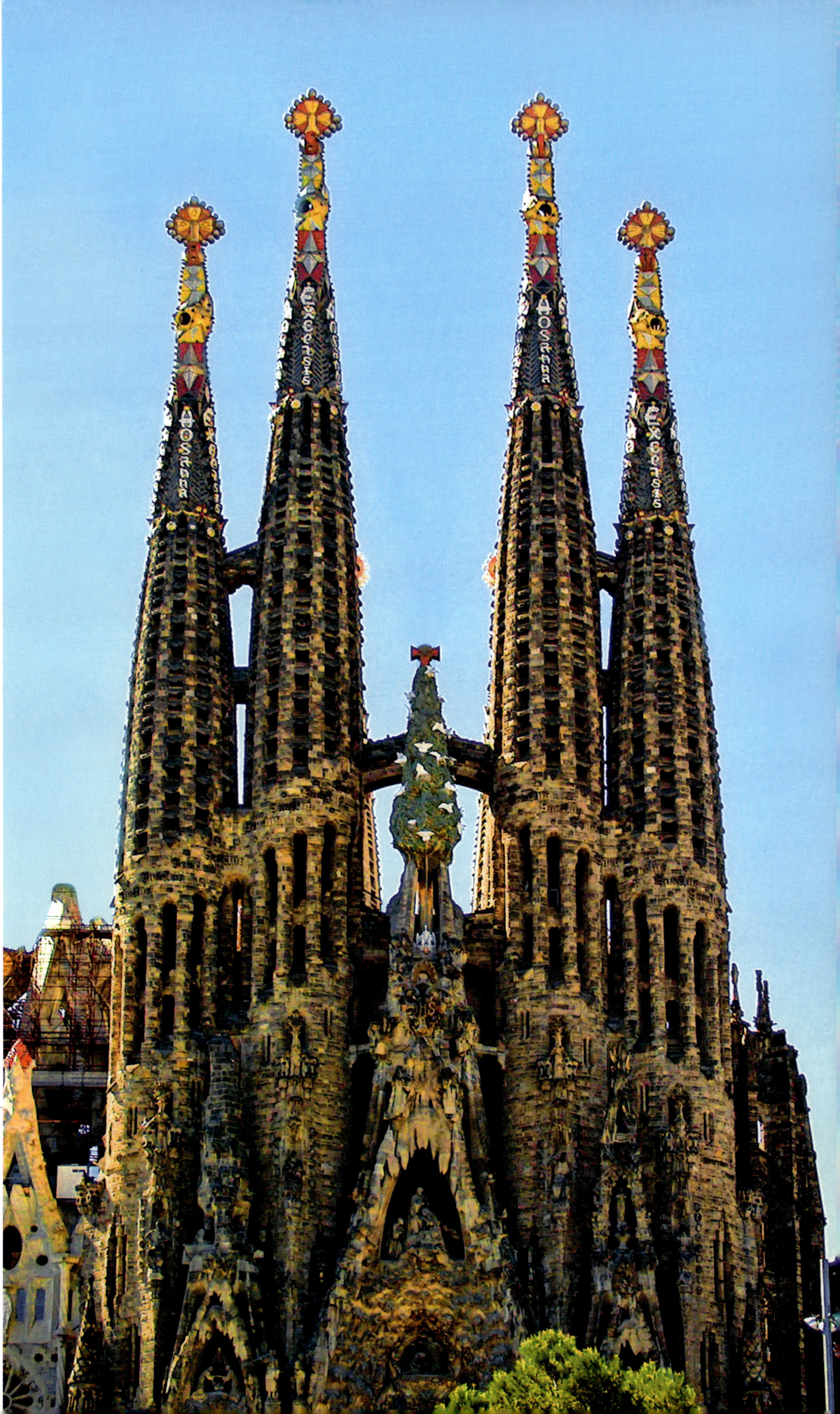

Double-Headed Eagles

We have seen examples of exaggerated symmetries in depicting the faces of important people. Another expression of exaggerated symmetry is doubled heads of the eagle that often occurs in coats of arms. Typical examples are those of the Austrian Habsburgs and the Russian Romanovs, but their use is not limited to large empires.

Top and bottom left: Toledo, Spain
Bottom right: Prague

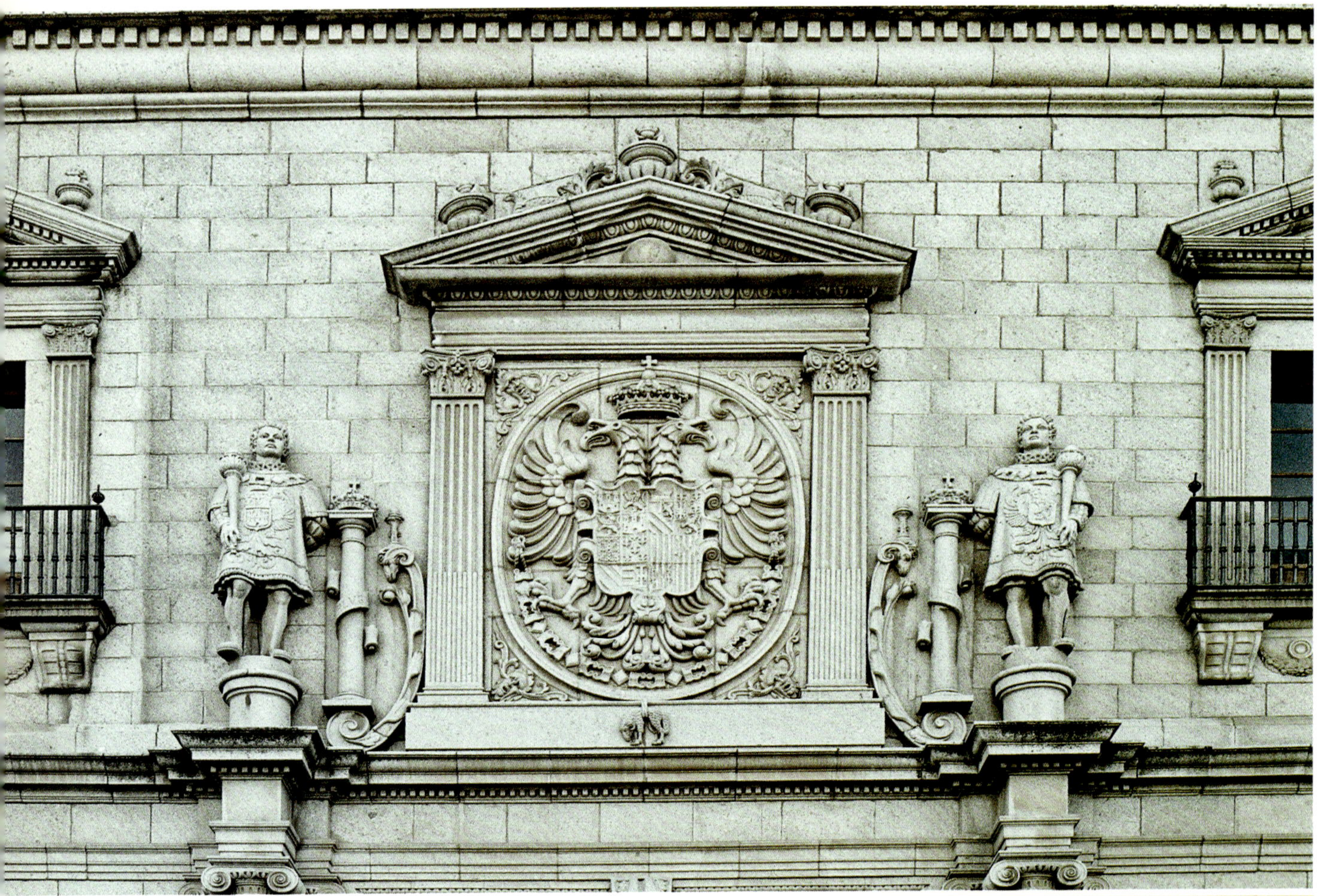

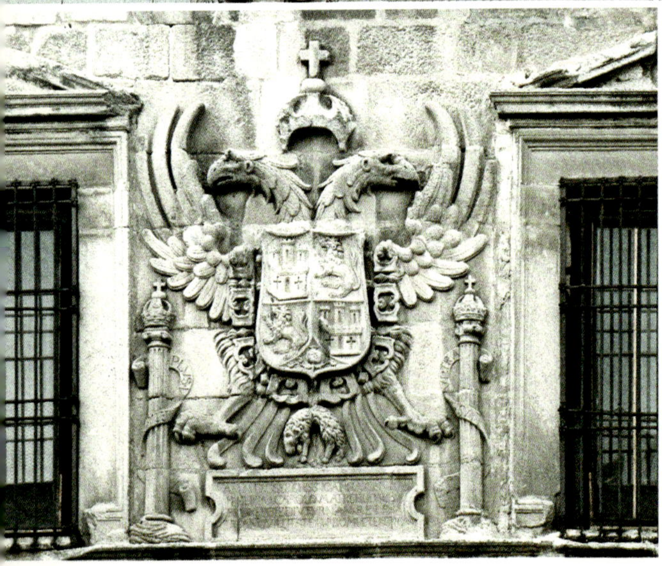

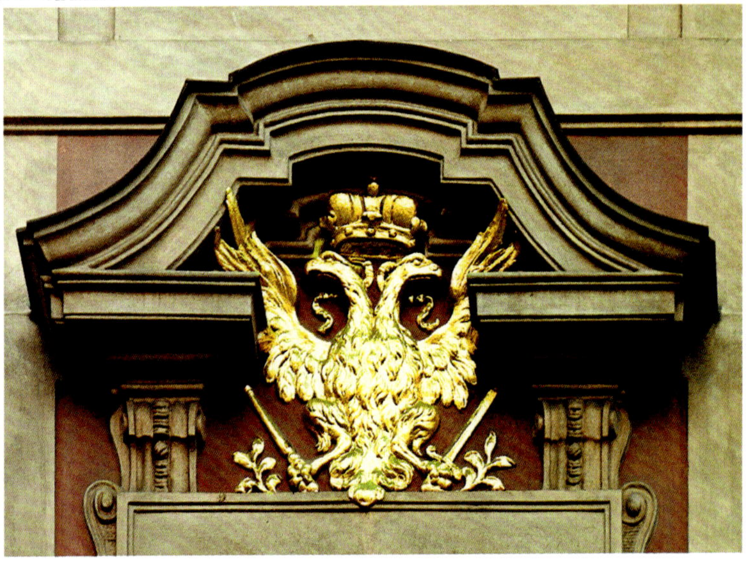

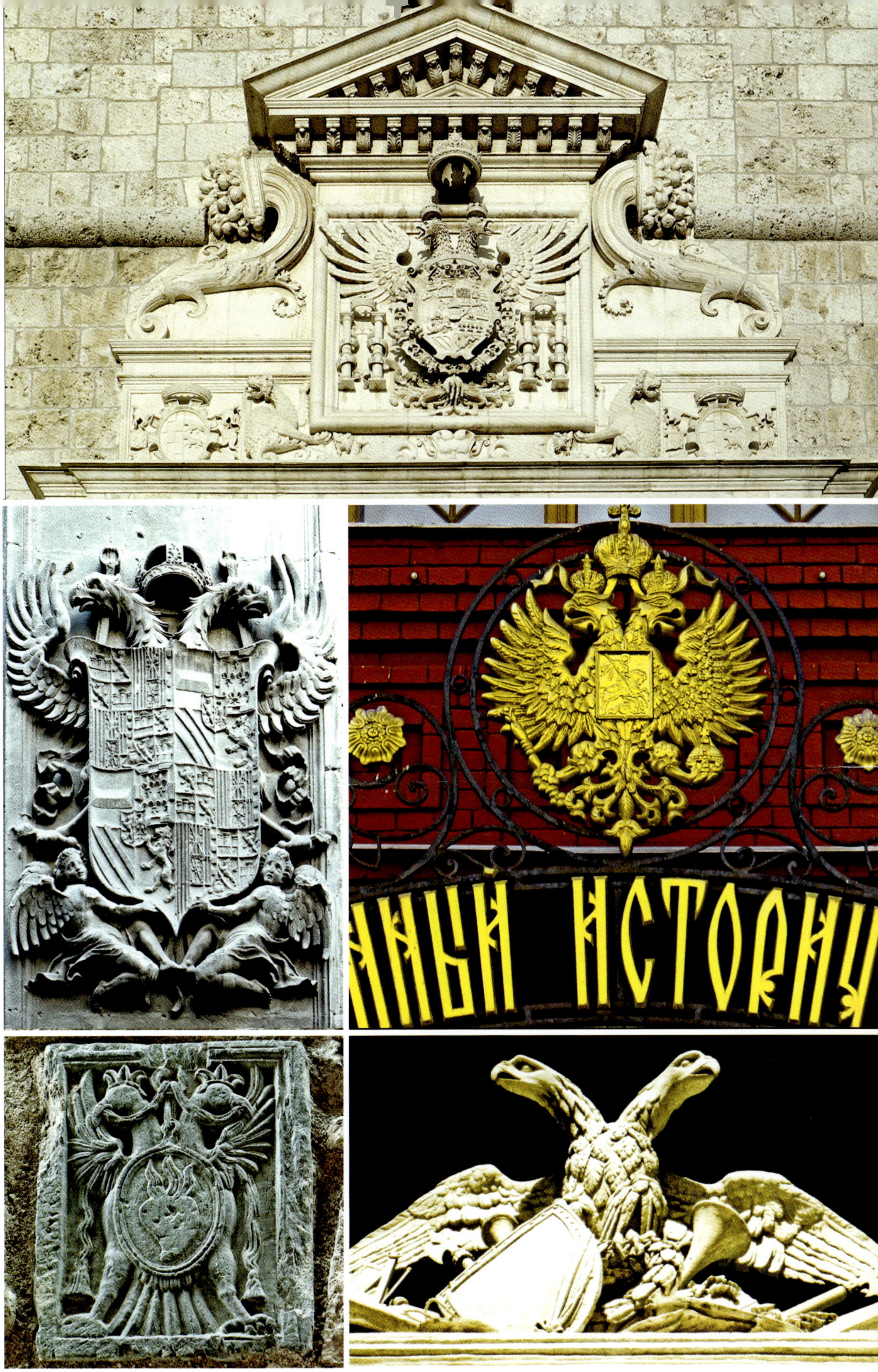

Top: L'Aquila, Italy
Middle row, from the left: Madrid; Moscow
Bottom: Toledo, Spain; Vienna, Austria

Music

Johann Sebastian Bach's *Contrapunctus XVIII* is a beautiful example of mirror symmetry in musical composition. In the detail depicted here, there is horizontal symmetry between the two parts.

Béla Bartók's *Microcosmos* is also represented by a detail only. It also shows horizontal mirror symmetry. When a music teacher, Mária Apagyi in a school in Komló, Hungary, asked her pupils to draw something inspired by listening to *Microcosmos*, the children invariably produced patterns of bilateral symmetry although the actual drawings came in great variations.

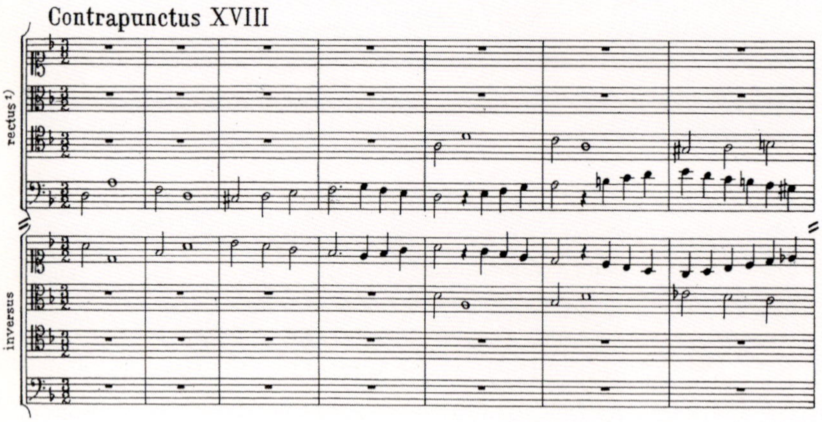

J. S. Bach, *Contrapunctus XVIII* (detail)

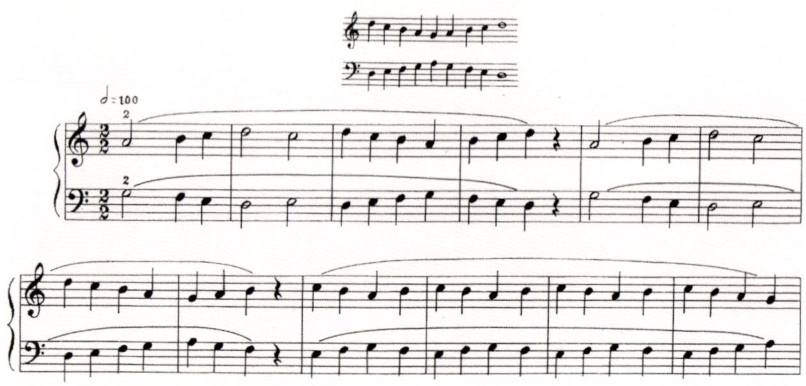

B. Bartók, *Microcosmos* (detail)

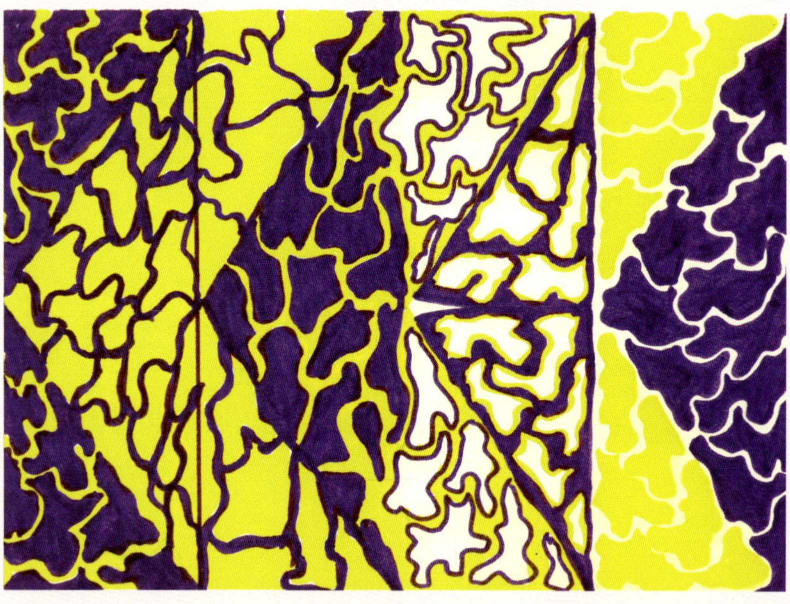

Kinga Aszódi, 12-year-old, 1982

Gábor Tóth, 13-year-old, 1982

Gábor Tóth, 13-year-old, 1982

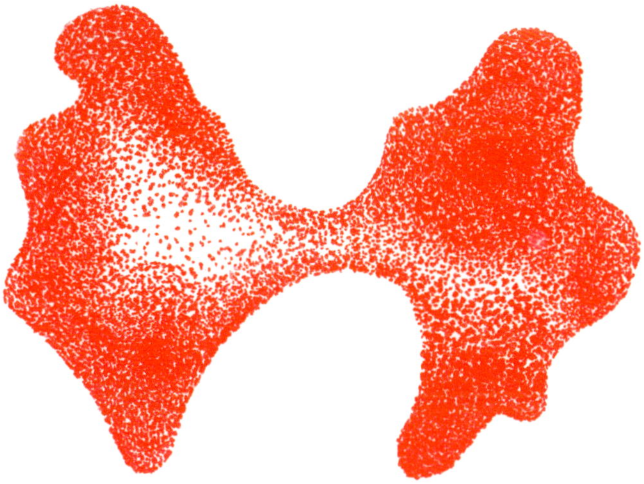

Csaba Király, 14-year-old, 1979

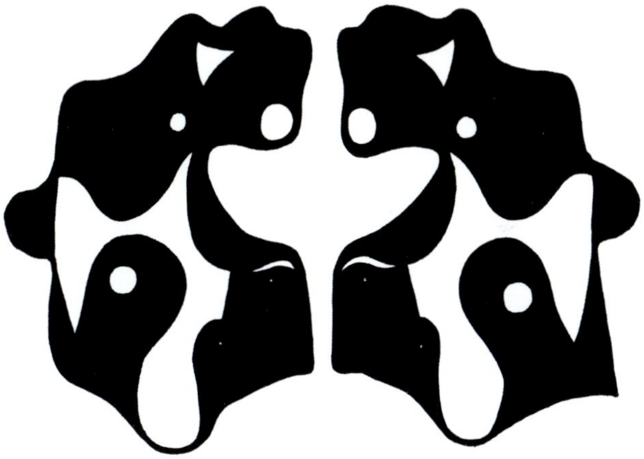

Erika Gaszt, 12-year-old, 1982

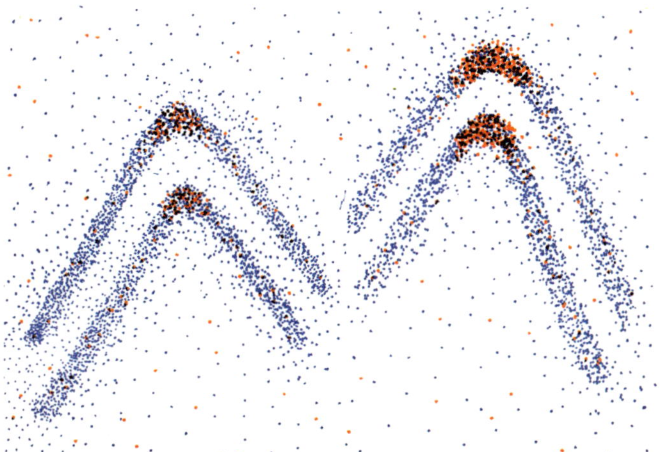

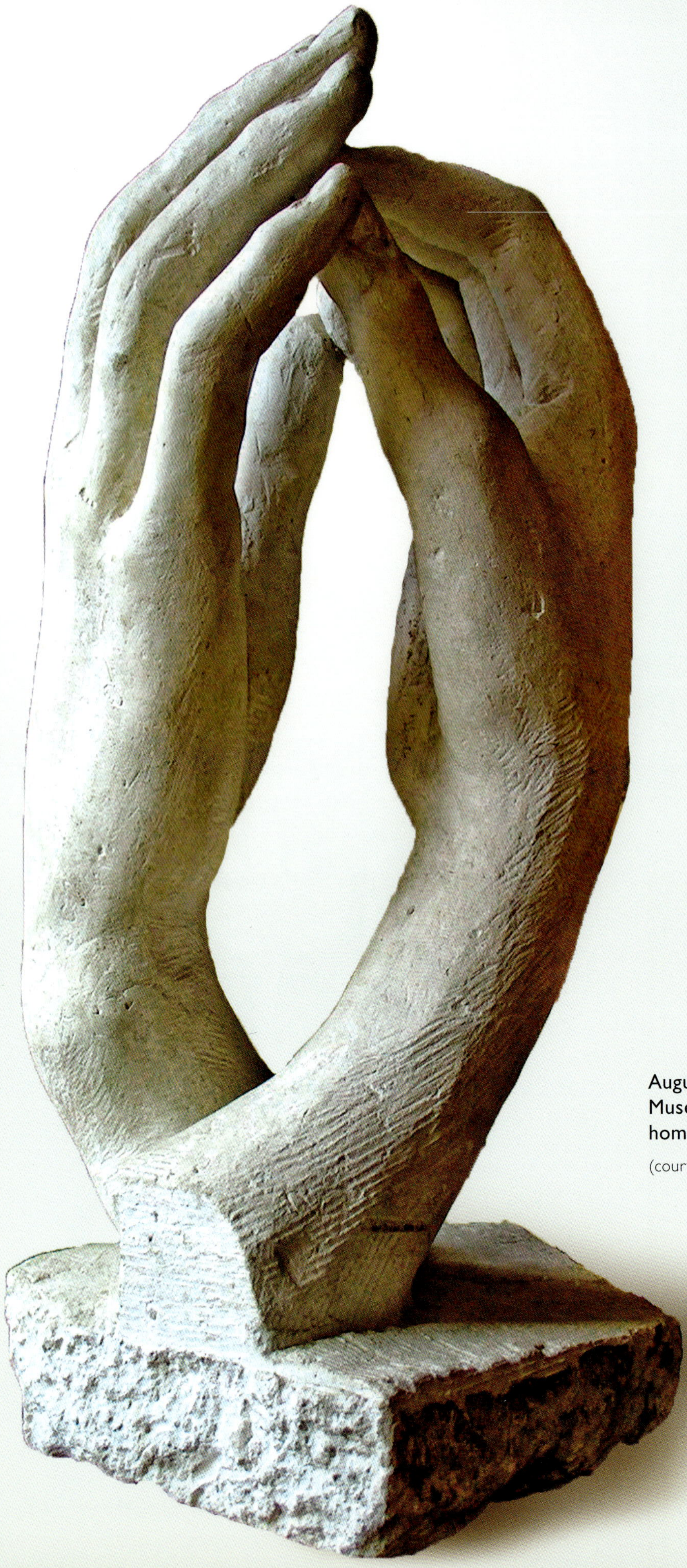

Auguste Rodin, *The Cathedral* in the Rodin Museum, Paris. This sculpture depicts two homochiral hands.

(courtesy of the Rodin Museum)

Chirality

Some objects may have no symmetry at all in themselves, but they can occur in pairs where the two are related by mirror symmetry. An obvious example is the human hand. Our two hands are not identical and they are not superimposable; that is, they cannot be brought into coincidence with each other. Each of our two hands has its own sense, or direction, and the two have opposite senses, or directions. For example, with one hand, the first finger is to the left of the thumb; with the other, it is to the right. This is "handedness" or "chirality" as the word chiral corresponds to hand in Greek. Our two hands have opposite senses; they are called heterochiral. Two hands with the same sense are called homochiral.

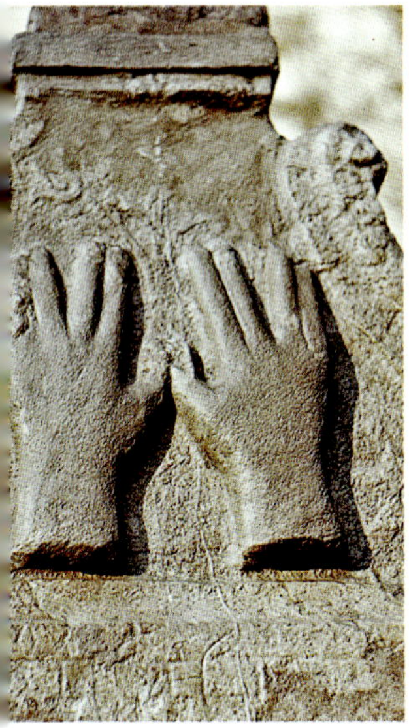
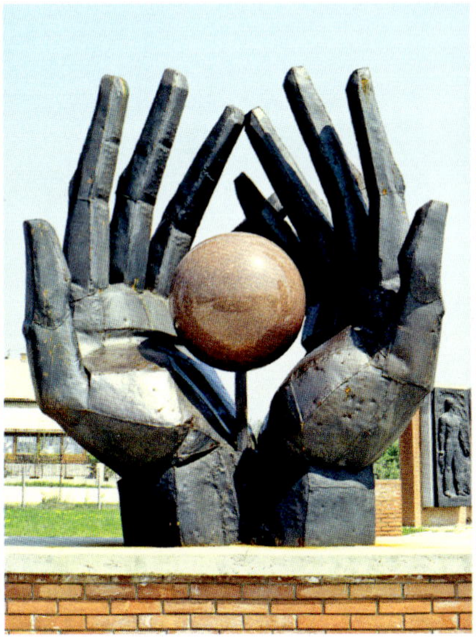
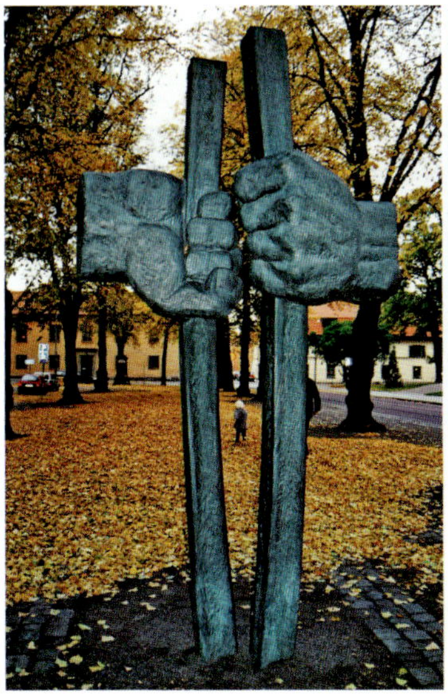
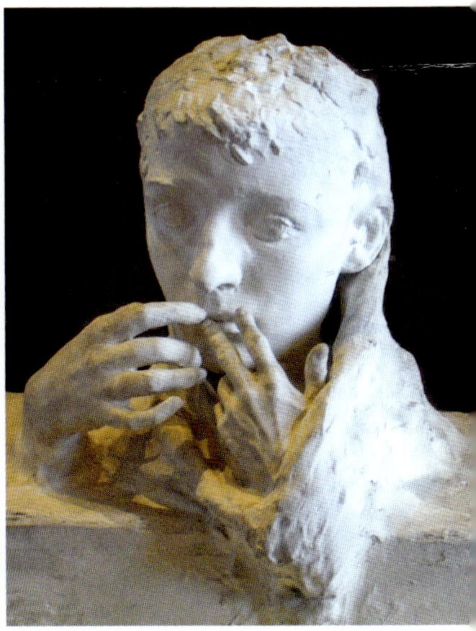

From the left:

Tombstone in the old Jewish cemetery in Prague

Sculpture in the Sculpture Park in Budapest
(sculptures associated with the communist era are collected in this park)

Sculpture in Uppsala, Sweden

Auguste Rodin, *L'Adieu* in the Rodin Museum, Paris
(courtesy of the Rodin Museum)

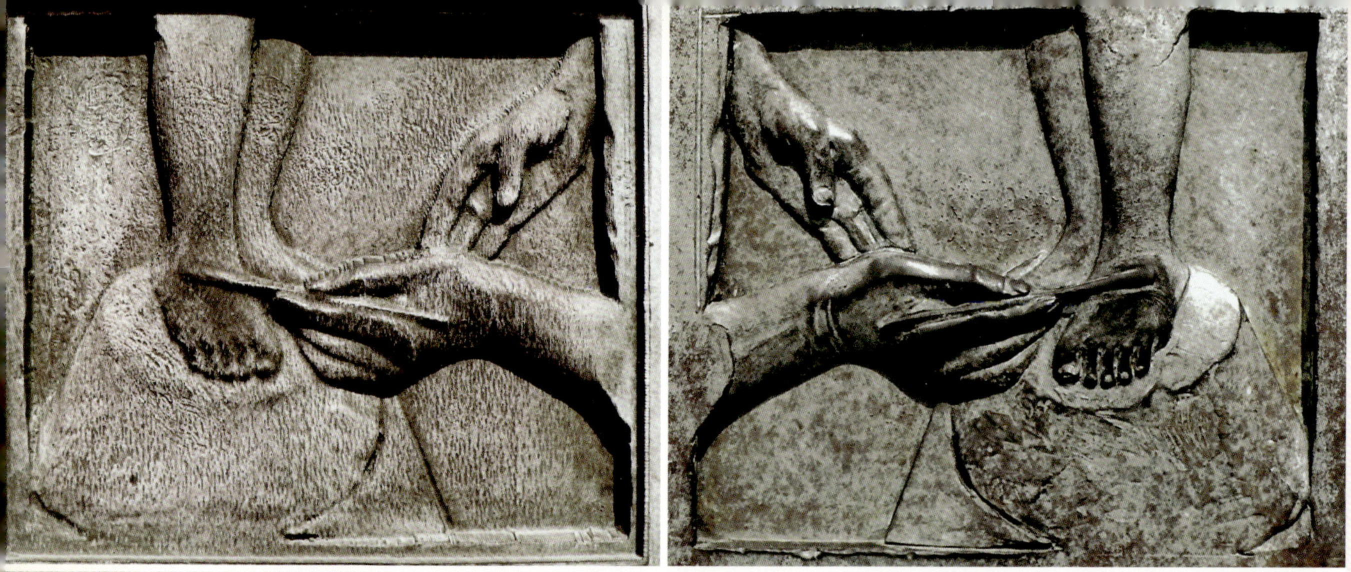

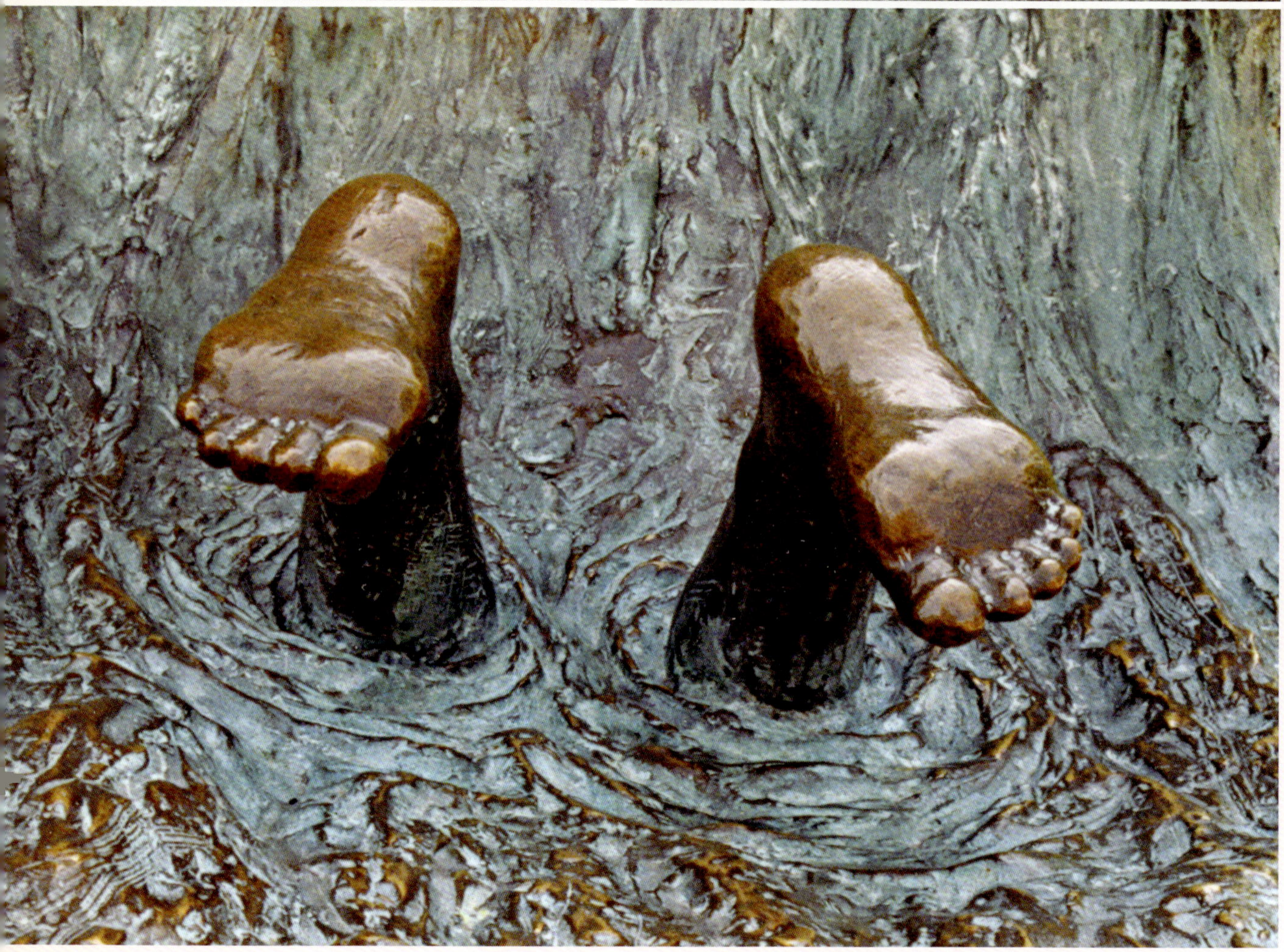

Top: Two images of the Franklin D. Roosevelt memorial in Washington, DC

Both show chiral pairs of hands as well as legs; in addition, the two images are related to each other by an approximate reflection plane.

Bottom: Kay Worden, *Wave* (detail) in Newport, Rhode Island

By analogy of chirality, when legs are involved in such pairs, this may be called podality.

There are two spiral staircases in the church below, forming a heterochiral pair.

The interior of the church St. Etienne du Mont in Paris

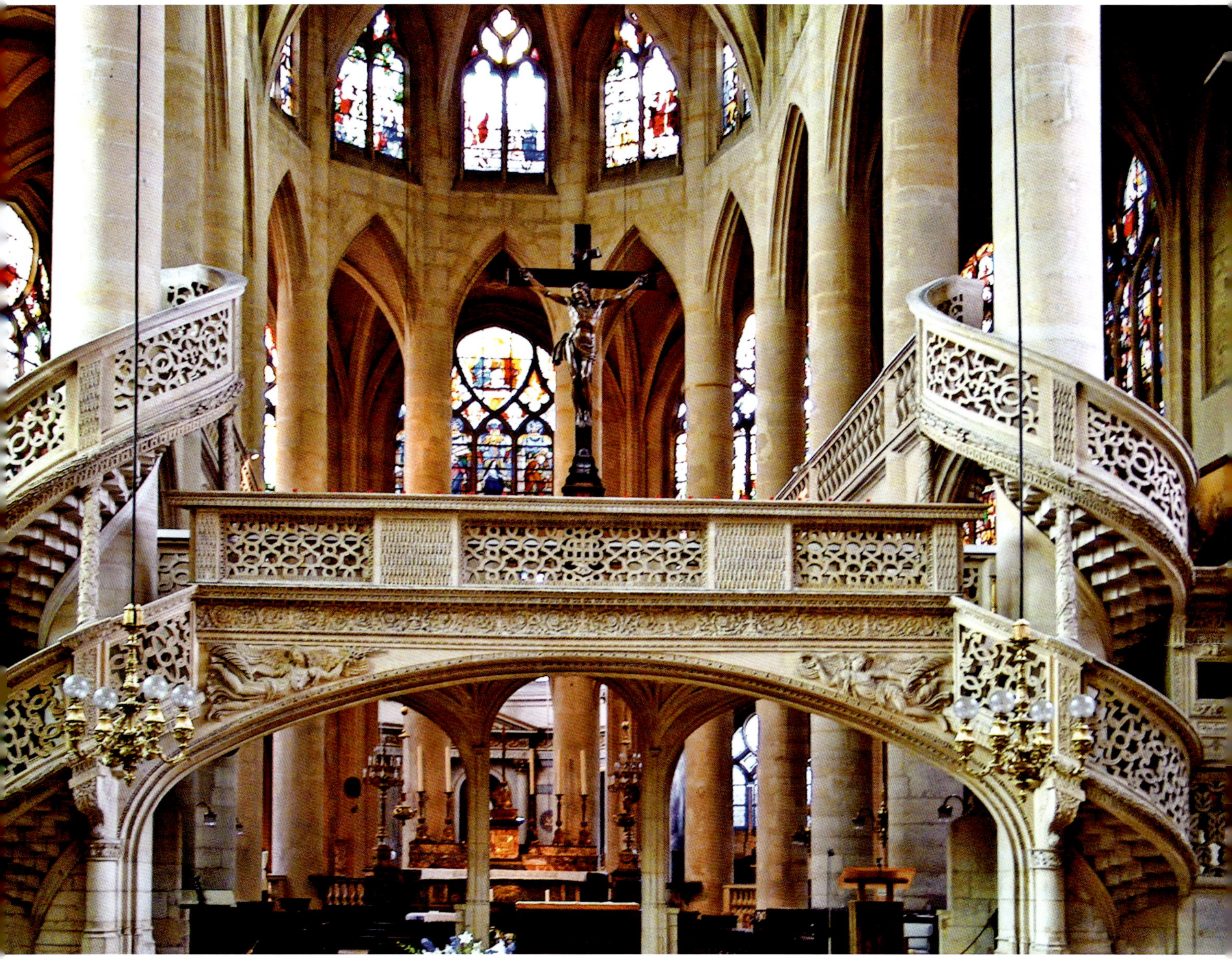

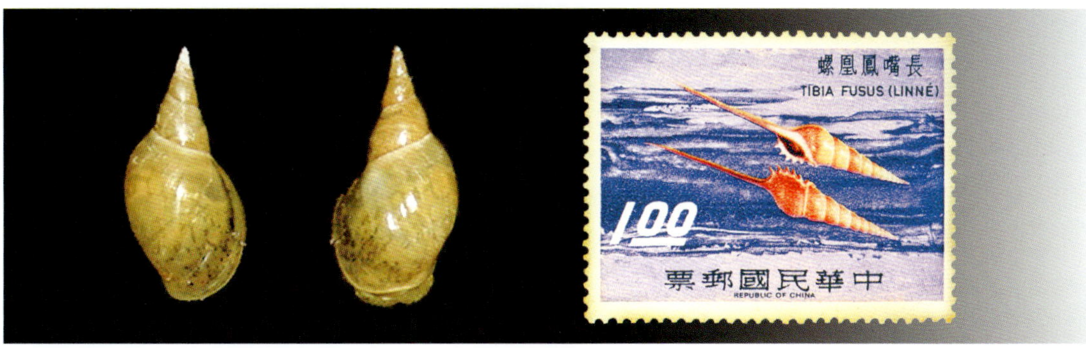

Hetereochiral pair of the snail *Lymnaea stagnalis*

(courtesy of Reiko Kuroda, Tokyo)

The houses of the vast majority of shells turn in the same direction, but there are exceptions, and heterochiral pairs of shells have been observed.

La Coupe du Roi

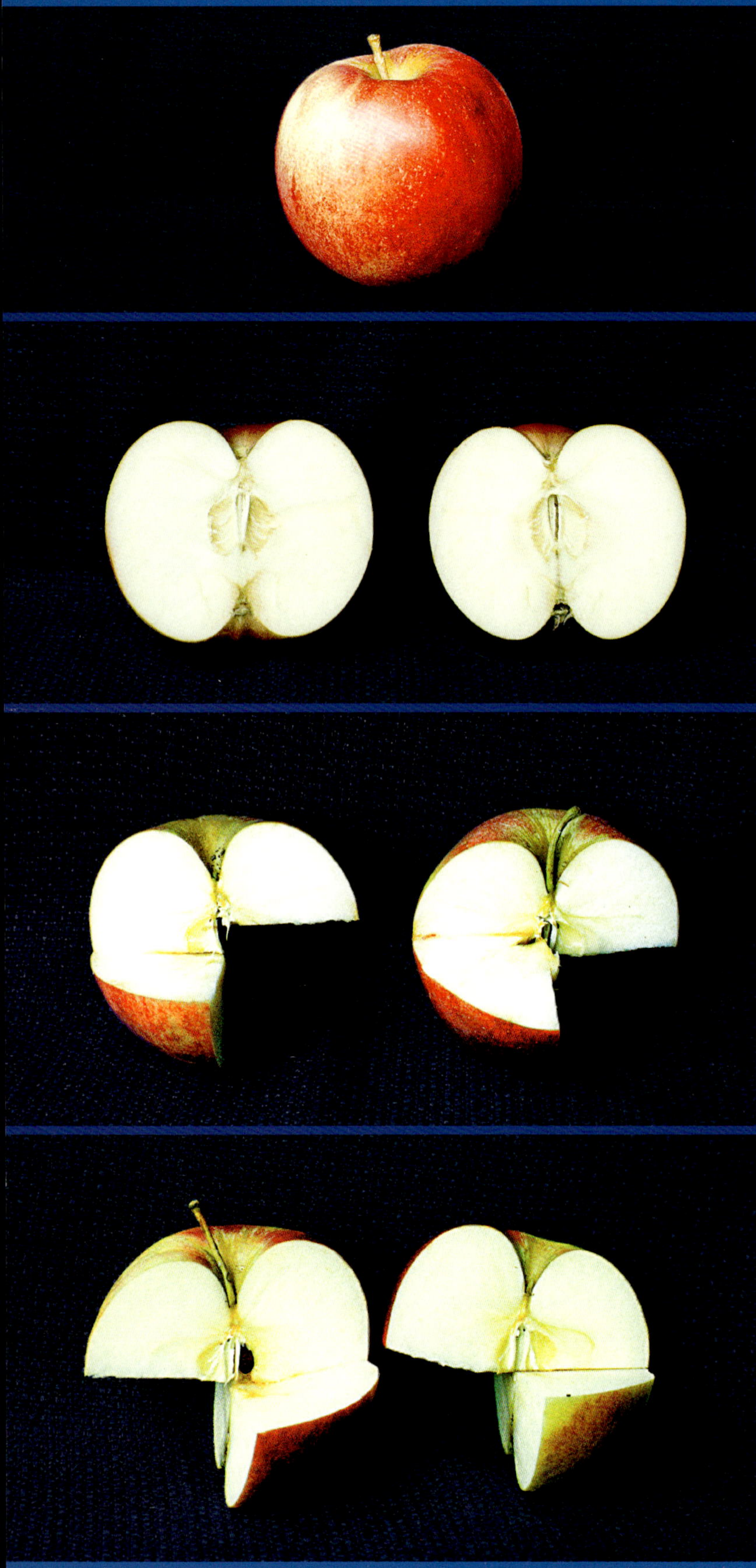

Cutting an apple in the usual way gives us two halves that are not only each other's mirror images, but are also superimposable. Thus, these two halves are not chiral; it is impossible to distinguish them as a "left-handed" and a "right-handed" half. So we may ask the question, Can an apple be dissected into left-handed and right-handed halves? However we try, we cannot dissect an apple into left-handed and right-handed halves. However, it is possible to dissect an apple into two homochiral halves. Thus, from two different apples we can produce two pairs of opposites: a pair of left-handed and a pair of right-handed halves. The different halves, however, cannot be combined into one apple.

The way to produce chiral halves is a French parlor trick and is called "La Coupe du Roi," "The Royal Cut." This is how it works:
Make two vertical half-cuts through the apple; one from its top to its equator, and the other, perpendicular to the first, from its bottom to its equator. Then make two nonadjacent quarter cuts along the equator of the apple. Following these cuts the apple should split into two homochiral halves. The recipe can be followed in two senses and thus produce two left-handed halves in one case and two right-handed ones in the other.

Crystals, Molecules, and Medicine

Crystals and molecules can also be chiral; in fact, the discovery of chirality happened in crystallization experiments by Louis Pasteur in 1848 when he observed the two kinds of crystals of the same substance. Molecules can also be chiral and the living organisms contain a large number of such substances. Almost all naturally occurring amino acids—the building elements of proteins—are chiral.

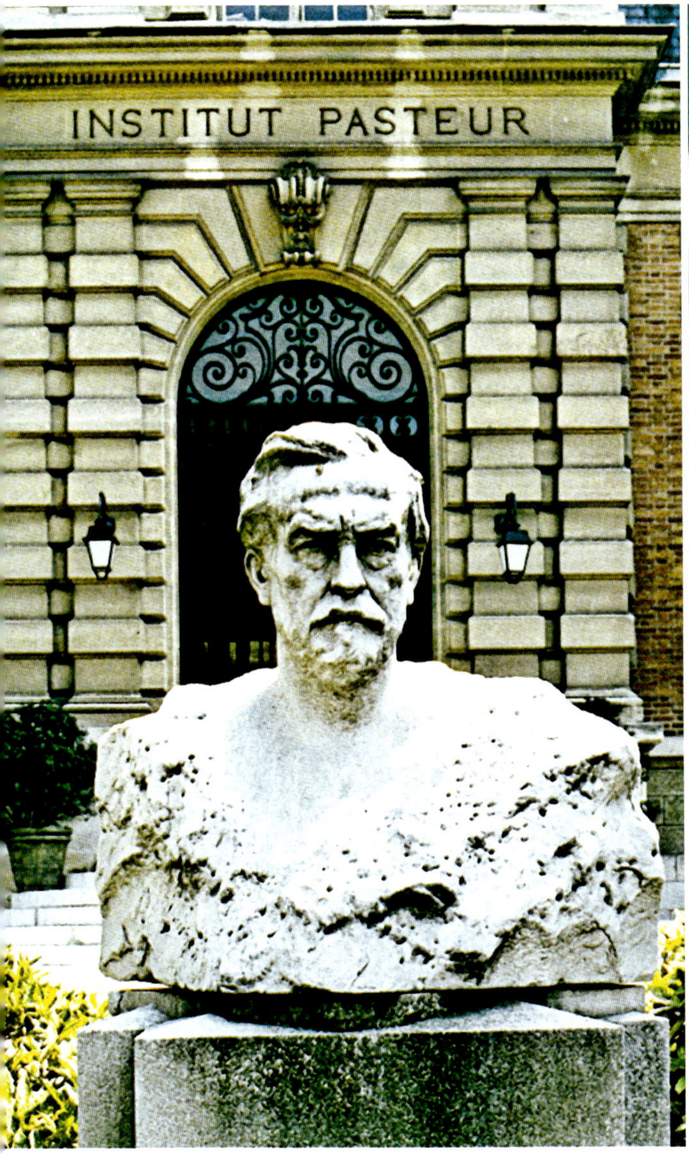

Louis Pasteur's bust in front of the *Institut Pasteur* in Paris, France

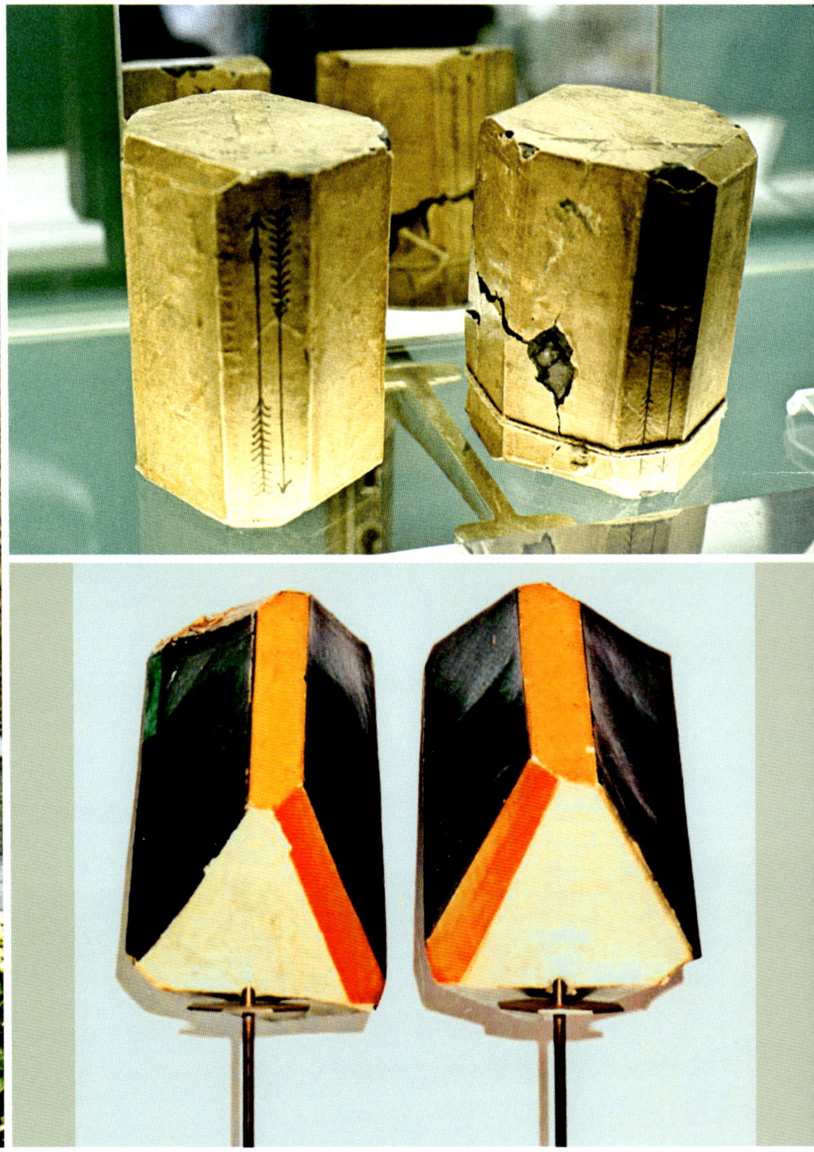

Pasteur's models in the Pasteur Museum at the *Institut Pasteur*

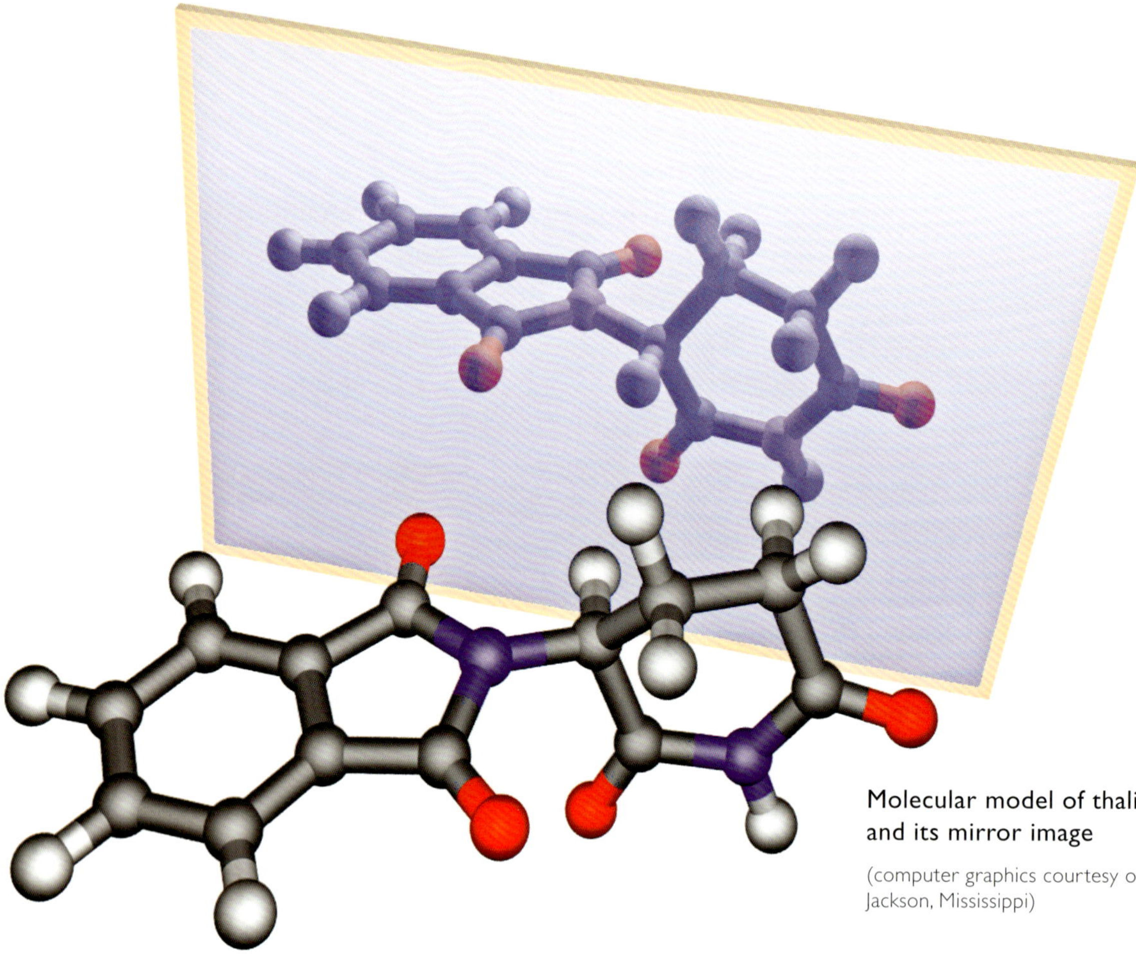

Molecular model of thalidomide and its mirror image

(computer graphics courtesy of Ilya Yanov, Jackson, Mississippi)

There is a unique situation in that all amino acids in living organisms occur as left-handed, but never right-handed. Other substances important for life, such as nucleotides—the building elements of nucleic acids—appear in right-handed versions only. That some substances occur always left-handed and others always right-handed is characteristic for all life processes, and is the same in humans, animals, plants, and microorganisms. Why this happens is a great puzzle, one that cannot be solved satisfactorily at this time. Nobel laureate Vladimir Prelog called this phenomenon a problem of "molecular theology!"

Nature has thus made a choice, but when these substances are produced in the laboratory, their left-handed and right-handed versions occur in equal amounts. However, the left-handed and the right-handed versions of the same substance may have very different properties. A tragic example was thalidomide, which was used in the 1950s in Western Europe (under the name Contergan) for alleviating the morning sickness of pregnant women in the first trimester of their pregnancy. The drug was marketed as a mixture of left-handed and right-handed molecules. By the early 1960s, many birth defects in Western Europe were associated with this drug (which was never approved and marketed in the United States). Research eventually discovered that one version of the molecule had beneficial effects, but its chiral pair acted as teratogen. Recent results indicate that the situation is more complicated because the molecules undergo rapid interconversion between the two chiral versions in the human organism. Today, in both the United States and the European Union any new substance must be thoroughly investigated for the possible differences in the effects of both their chiral versions. Here we list a few substances, pharmaceuticals as well as other substances, indicating the actions of both versions of their molecules.

Right-hand version		Left-hand version
	Ethambutol	
Treats tuberculosis		Causes blindness
	Penicillamine	
Treats joints		Very toxic
	Naproxen	
Reduces pain, inflammation (but risks heart disease)		Toxic for the liver
	Propoxyphene*	
Pain reliever	Darvon / Novrad	Cough medicine
	Norgestrel	
No harmful effect		Contraceptive
	Asparagine	
Bitter		Sweet
	Carvon	
Caraway smell		Spearmint smell
	Limonene	
Orange smell		Pine smell

* Note that the names of the respective medications are each other's mirror images, Darvon® versus Novrad®.

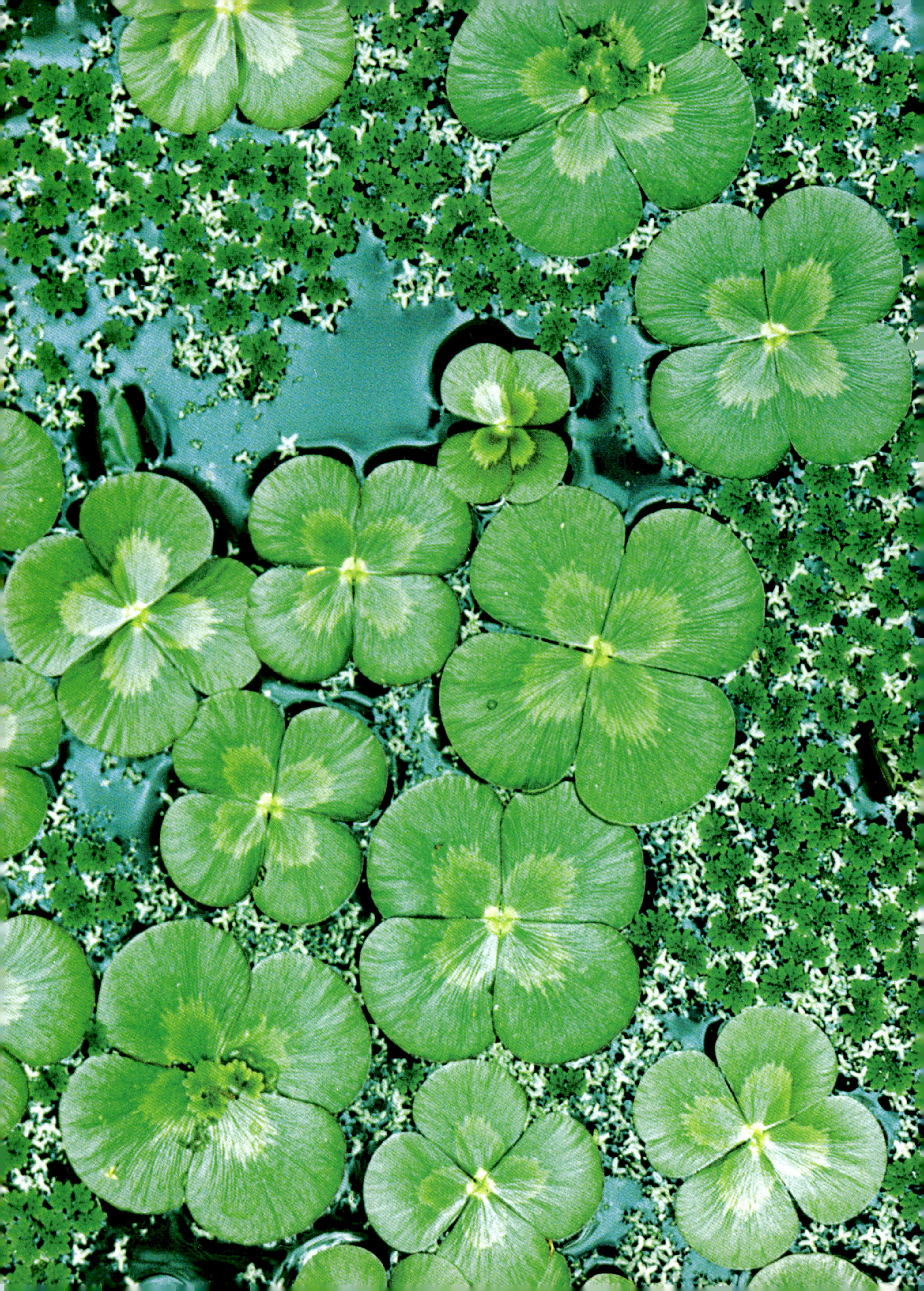

Animals seldom have more than one symmetry plane whereas plants and especially flowers often do.

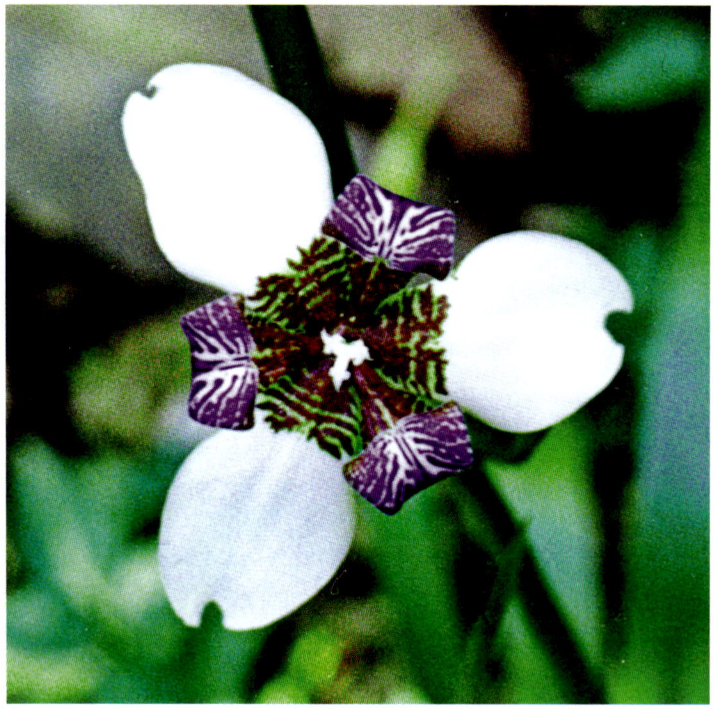

A total of four symmetry planes can be placed across and between the leaves of the water plant and three symmetry planes, one across each petal of the flower with three petals, above. The crossing lines of the symmetry planes are axes of rotational symmetry, so we may say that multiple mirror symmetries generate rotational symmetry in addition. We will discuss rotational symmetry in a subsequent chapter.

Multiple Mirrors

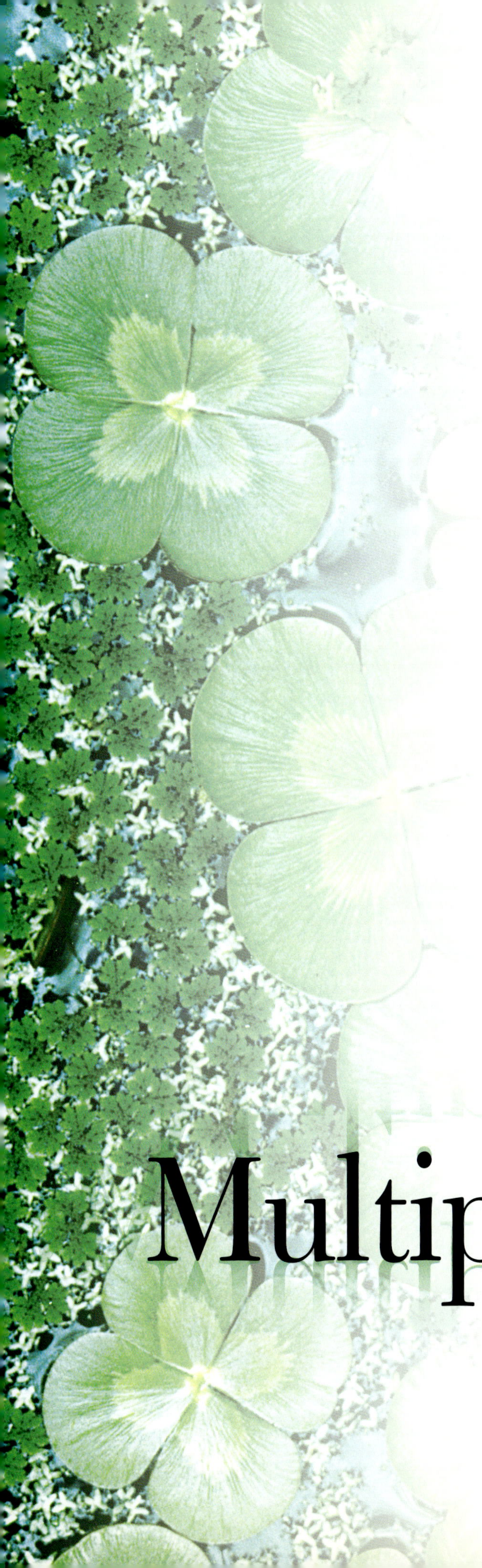

Water plant from Oahu, Hawaii

Flowers and Leaves

Flowers with three petals are rather rare whereas flowers with five petals are quite common. For a five-petal flower five symmetry planes can be placed across the flower, one across each petal. The five planes then generate fivefold rotational symmetry with the axis of rotation along the line where the symmetry planes cross—usually coinciding with the stem of the flower. Most of the flowers presented here are from Oahu, Hawaii, and display threefold, fourfold, and fivefold symmetries.

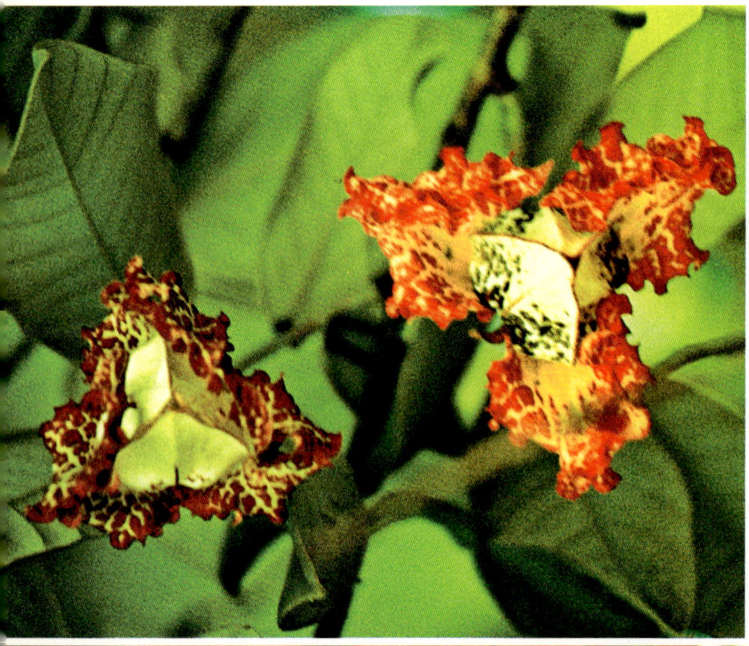

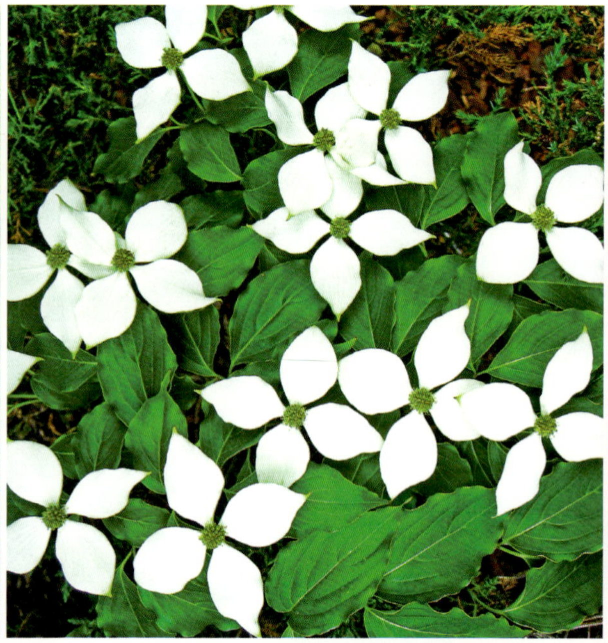

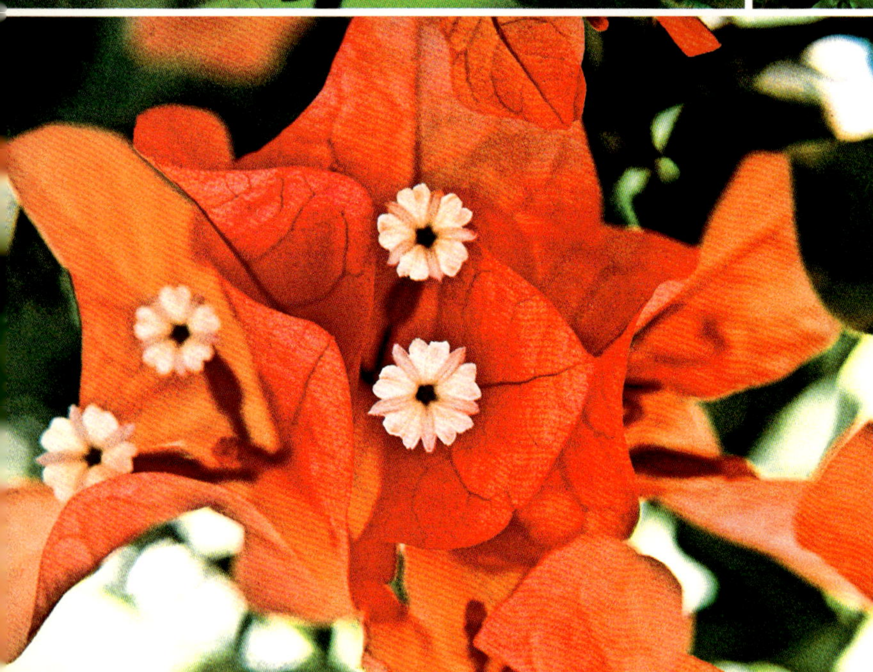

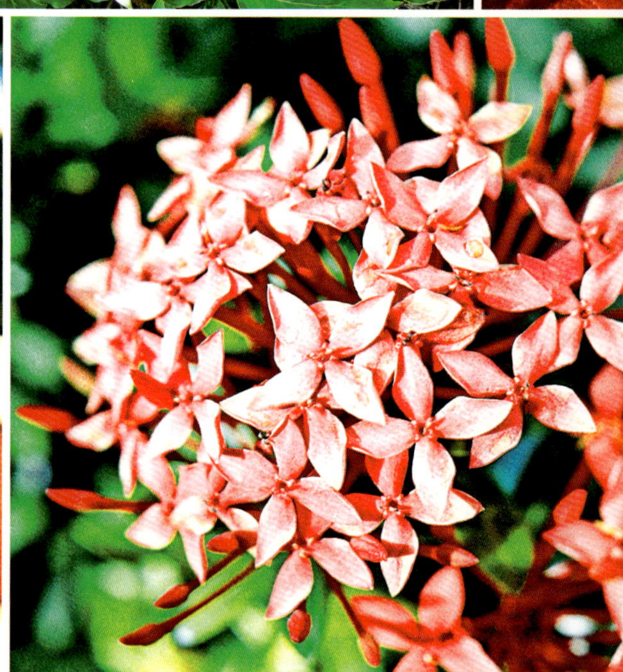

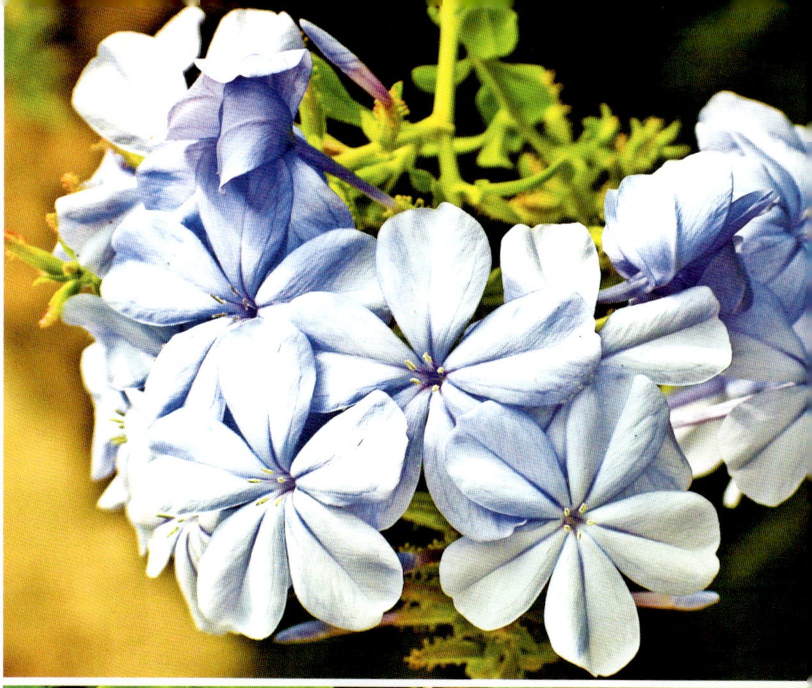

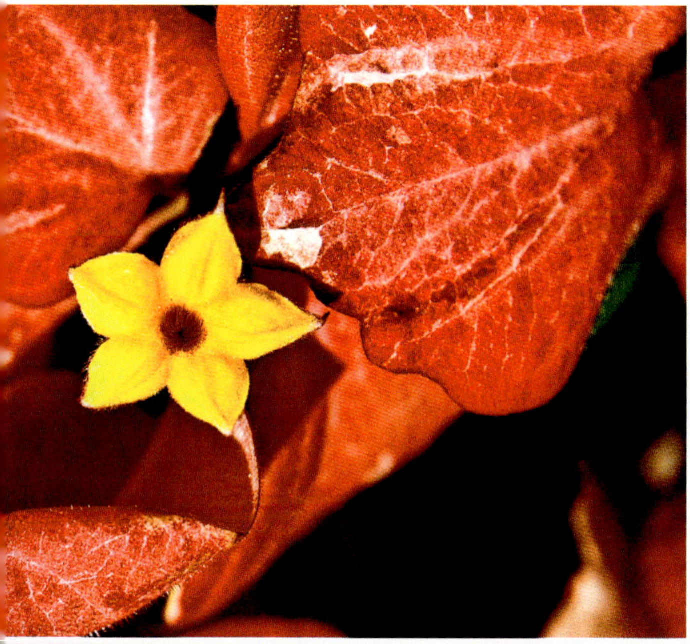

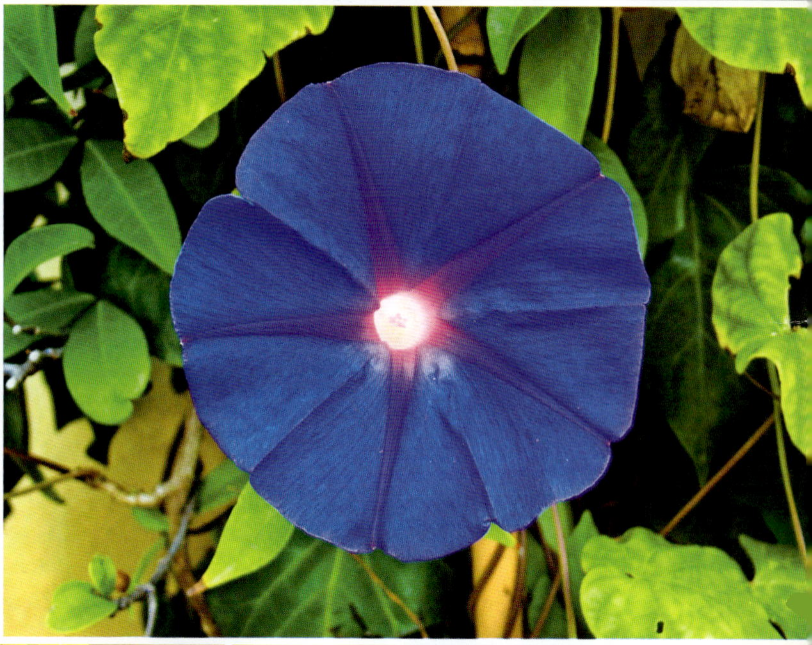

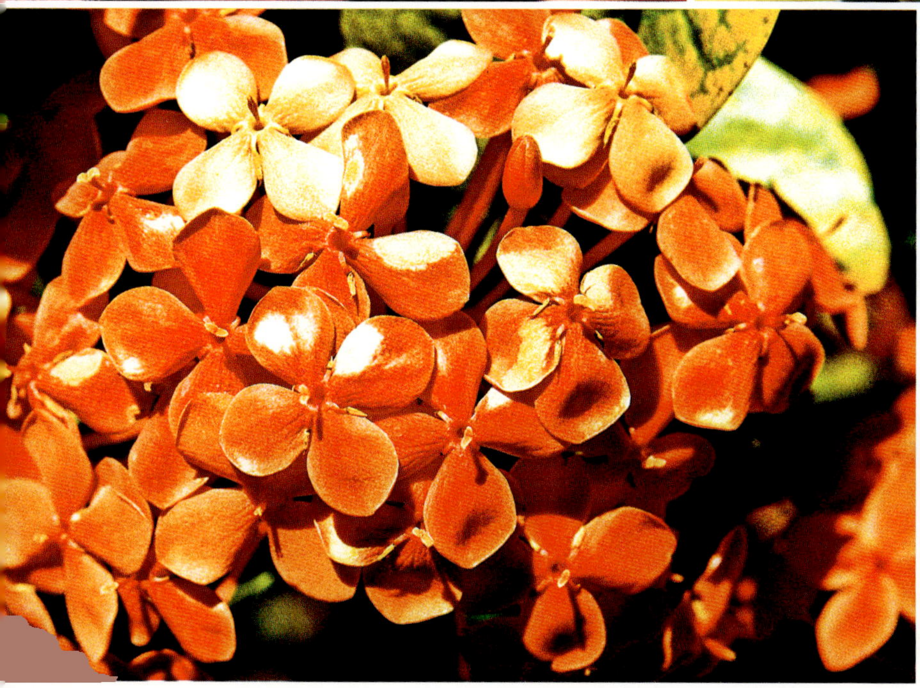

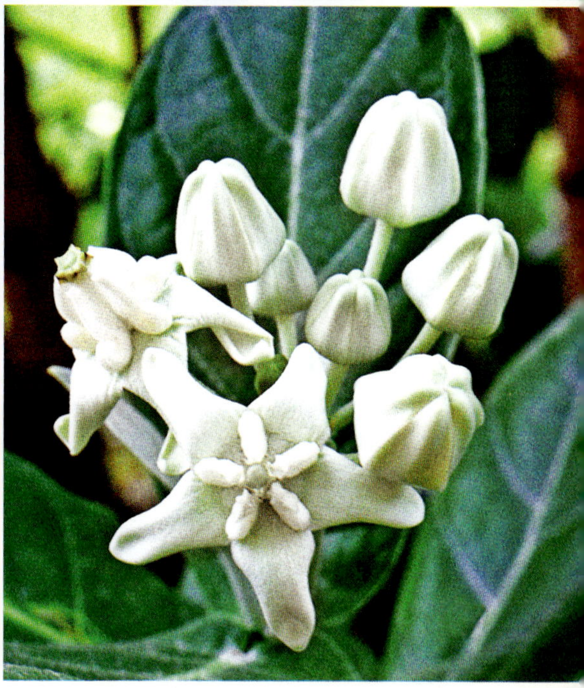

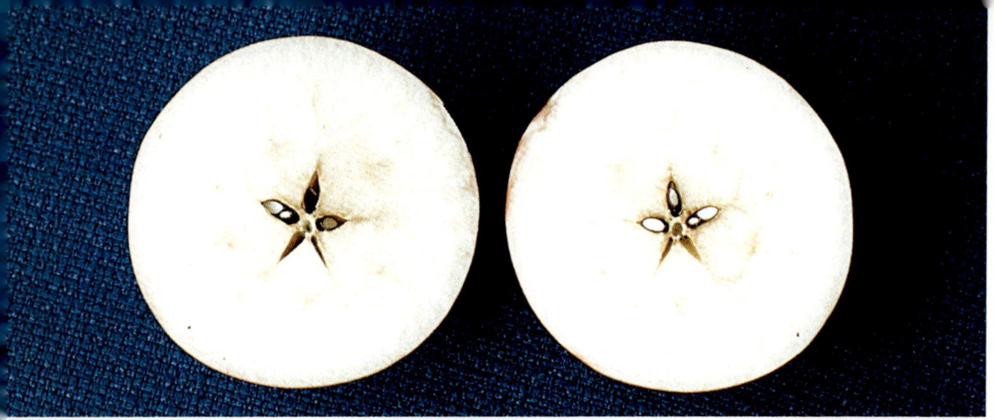

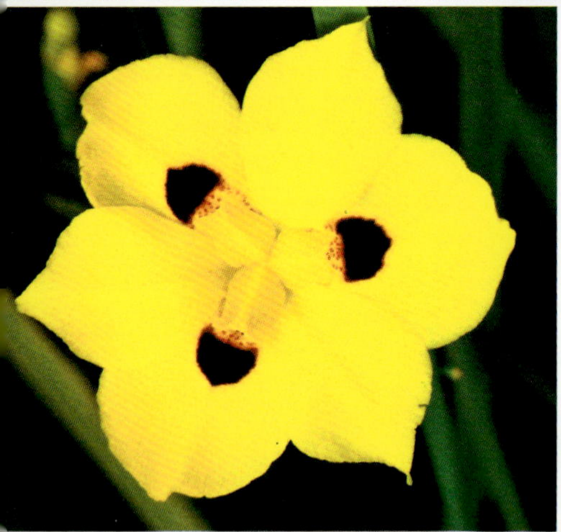

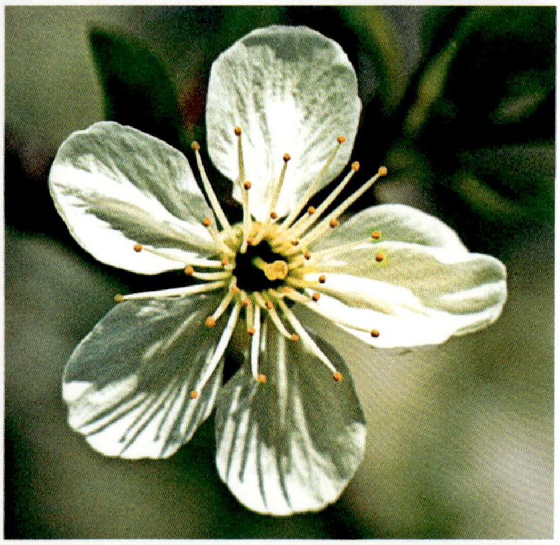

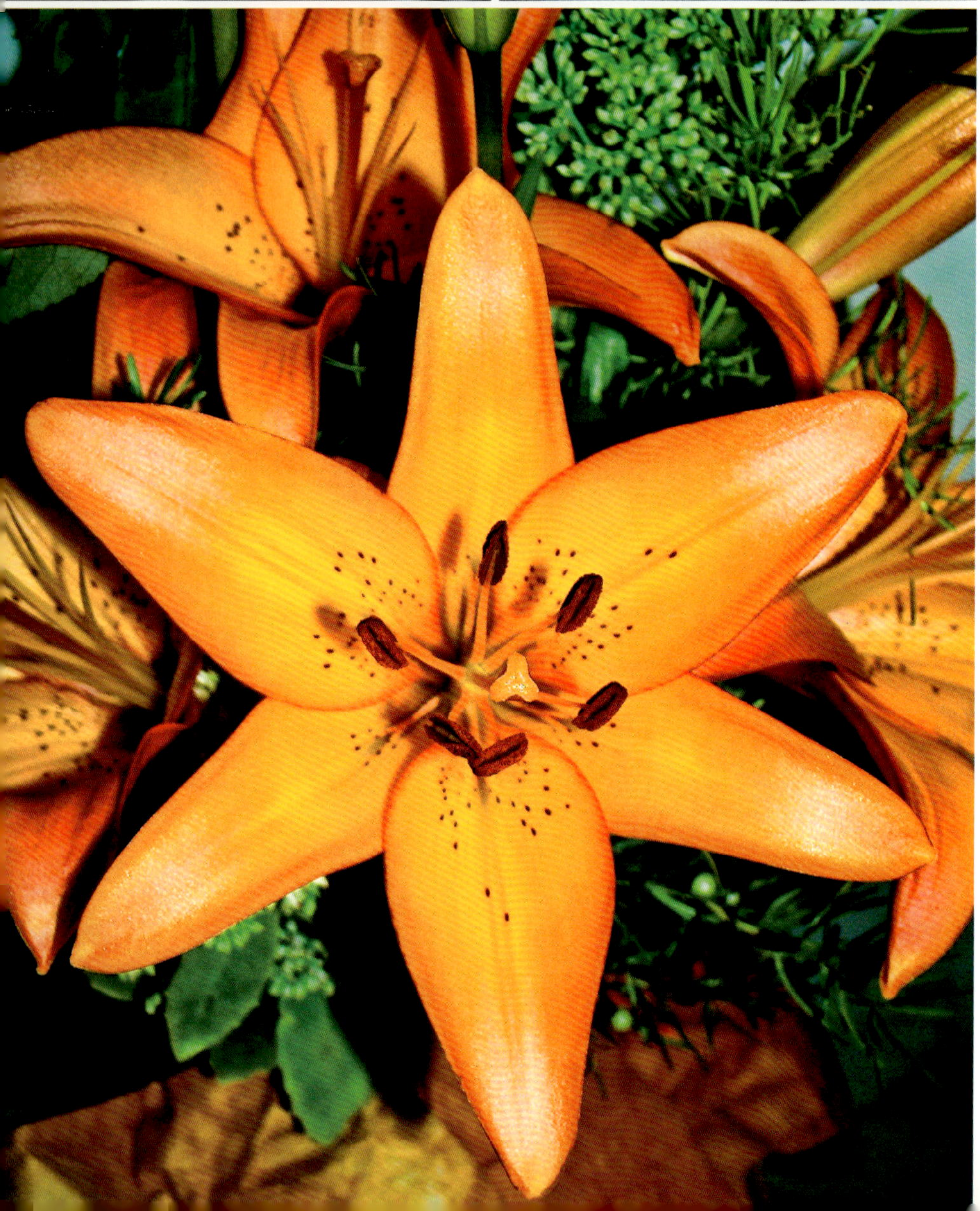

The apple core when the apple is dissected in the plane perpendicular to its stem shows fivefold symmetry and so does the apple blossom (below, right).

The flower at the left and the one at the bottom of this page have six petals, but their symmetry is only threefold. It is only possible to place three symmetry planes across these flowers because their six petals are not in the same plane; only three of them are in the same plane in each case.

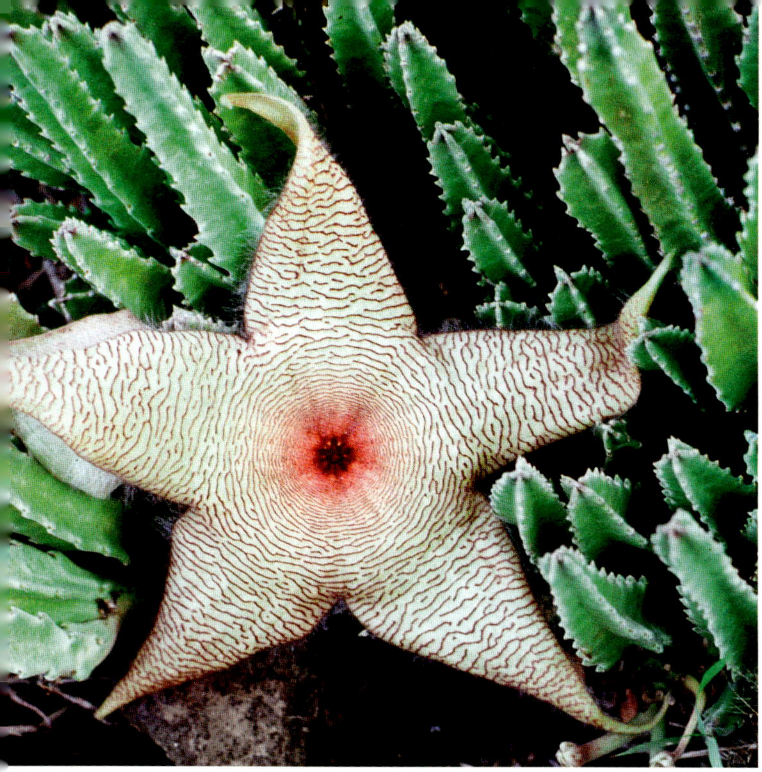

This cactus-like flower is not a cactus at all. It resembles the starfish, but it is not one. It came to Hawaii from southern Africa.

Carrion flower
(Stapelia gigantea)

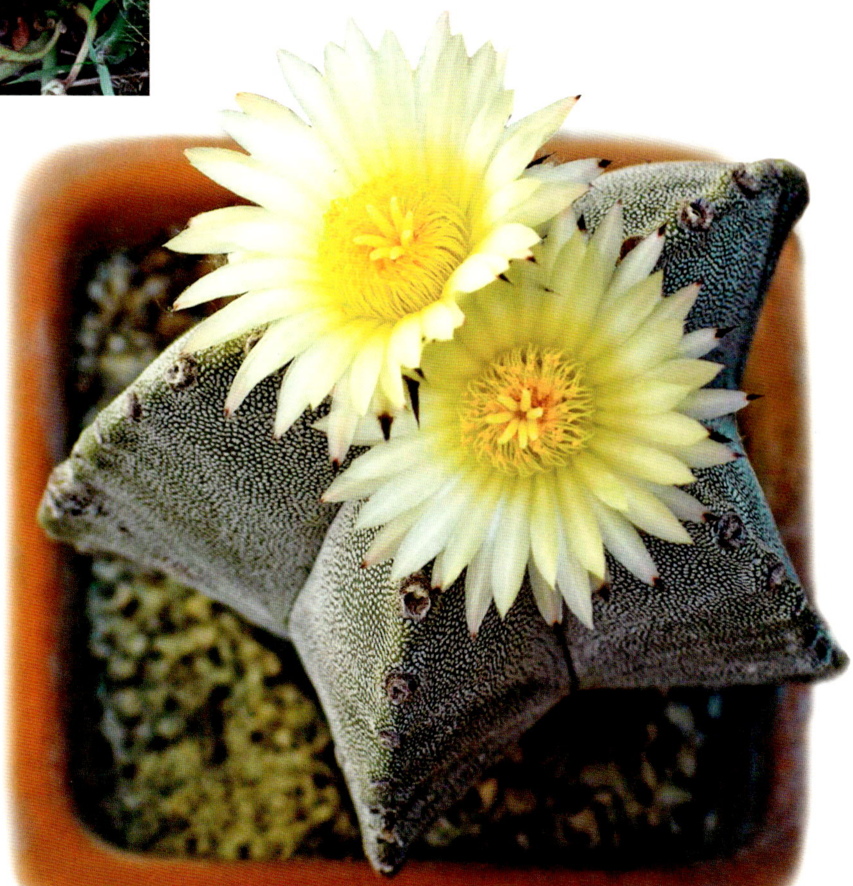

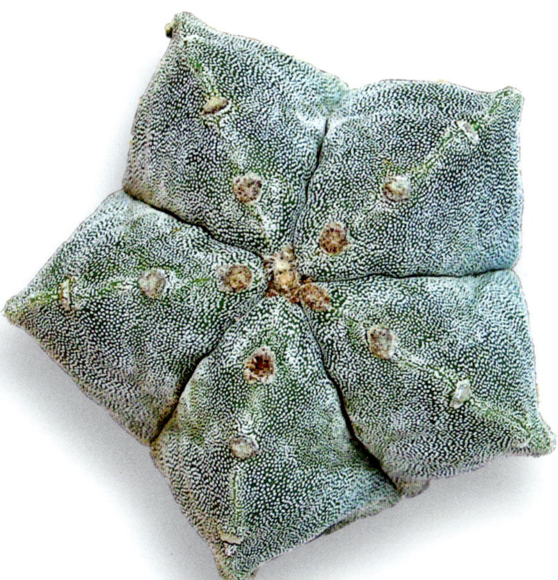

The cactus in the picture also has fivefold symmetry. Above it is shown blossoming. Taken together with its flowers, there is no symmetry. The flowers by themselves are symmetrical, but it is not easy to count their many petals—they have many-fold symmetry. The number of pistils is seven, so that part has sevenfold symmetry.

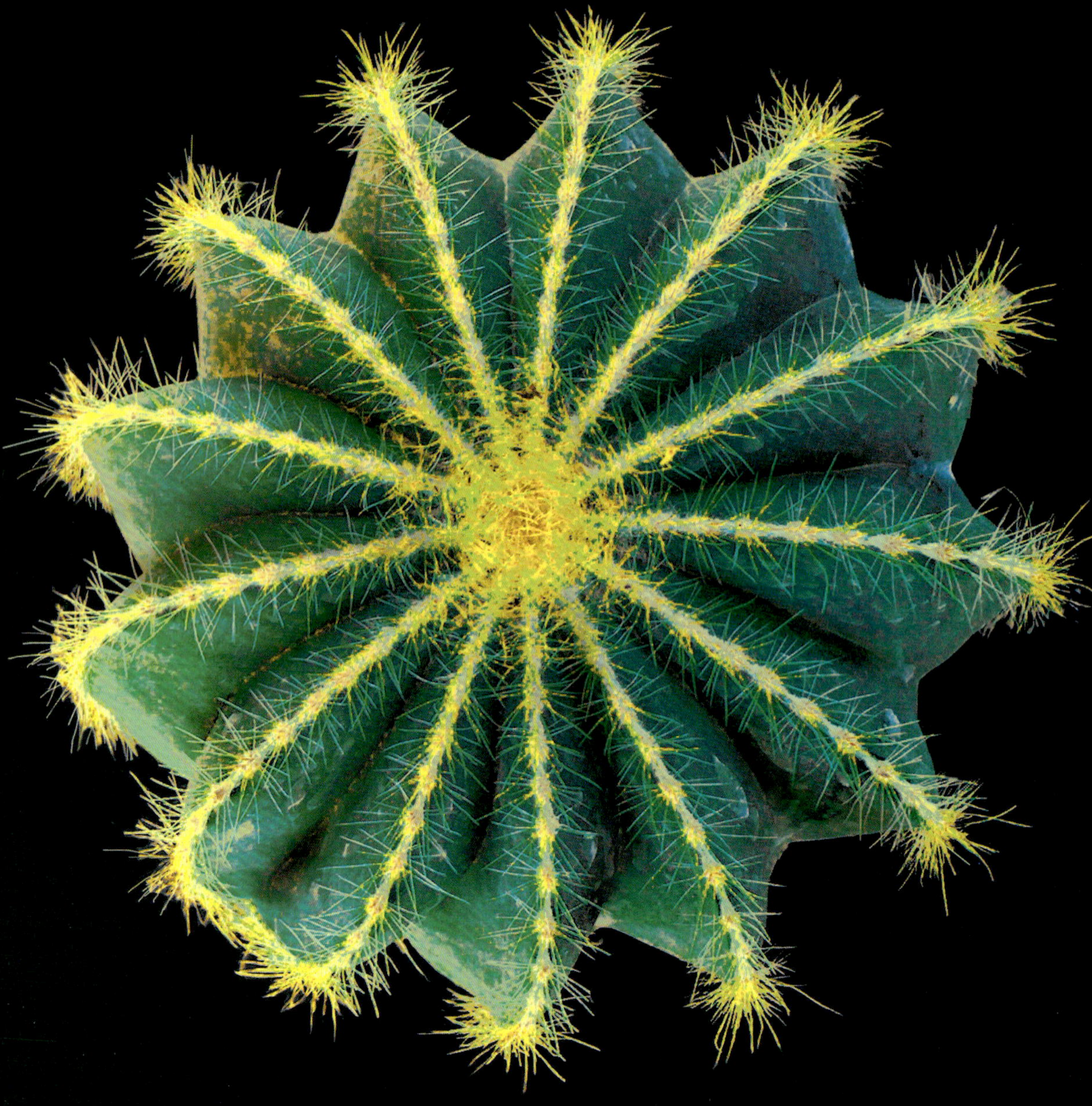

Cacti are rich in symmetries, especially in multiple mirror symmetries.

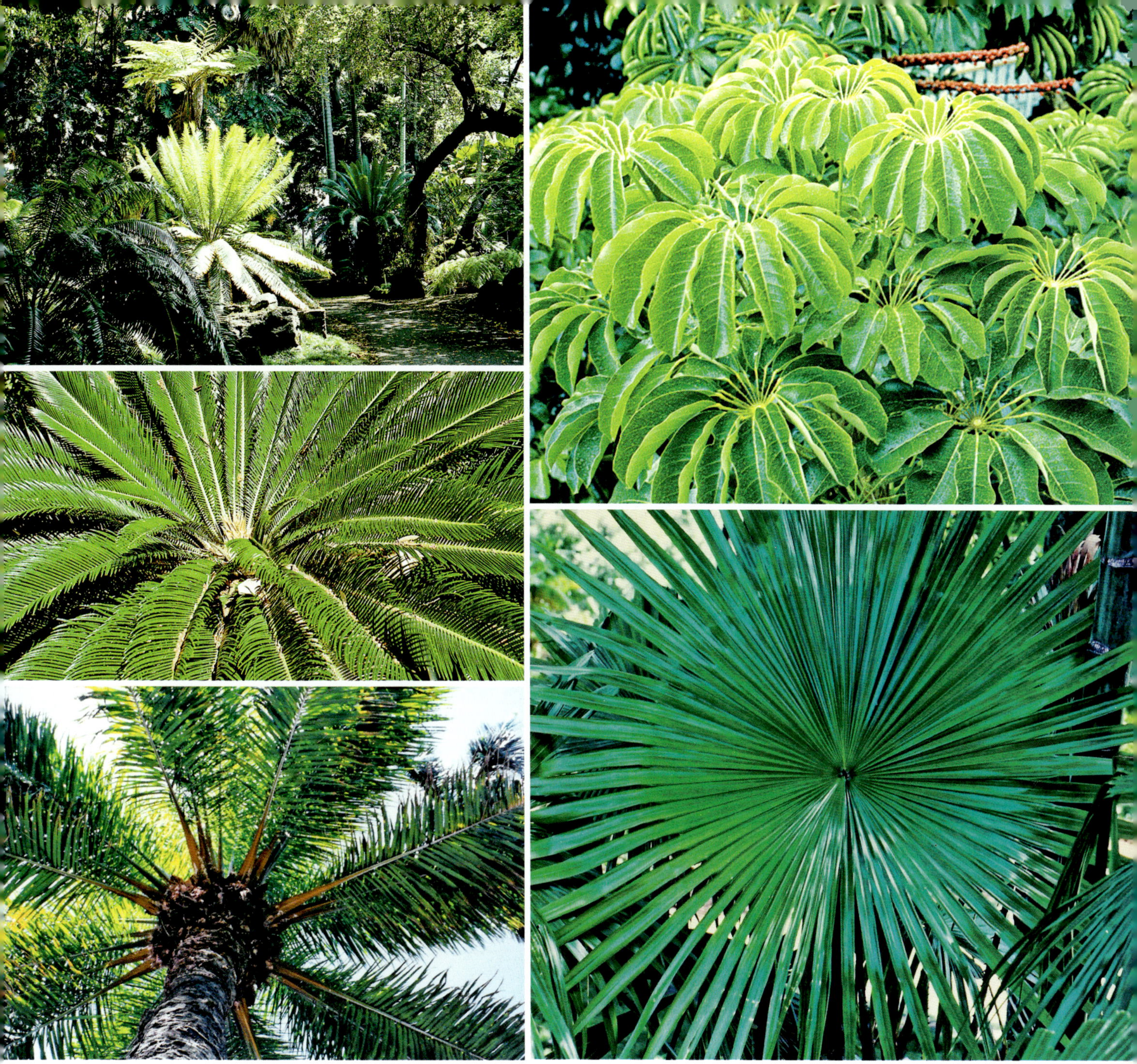

Multiple mirror symmetries often occur in the world of plants. All these examples are from Oahu, Hawaii.

Man-made Objects

Symmetry planes induce the feeling of calmness and security and occur often in artifacts. From 2 symmetry planes to 16, there is a variety of multiple symmetry planes in these examples.

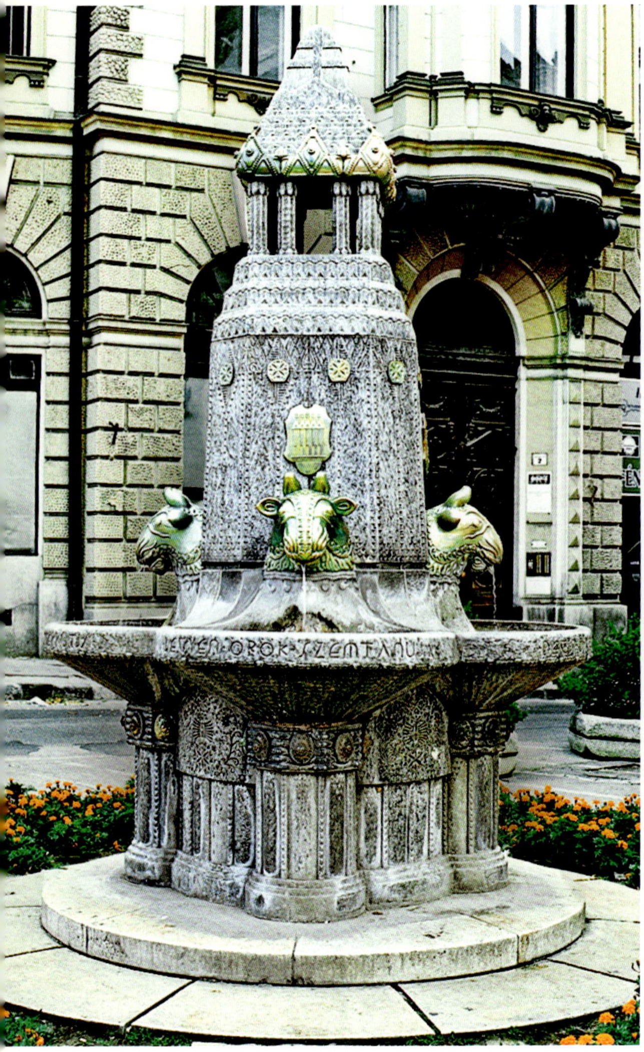

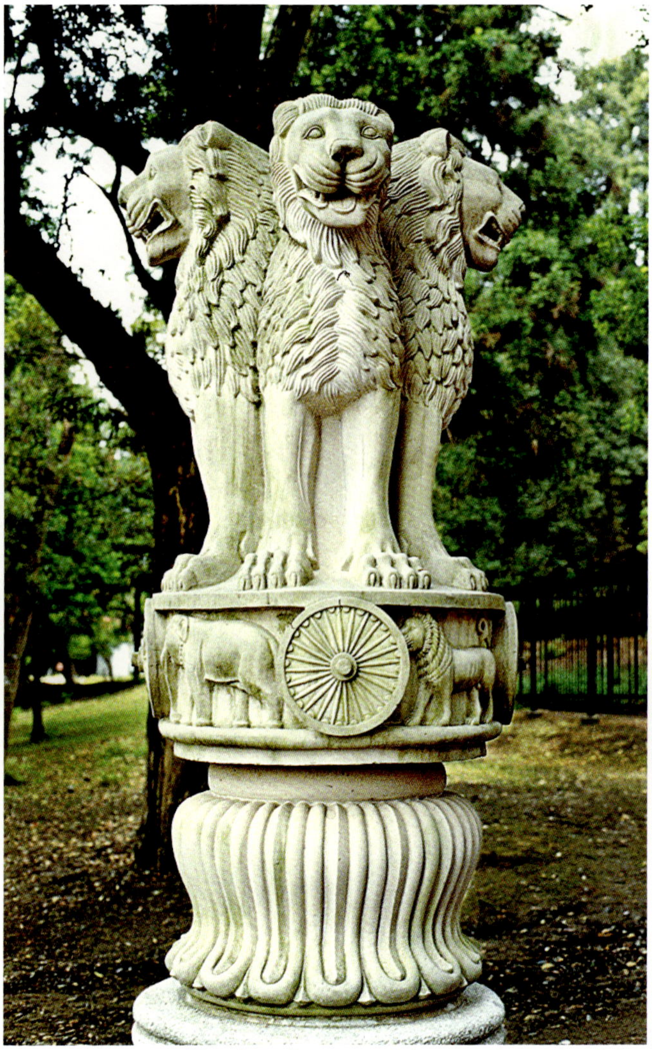

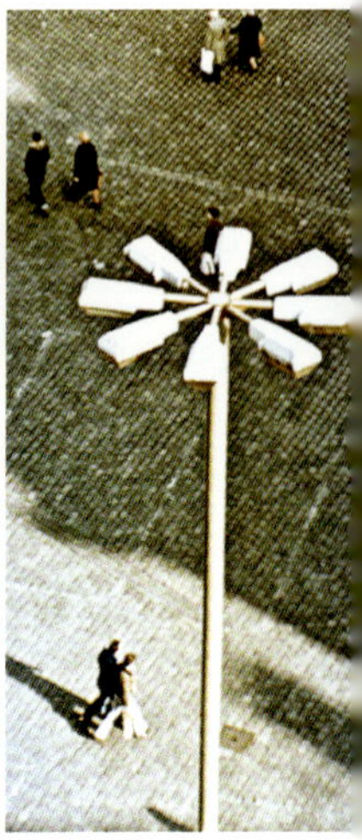

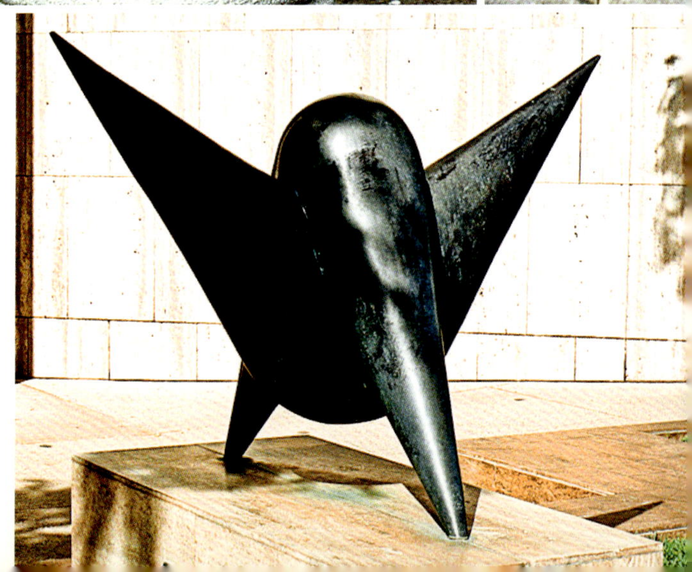

From upper left clockwise:

Zsolnay Fountain in Pécs, Hungary
Four lions in a monument honoring India in Paris
Street lamp in Prague
Sculpture in Honolulu, Hawaii

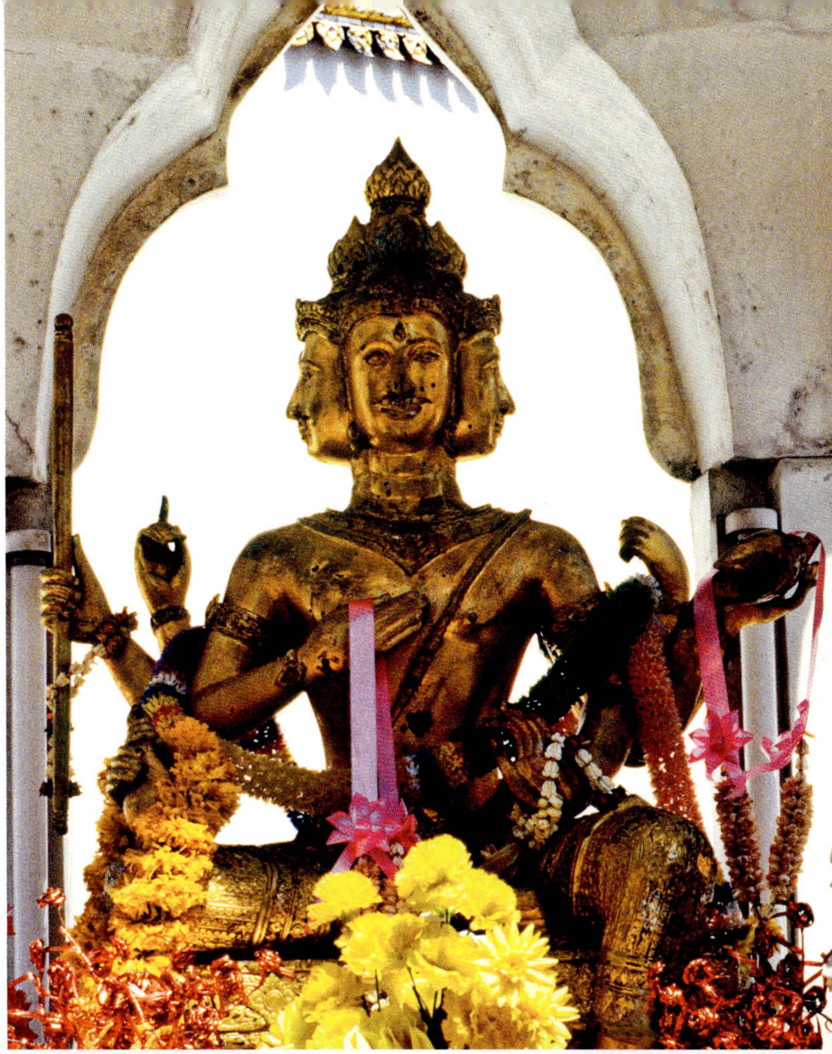

Buddha statue in Bangkok

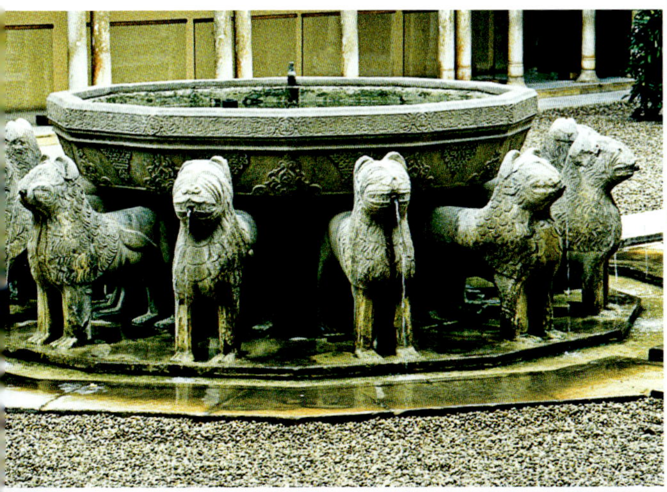

The Lion Fountain at the Alhambra
Palace in Granada, Spain

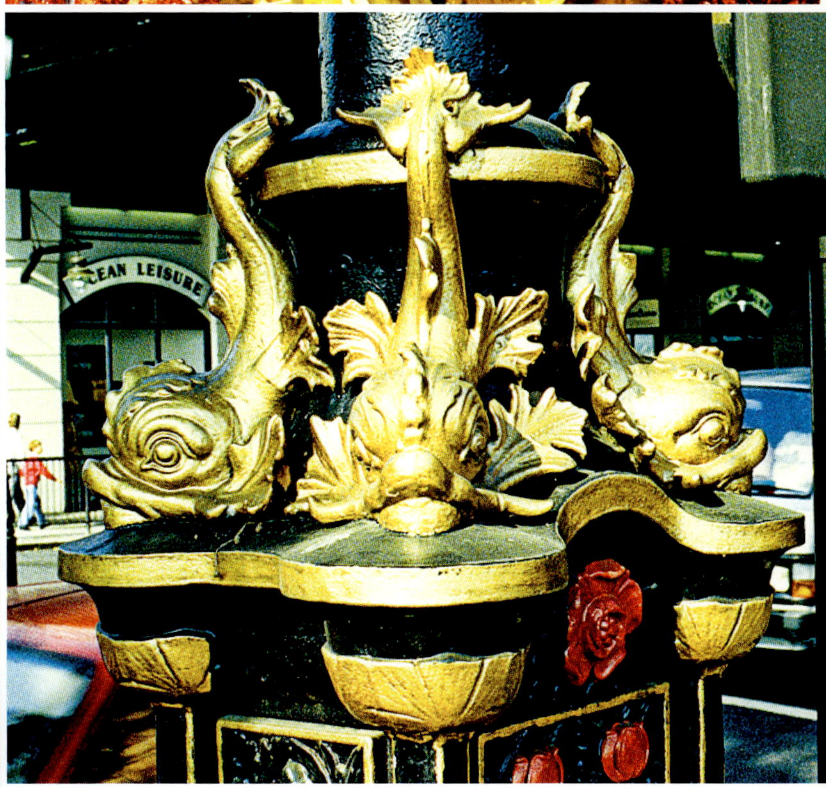

Lamp stand (detail) in London

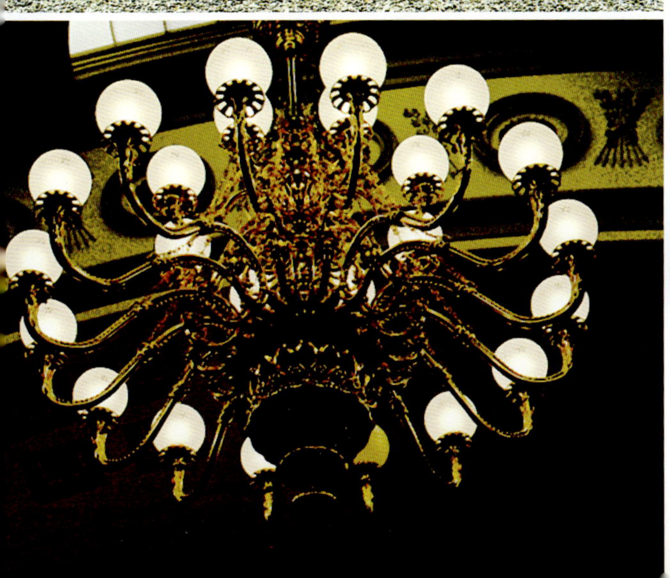

Chandelier in the State Capitol,
Austin, Texas

Architecture

Mirror planes are common in architecture although too much symmetry in buildings may hinder easy orientation. Certain symmetries, however, follow from the functions of constructions. It is natural to have two perpendicular symmetry planes for bridges, for example.

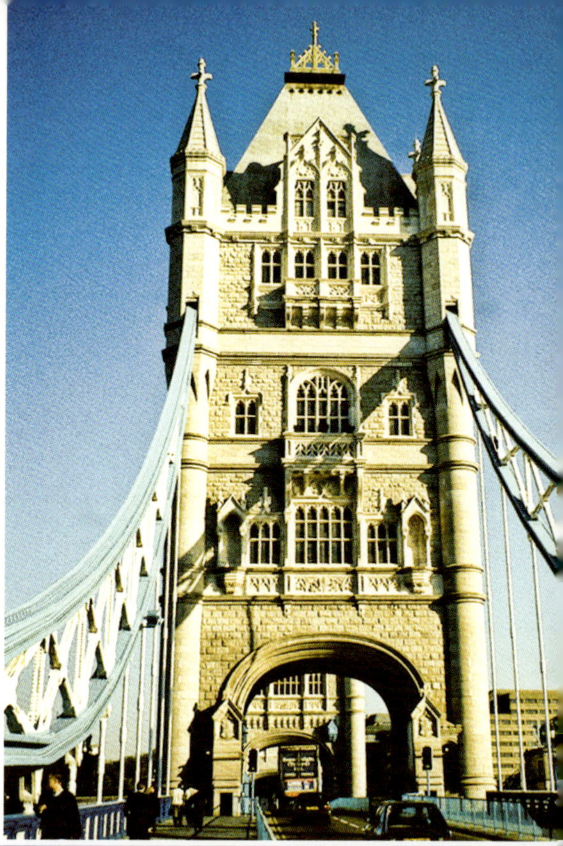

Tower Bridge in London in two views

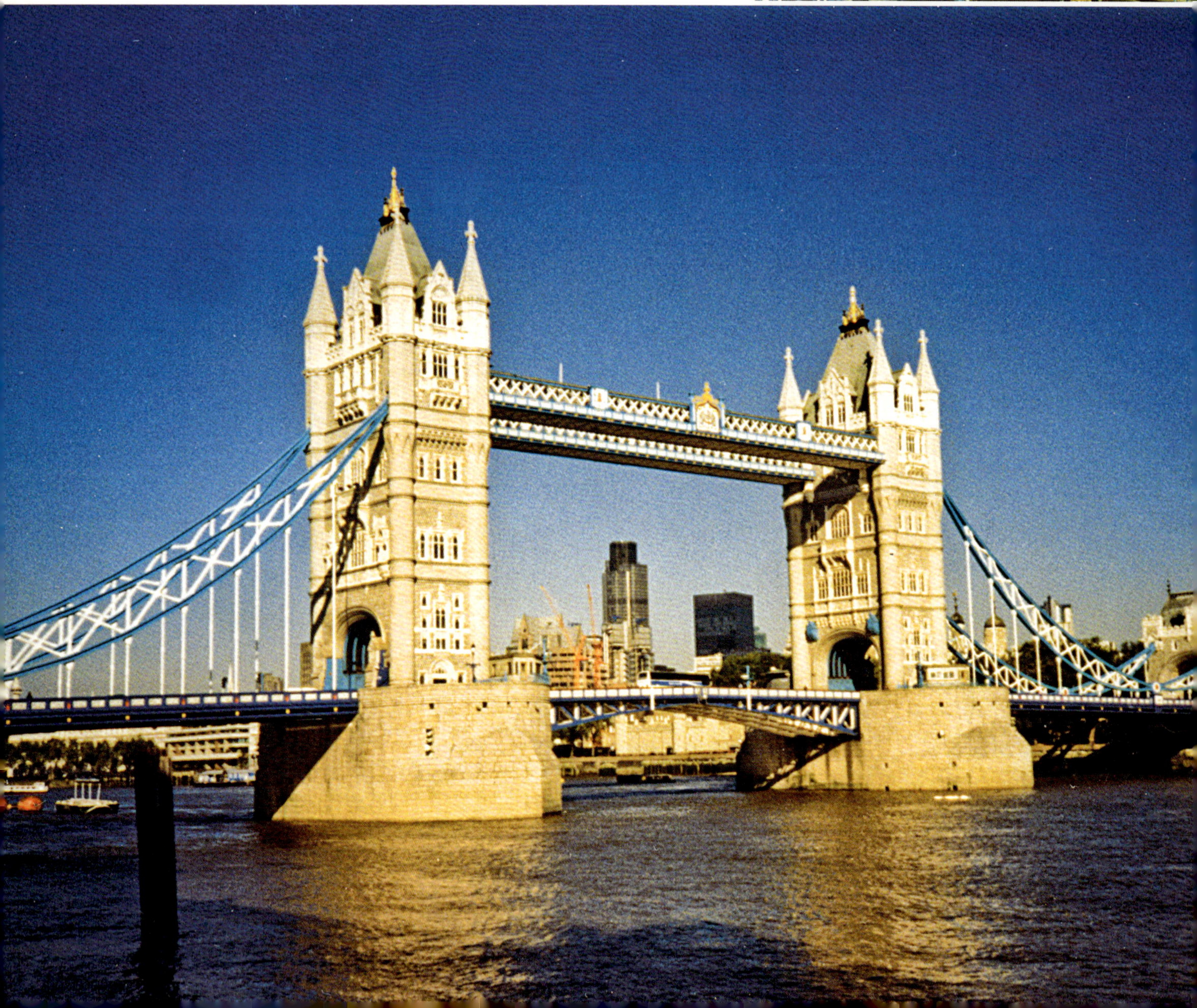

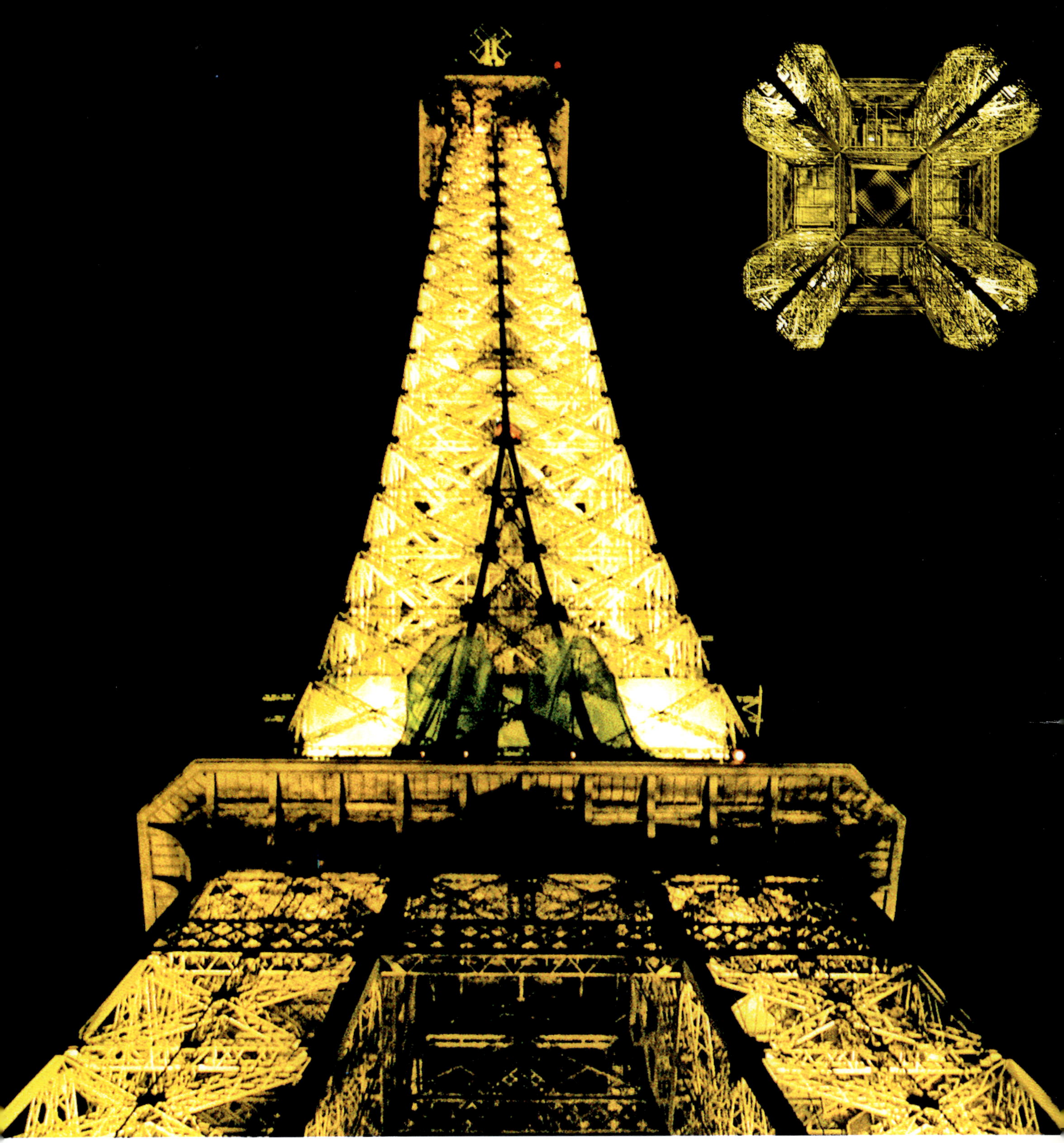

The Eiffel Tower in Paris from the side and from below, both at night

The Eiffel Tower in Paris has four symmetry planes, two across its sides and two connecting its corners though its square outline and fourfold symmetry is not directly seen from the side. The symmetries of buildings and big structures would be easier to determine by flying above them, but we have seldom the opportunity of doing so. In case of the Eiffel Tower, however, it is possible to experience its outline directly from below. The images of the Eiffel Tower presented here were recorded from the side and from below, both at night.

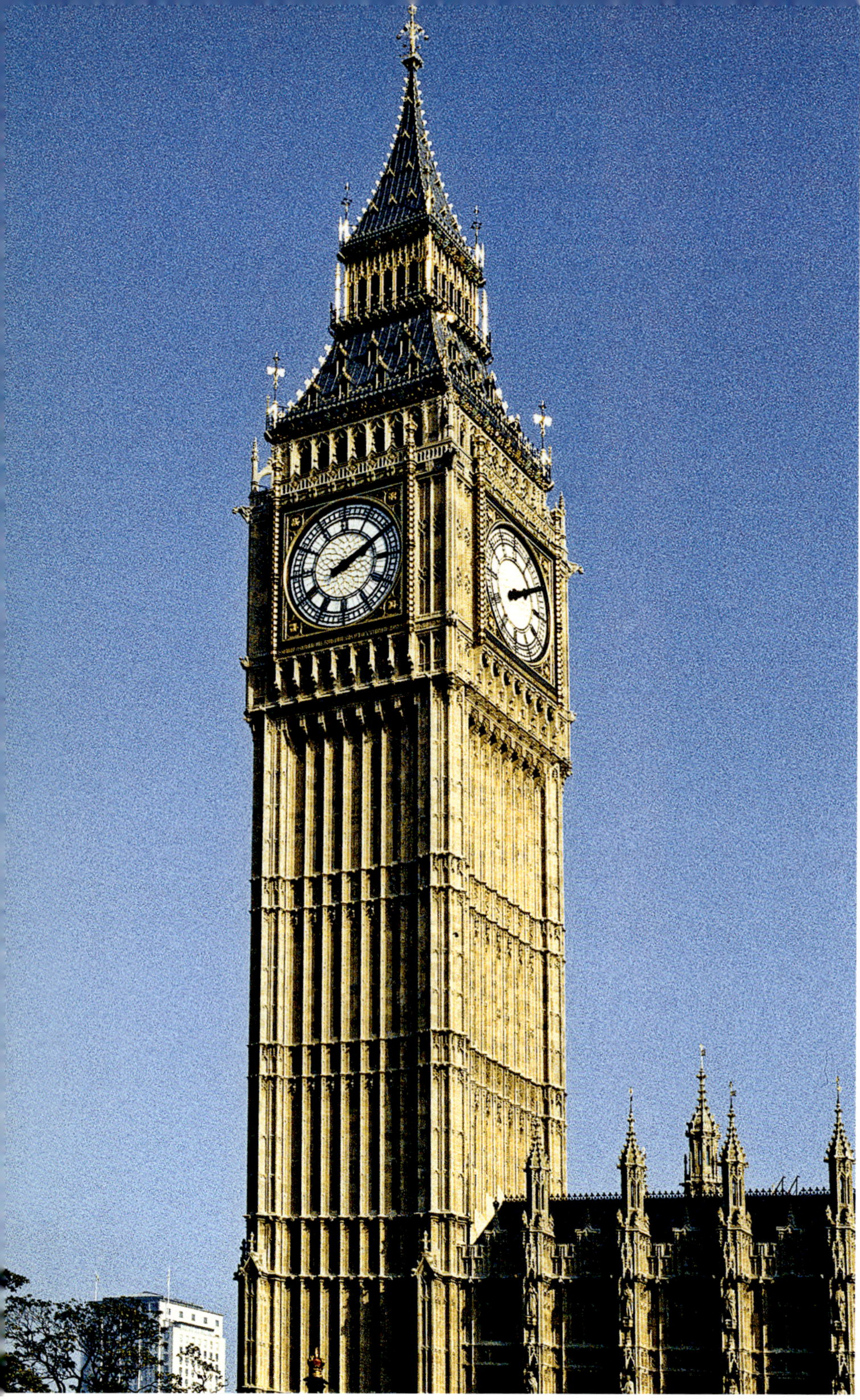

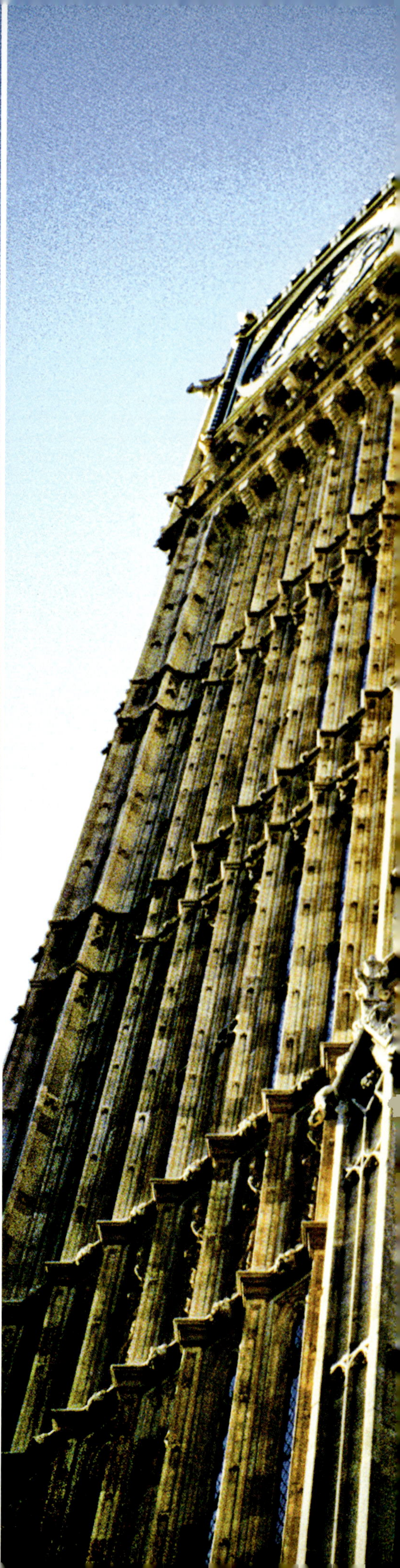

The Big Ben in London in two views; one, the usual, and the other, somewhat more special

Many other famous towers also have fourfold symmetries.

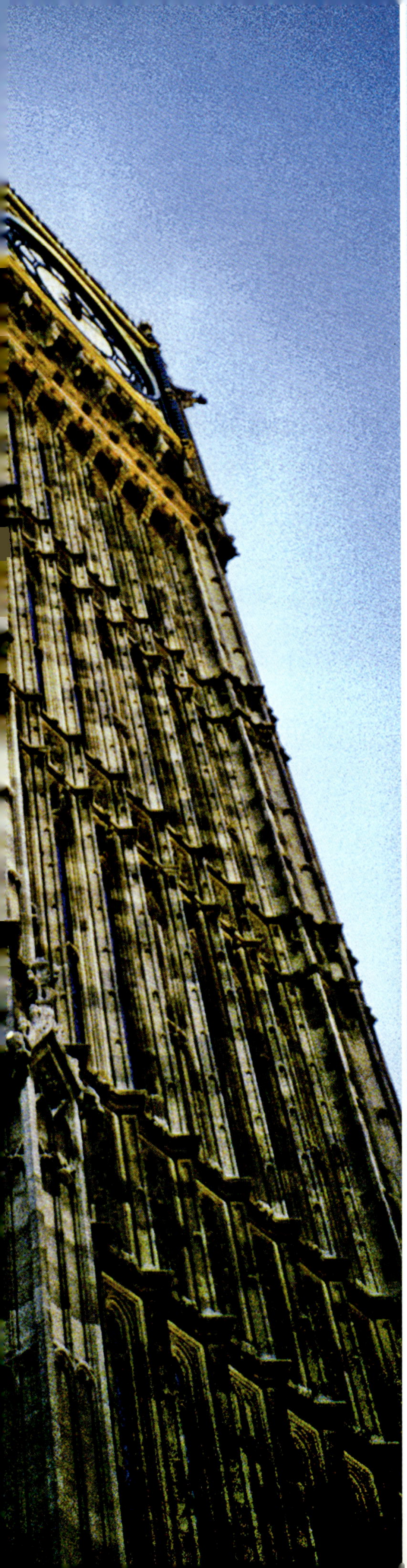

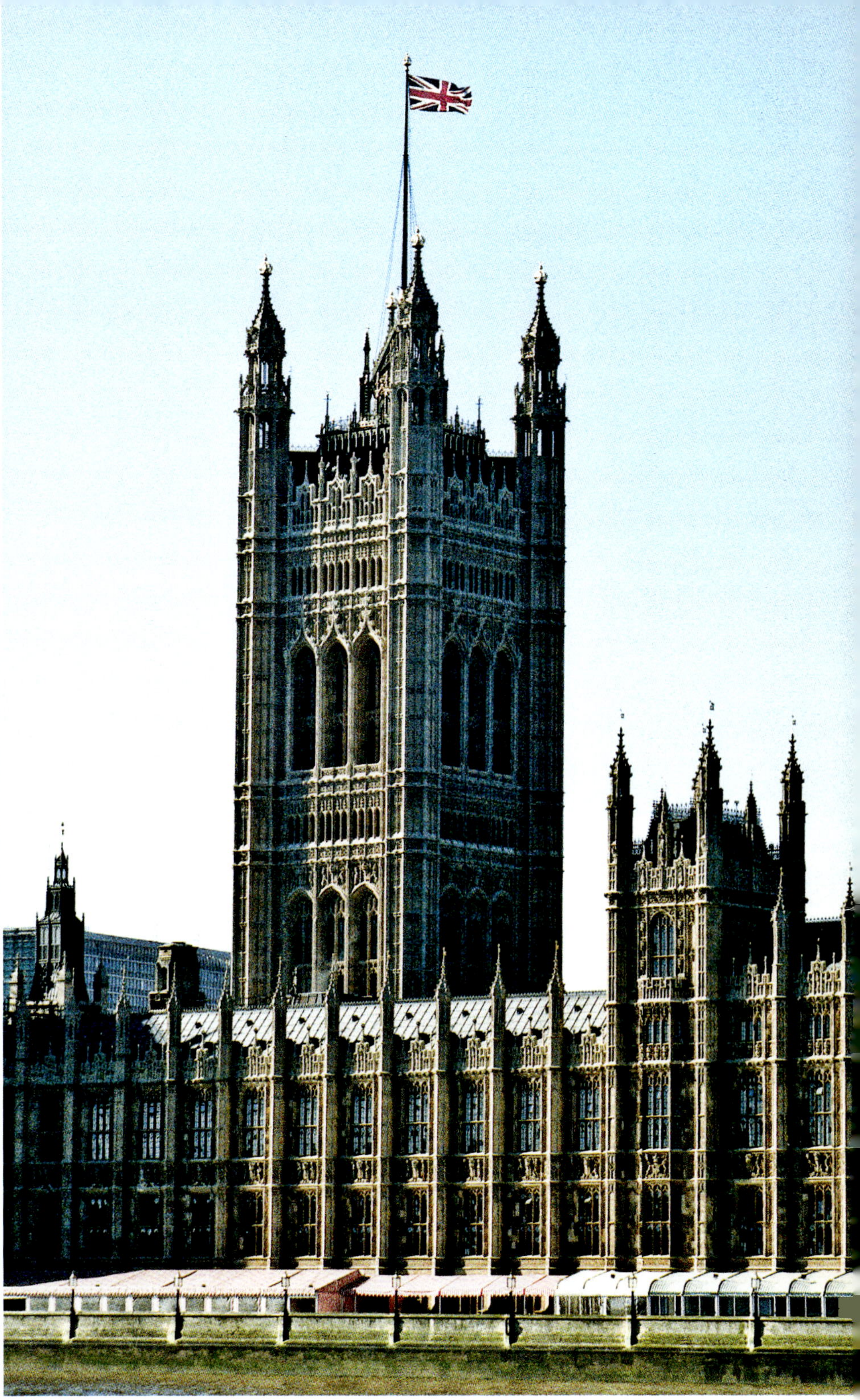

The Parliament Tower in London

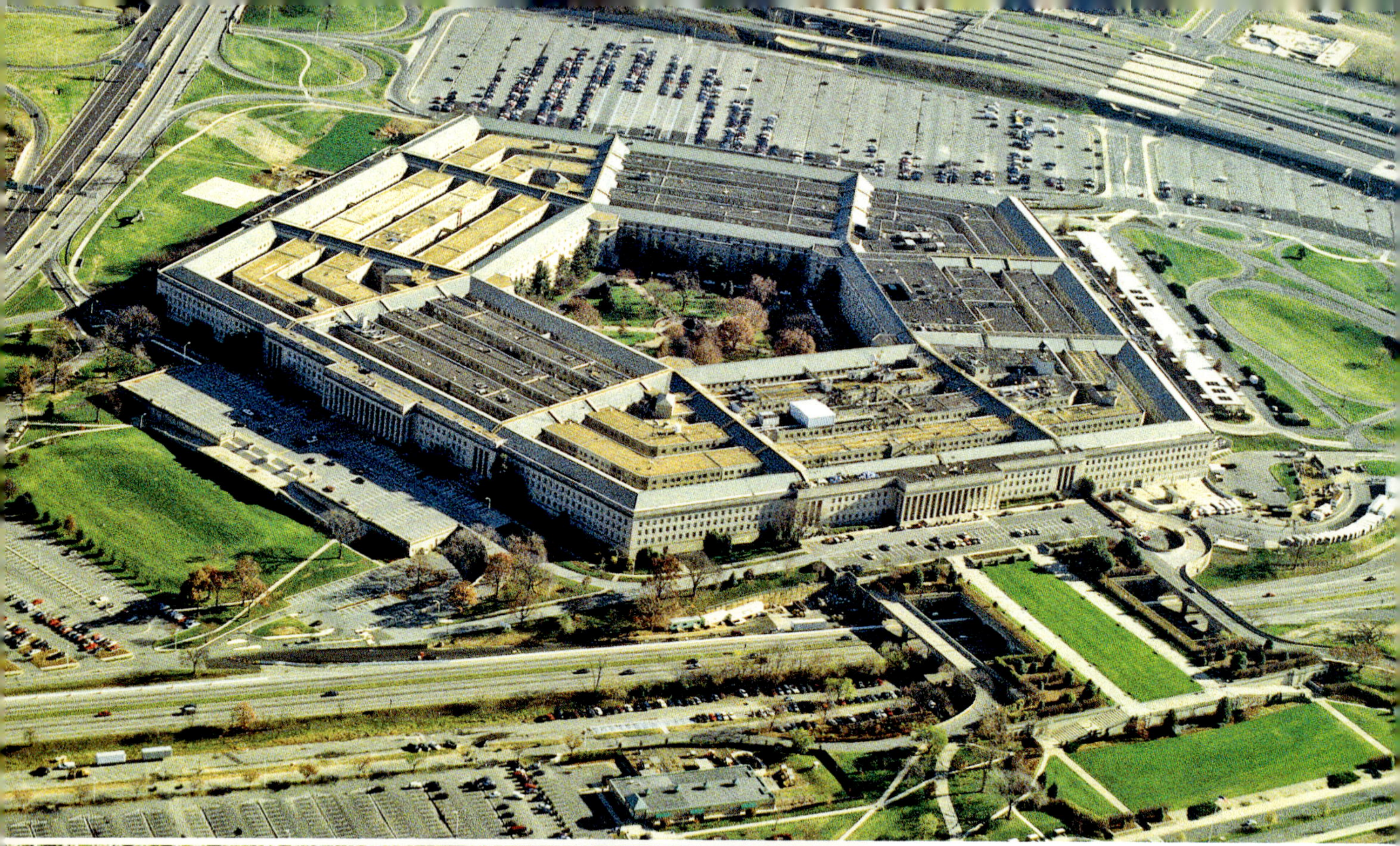

The Pentagon, the building of the U.S. Department of Defense in Washington, DC

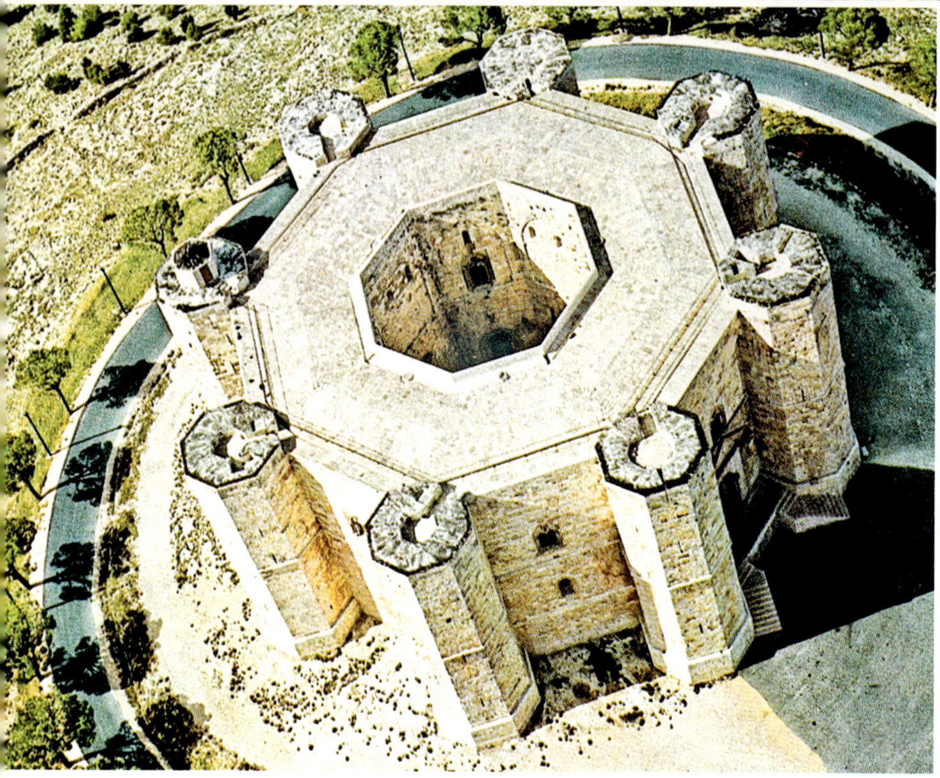

Castel del Monte in Apulia, Italy

The Pentagon, as suggested by its name, has fivefold symmetry. In this case, we were lucky enough to fly close to it to take an aerial snapshot.

The famous Castel del Monte in southern Italy has eightfold symmetry. The main building has this symmetry and so do the adjacent towers at each of its eight corners.

The famous corncob-like buildings in Chicago, the Leaning Tower in Pisa, Italy, and Hotel Budapest in Budapest all have multiple mirror planes. These buildings can also be described with cylindrical symmetry that will be discussed a little later.

Top: Leaning Tower in Pisa, Italy

Bottom: Hotel Budapest in Budapest

Corncob-like skyscrapers in Chicago

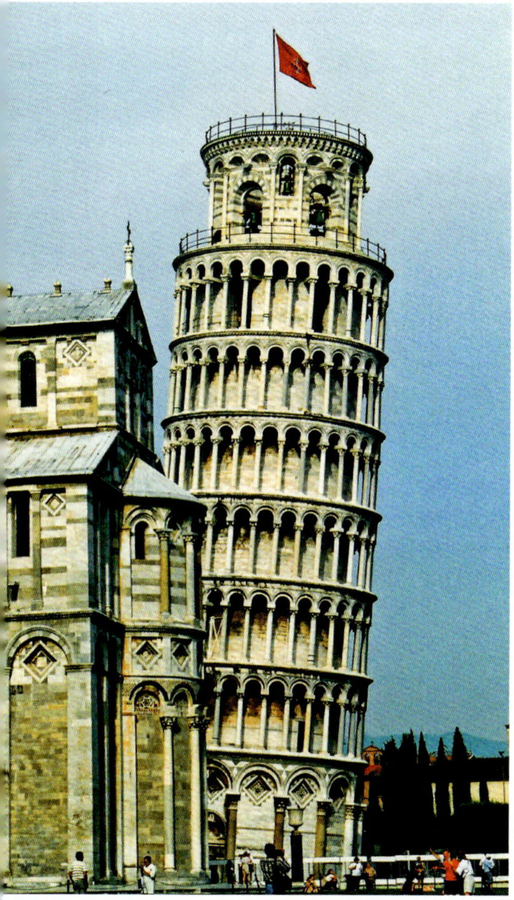

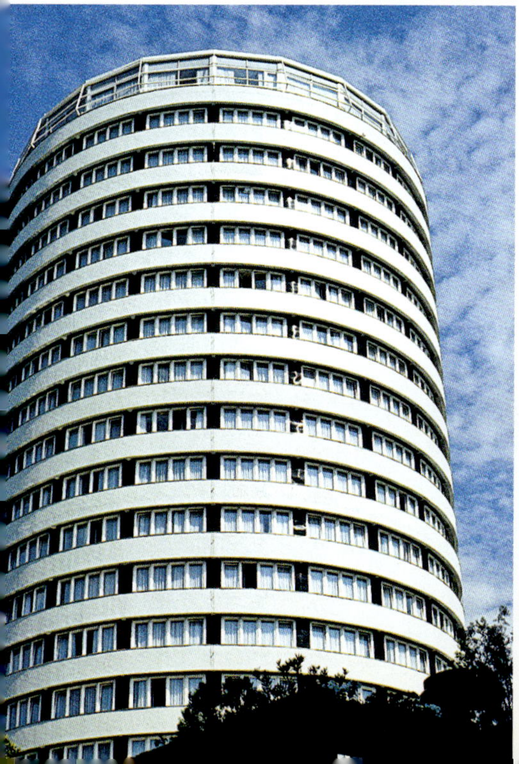

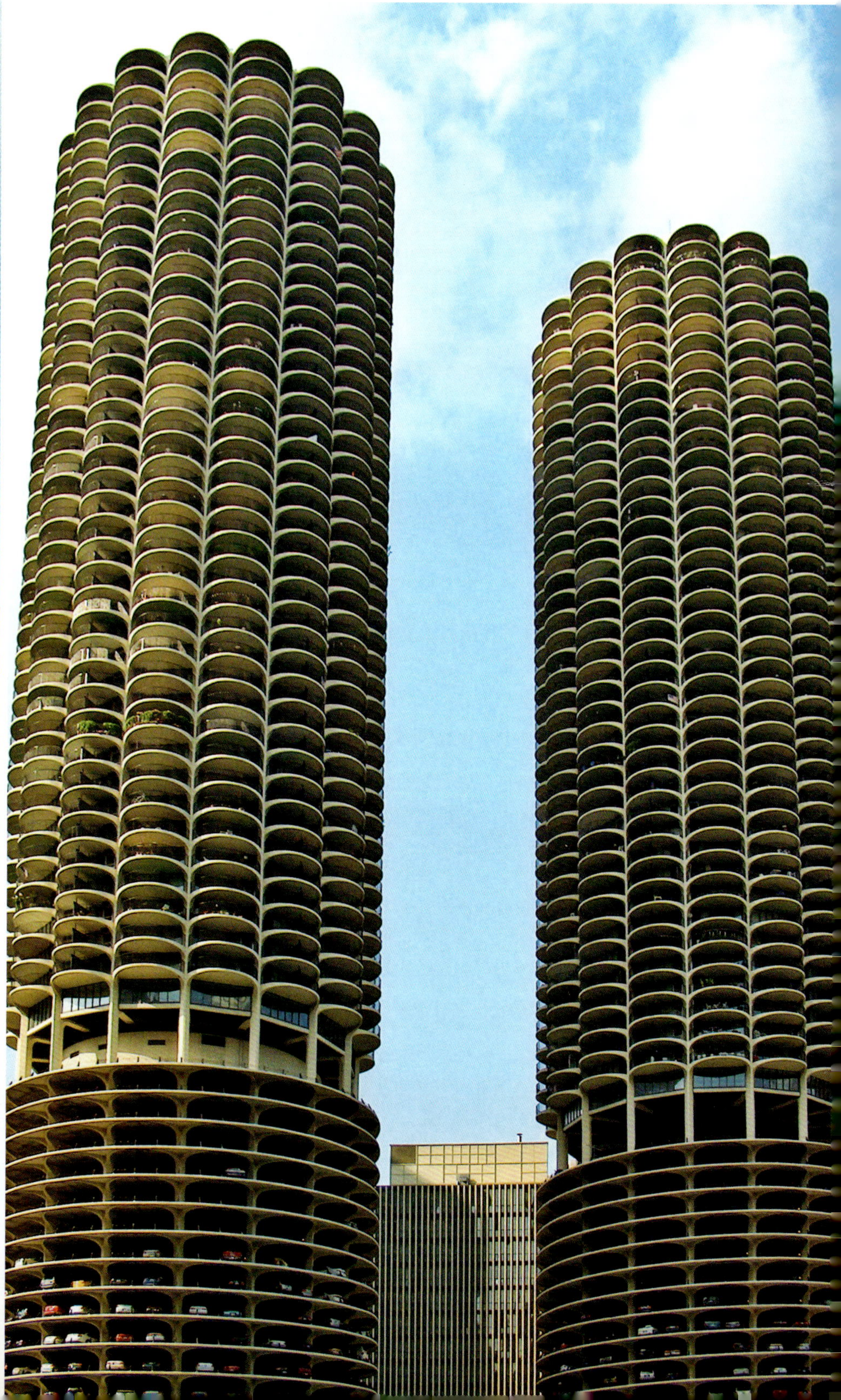

Rosettes have multiple mirror symmetries as indicated by a few examples:

Window of the Notre Dame Cathedral in Paris from the outside and inside

Rosettes in Manhattan, New York (below) and in Erice, Sicily (right)

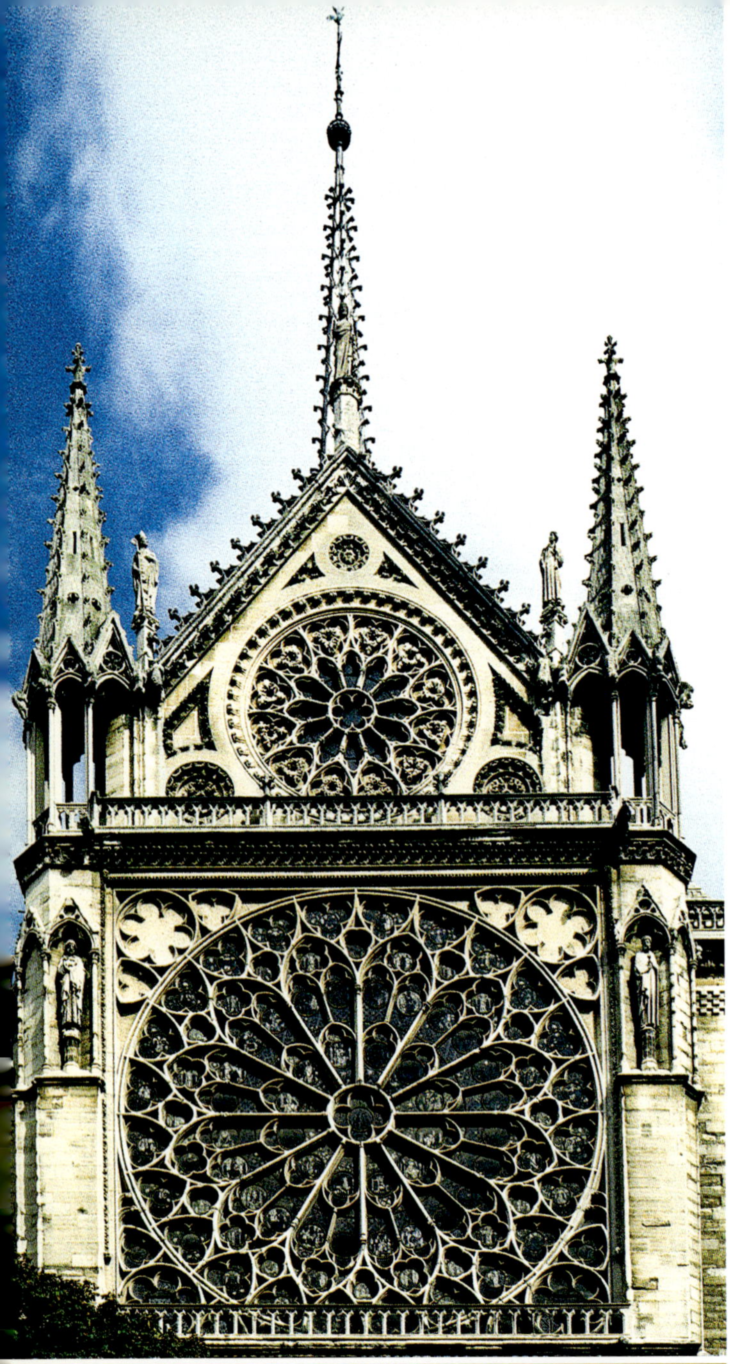

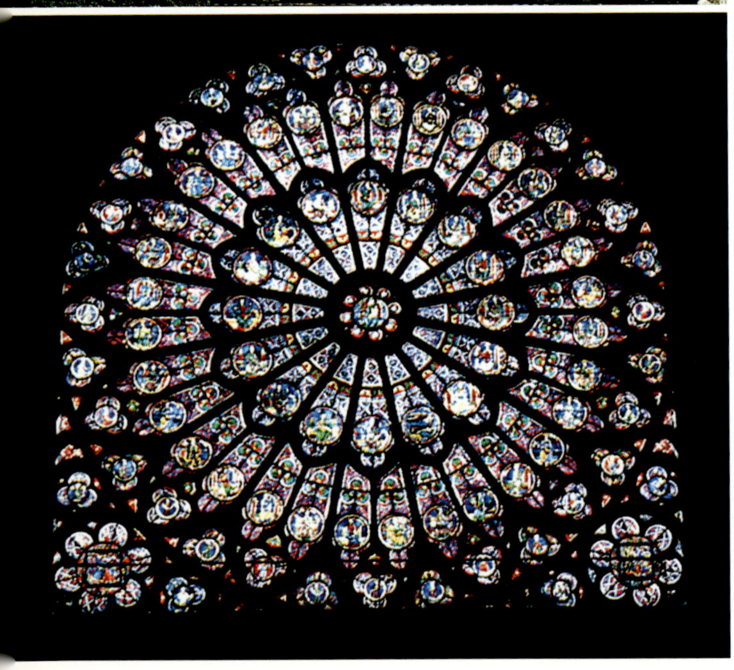

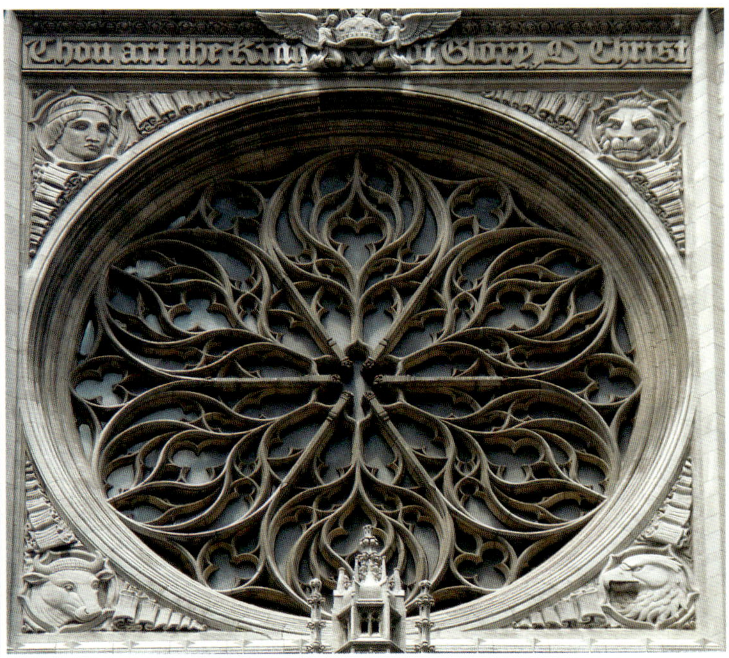

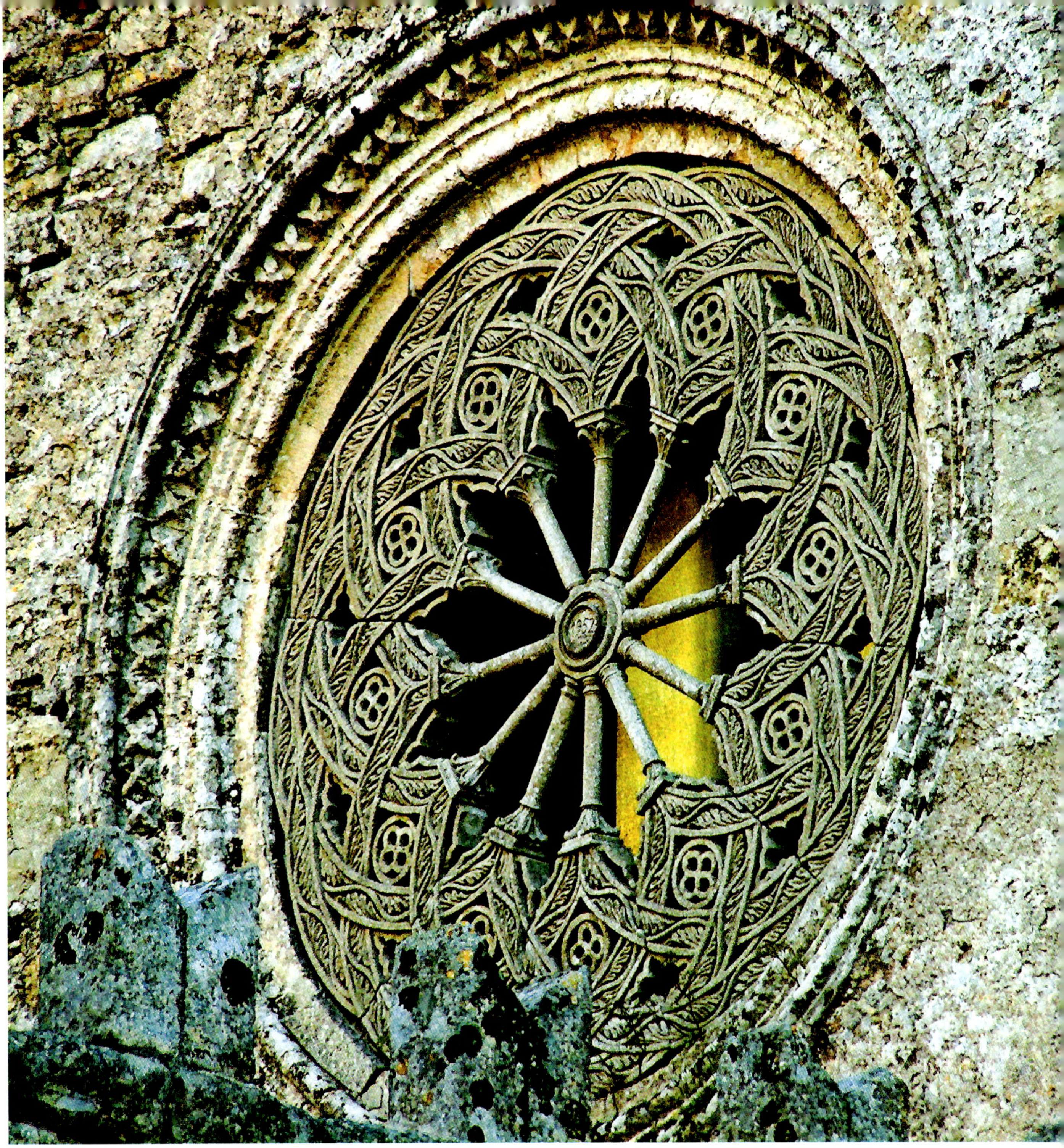

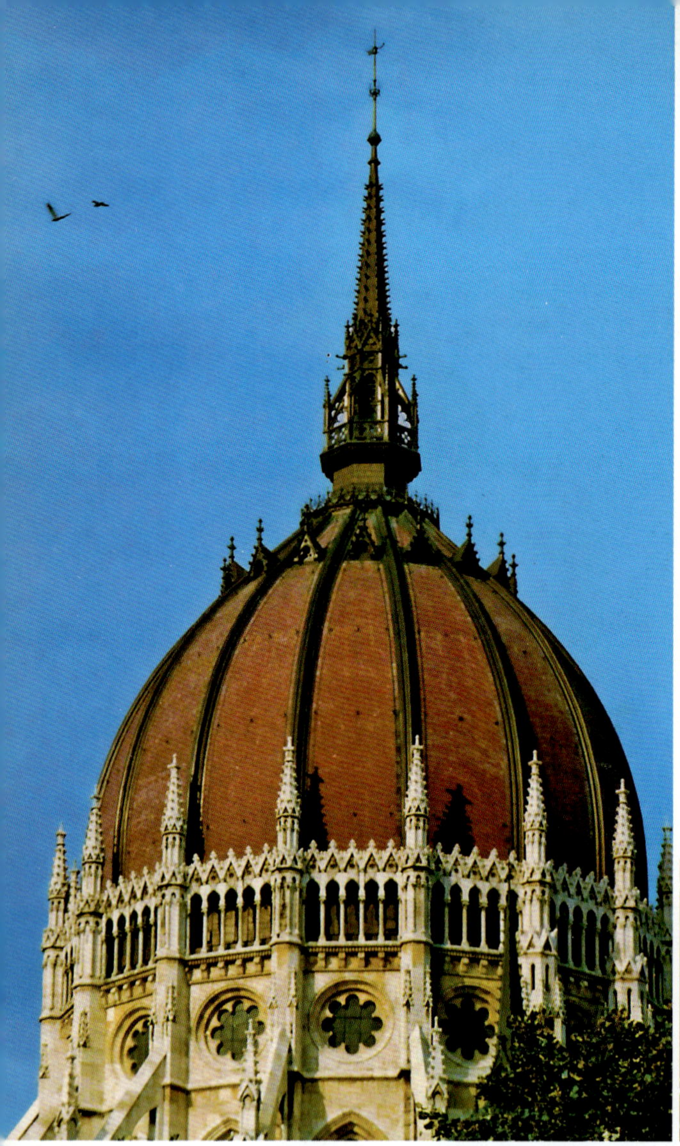
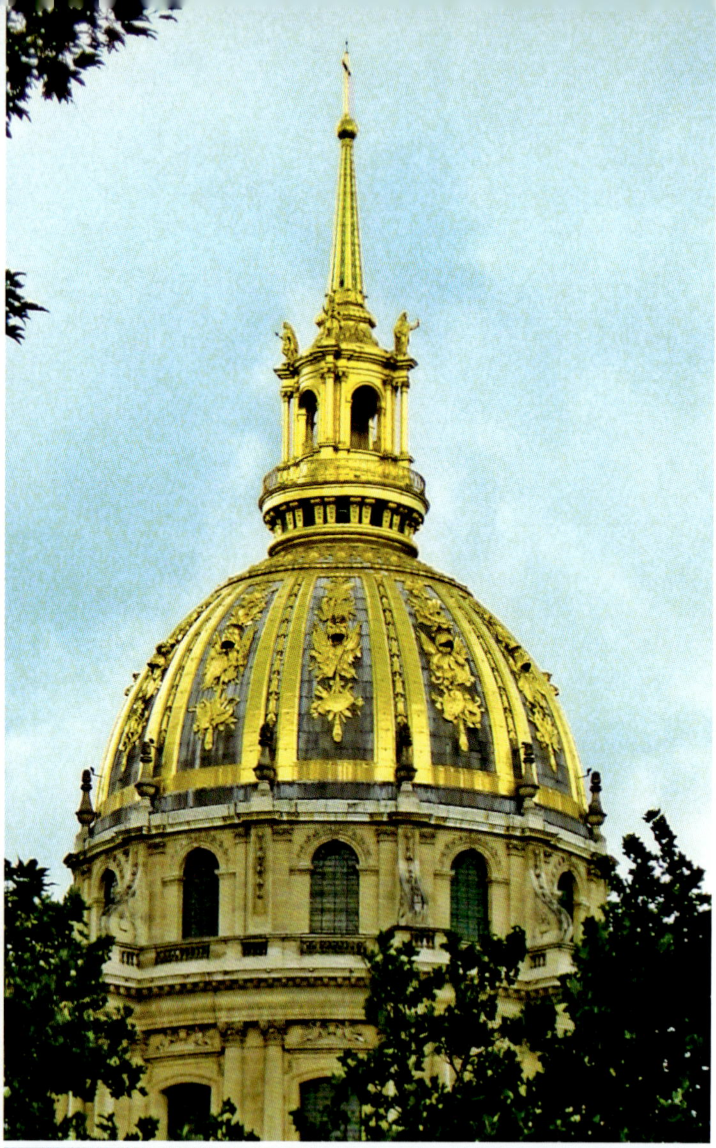
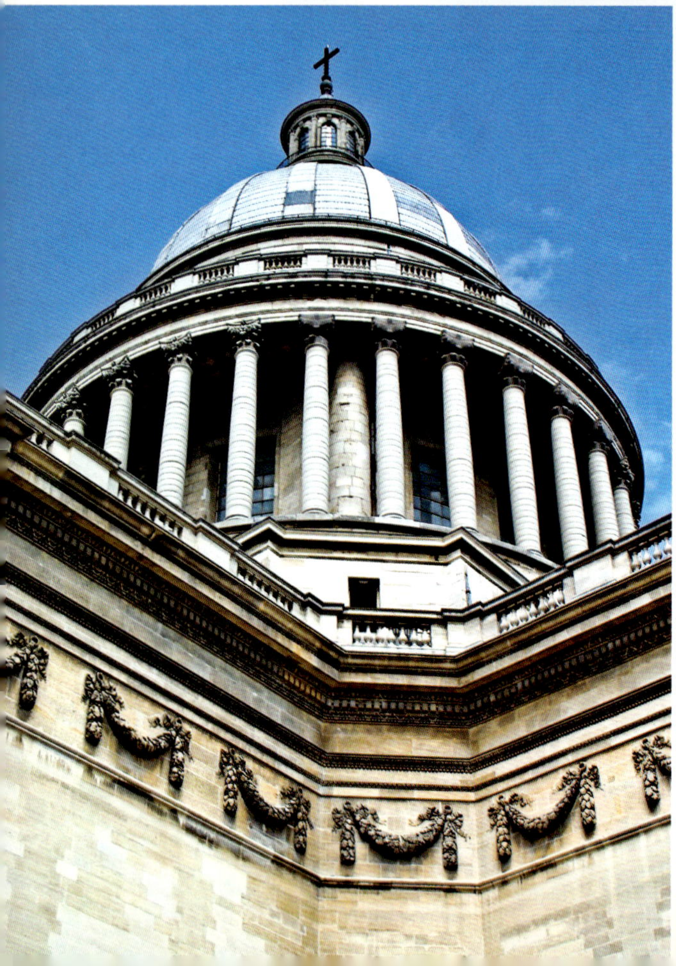
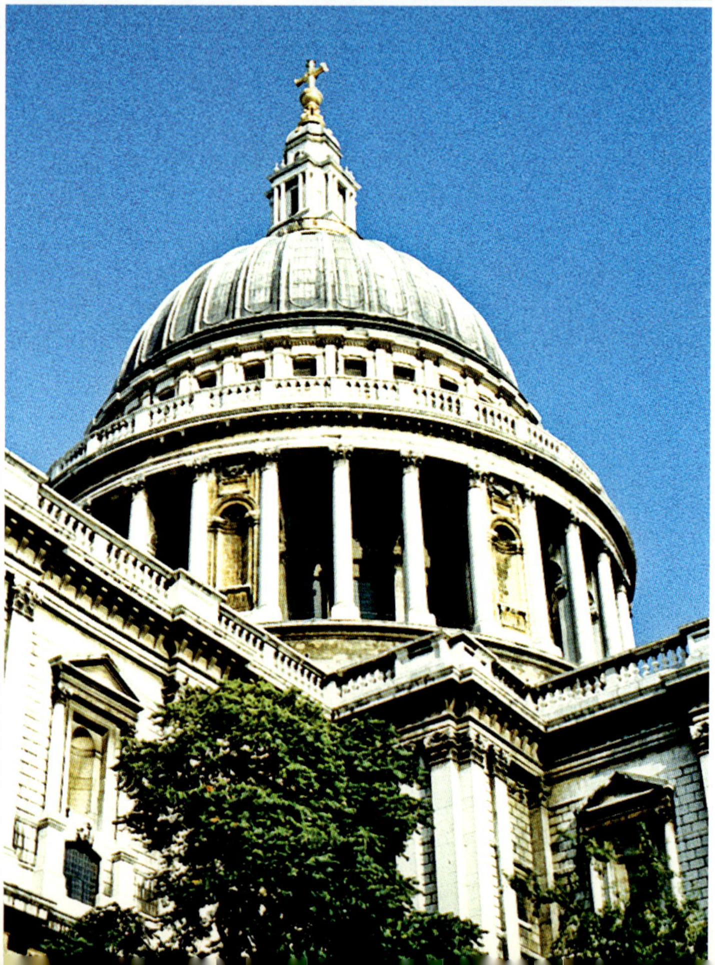

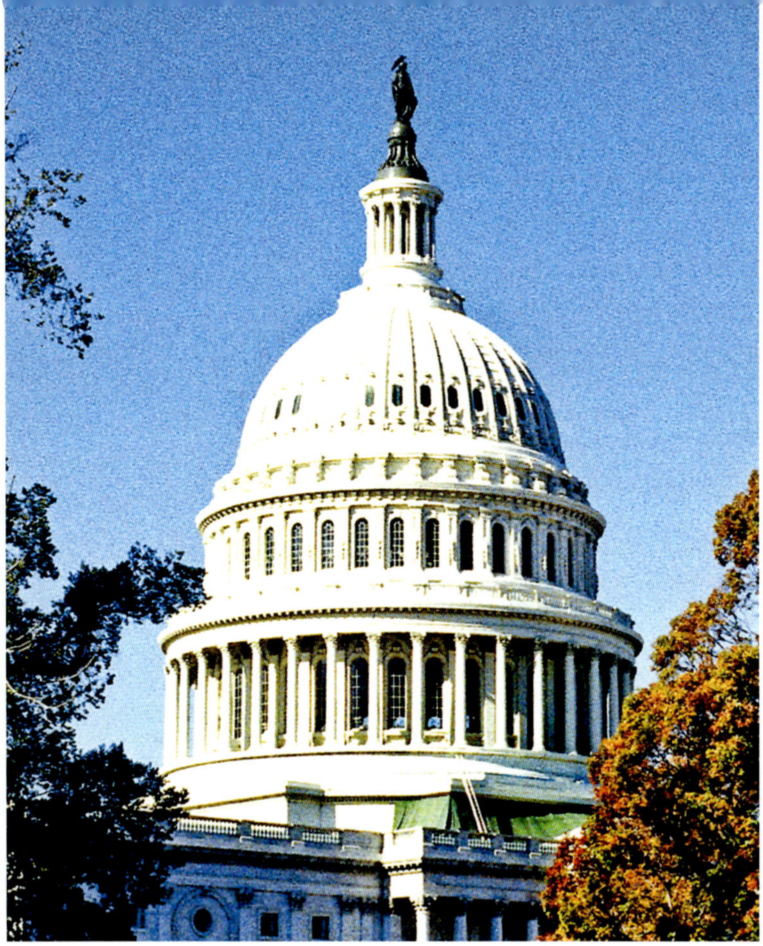

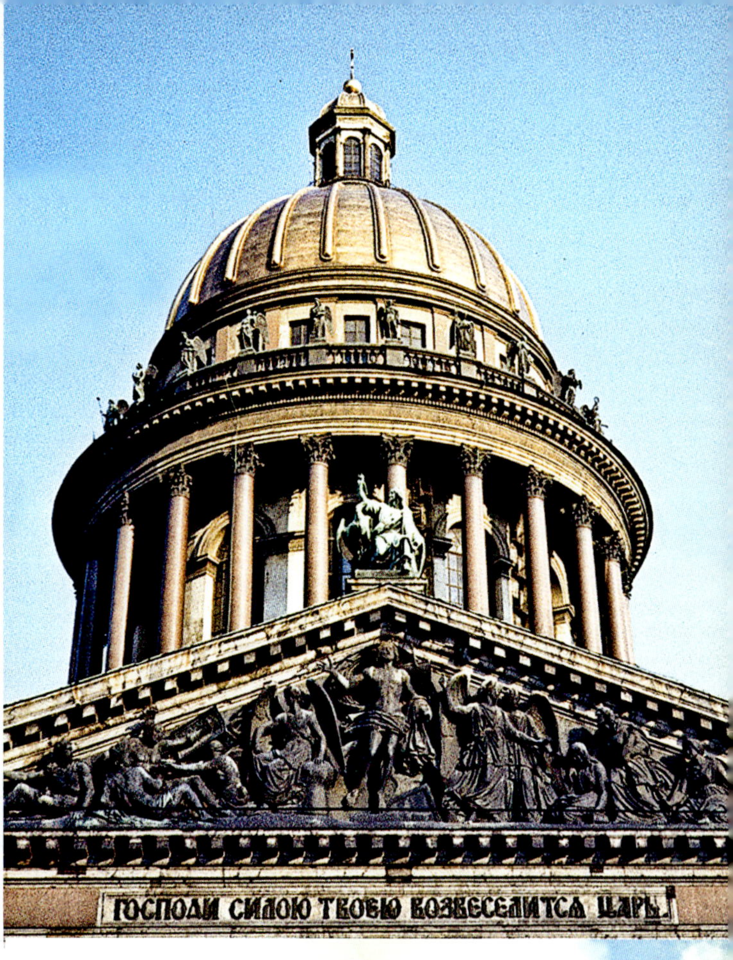

The cupolas of many important buildings, among them many state capitols in the United States, have multiple mirror planes.

Page 56

From upper left clockwise:

Parliament building in Budapest
Invalids' Church in Paris
St. Paul's Cathedral in London
Pantheon in Paris

Page 57

From upper left clockwise:

Capitol Dome in Washington, DC
St. Isaac's Cathedral in St. Petersburg
St. Peter's Basilica in Vatican City

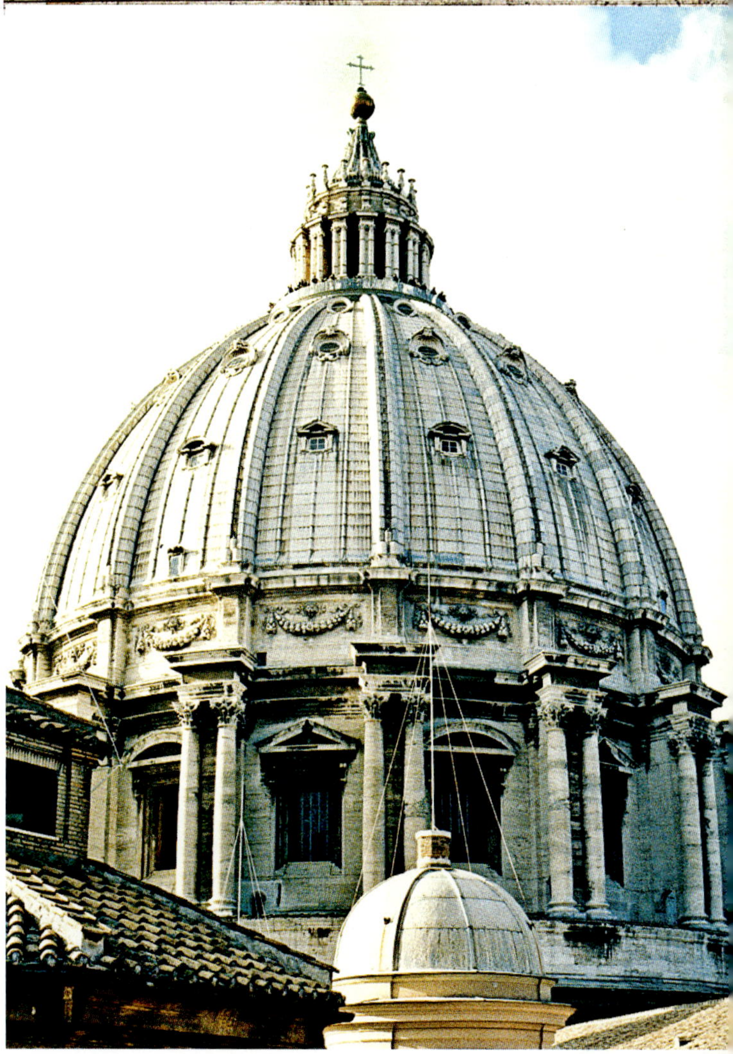

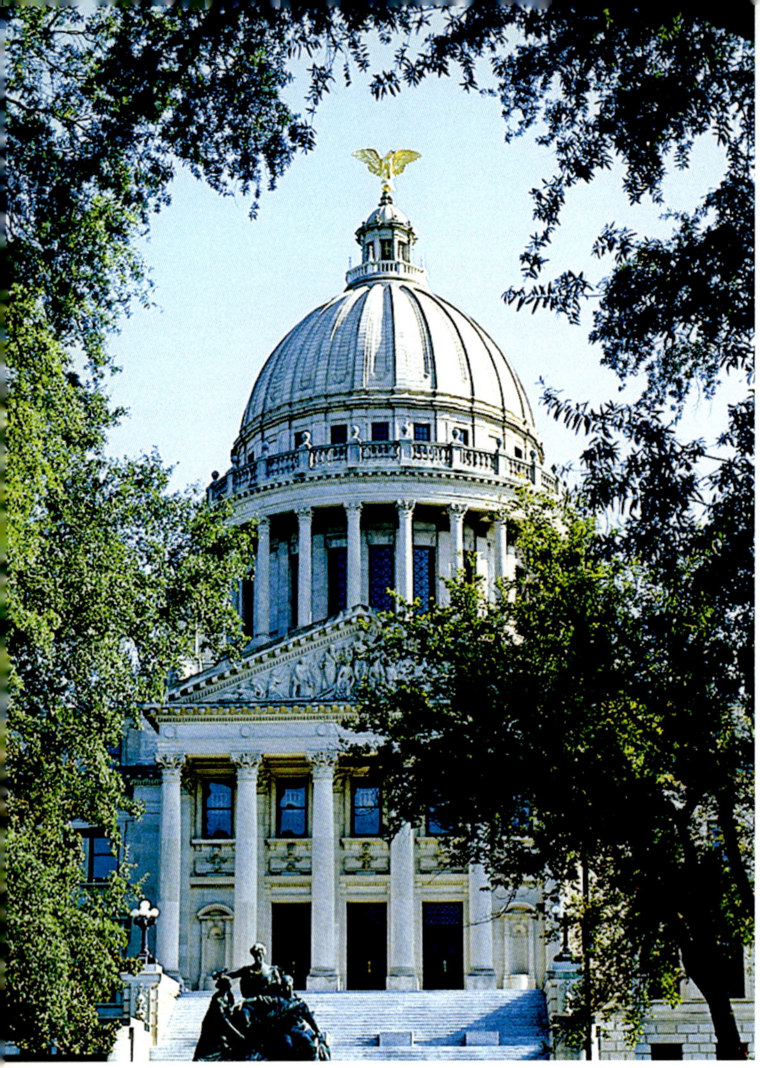

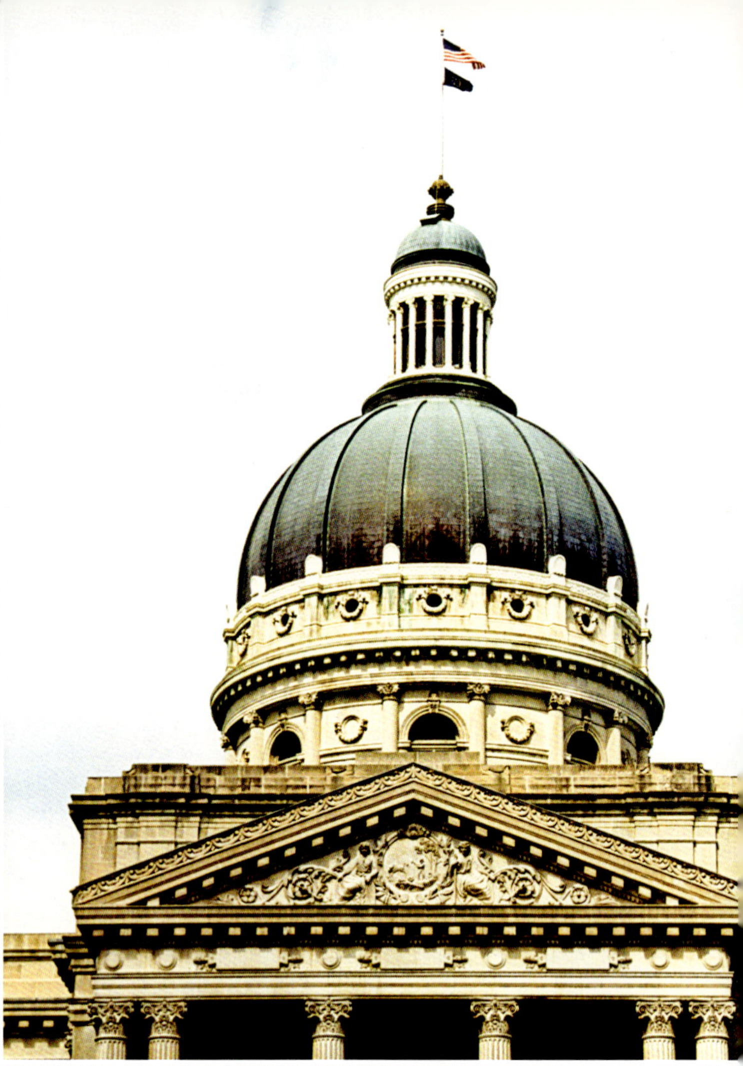

A sampler of American state capitols.

From upper left clockwise:

Jackson, Mississippi
Indianapolis, Indiana
Providence, Rhode Island

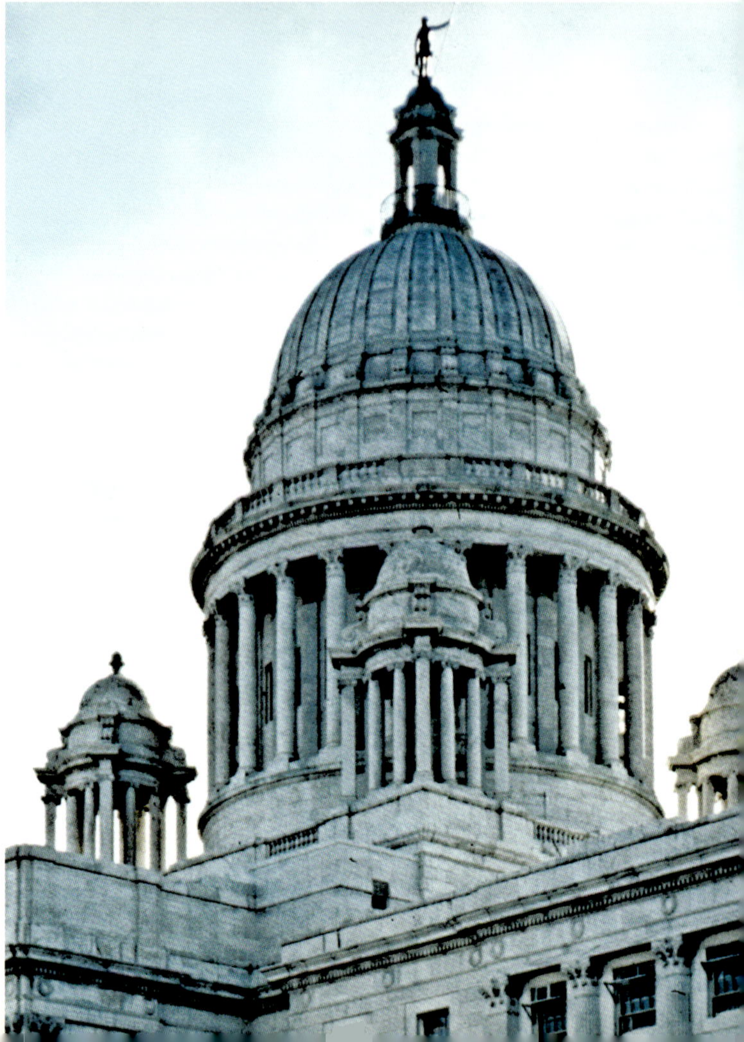

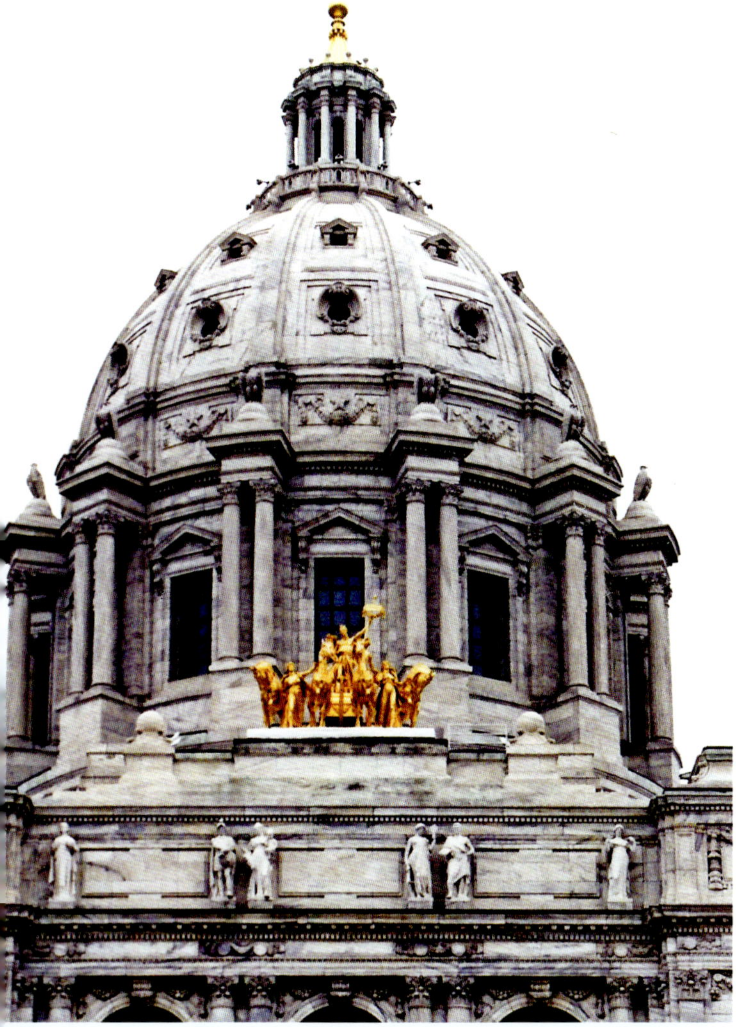

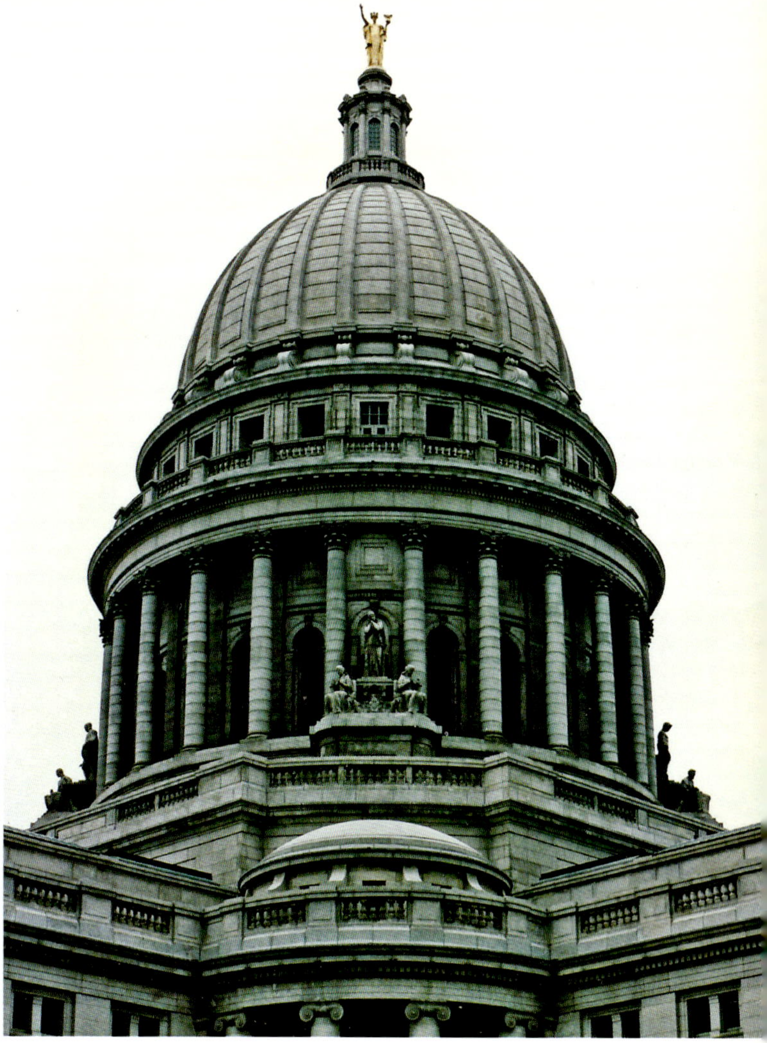

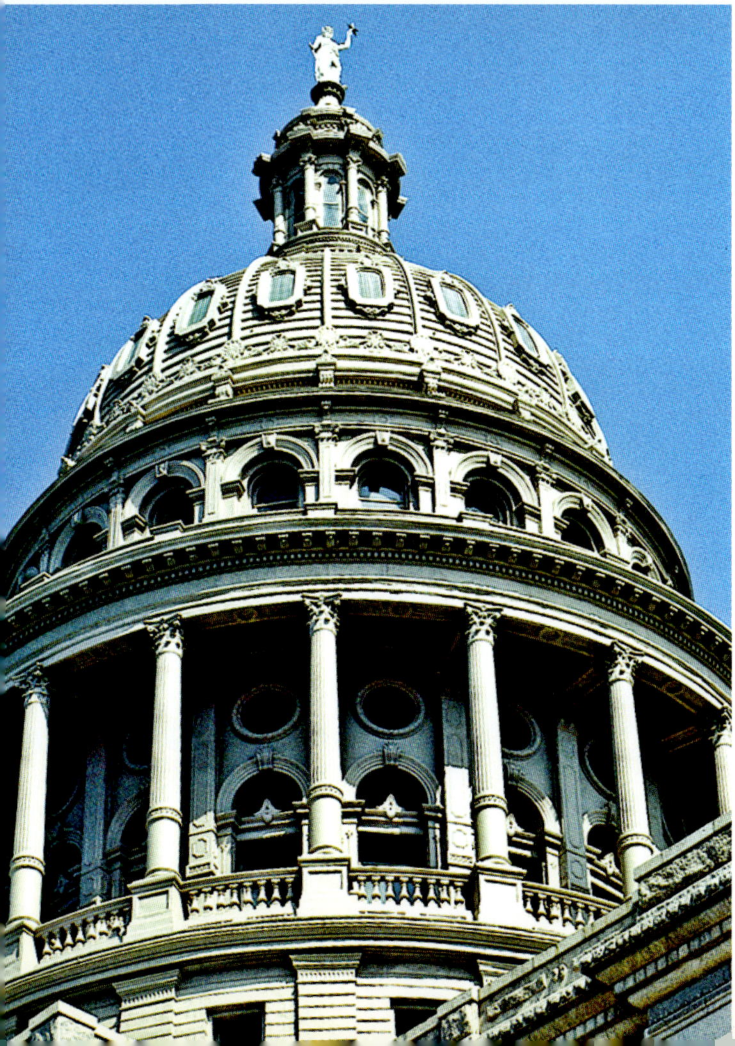

From upper left clockwise:

St. Paul, Minnesota
Madison, Wisconsin
Austin, Texas

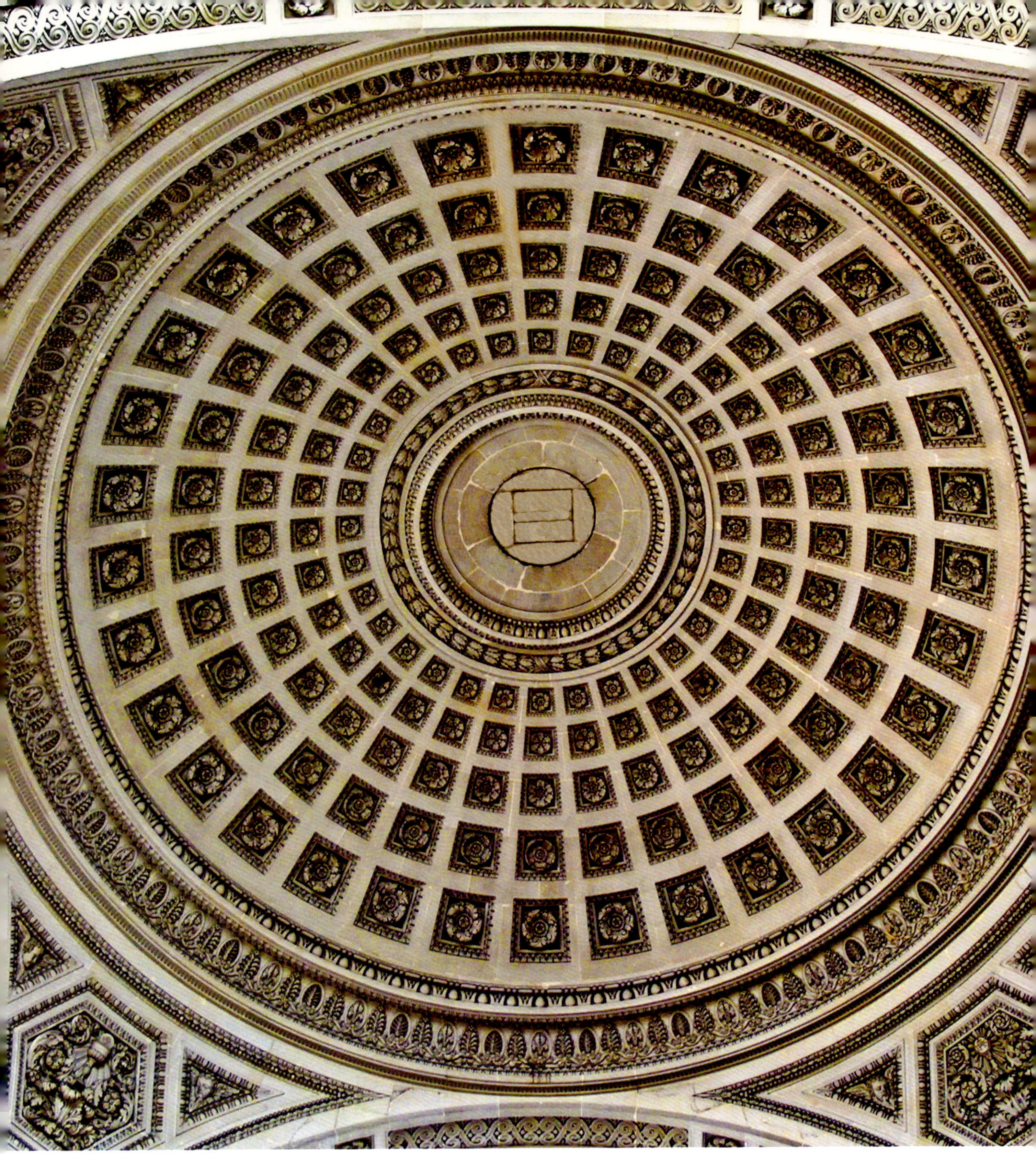

Pantheon in Paris

The interiors of these building are also rich in multiple symmetries.
The two examples here show the decorations of cupolas from the inside.

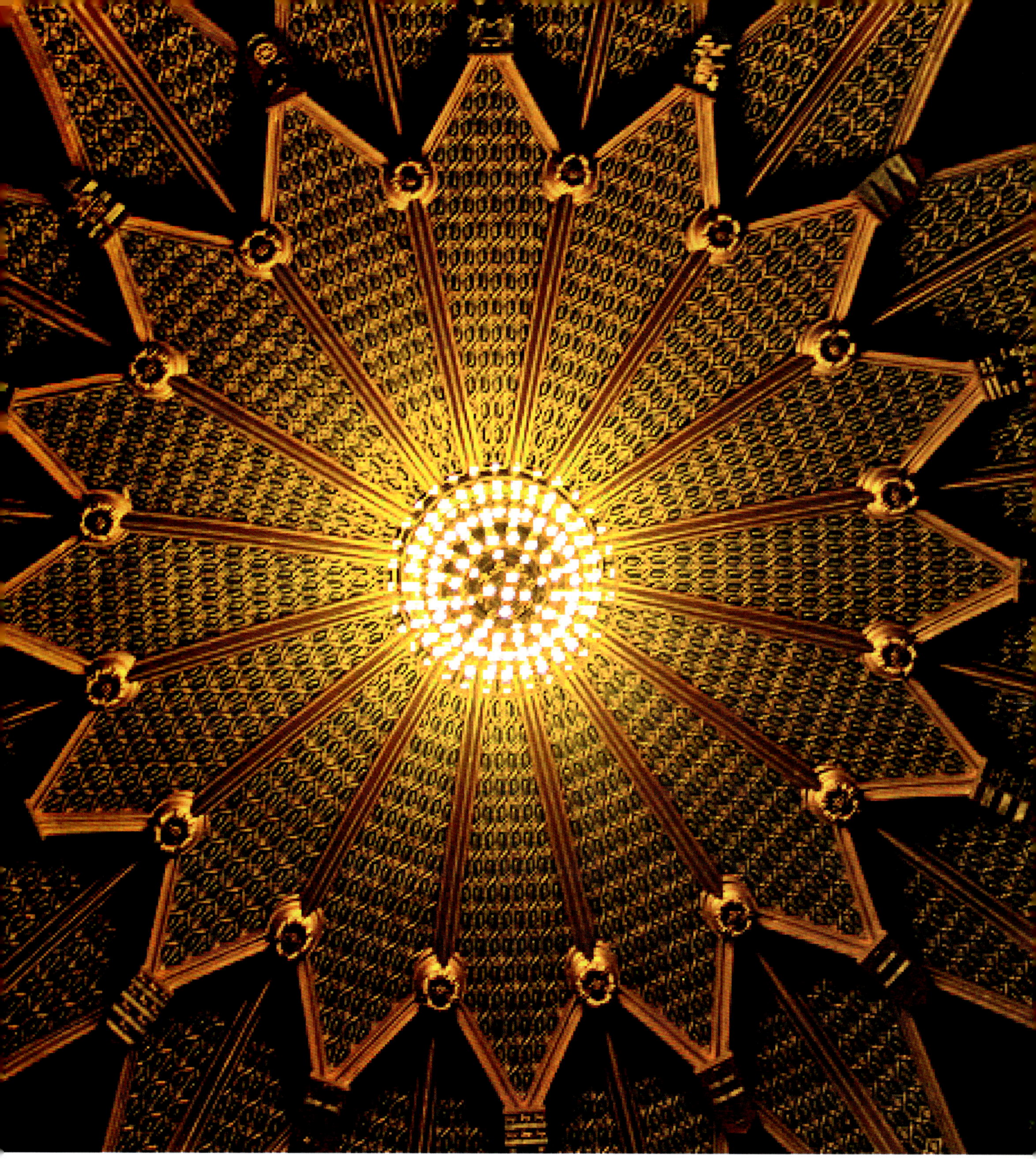

Cupola Hall in the Parliament building in Budapest

(photograph by and courtesy of Eszter Hargittai, Evanston, Illinois)

Snowflakes

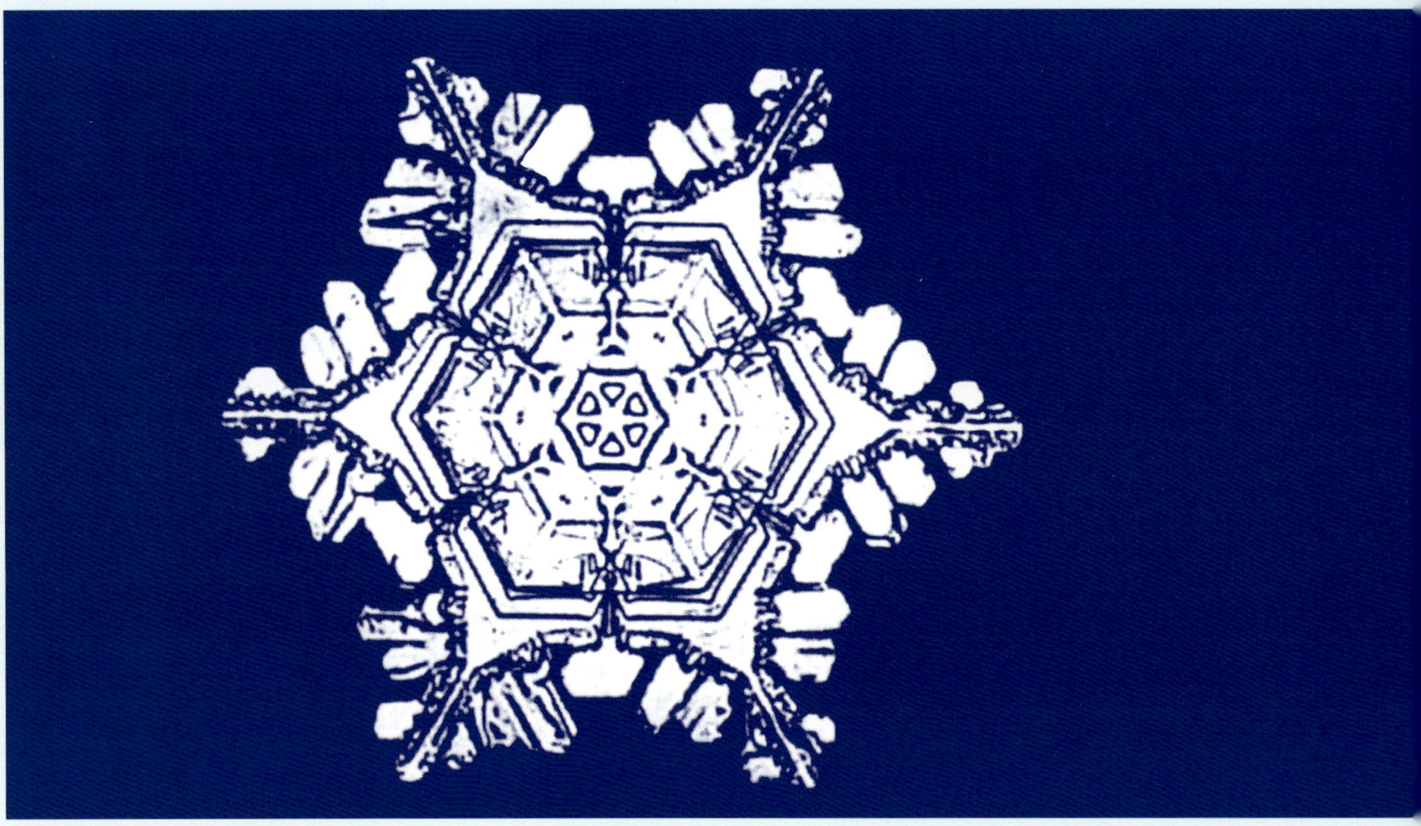

Thomas Mann describes the beauty of snowflakes in his famous novel *The Magic Mountain*:

"... the exquisite precision of form displayed by these little jewels, insignia, orders, agraffes — no jeweler, however skilled, could do finer, more minute work ... And among these myriads of enchanting little stars, in their hidden splendour that was too small for man's naked eye to see, there was not one like unto another; an endless inventiveness governed the development and unthinkable differentiation of one and the same basic scheme, the equilateral, equiangled hexagon ..."

Some of the most beautiful examples of reflection and rotation in nature can be seen in snowflakes. The conditions must be cold and dry to observe the falling snowflakes individually. It is a unique and captivating experience. The snowflakes are not only beautiful jewels but each is unique unto itself.

As Thomas Mann noted, the snowflakes have hexagonal symmetry. First of all, they have six reflection planes of which three go through the branches and three between them. There is an axis of six-fold rotational symmetry in the intersection of the reflection planes going through the center of the snowflake. There is then yet another mirror plane perpendicular to the other symmetry planes and this symmetry plane is in the plane of the snowflake itself, as if slicing the snowflake into two thinner snowflakes. There was no such additional symmetry plane in the flowers or, for another example, in the eight-branched streetlight.

The hexagonal symmetry of the snowflakes follows from their internal structure. The snowflakes consist of water molecules linked to each other in a regular hexagonal pattern. The links between the water molecules are not very strong, but are strong enough for maintaining this order. In such a link, the slightly positively charged hydrogen atoms of one water molecule are joined by the slightly negatively charged oxygen atom of the adjacent water molecule and countless water molecules form a virtually endless network in the same manner. The

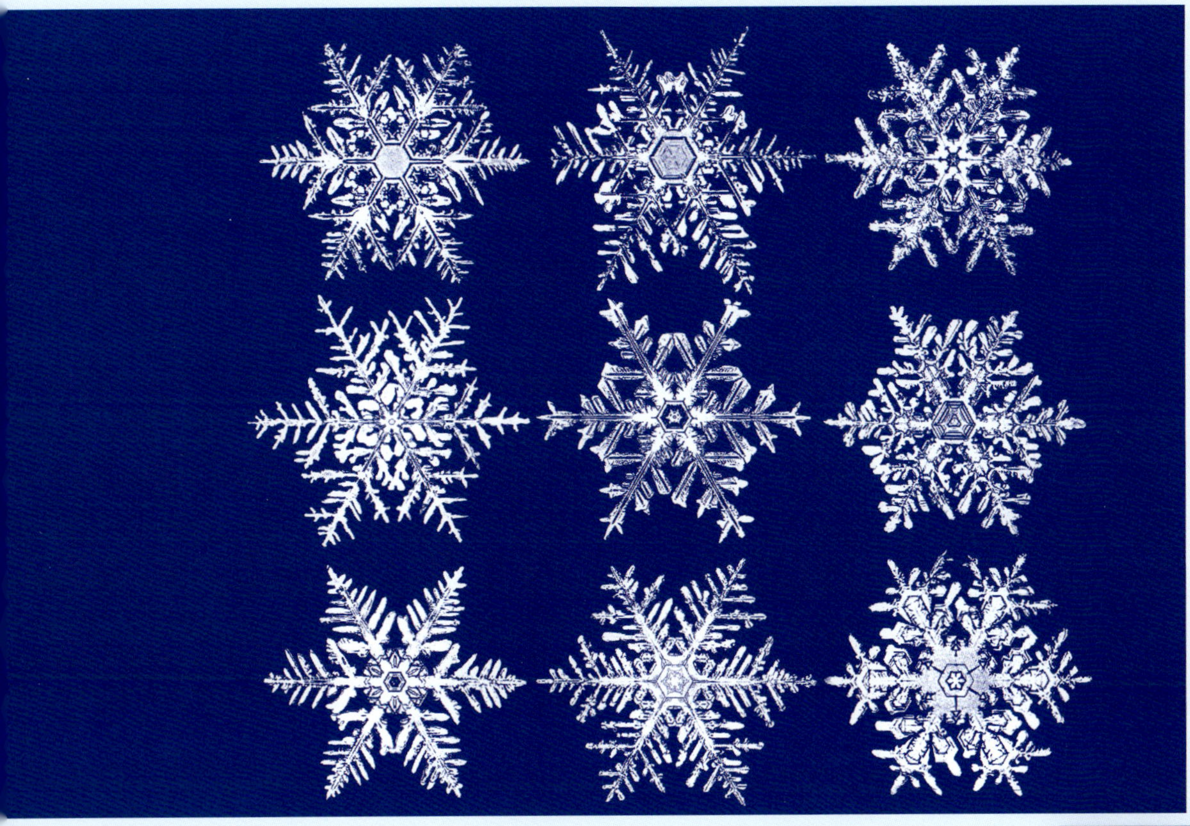

snowflake is an ice crystal and it grows uniformly in six directions if there is nothing that would disturb this uniformity. If there are impurities or other differences in the local conditions, then the snowflake will not grow in a regular way. There are no two snowflakes exactly the same because there are no two snowflakes that would have exactly the same conditions around them when they grow. This is why the snowflakes come in an endless variety.

The oldest known recorded statement on snowflake forms dates back to the second century BCE (that is, about two thousand two hundred years ago) in China. According to Joseph Needham and Lu Gwei-Djen [*Weather* 1961, 16, 319], it was stated as early as 135 BCE that "Flowers of plants and trees are generally five-pointed, but those of snow are always six-pointed." Needham and Lu quote a physician from 1189:

"... the reason why double-kernelled peaches and apricots are harmful to people is that the flowers of these trees are properly speaking five-petalled yet if they develop with sixfold [symmetry], twinning will occur. Plants and trees all have the fivefold pattern; only the yellow-berry and snowflake crystals are hexagonal. This is one of the principles of Yin and Yang. So if double-kernelled peaches and apricots with an (aberrant) sixfold [symmetry] are harmful, it is because these trees have lost their standard rule."

Six was a symbolic number for water in many classical Chinese writings. The examination of snowflake shapes and their comparison with other shapes was

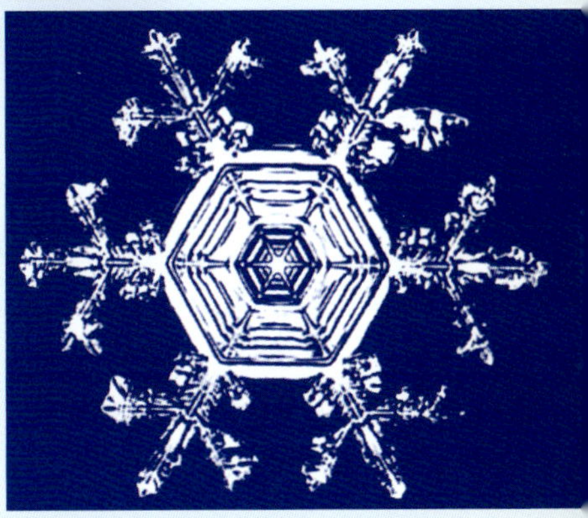

All snowflake illustrations from W. A. Bentley, W. J. Humphreys, *Snow Crystals*. Dover, New York, 1962

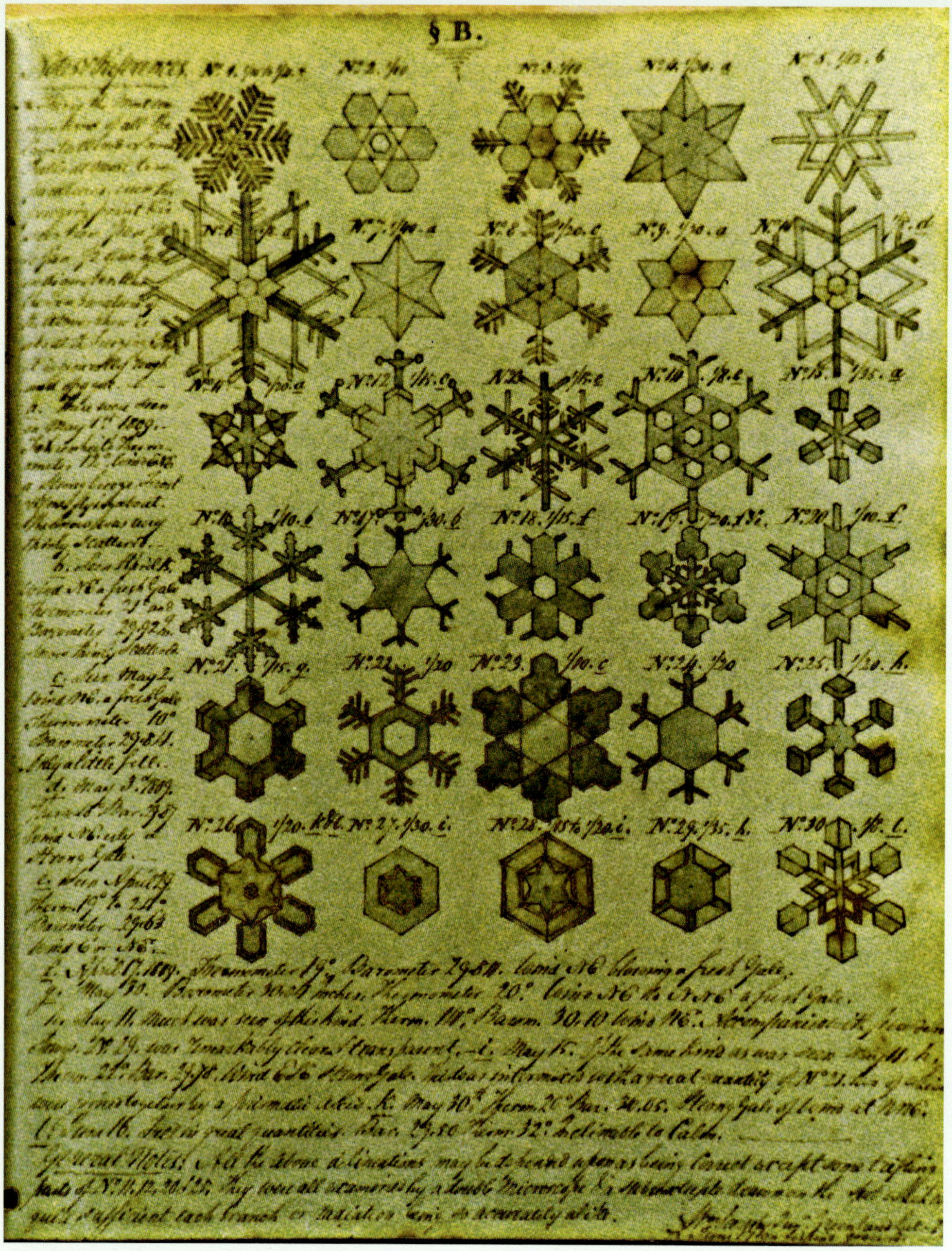

A page from William Scoresby's log book in 1806 (W. Scoresby, *Arctic Scientist*. Caedmon of Whitby Publishers, Whitby, UK, 1976)

considered to be of great importance in East Asia. As a forerunner of modern investigations into the correlation between snowflake shapes and meteorological conditions, the following was noted in the thirteenth century:

"The Yin embracing Yang gives hail,
the Yang embracing Yin gives sleet.
When snow gets six-pointedness,
it becomes snow crystals. When hail gets
three-pointedness, it becomes solid."
(Joseph Needham and Lu Gwei-Djen, *Weather*)

William Scoresby was an Arctic scientist, who made drawings. He was a young man, who at the age of 16 travelled with his father to the whale fisheries in Greenland in 1806. In addition to his sketches, he made careful observations and recorded them in his log book.

Johannes Kepler was the first European scientist, who recognized the hexagonal symmetry of snowflakes. He published a small book in 1611 with the title *The Six-cornered Snowflake*. Kepler supposed that the hexagonal shape originates from the internal arrangement of the building water particles of the snow crystal. In this way he operated with the notion of what we would call molecule today.

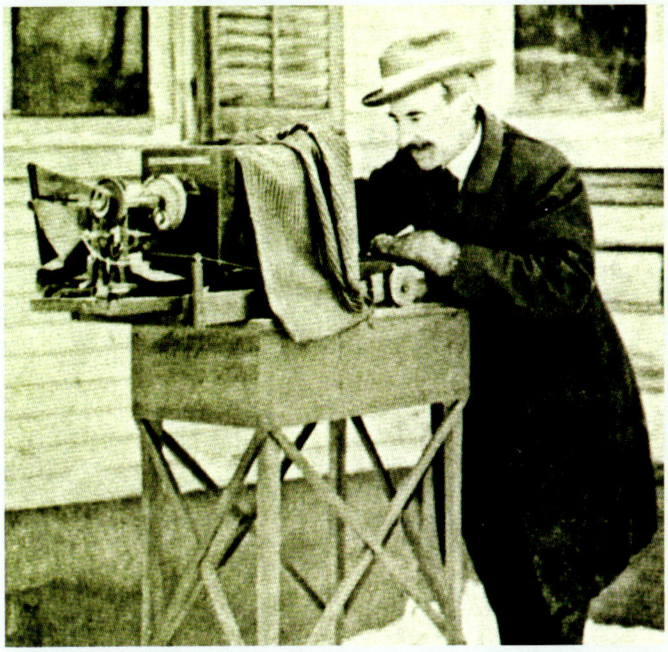

W. A. Bentley photographing snow crystals

The best conditions for observing snowflakes are dry and cold weather. Under such circumstances the snowflakes can be kept on a surface and can be photographed. Under humid and not so cold conditions the snowflakes melt fast and their examination is hardly possible.

The most famous book on snowflakes, *Snow Crystals*, by W. A. Bentley and W. J. Humphreys, appeared first in 1931. Bentley was the photographer, who devoted his life to taking photographs of snow crystals and collected at least 6,000 images in his workshop at Jericho, Vermont, in the northeastern United States. Humphreys wrote the text in their joint book.

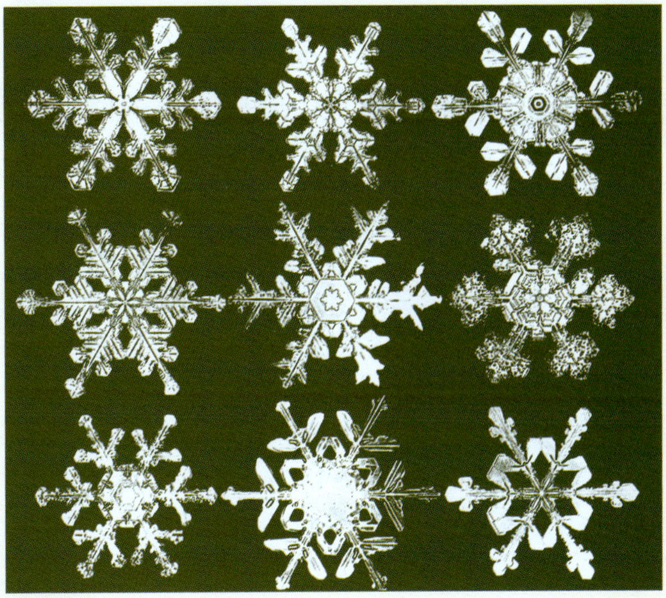

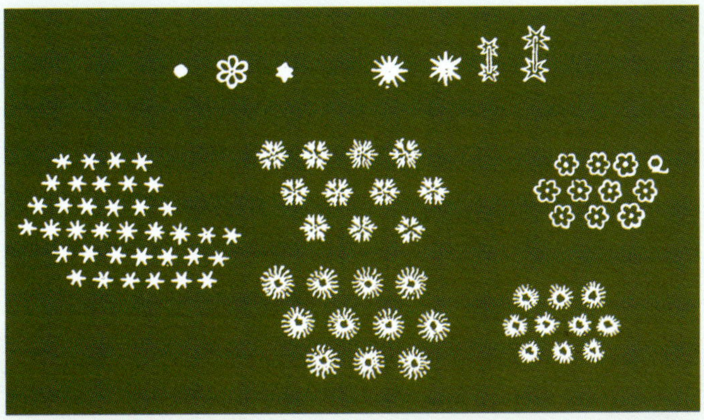

Drawings of snowflakes by Descartes

The great French mathematician René Descartes observed the shapes of snowflakes and drew them in 1635. His drawings were rather sketchy, but the hexagonal symmetry appears in them unambiguously.

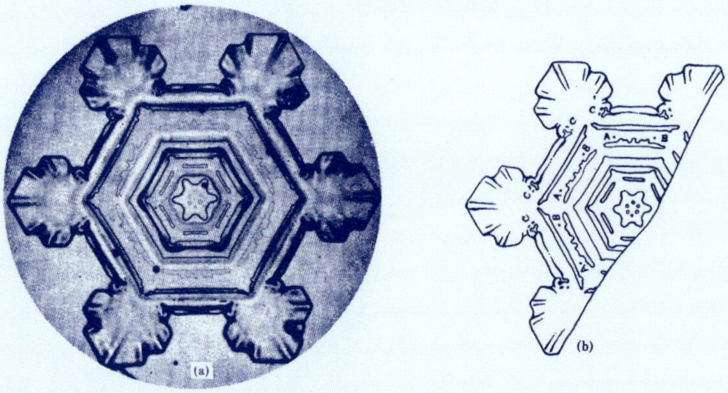

Photomicrograph of a snowflake and sketch of part of the crystal by Ukichiro Nakaya

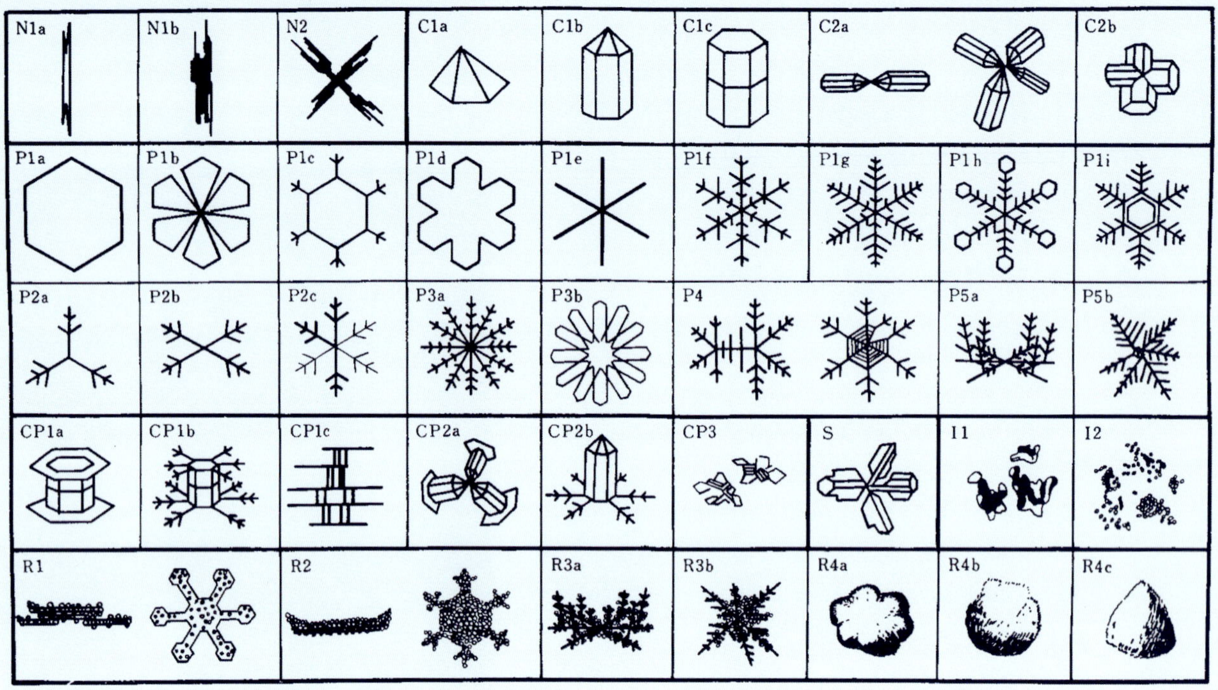

Nakaya's drawings of snowflakes

The booklet entitled *Snow* by Ukichiro Nakaya appeared in Japanese in 1938, and it has since been reprinted dozens of times. Working in Hokkaido, the northernmost big island of Japan, Nakaya recorded naturally occurring snow crystals, classified them, and investigated their properties. He also worked out techniques to produce artificial snowflakes and succeeded in obtaining different types of snowflakes under different conditions of formation.

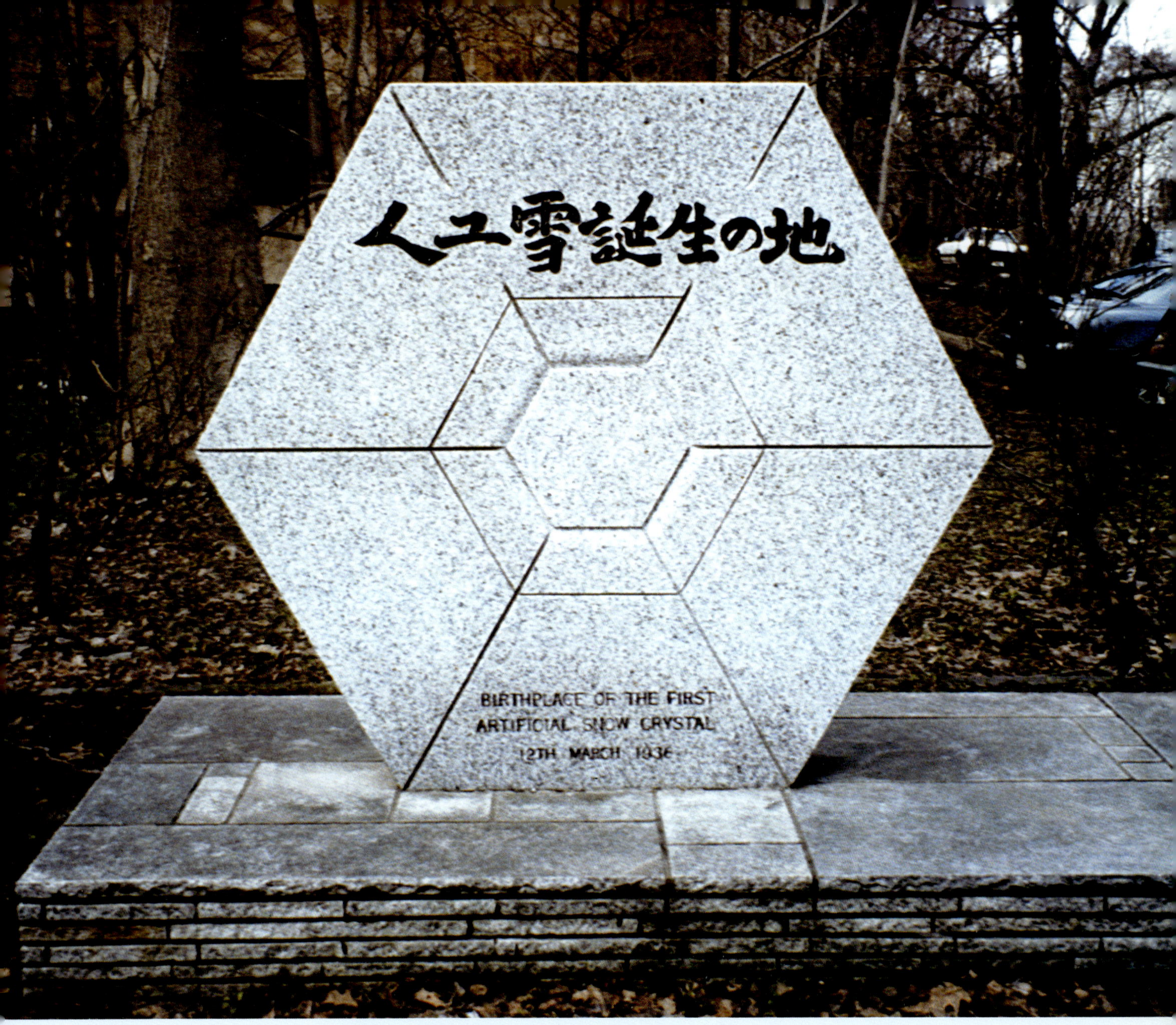

Sculpture on the campus of Hokkaido University in Sapporo, Japan, honoring Nakaya and commemorating the birth of the first artificial snowflake in 1936

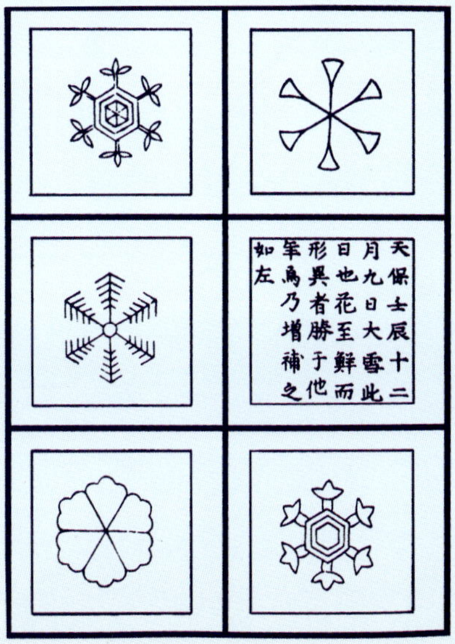

Sekka Zusetsu of Doi produced a series of beautiful sketches of snowflakes in Japan in 1832, communicated by Nakaya in his book.

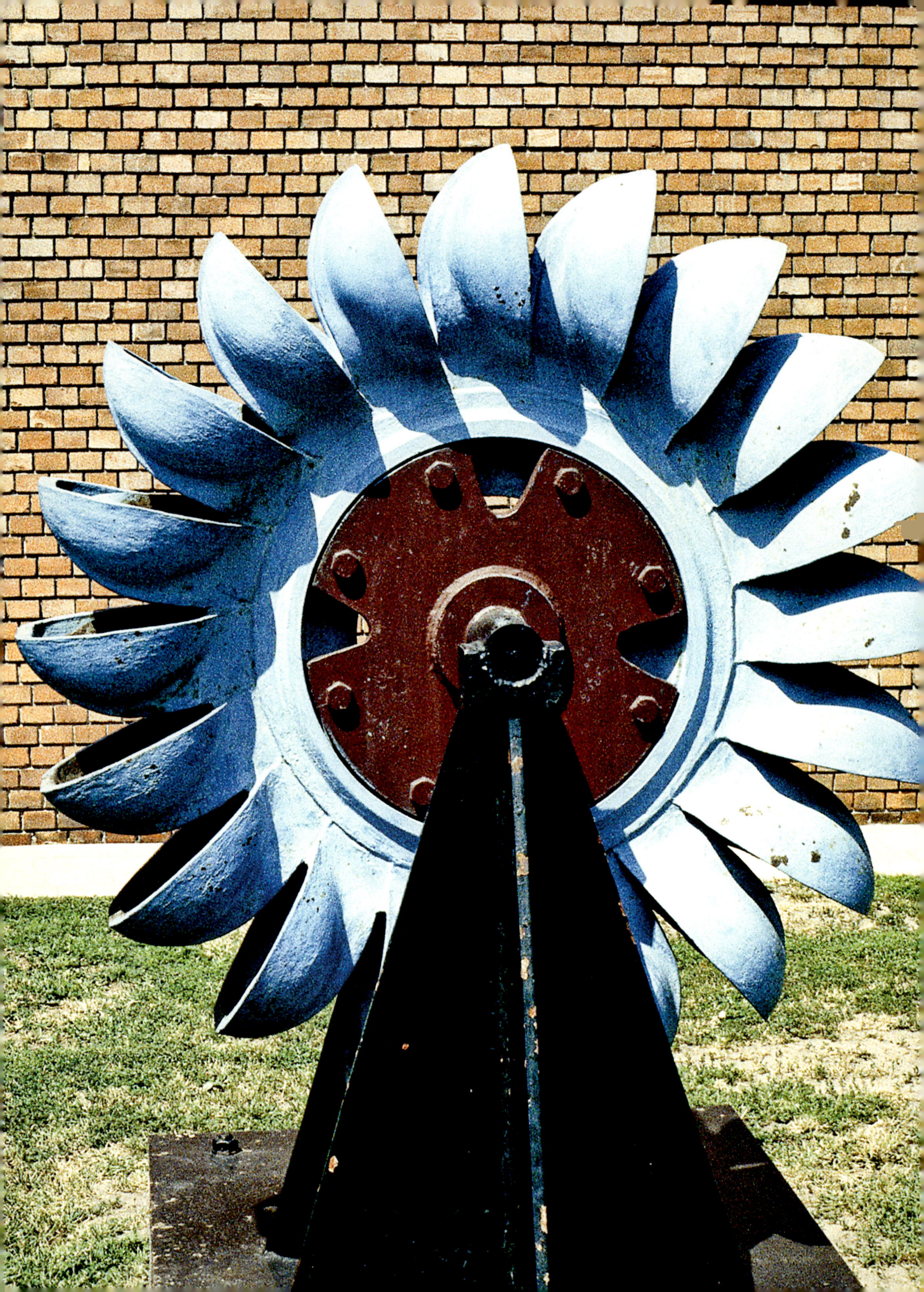

Rotational Symmetry

We talk about rotational symmetry when an object is rotated around its axis, and it appears in the same position two or more times during a complete revolution.

In the previous chapter, we have already seen rotational symmetry when it was generated by the simultaneous presence of two or more mirror planes. The axis of rotation was along the line of intersections of the mirror planes. This is why we could say that in that case the rotational symmetry was a "byproduct" of mirror symmetries.

Rotational symmetry may occur by itself as well, that is, without the presence of mirror planes.

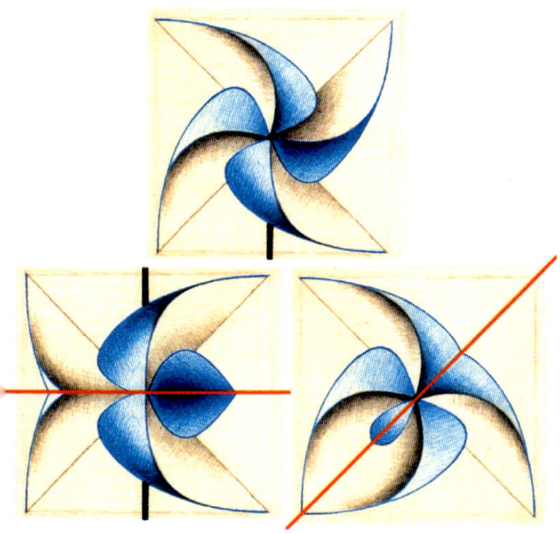

The pinwheel has rotational symmetry. We can easily rotate the pinwheel by blowing air at it gently, or let the wind rotate it. When the pinwheel rotates slowly, we can see that at each quarter turn it is in a new position, yet it looks exactly the same as it did at the start. The pinwheel has fourfold rotational symmetry because its original view appears four times during a complete rotation.

On the other hand, the pinwheel has no mirror symmetry. We find it impossible to place a mirror on it in such a way that it would reproduce the shape of the pinwheel. Thus the pinwheel has no mirror symmetry, *only* rotational symmetry.

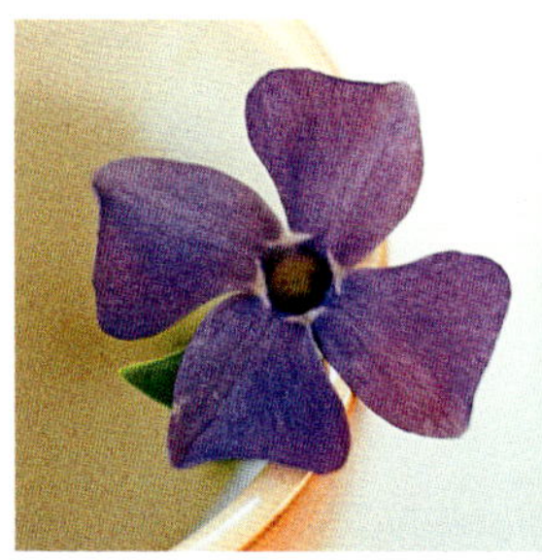

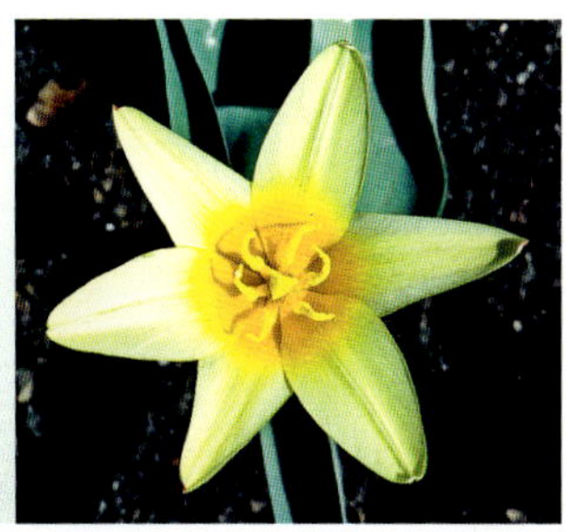

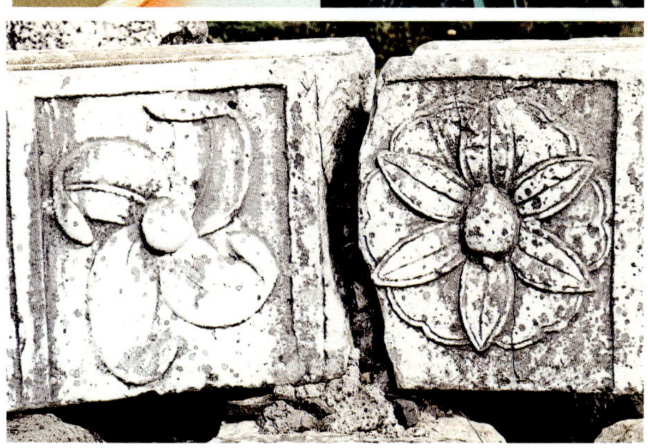

Of these two flowers, the left one has only rotational symmetry and the right one has mirror symmetries *and* rotational symmetry. The two pieces of stone carving from the ancient road Via Appia Antica in the outskirts of Rome, Italy, seem to reproduce the two kinds of symmetry.

Rotating waterwheel in front of the Technical Museum in Budapest

Flowers

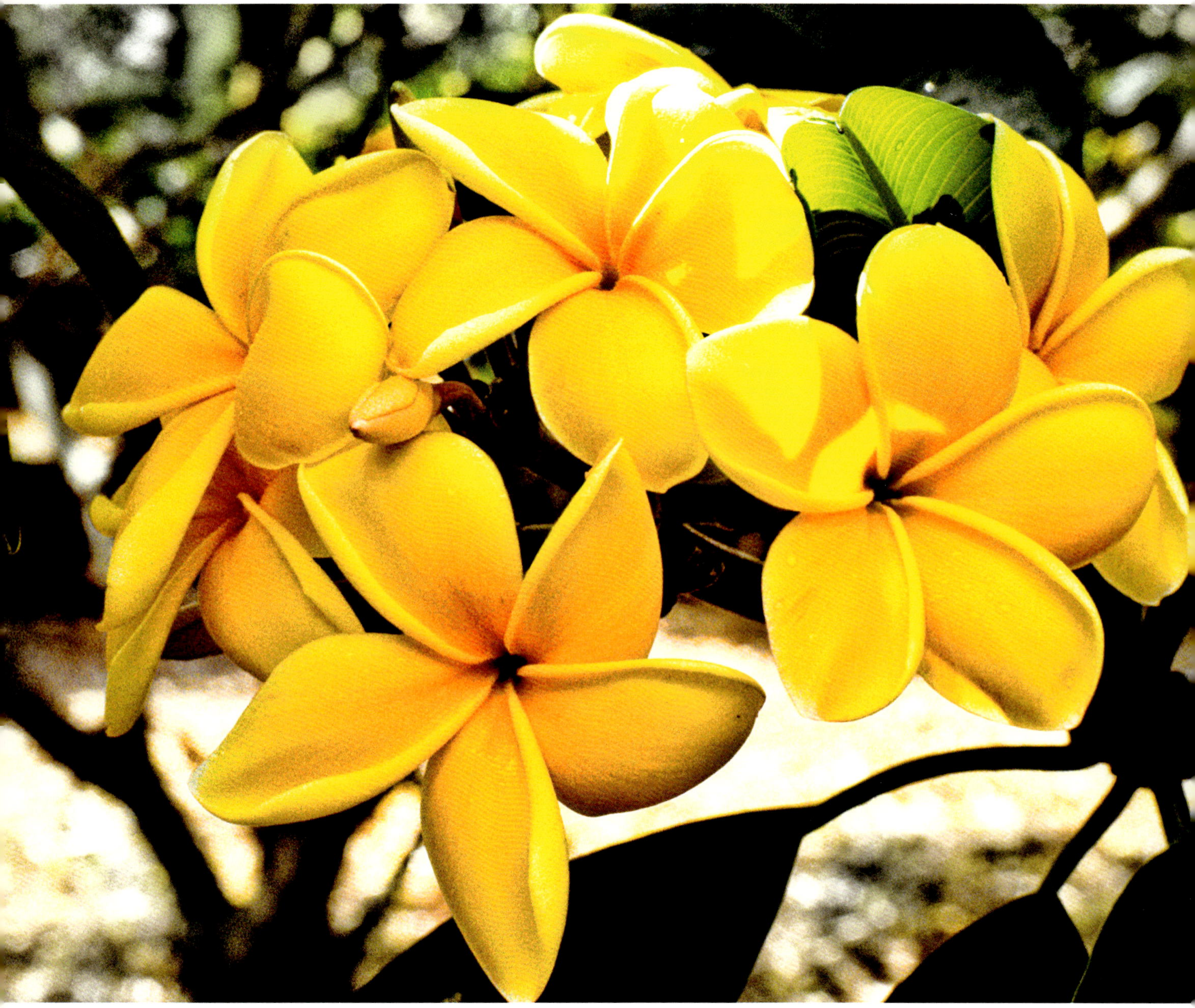

There are many flowers with rotational symmetry only. The most frequent ones have fivefold rotational symmetry. Most of the examples presented here are from Hawaii. The Tiare (*Gardenia Taitensis*) is interesting because it appears in versions with 6-, 7-, and 8-fold symmetries.

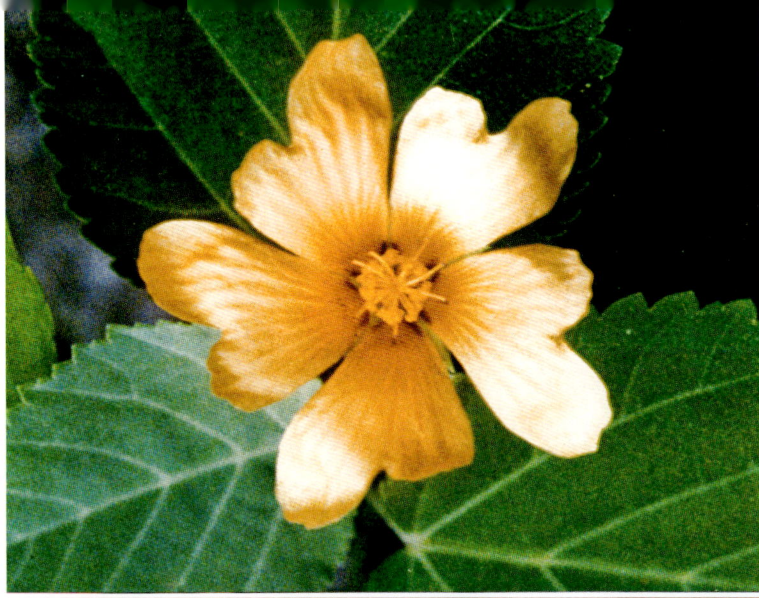

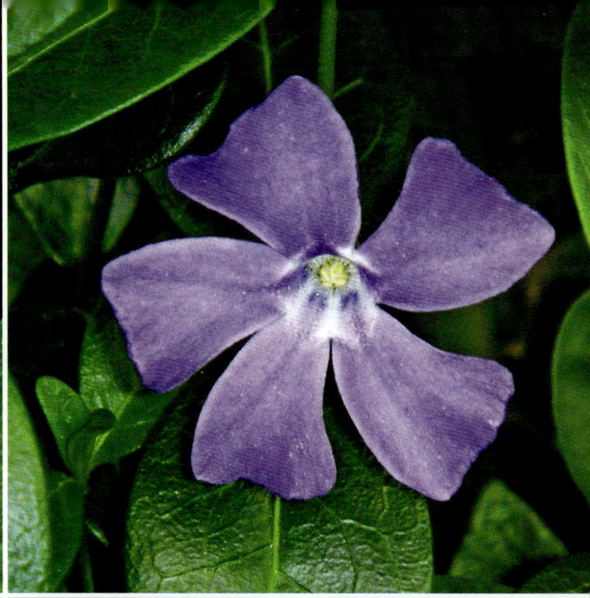

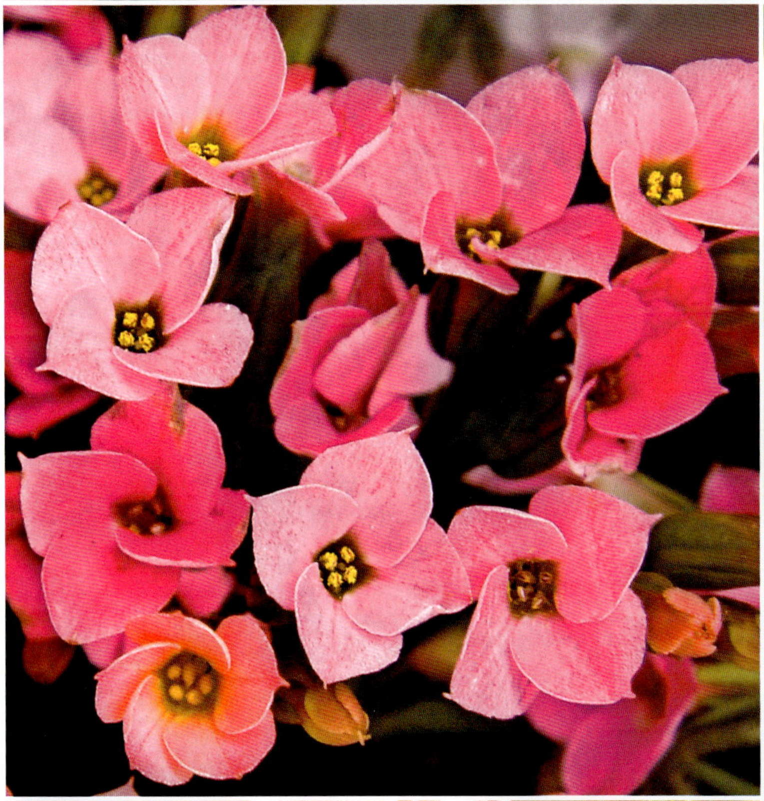

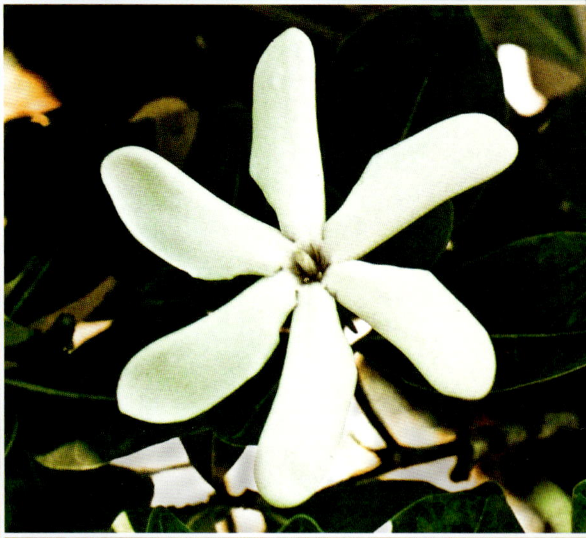

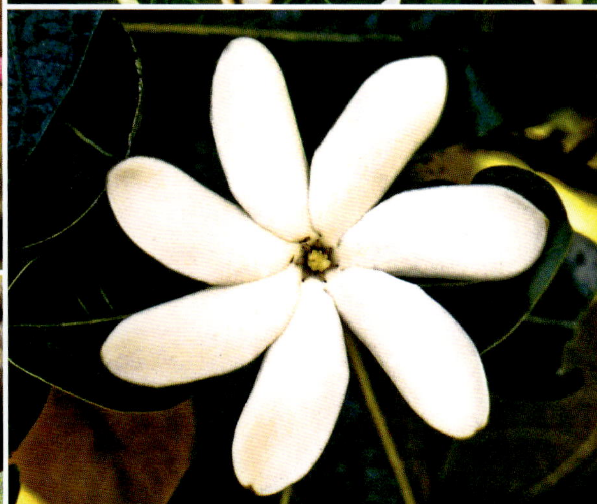

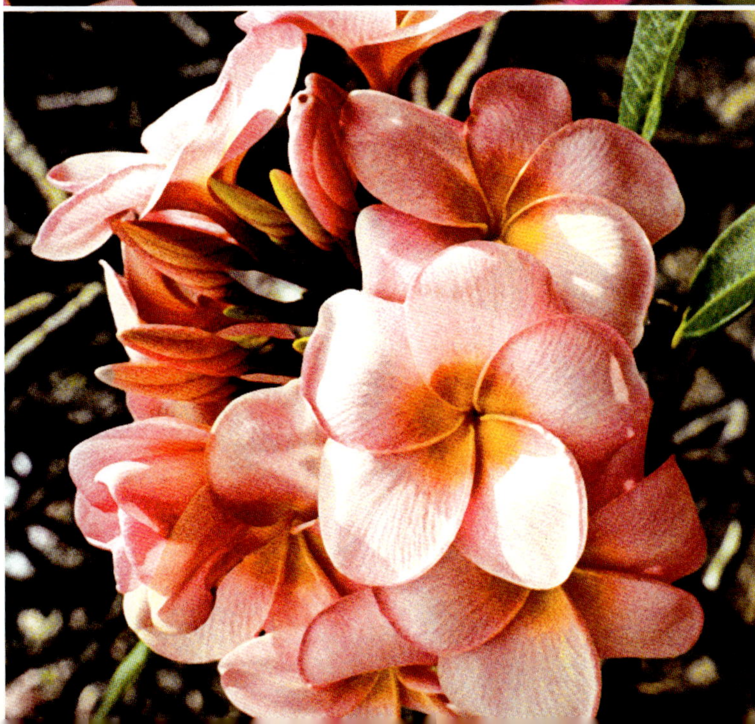

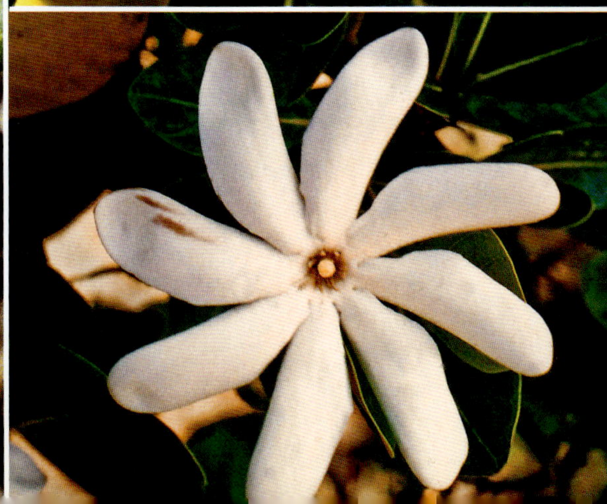

Rotating Blades

Rotating parts of various machines, such as propellers, have rotational symmetry only. The familiar windmills in Holland with four blades are very much like pinwheels. In the old days, wind power was used to rotate huge mill stones for grinding grain into flour. Rotational symmetry (without mirror planes) means that all the blades curve in the same direction, thereby facilitating catching the wind.

Recently, windmills have been utilized extensively as an alternative means of producing electricity without the need to burn nonrenewable fuels, such as oil and coal.

Two old rotating waterwheels are shown here that are exhibited in two different Norwegian cities. They are motionless now because they are no longer in operation.

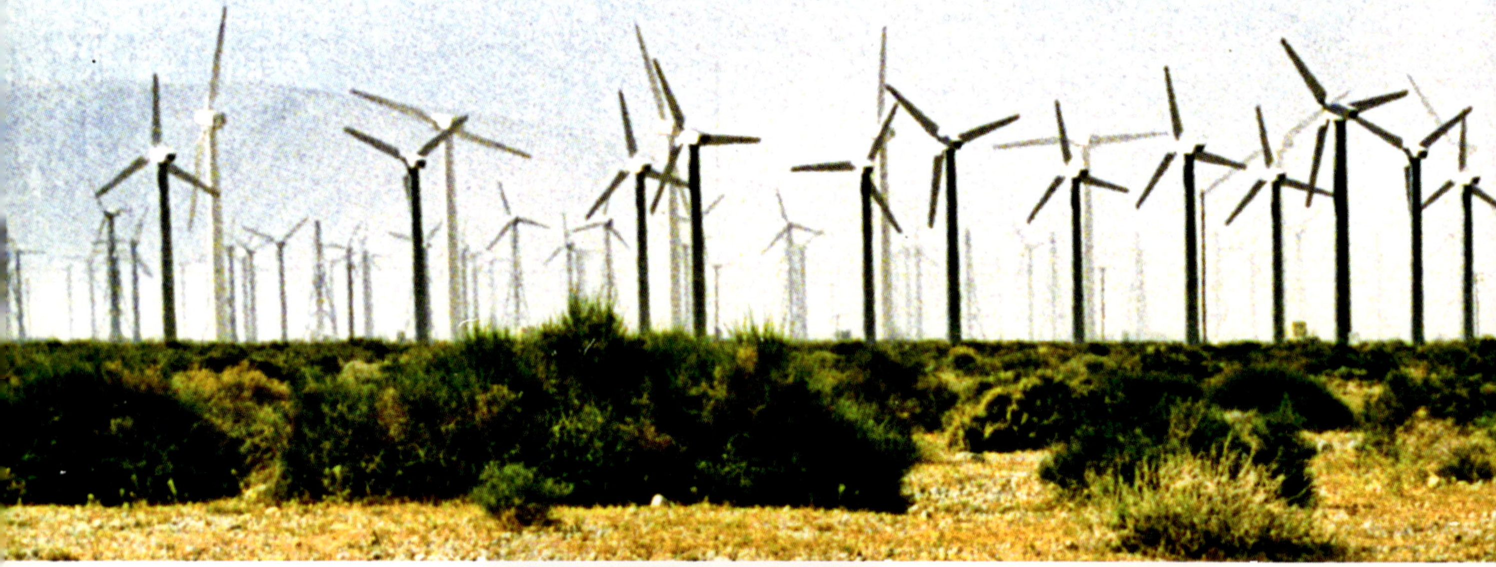

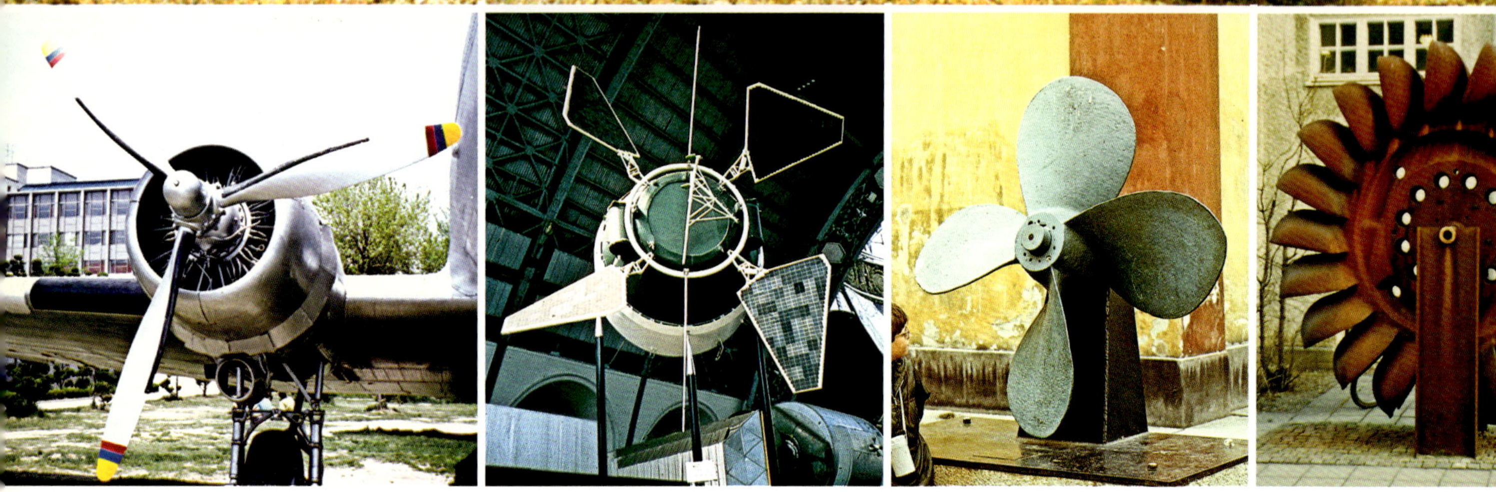

Top: Electricity-generating windmills (photograph courtesy of Lloyd Kahn, Bolinas, California)

Bottom: Propeller of airplane

Satellite (Space Museum, Washington, DC)

Propeller of a ship (Foundry Museum, Budapest)

Waterwheel in Trondheim, Norway

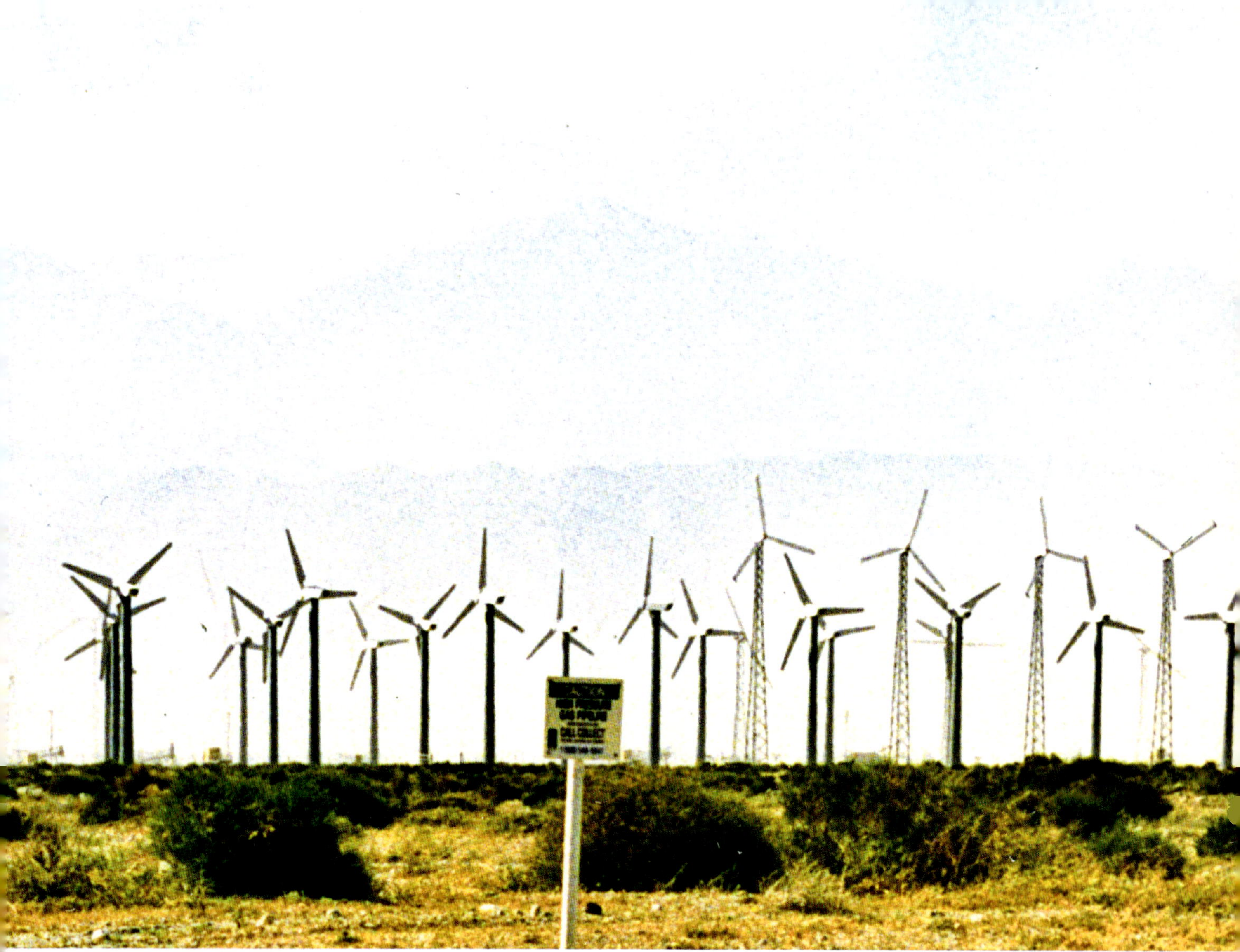

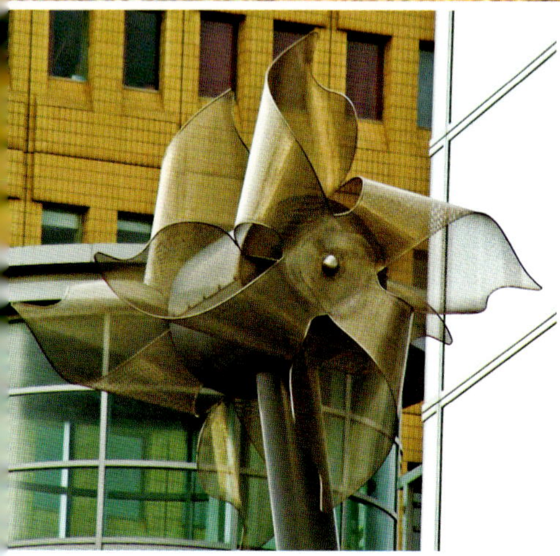

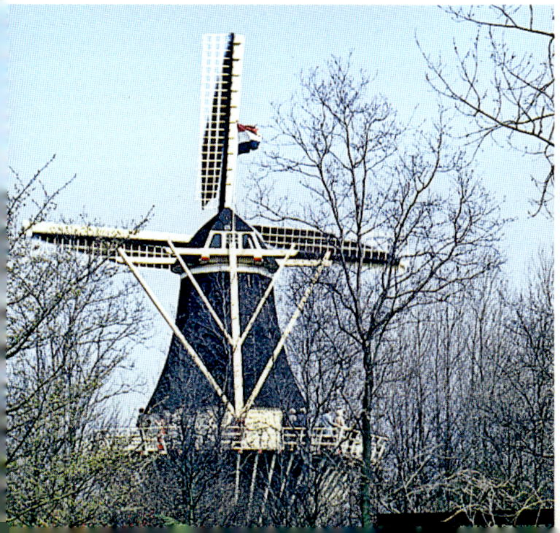

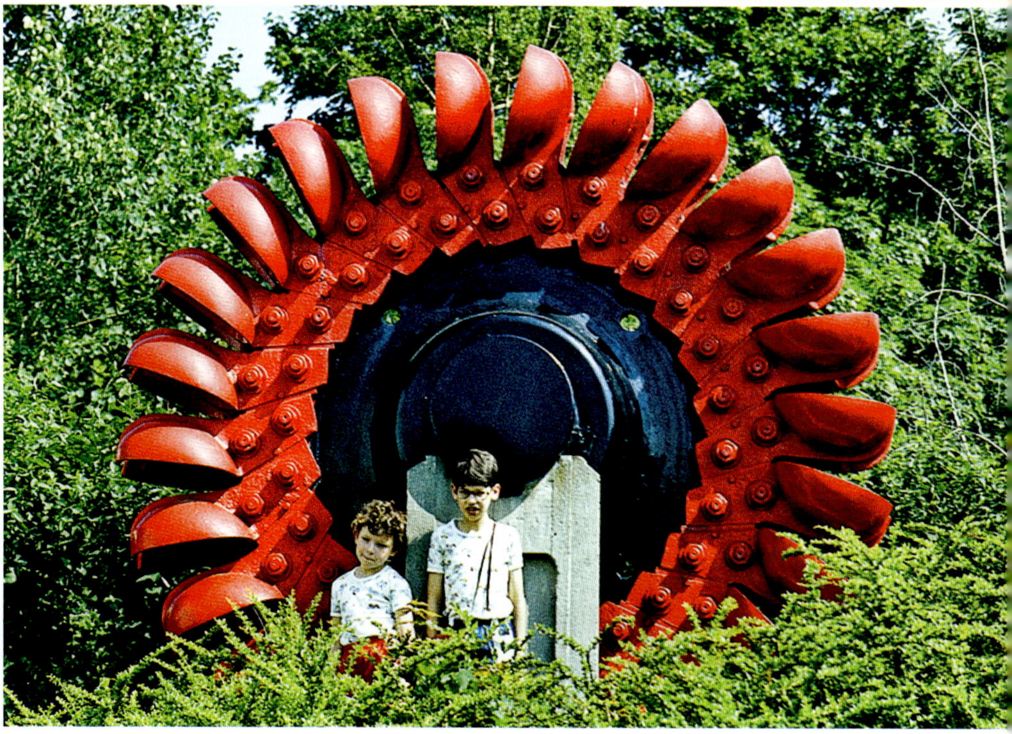

"Pinwheel" in front of a building in Manchester, England

Windmill in Holland

Waterwheel in Oslo, Norway

Logos

Transportation company logos

The logos of railway systems as well as many other companies involved in transportation often have twofold rotational symmetry and no symmetry plane. These logos imply the act of traveling somewhere and returning.

Left column	Right column
Korean	Dutch
British	Spanish
Austrian	Israeli
	American
	Hungarian
	Austrian

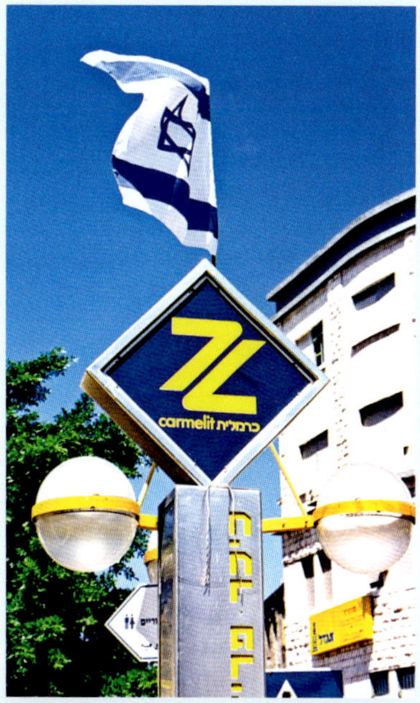

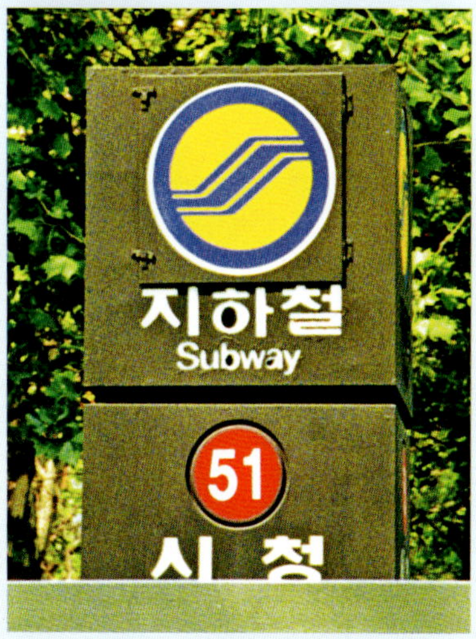

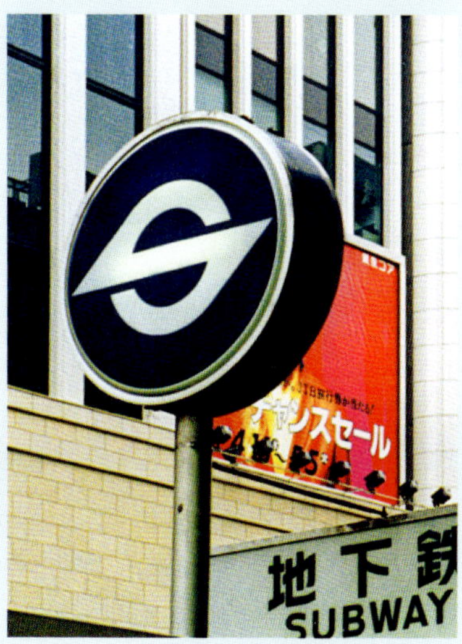

Left column

Italian
Austrian
American
Spanish
German

Right column

Israeli
Korean
Japanese

Recycling logos

Recycling logos most often have threefold rotational symmetry, but other kinds also appear. What is common among them is that they all have rotational symmetry and no mirror planes.

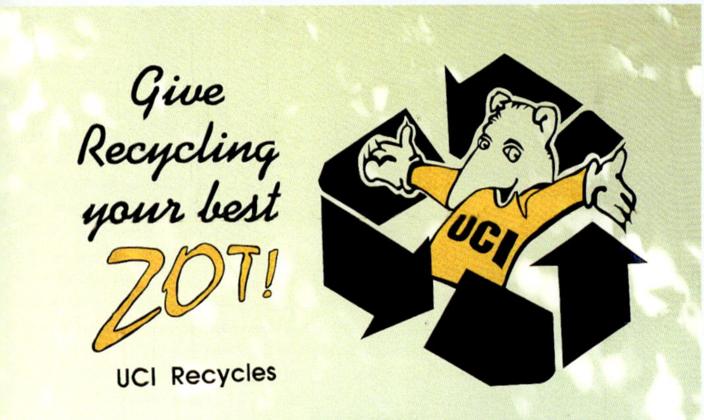

Our examples come from many different countries (United Kingdom; Italy; Portugal; and several states in the United States).

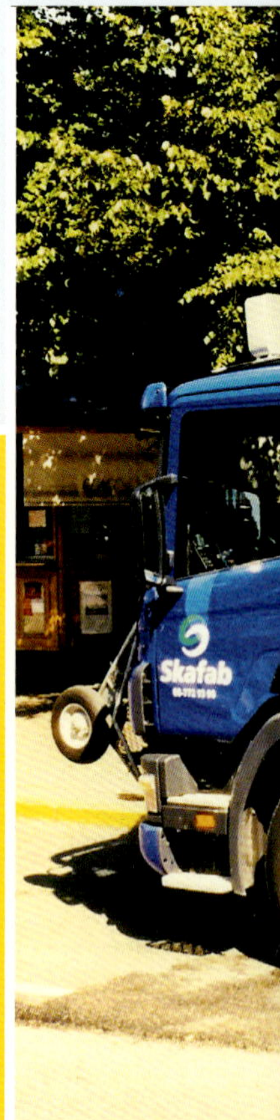

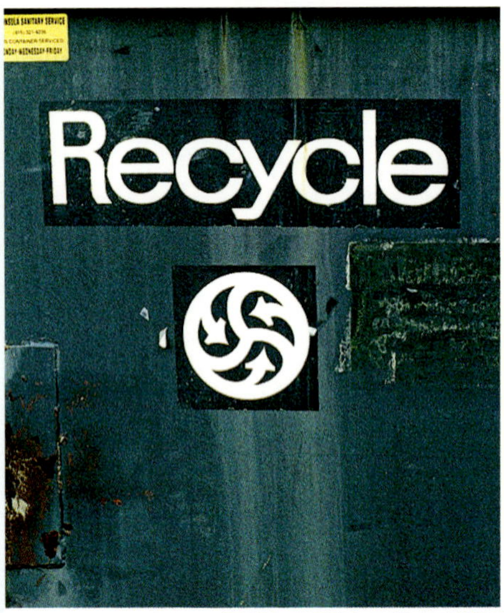

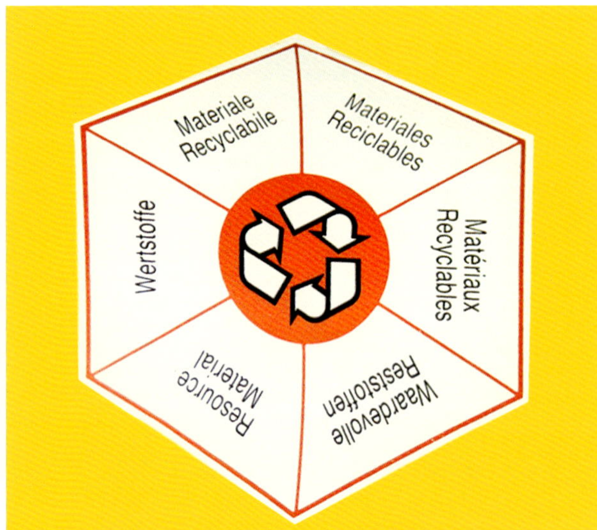

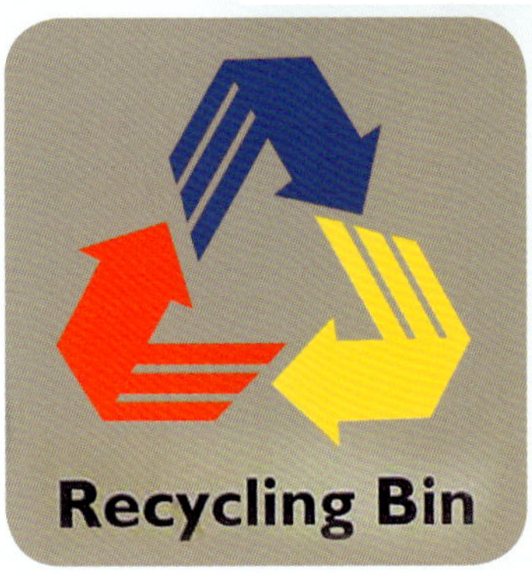

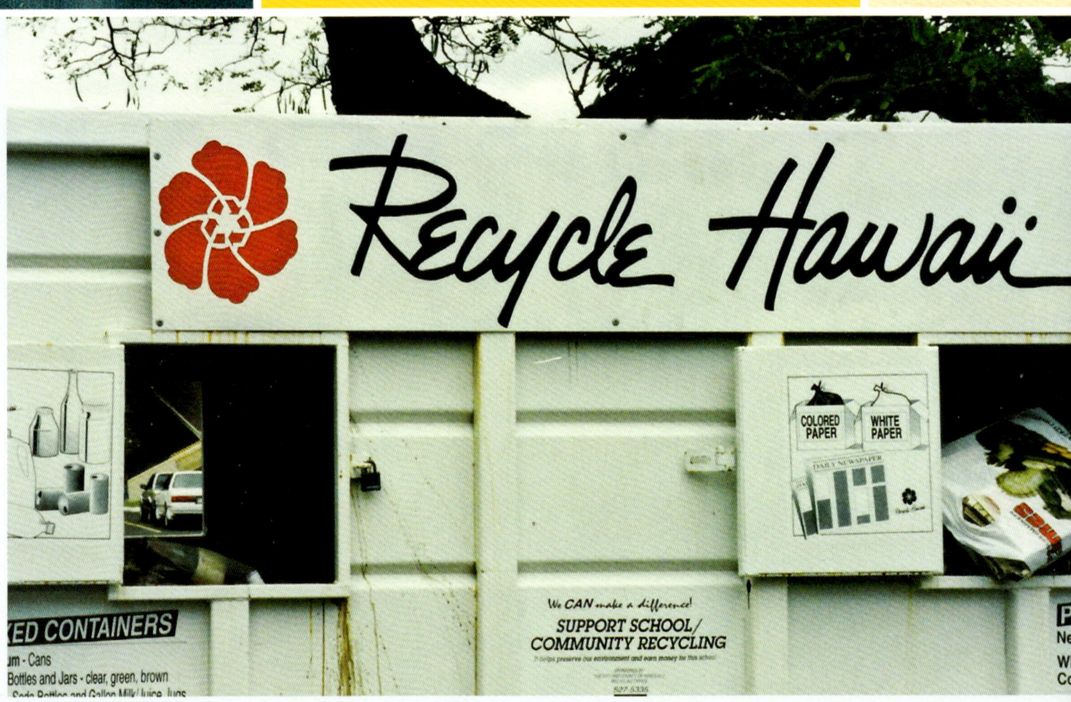

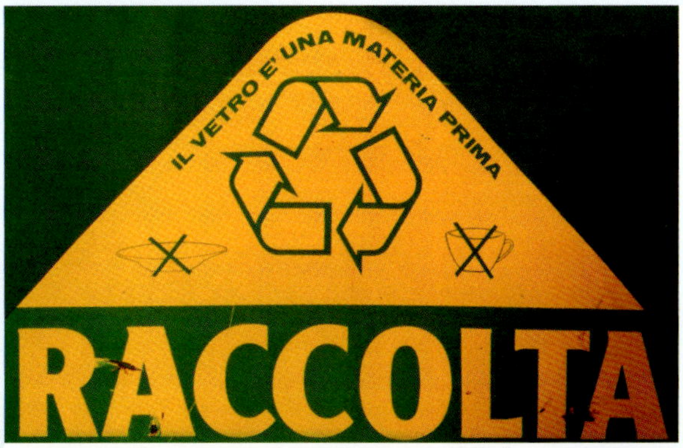
IL VETRO E' UNA MATERIA PRIMA
RACCOLTA

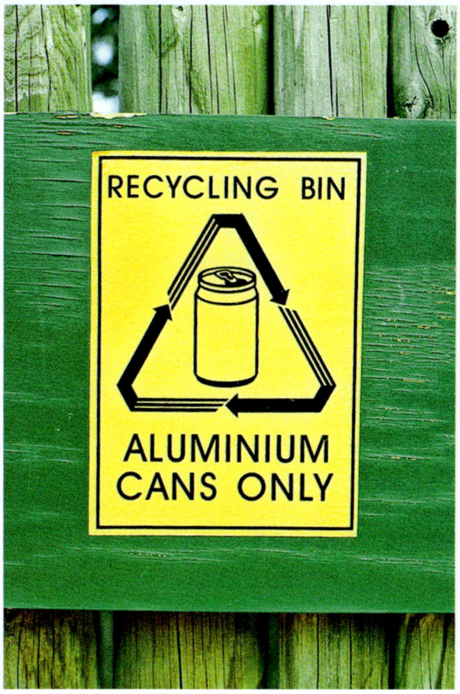
RECYCLING BIN
ALUMINIUM CANS ONLY

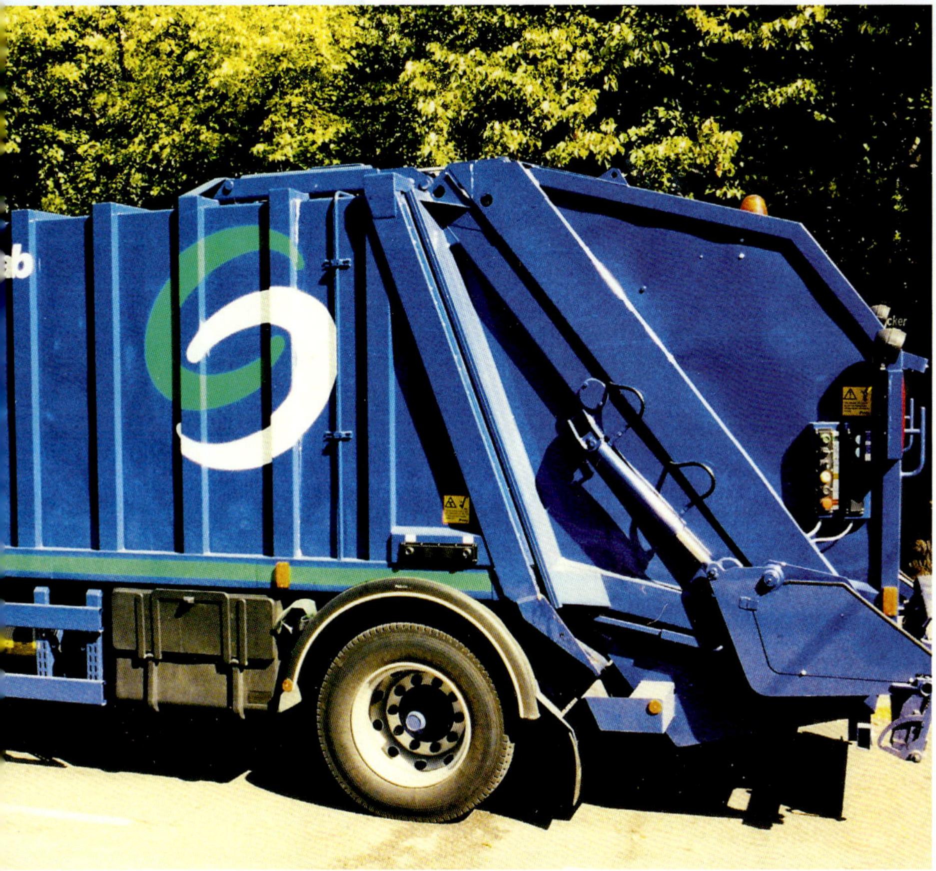

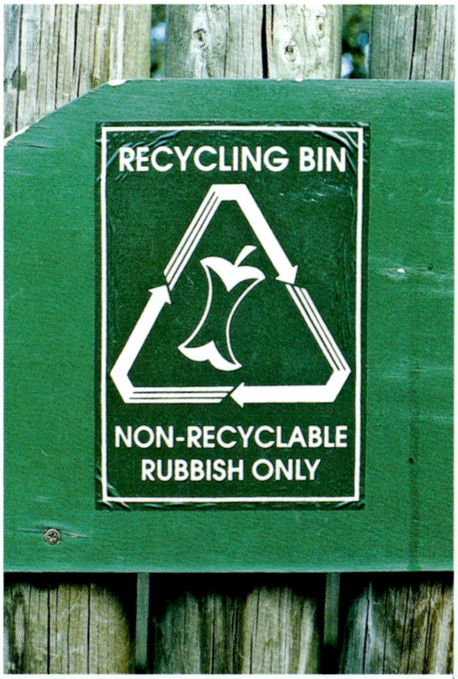
RECYCLING BIN
NON-RECYCLABLE RUBBISH ONLY

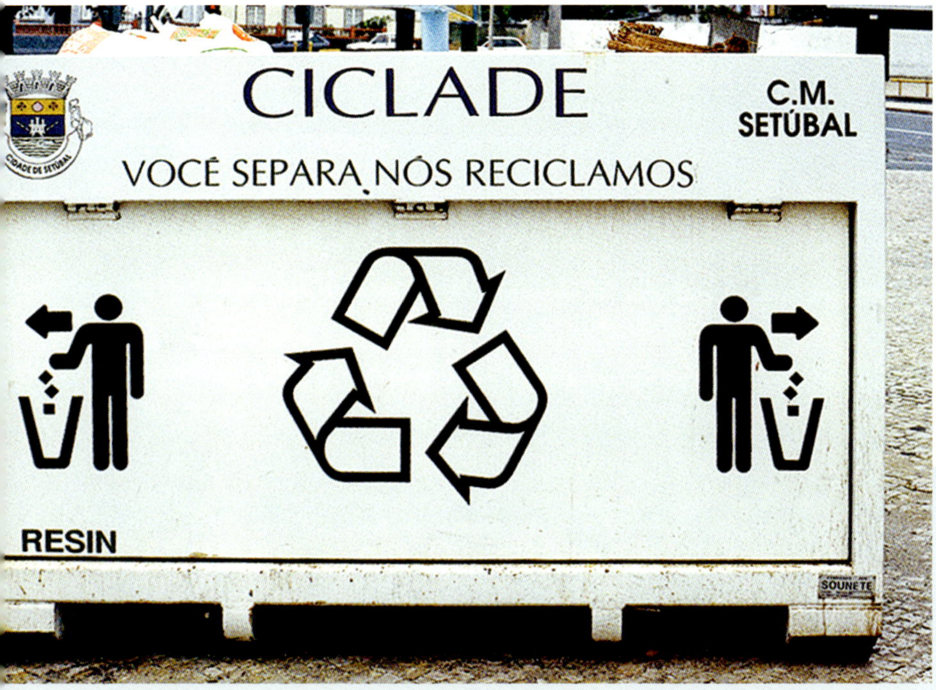
CICLADE
C.M. SETÚBAL
VOCÊ SEPARA NÓS RECICLAMOS
RESIN

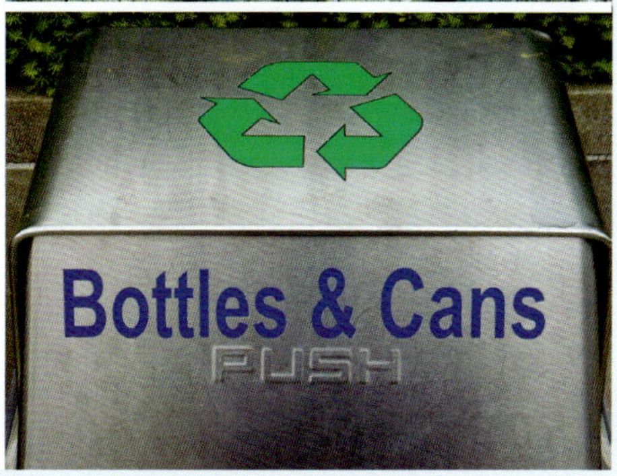
Bottles & Cans
PUSH

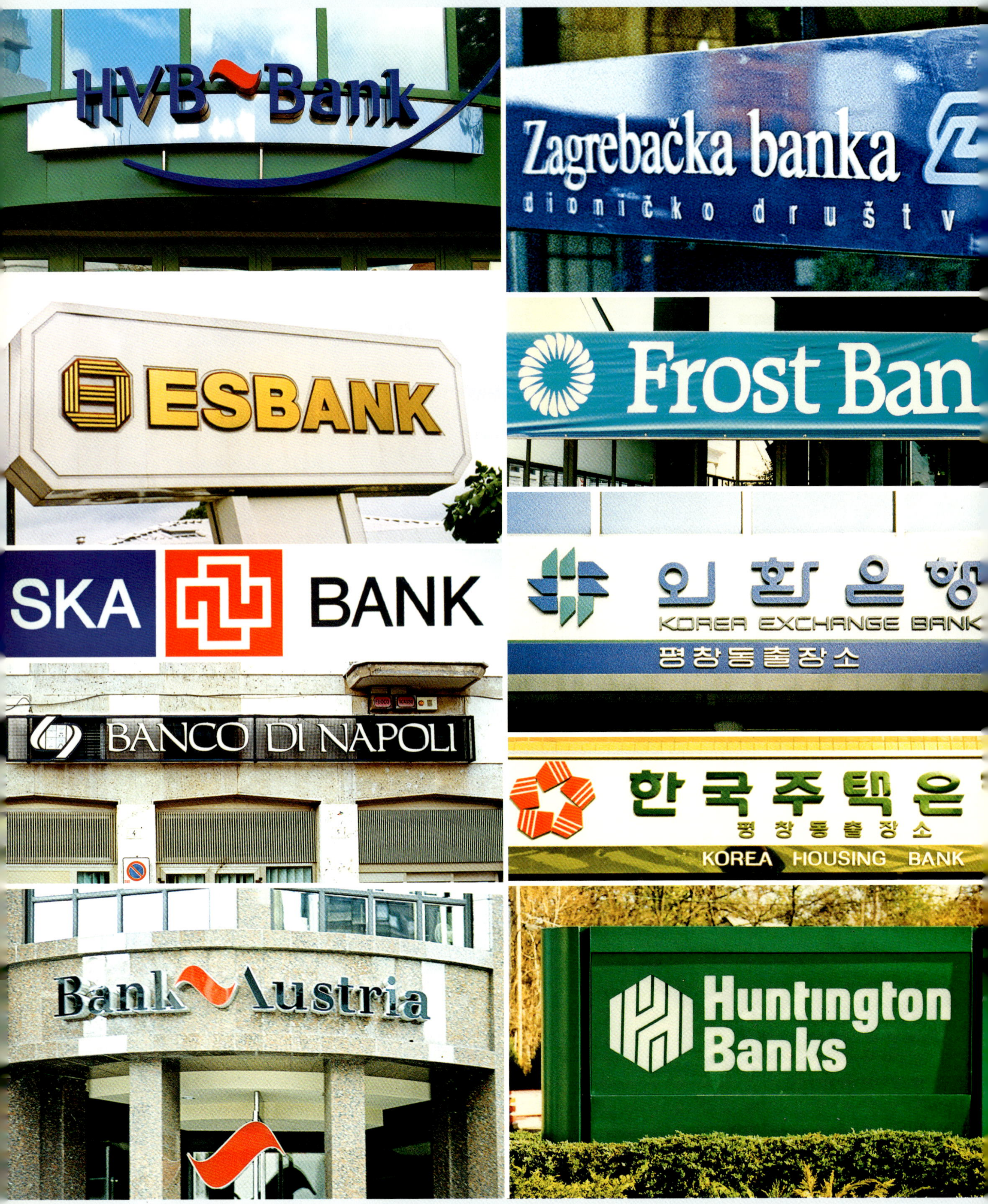
HVB Bank
Zagrebačka banka
dioničko društv
ESBANK
Frost Ban
SKA BANK
외환은행
KOREA EXCHANGE BANK
평창동출장소
BANCO DI NAPOLI
한국주택은
평창동출장소
KOREA HOUSING BANK
Bank Austria
Huntington
Banks

Bank logos

Banks, exchange offices, and other financial institutions often have logos with rotational symmetry, again, without a mirror plane. Their predilection for rotational symmetry may be interpreted by the character of their activities in that they move the money around. They display a variety of rotational symmetries.

Our examples come from American; Austrian; British; Croatian; Hungarian; Italian; Korean; Russian; Serbian; Swiss; and Turkish banks.

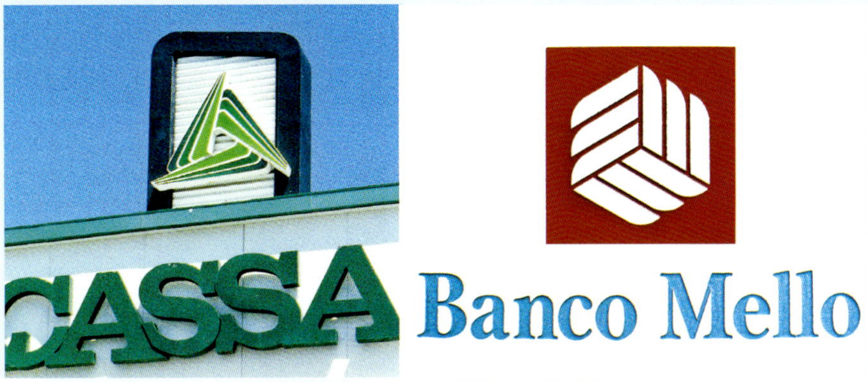

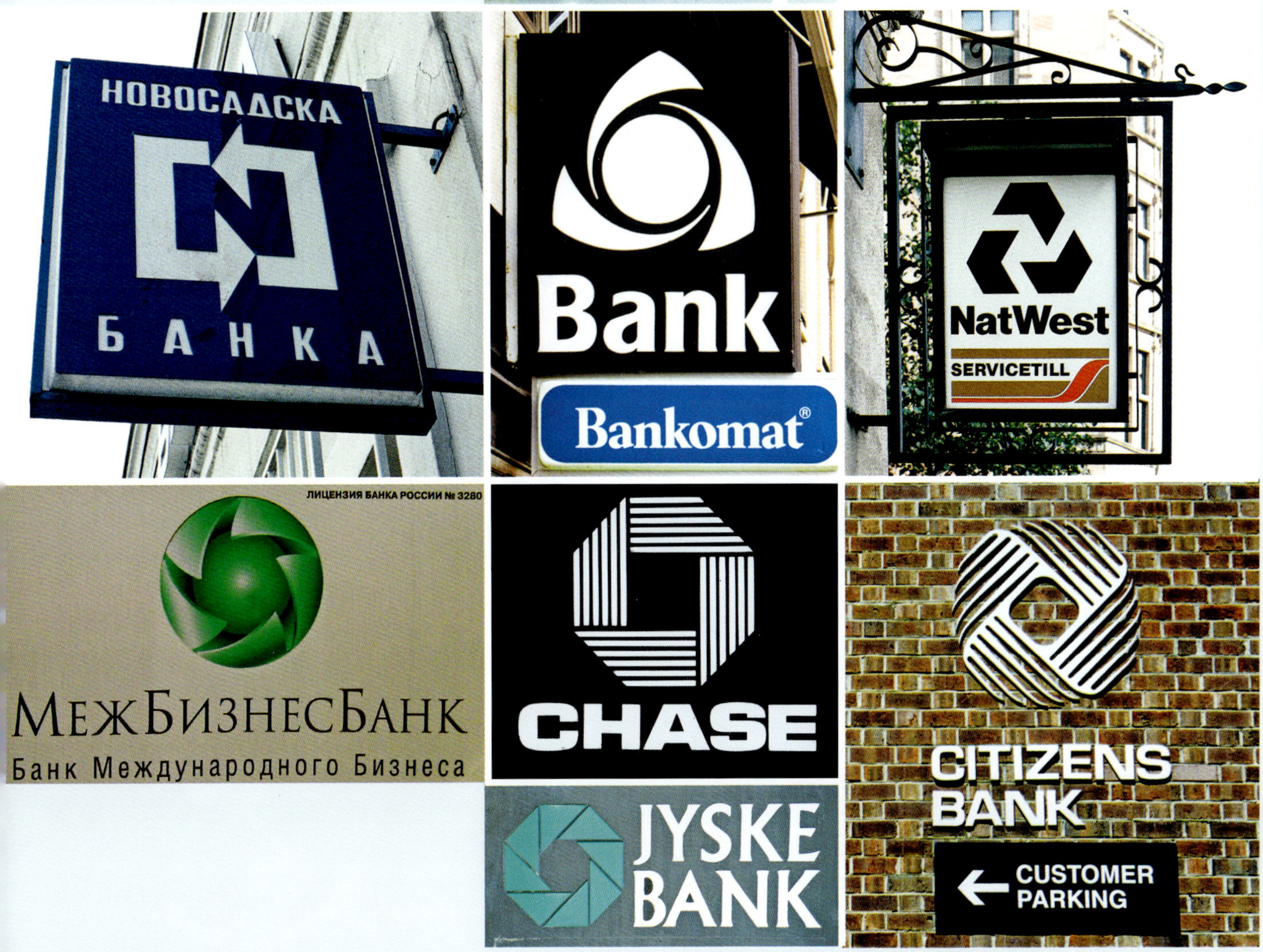

Hubcaps

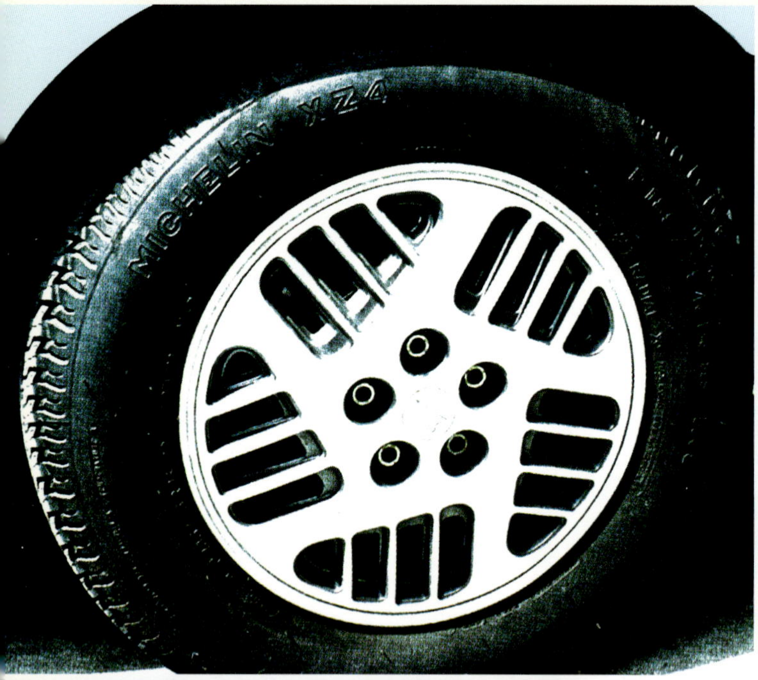

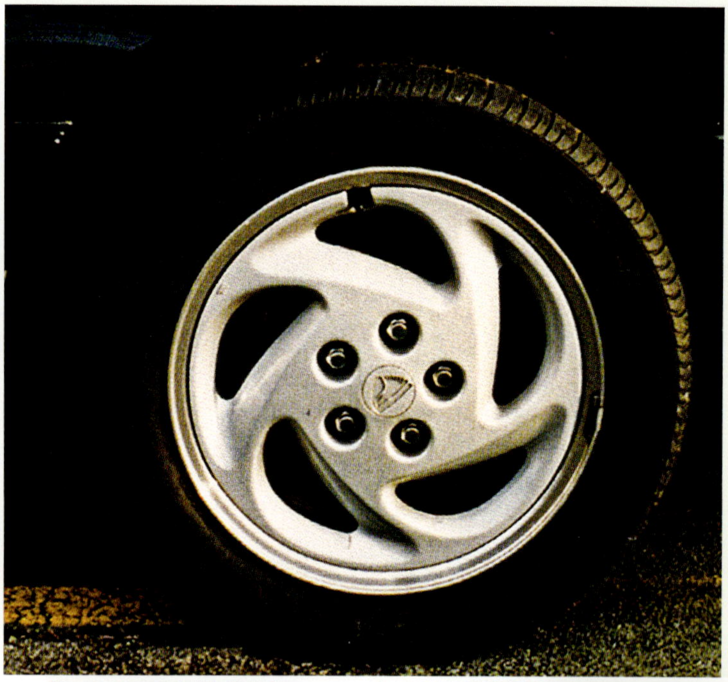

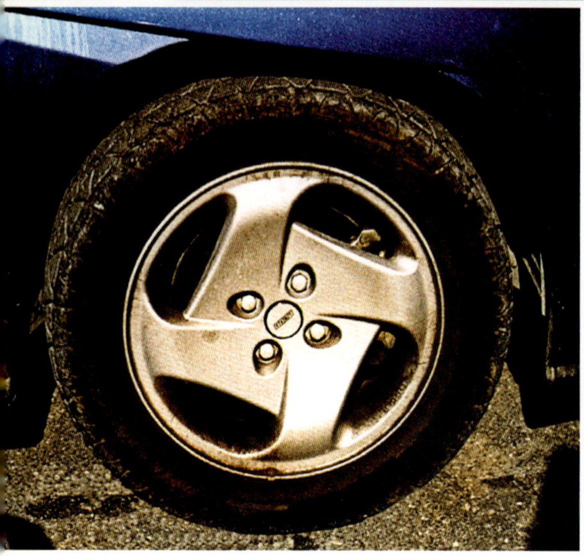

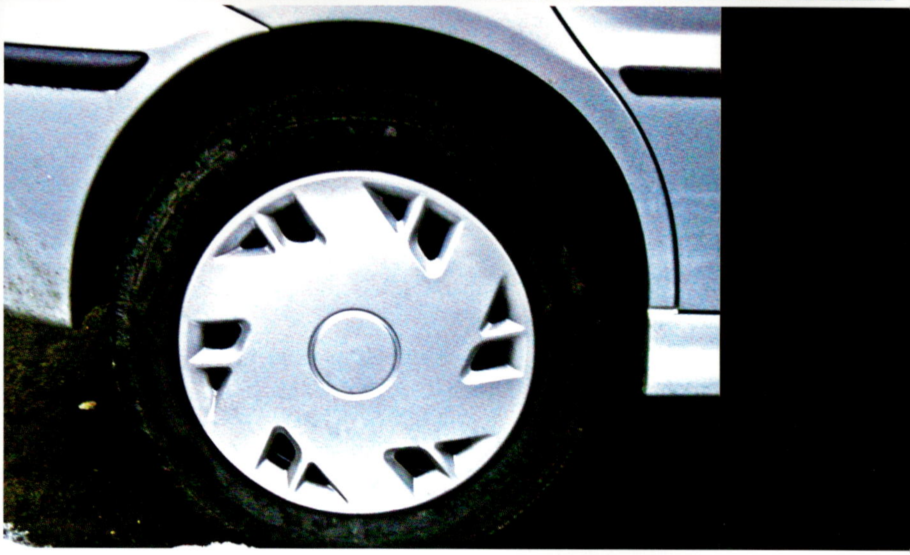

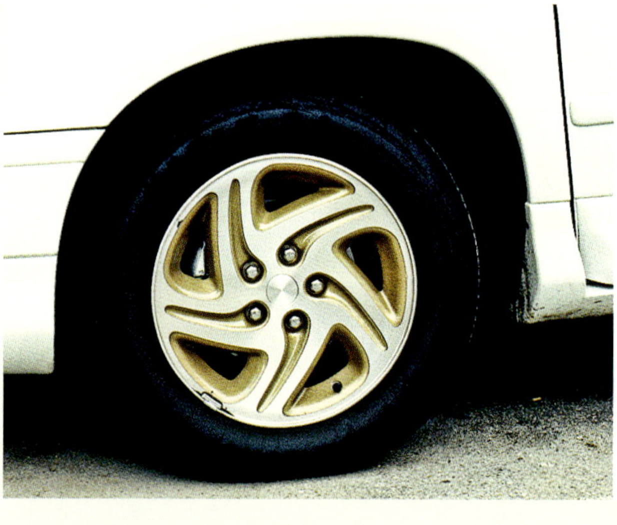

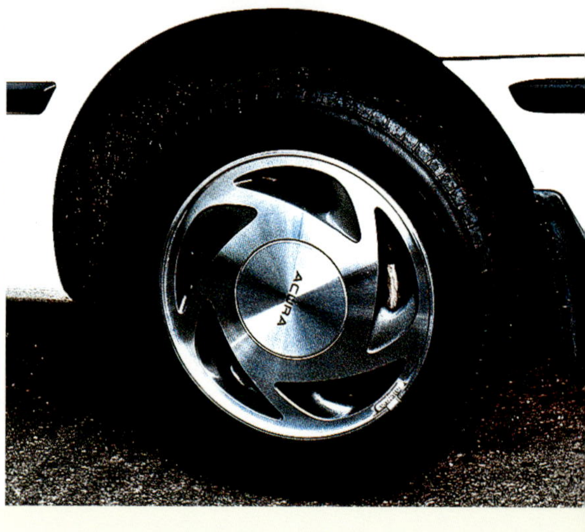

There are hubcaps with patterns of rotational symmetry only and others whose patterns display rotational symmetry and mirror planes as well. They both rotate regardless of their patterns, but our perception is that those with rotational symmetry only might be better suited for the rotational motion. Here we present a sample with exclusively rotational symmetries.

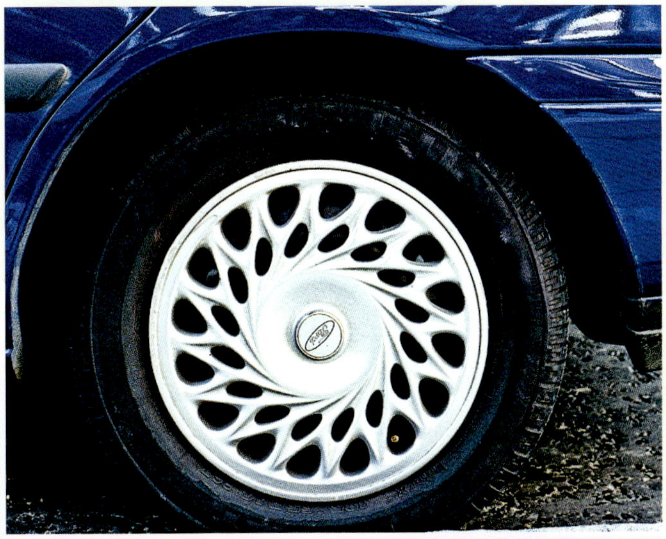

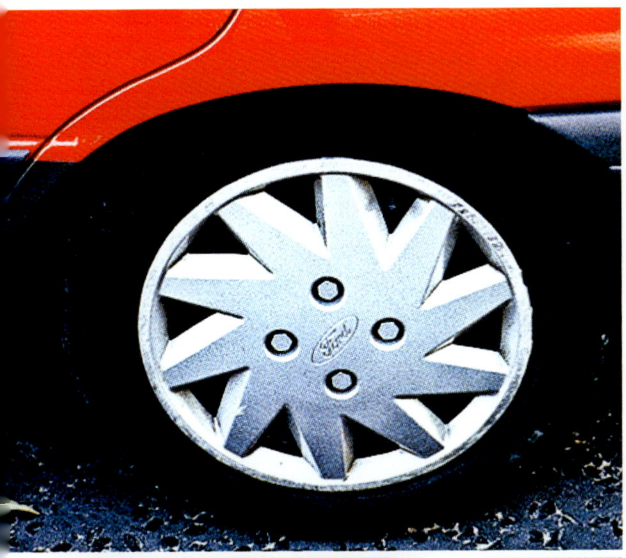

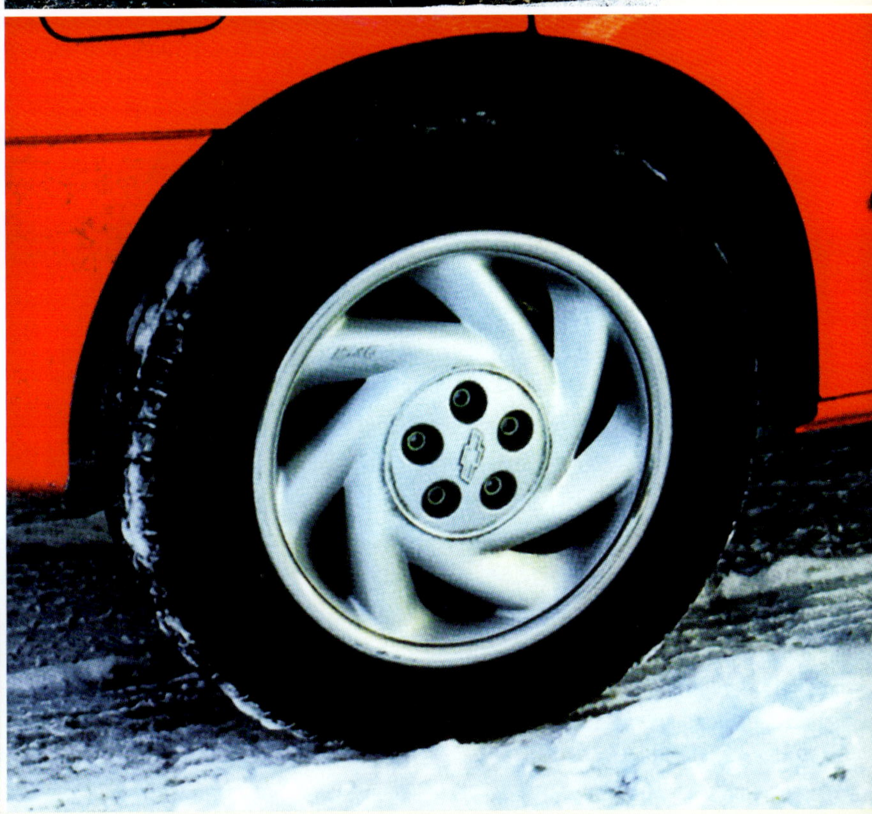

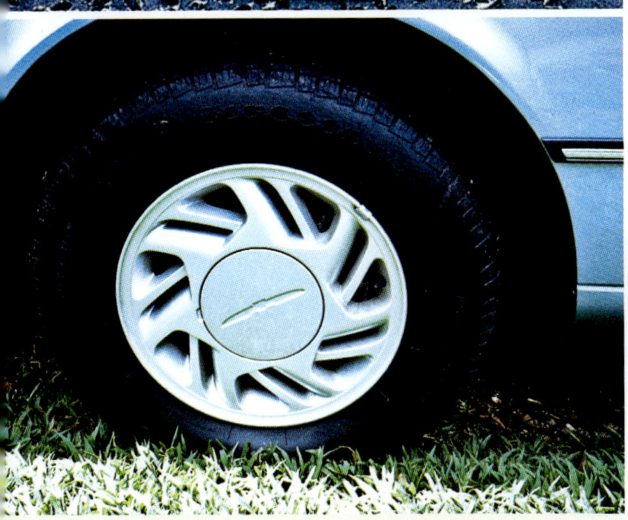

Sculptures

Rotational symmetry occurs in sculptures of some special kind, often representing fish or dolphins coiling around each other. They appear in different parts of the world.

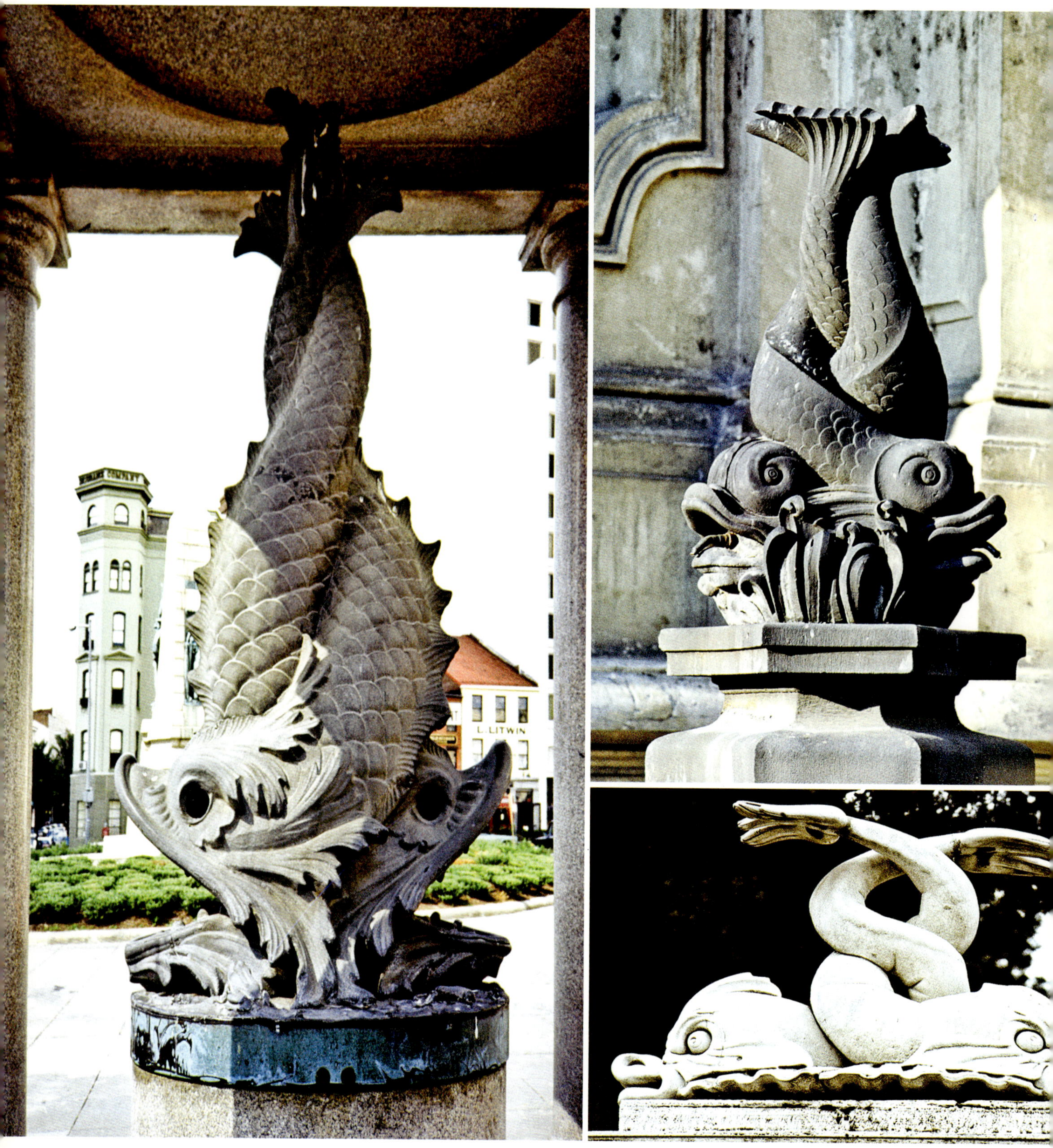

From left clockwise: Washington, DC; Prague; Rome

The sculptures in Washington and Rome have twofold symmetry; the Prague and Brighton sculptures have threefold symmetry; and the Linz sculpture has fourfold symmetry.

There is then Charles Perry's modern sculpture displaying threefold symmetry.

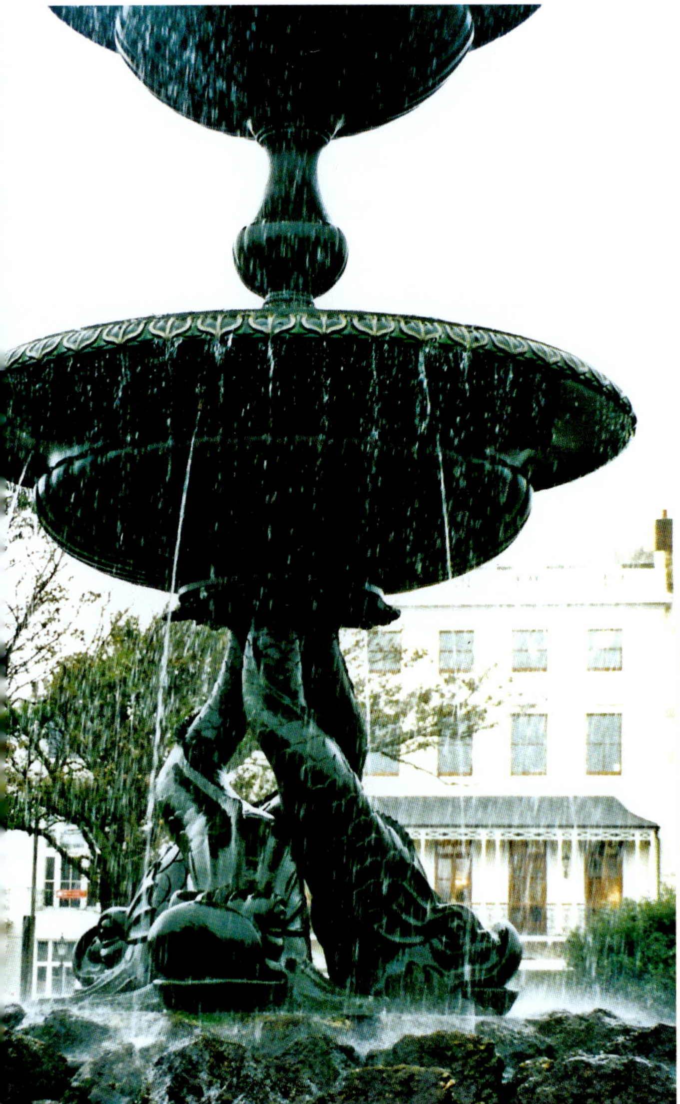

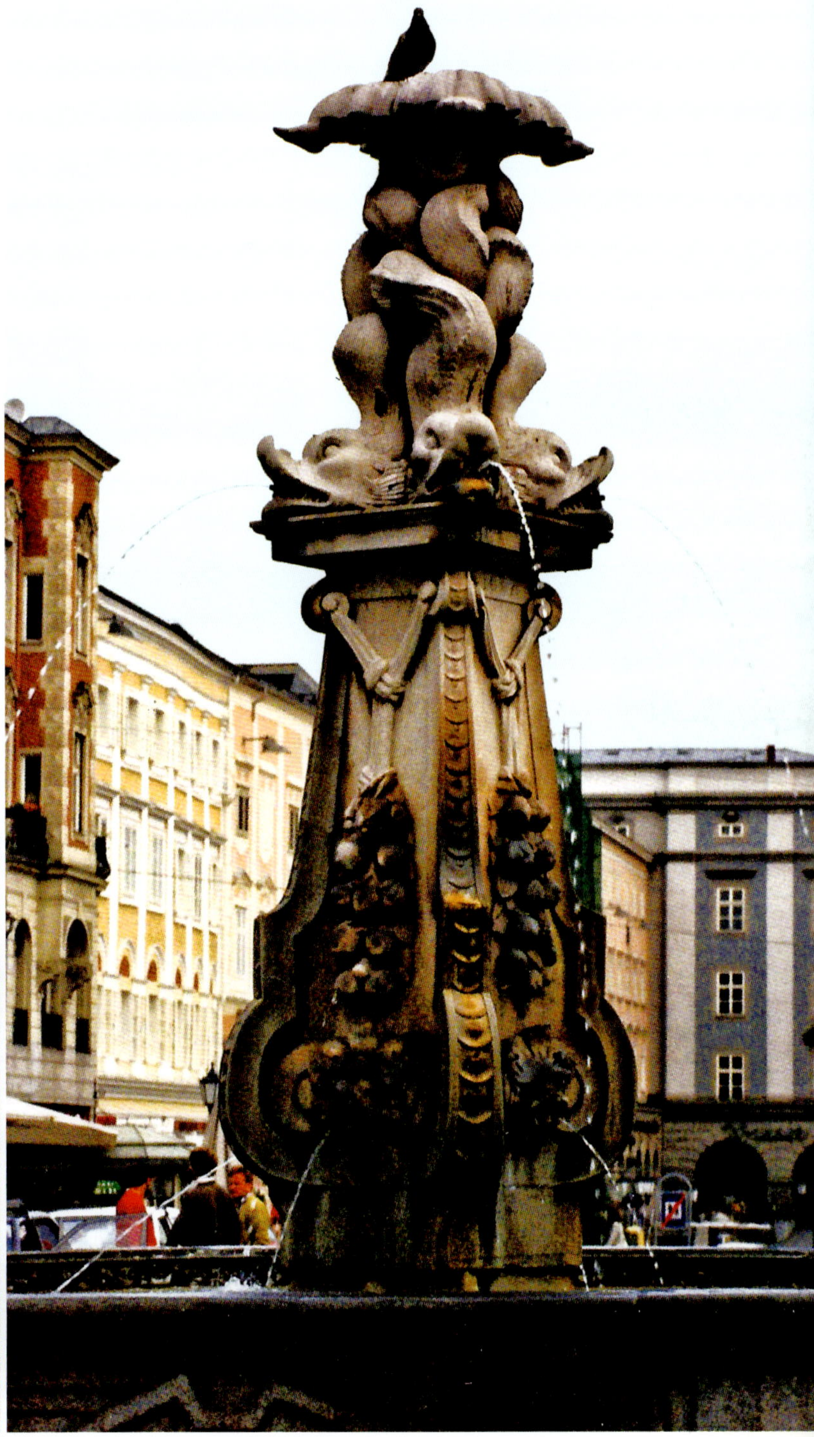

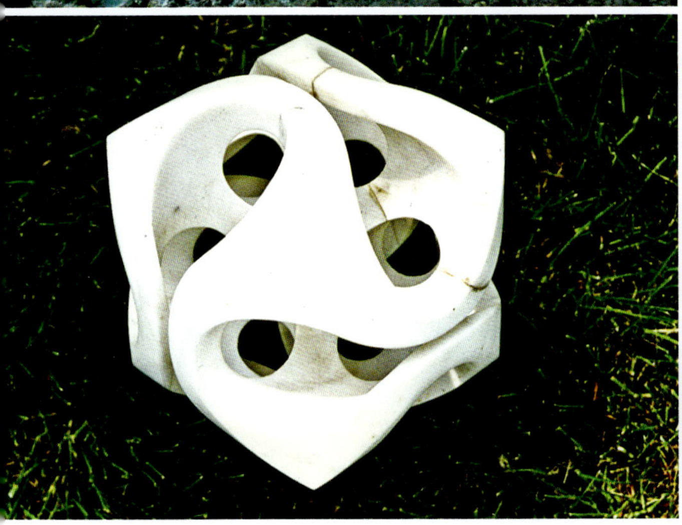

From upper left clockwise: Brighton, England; Linz, Austria; Charles Perry's (Norfolk, Connecticut) sculpture

Architecture

Churches and old buildings often have decorations with patterns of rotational symmetry as illustrated by these examples of European church windows.

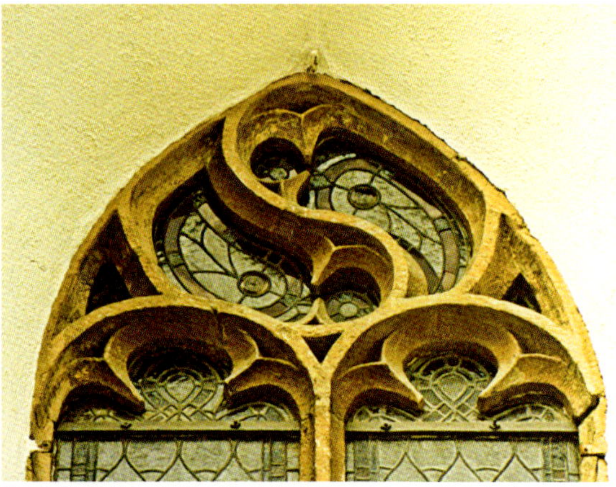

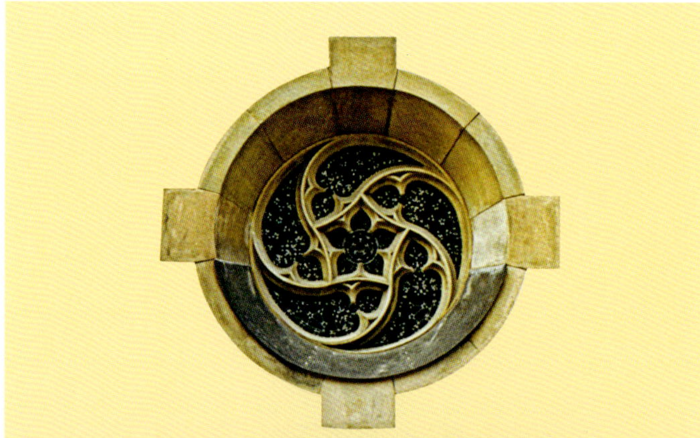

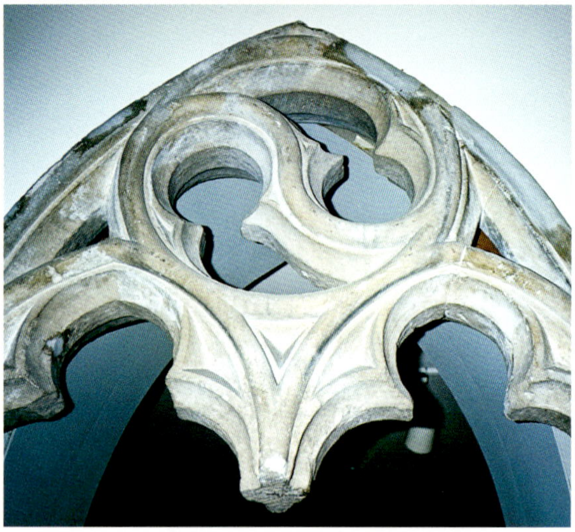

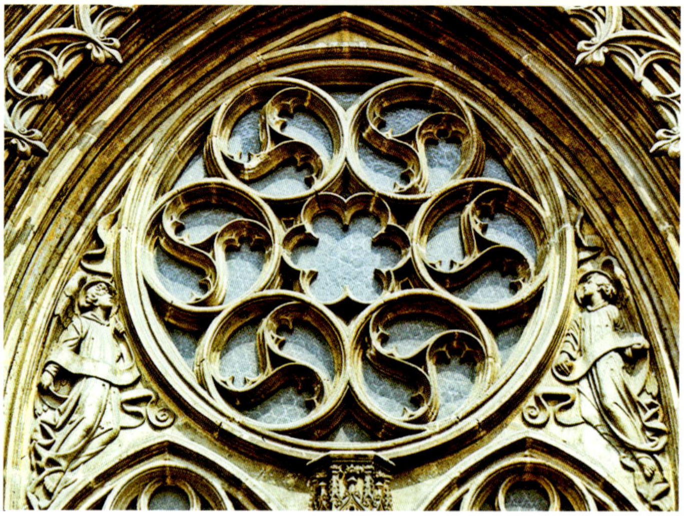

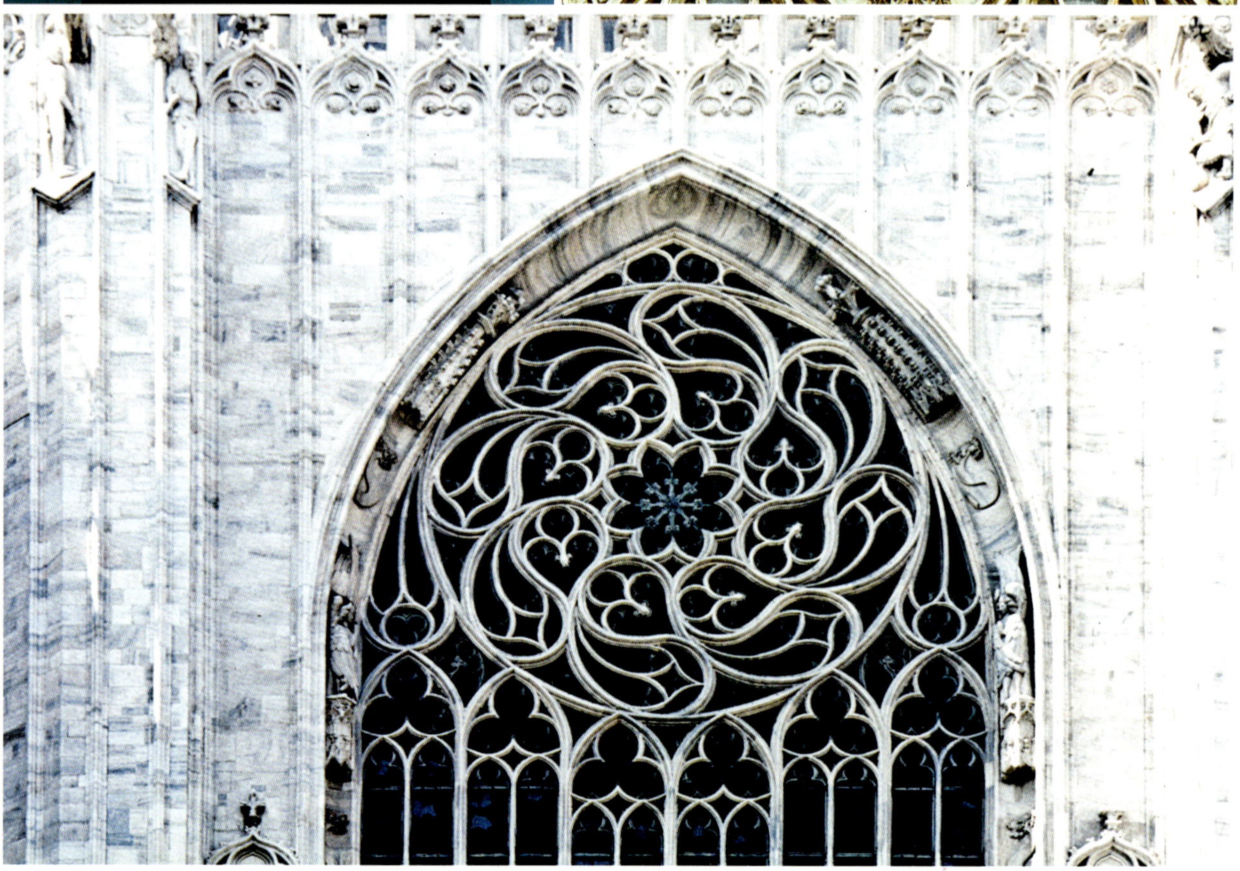

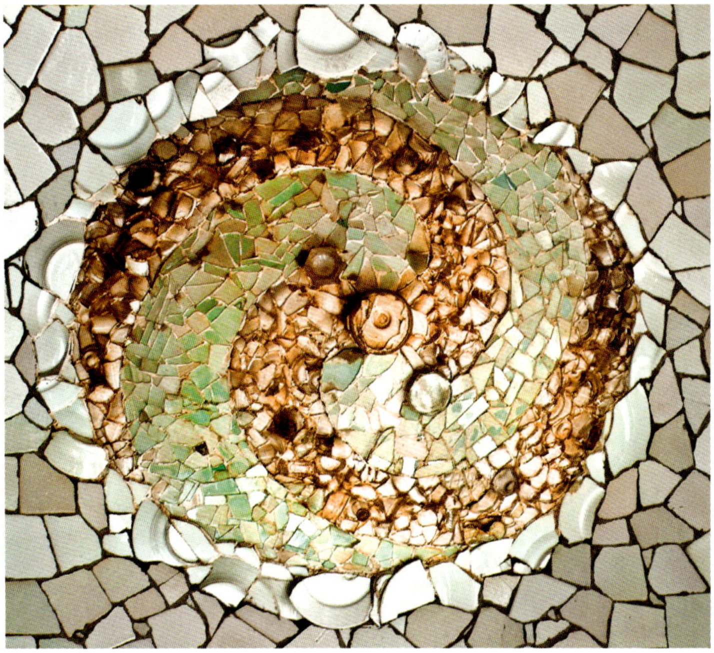
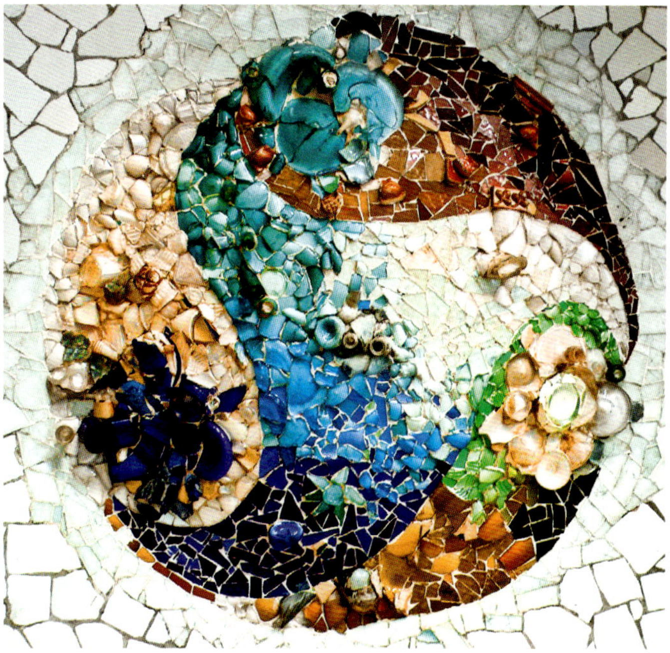
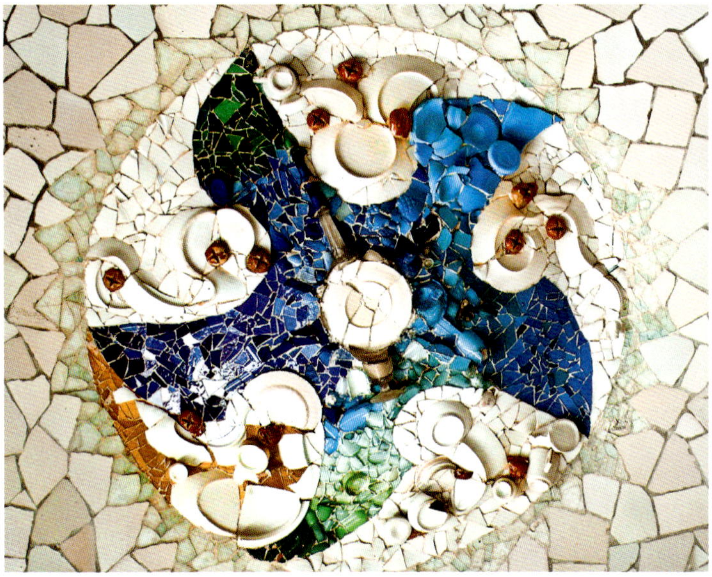
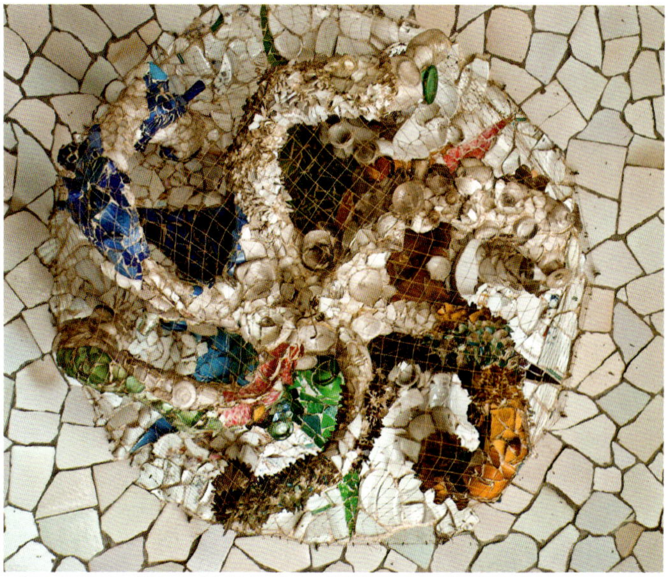
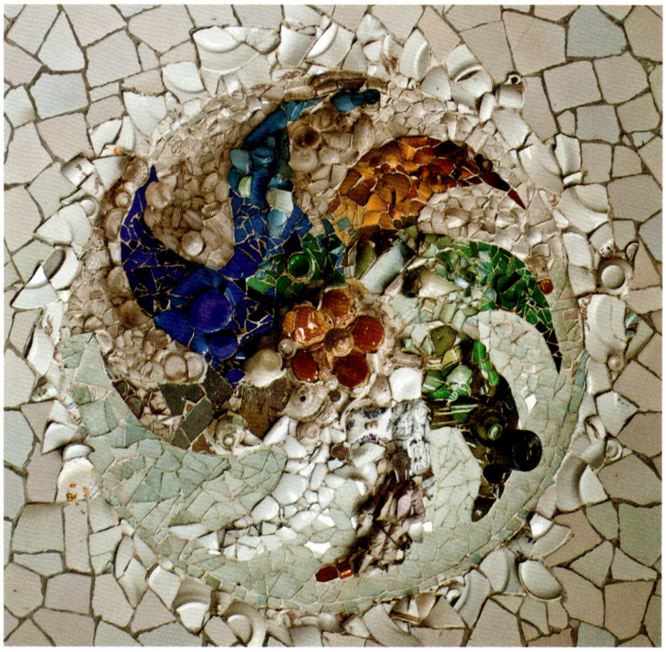

These mosaics decorating a ceiling in Park Güell in Barcelona, Spain, are the creations of Antoni Gaudi. The symmetries are rotational in all these decorations although they are rather approximate and are further masked by other decorative elements.

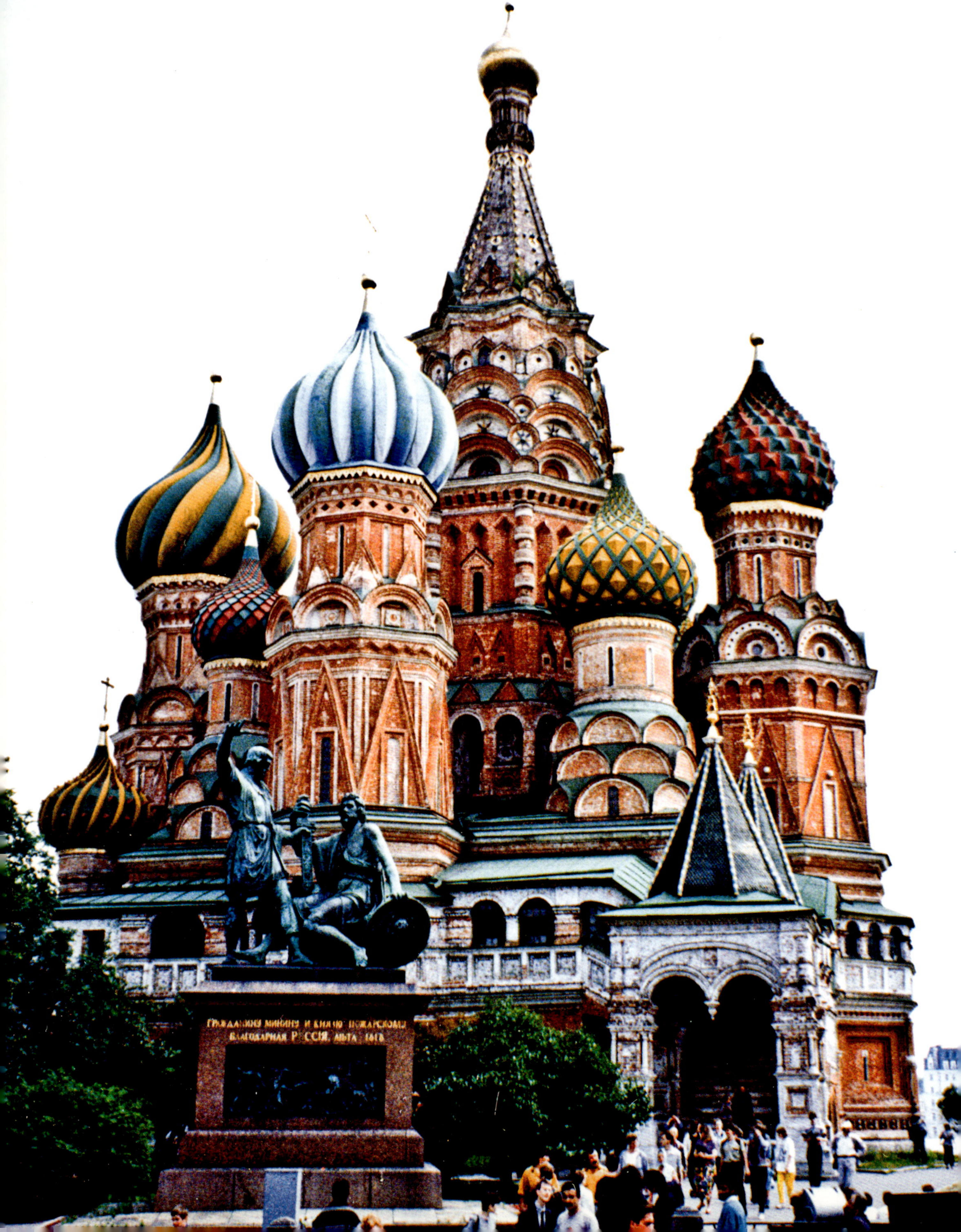

The magnificent towers display both rotational and combined rotational and mirror symmetries.

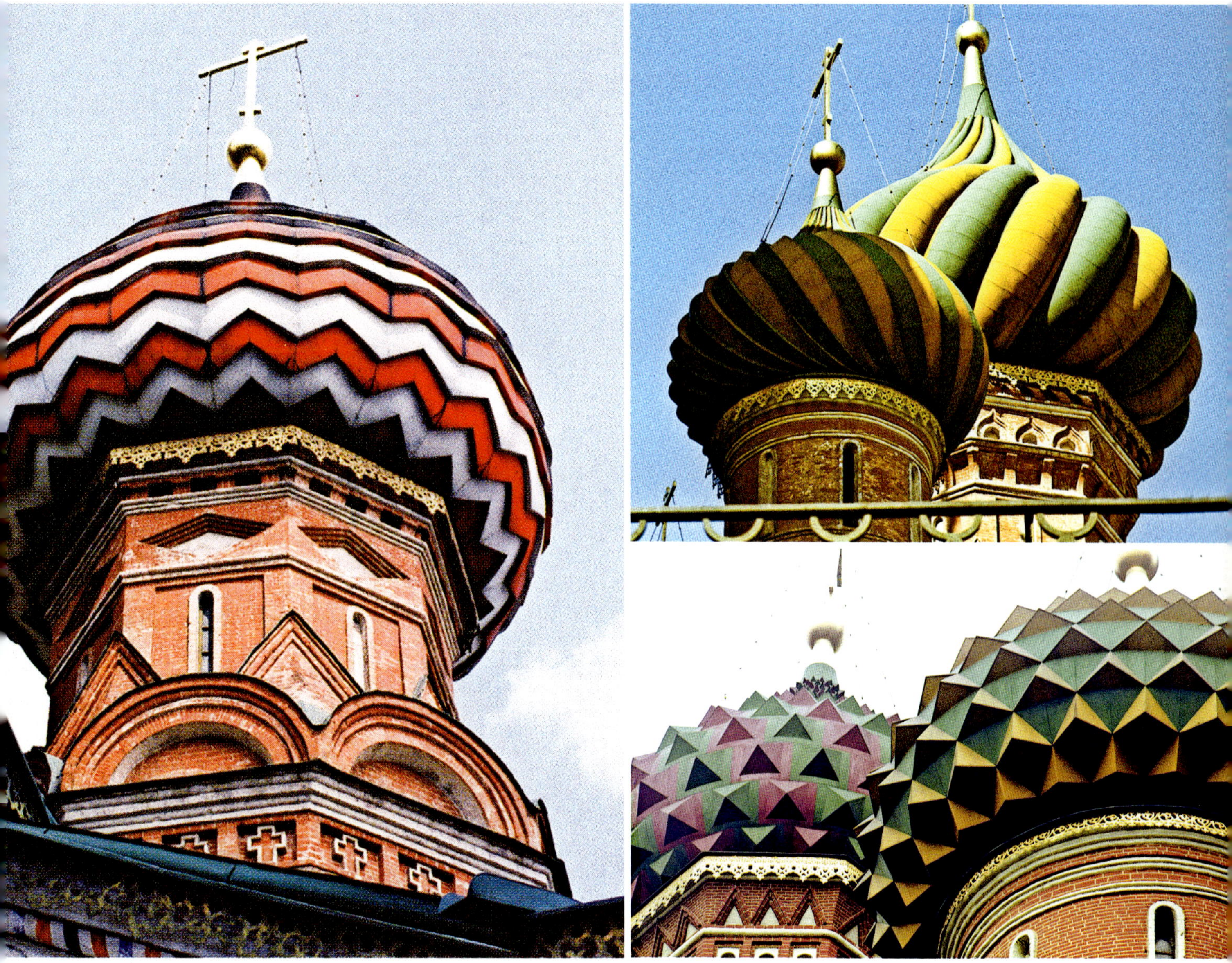

The Vasilii Blazhennii Cathedral in Moscow

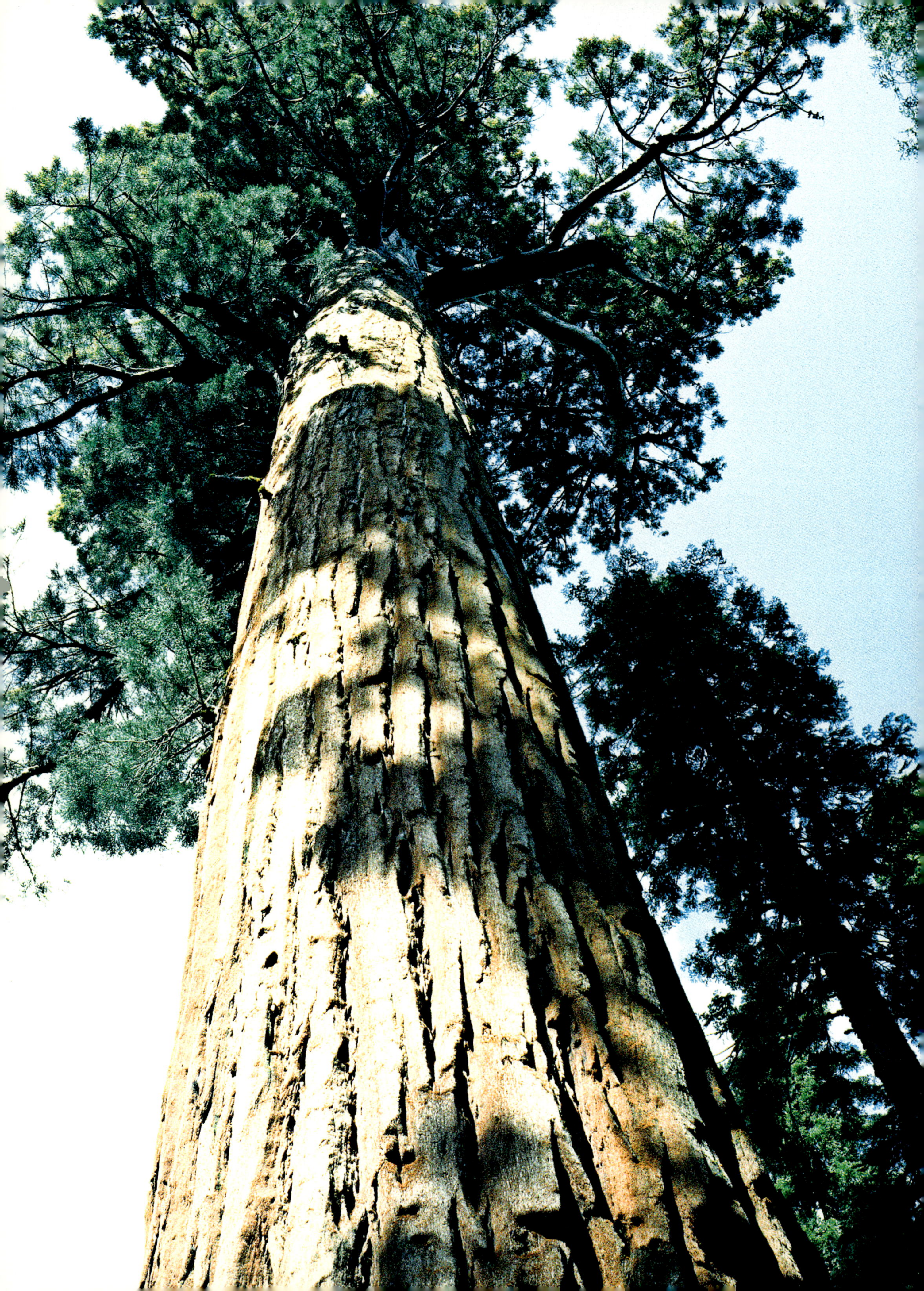

Shape and Movement

The bodies of most animals—as that of humans—have bilateral symmetry. This means that most animals (though not all) and humans have a left side and a right side that are each other's equivalents. But front and behind are not related in such a way, neither are top and bottom. Their bilateral symmetry conforms to their mode of motion. They move mainly forward as they walk, run, fly, swim, or crawl.

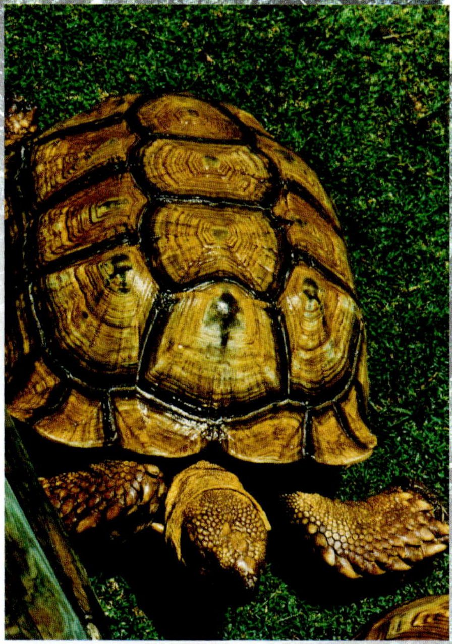

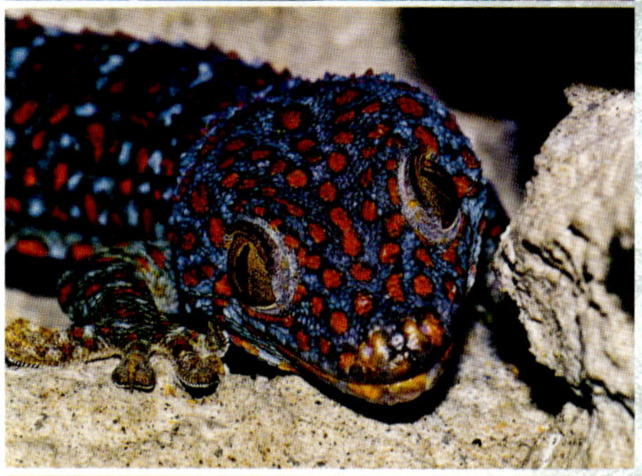

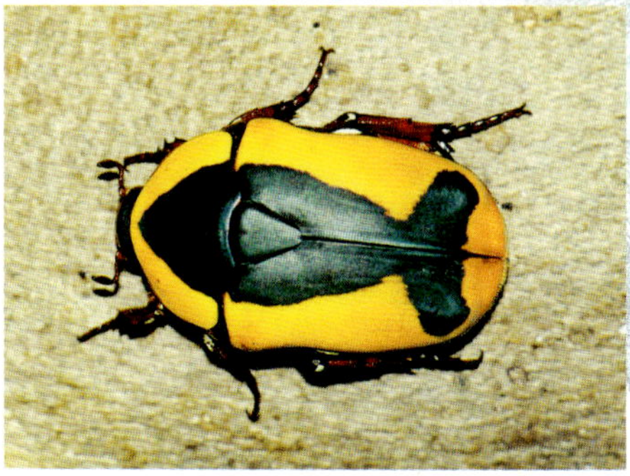

Giant tree in the Yosemite Park in California

More animals with bilateral symmetry.

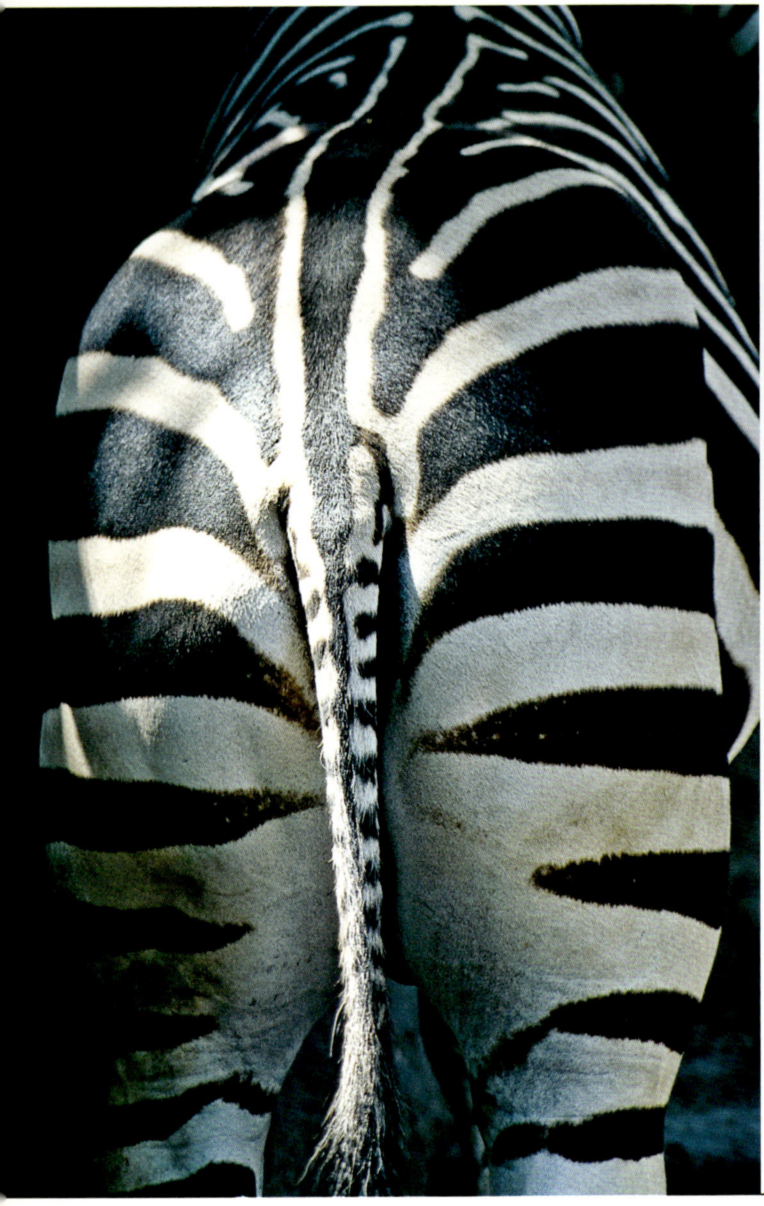

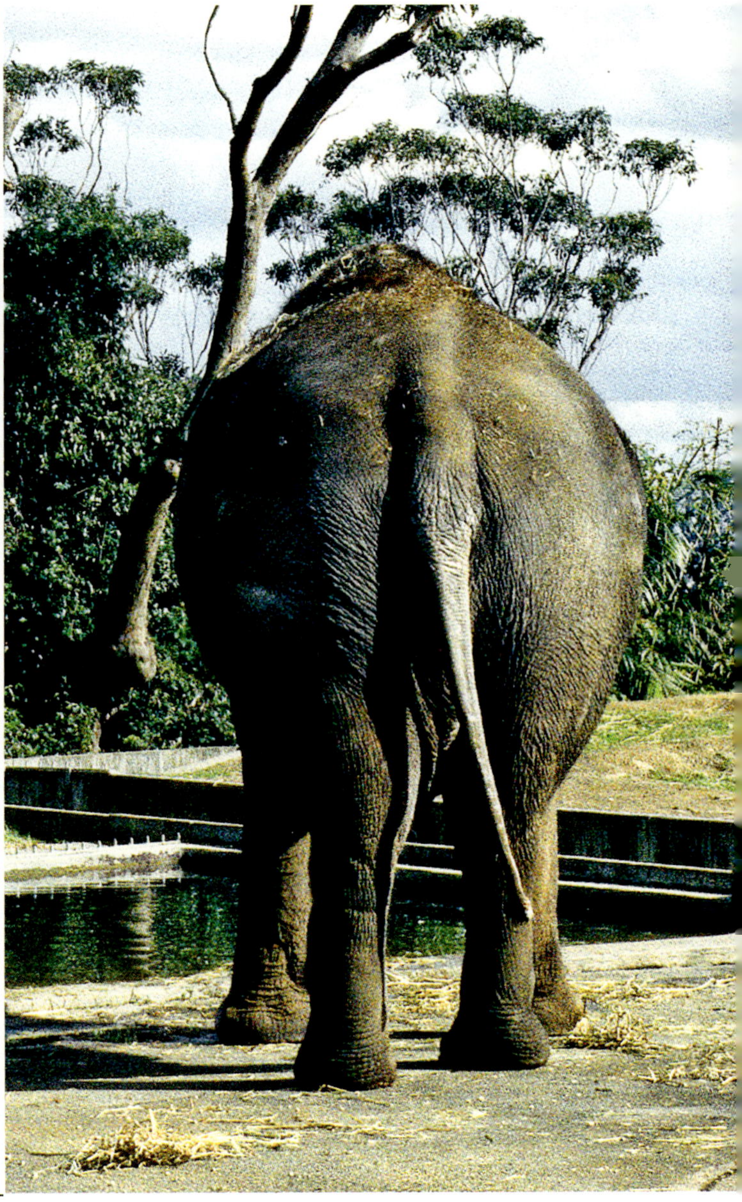

The animal images on page 89, the zebra on page 90, and the cat on page 91 are snapshots by Zoltán Bagosi of the Budapest Zoo (used by permission).

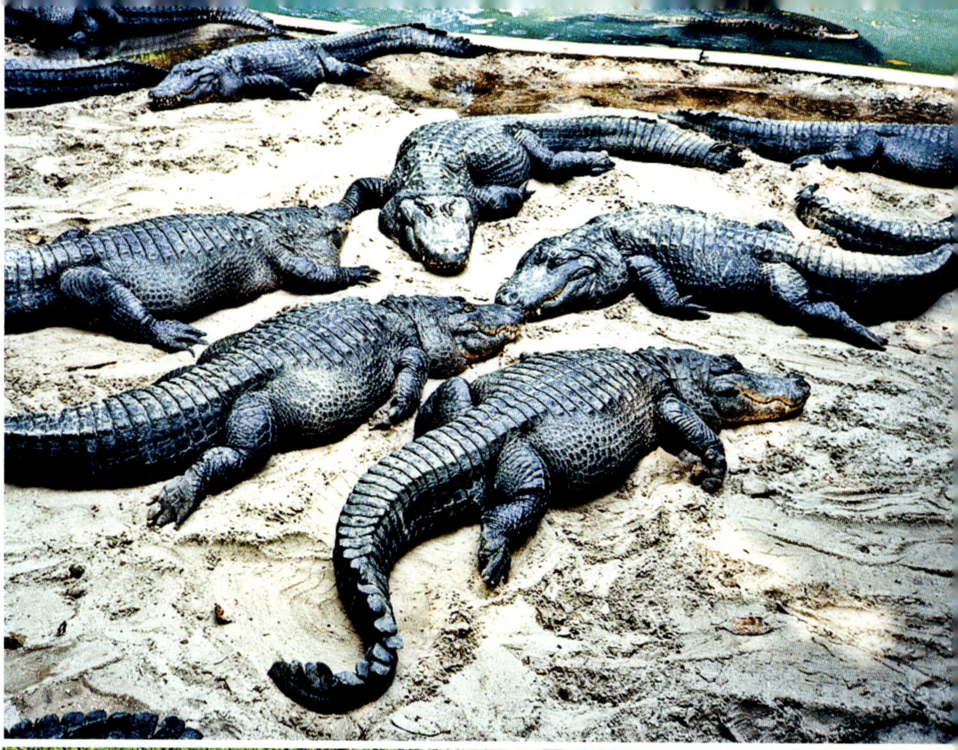

Three images from the Honolulu Zoo

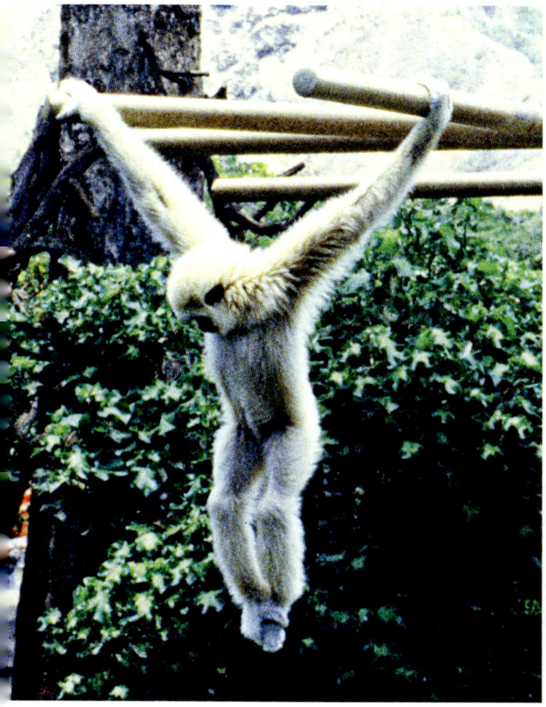

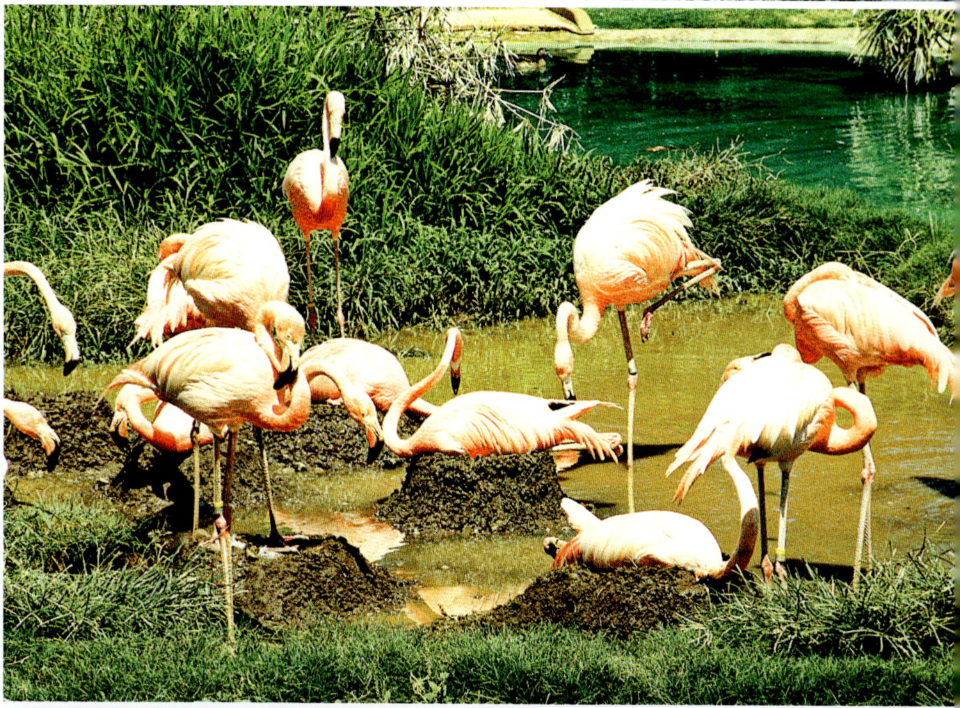

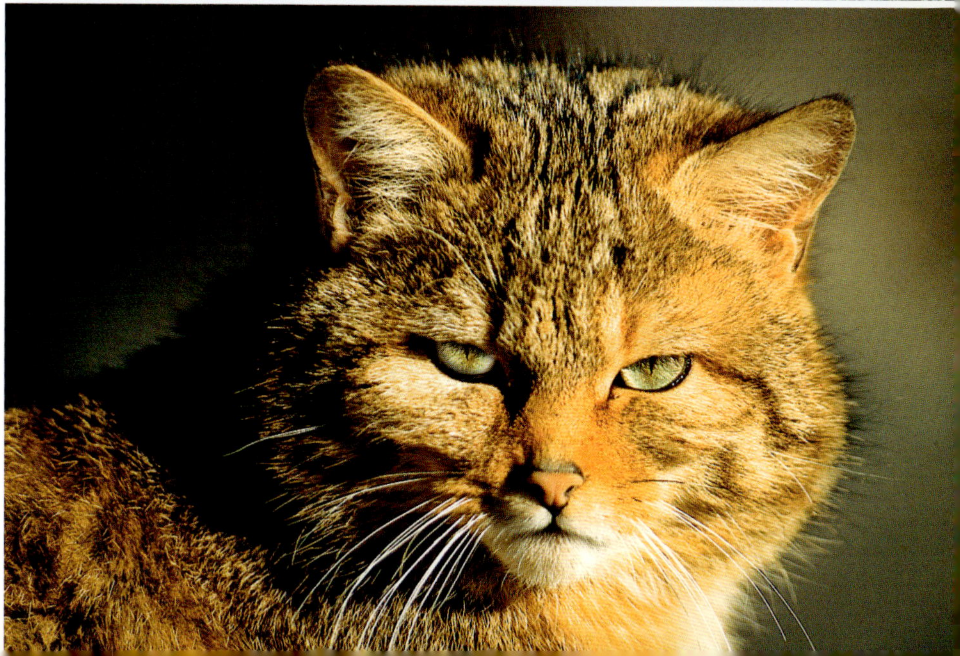

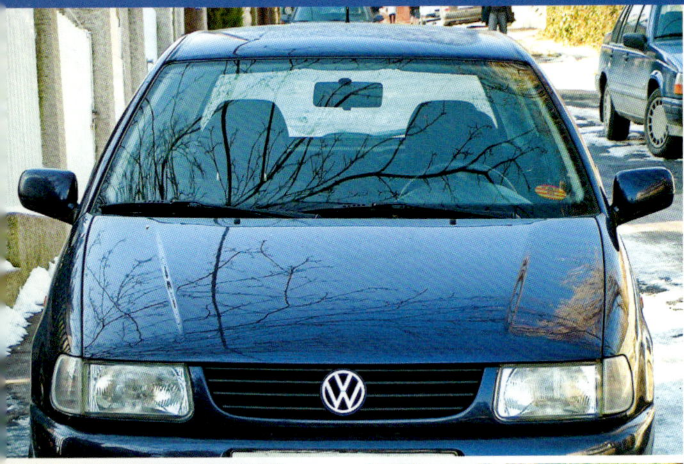

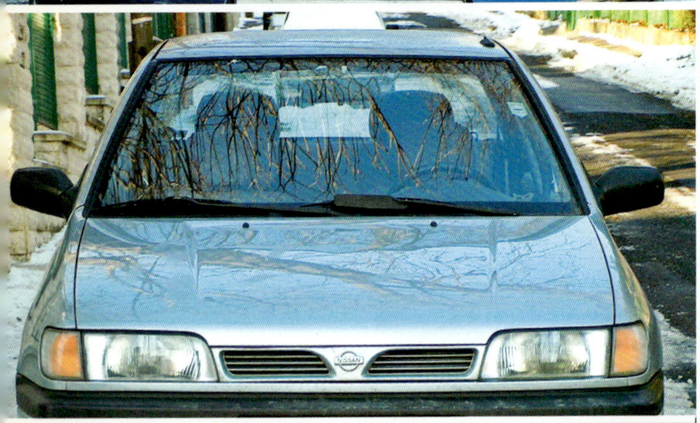

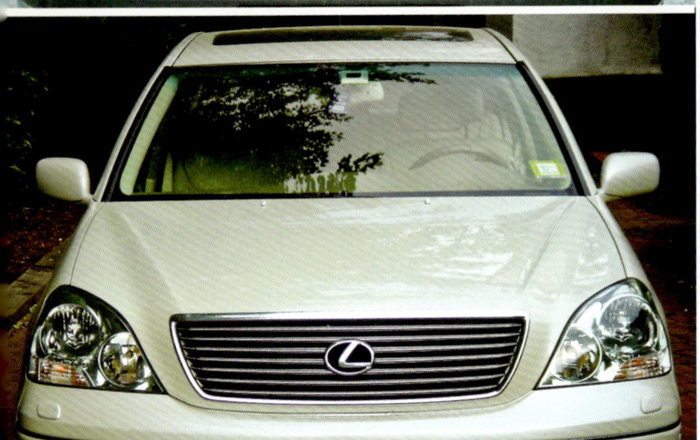

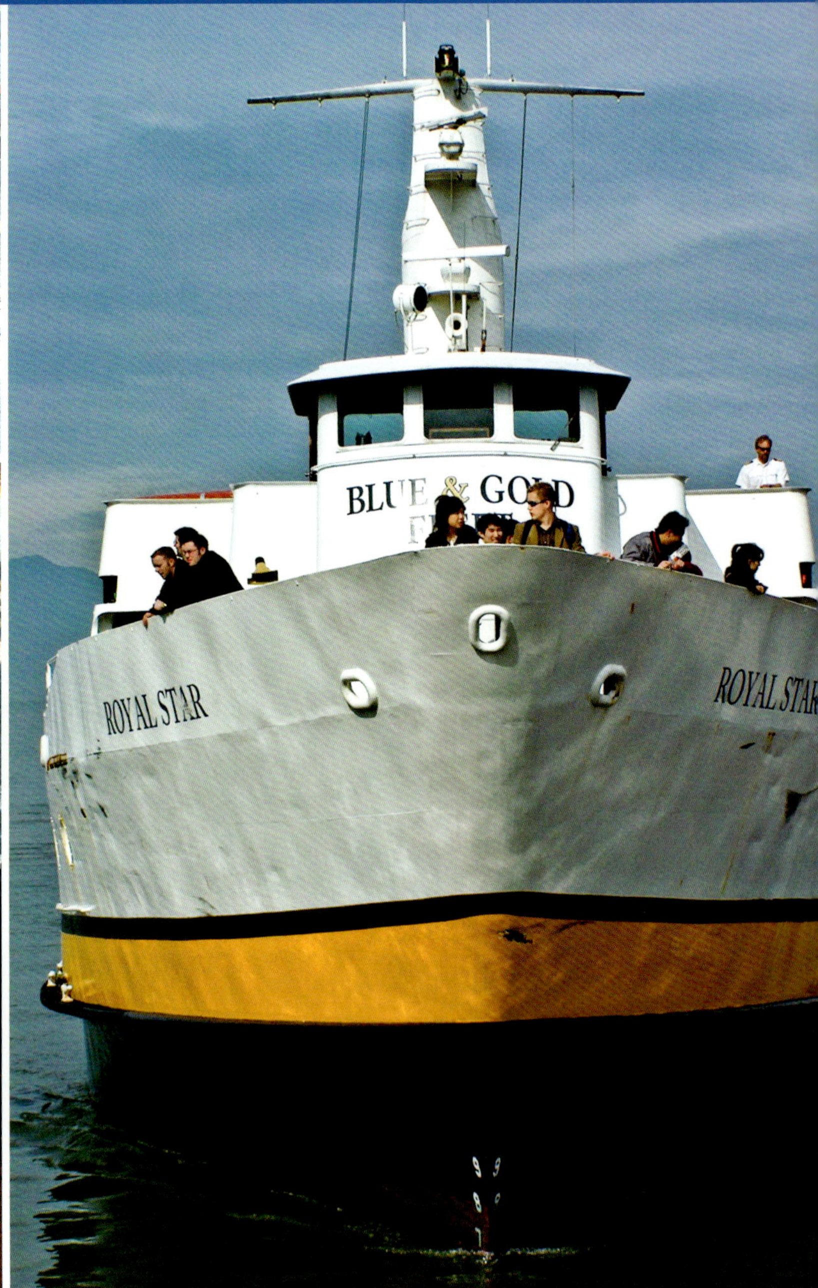

Almost all vehicles have bilateral symmetry. They are human-made objects and their bilateral symmetry is created for mobility. For efficient movement forward, the two sides must be balanced. This applies mostly to their external appearance; internally, there are differences between the two sides. For example, the driver side is different from the passenger side in cars.

Airplanes are also built to have bilateral symmetry. The images presented on this page were taken at an American air show by Michael Bartell, Atlanta, Georgia (used by permission).

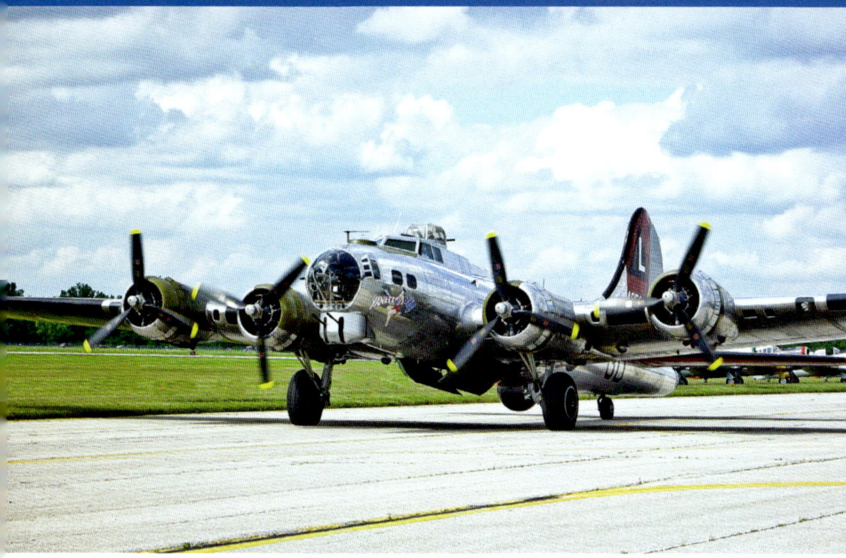

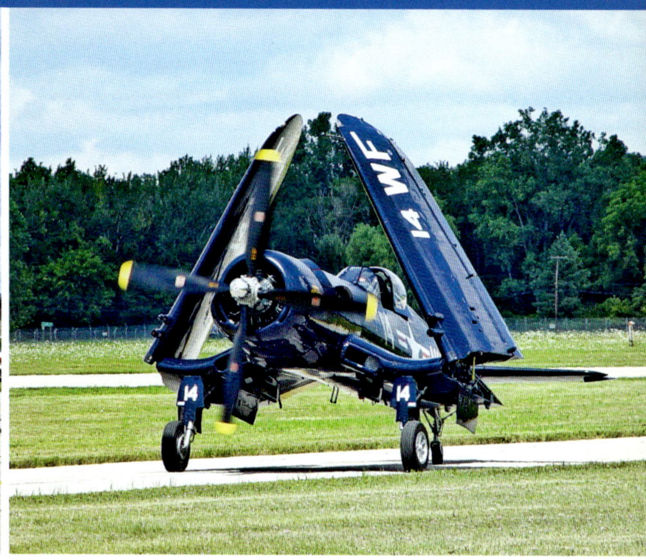

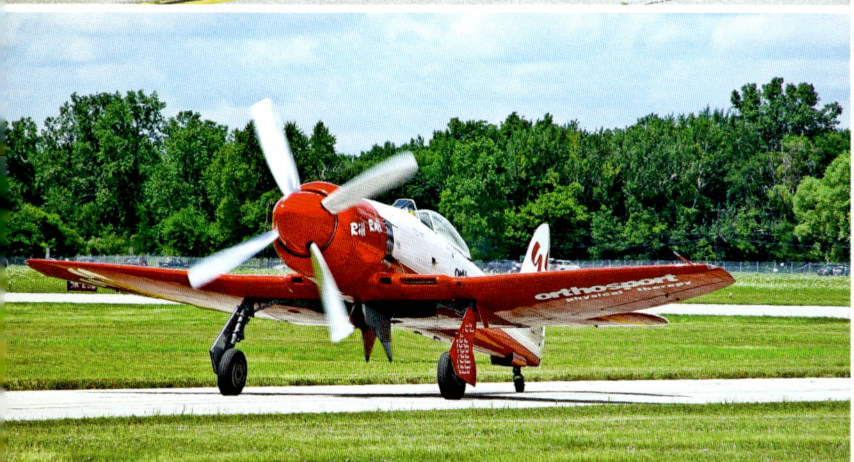

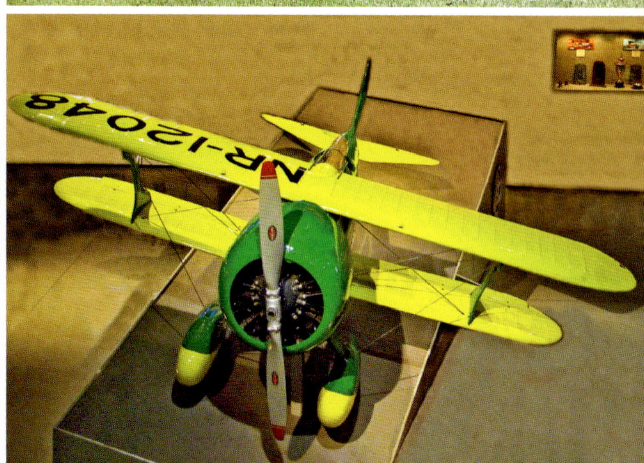

From the top left, clockwise:

B176 Boeing, the "Flying Fortress" (much used in World War II)

Vought-corsair (with folding wings to allow its transportation in larger aircraft)

"Laird Super Solution" (model)

"Howker Sea Fury"

Rocket at the Kennedy Space Center in Florida

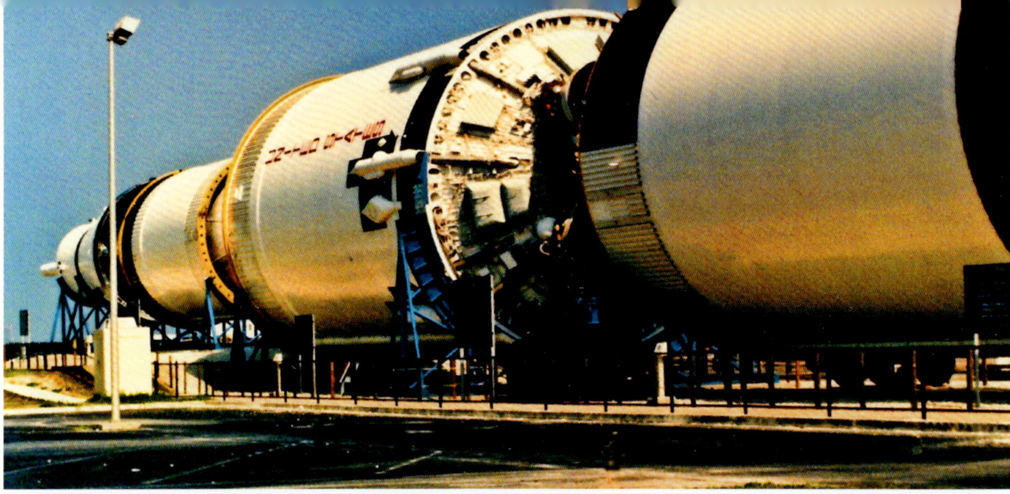

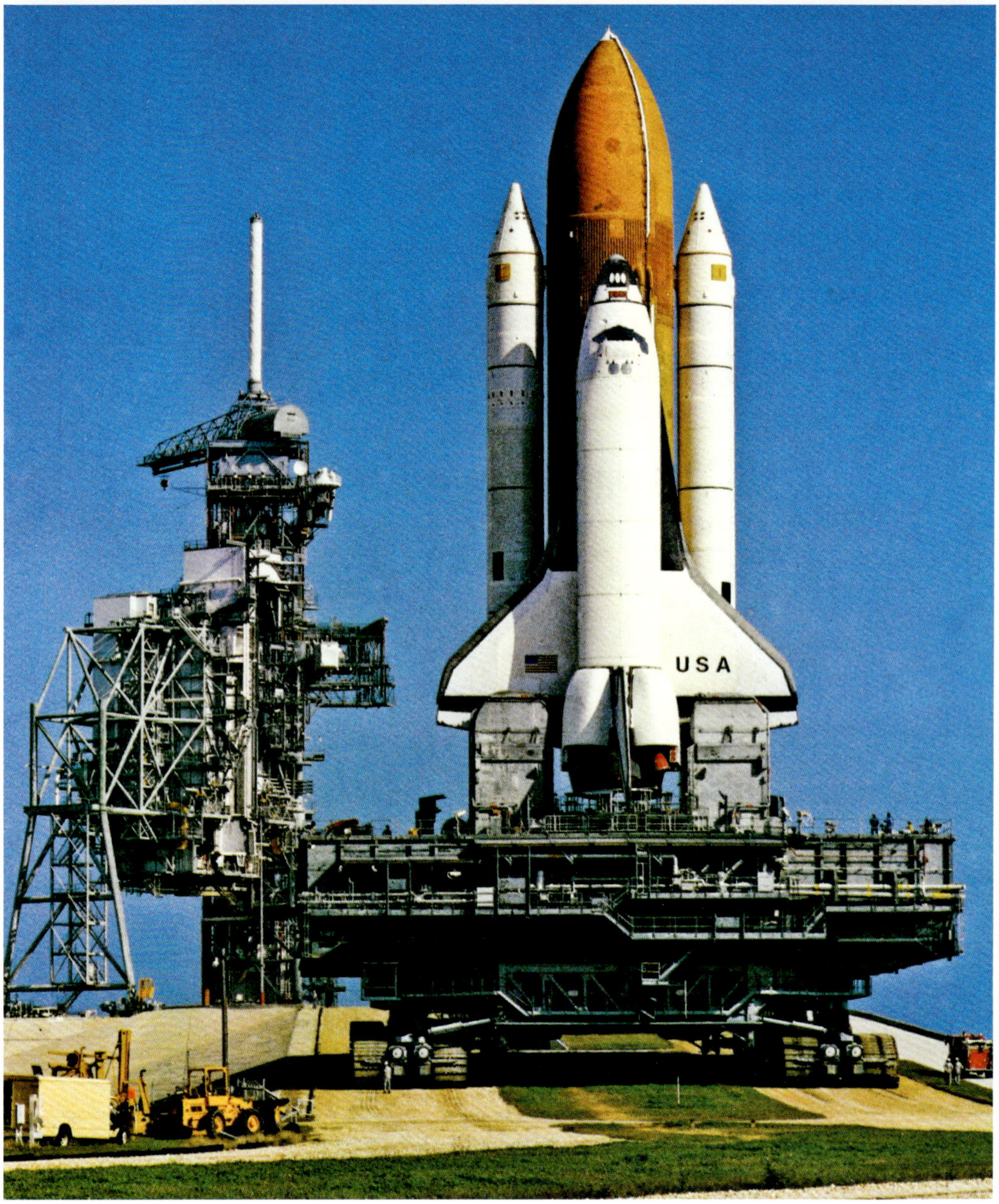

As rockets travel vertically from the surface of the Earth, there is no reason for them to have bilateral symmetry; they have cylindrical symmetry.

In cylindrical symmetry, everything is the same all around the vertical axis, but the top and the bottom are different.

Two pictures from Laser Photo Art, NASA Space Program Collection, Kennedy Space Center, Florida (reproduced with permission)

The NASA Space Shuttle has bilateral symmetry because upon its returning to the Earth it operates like an ordinary airplane. Its carrier rocket, however, has cylindrical symmetry.

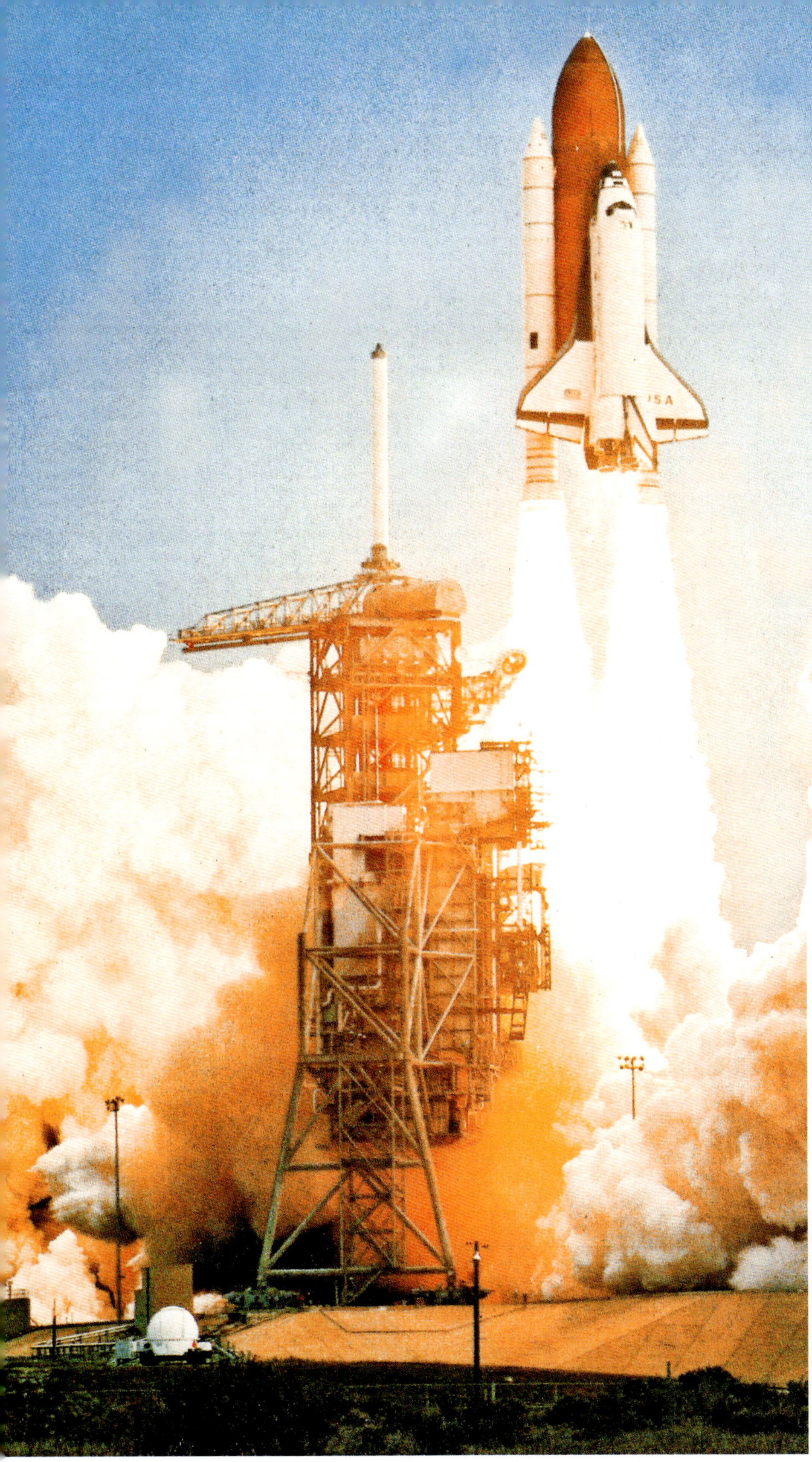

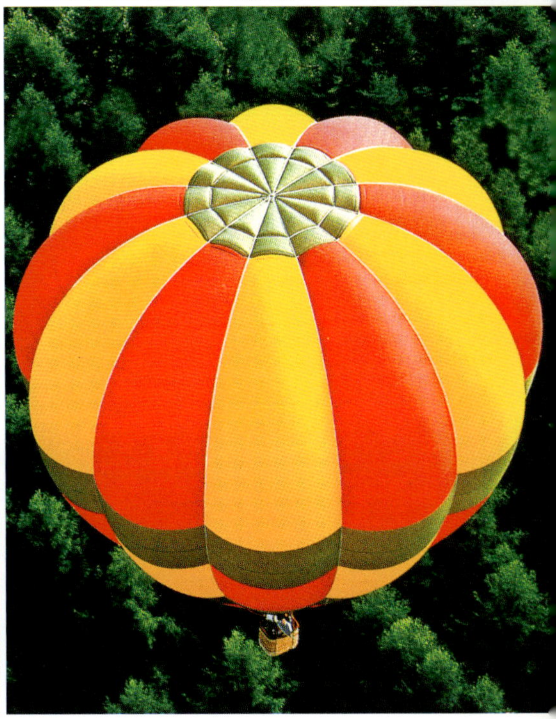

Balloons also have cylindrical symmetry.

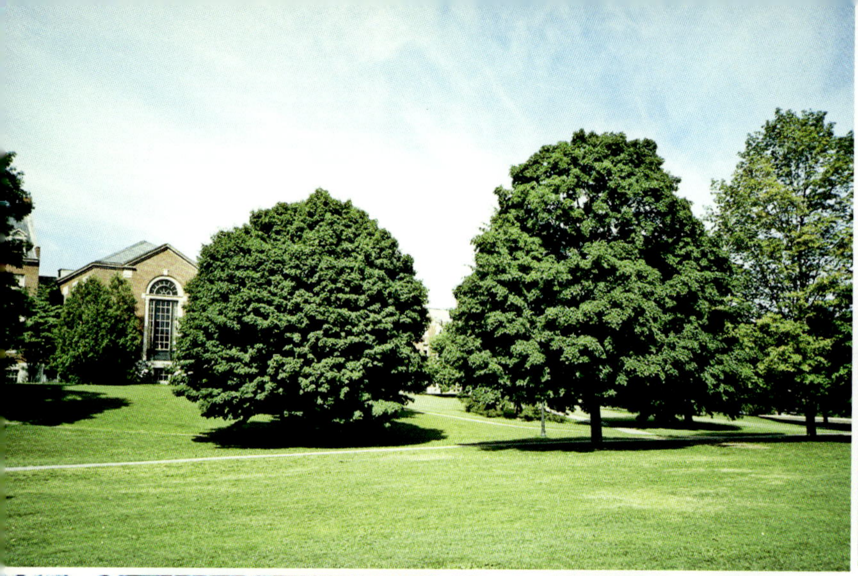
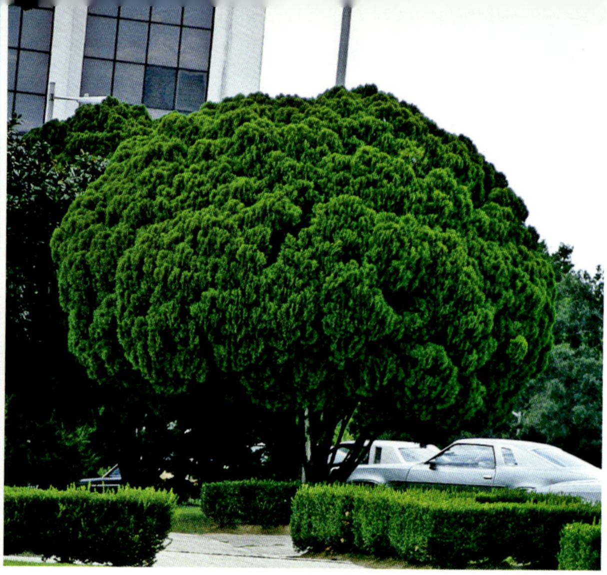
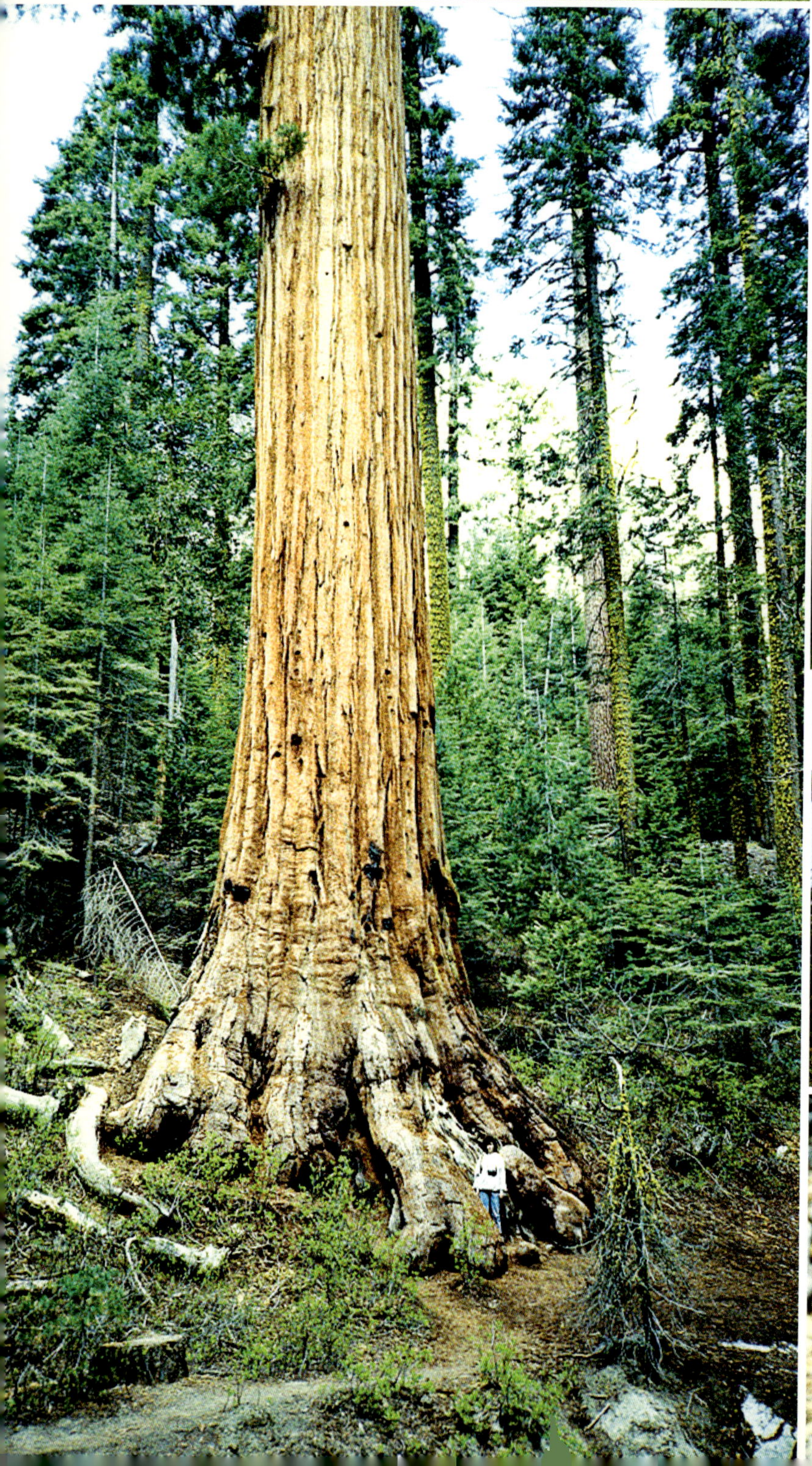
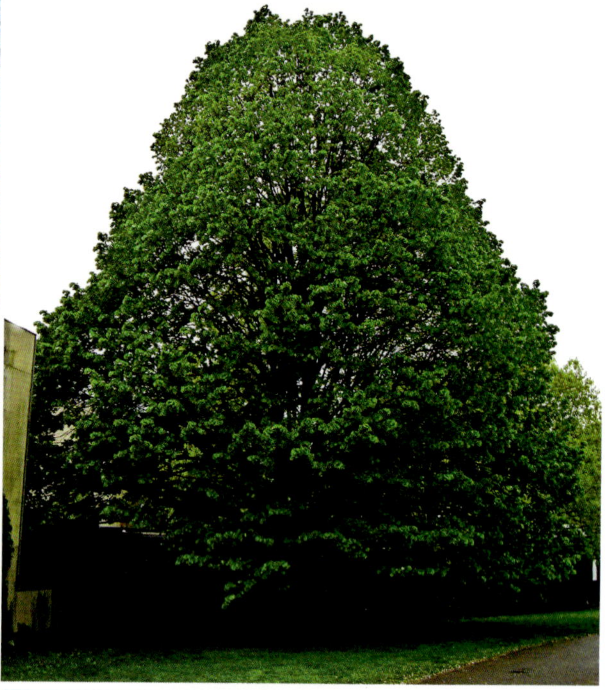
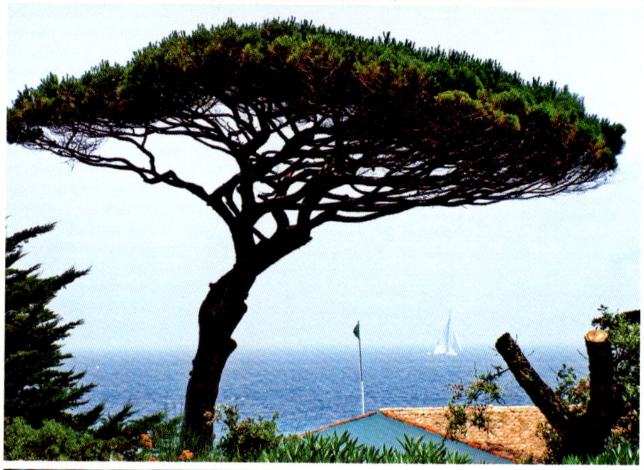
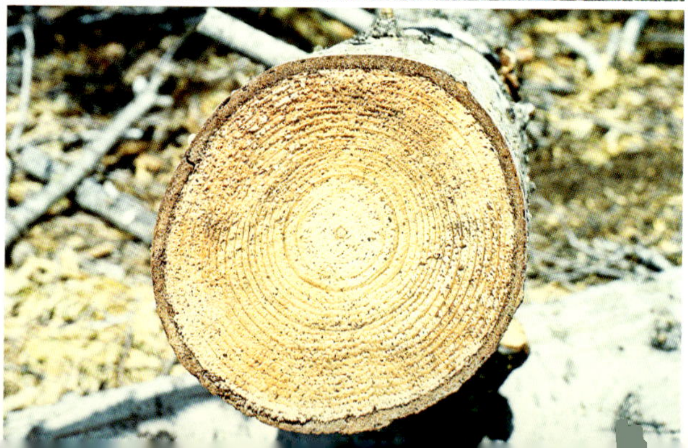

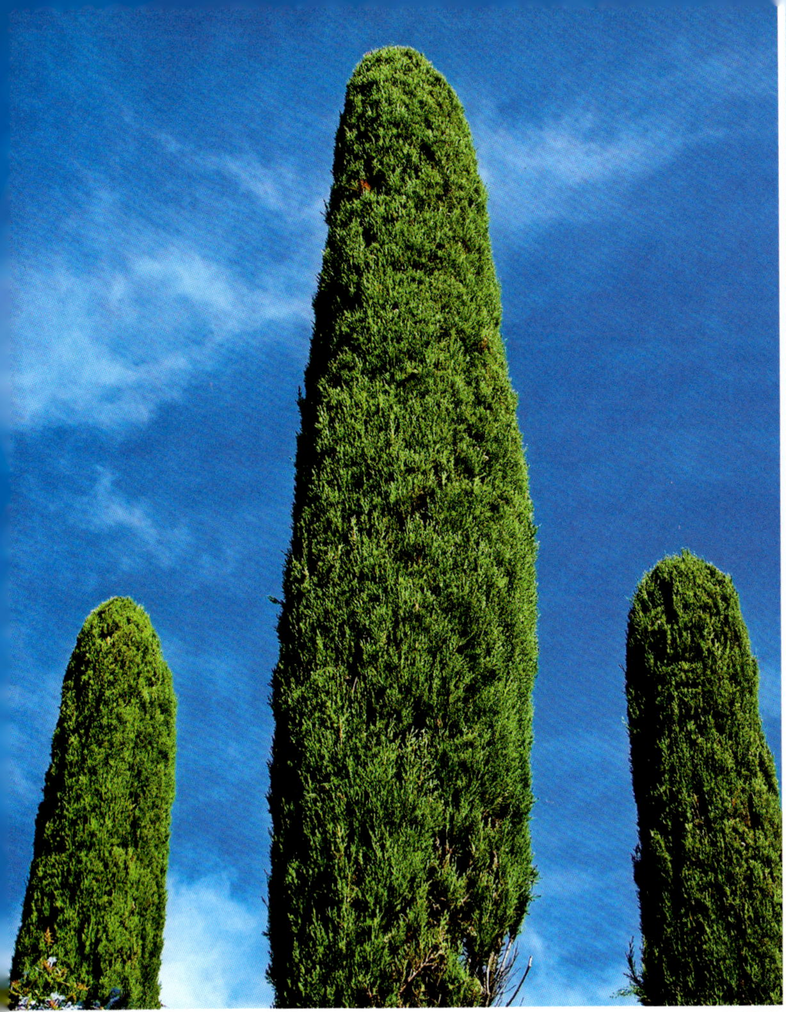

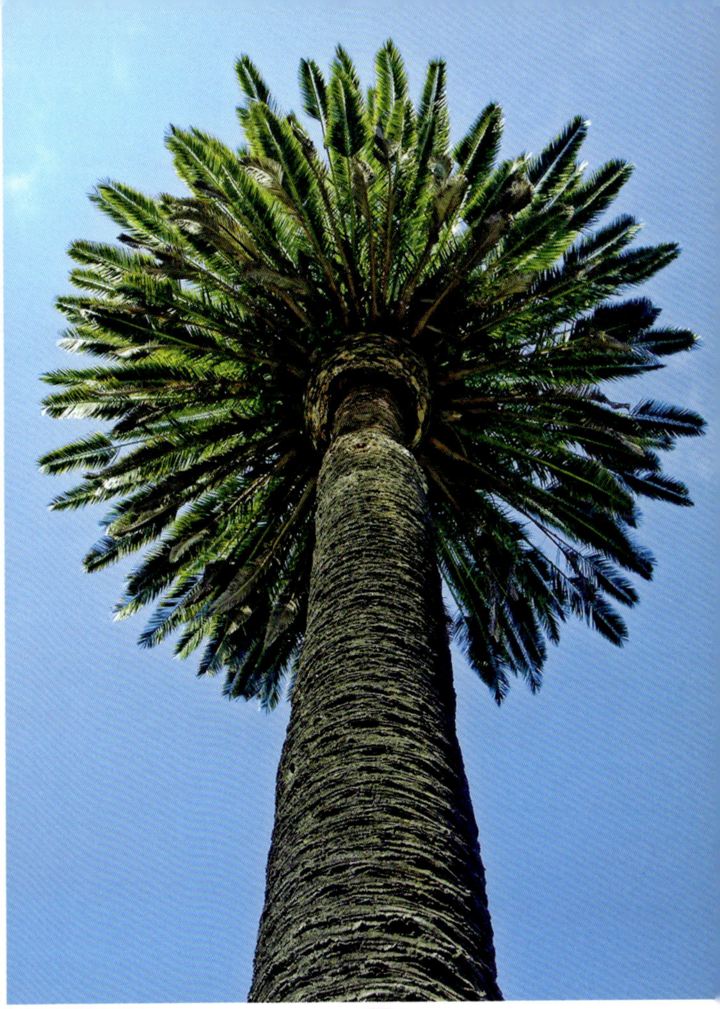

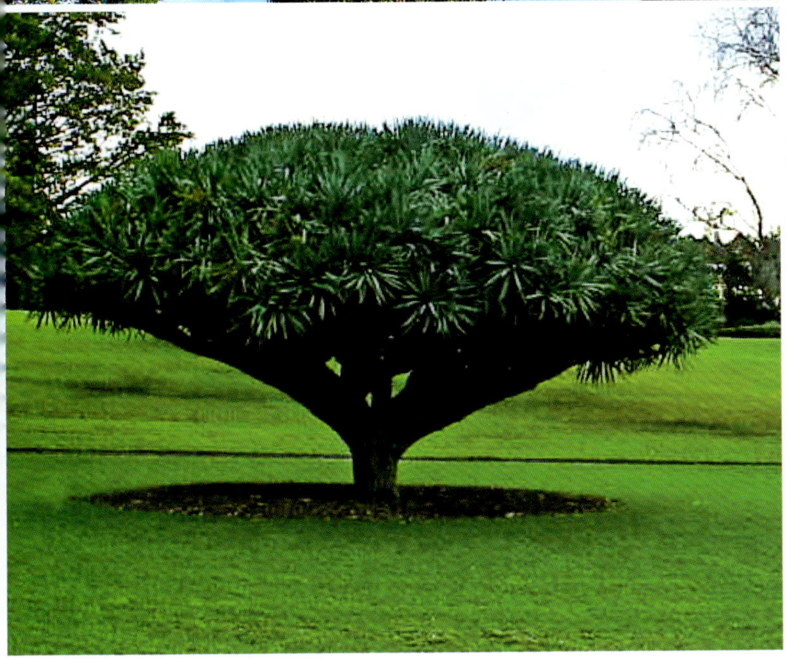

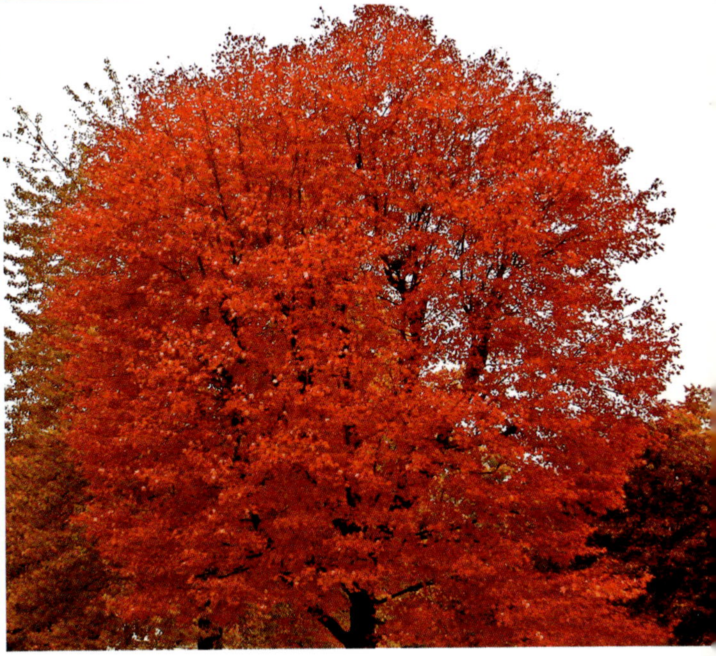

Cylindrical symmetry is characteristic for trees as well, although they appear in many different shapes. The most conspicuous is the cylindrical symmetry of the tree trunk, which grows an additional ring each year. These rings may be perfect circles, but may also be less than perfect depending on the conditions of growth. When the trunk is cut through, perpendicular to its axis, the annual rings become visible.

The size of the huge tree trunk in the Yosemite Park can be appreciated by comparing it to the person standing next to it.

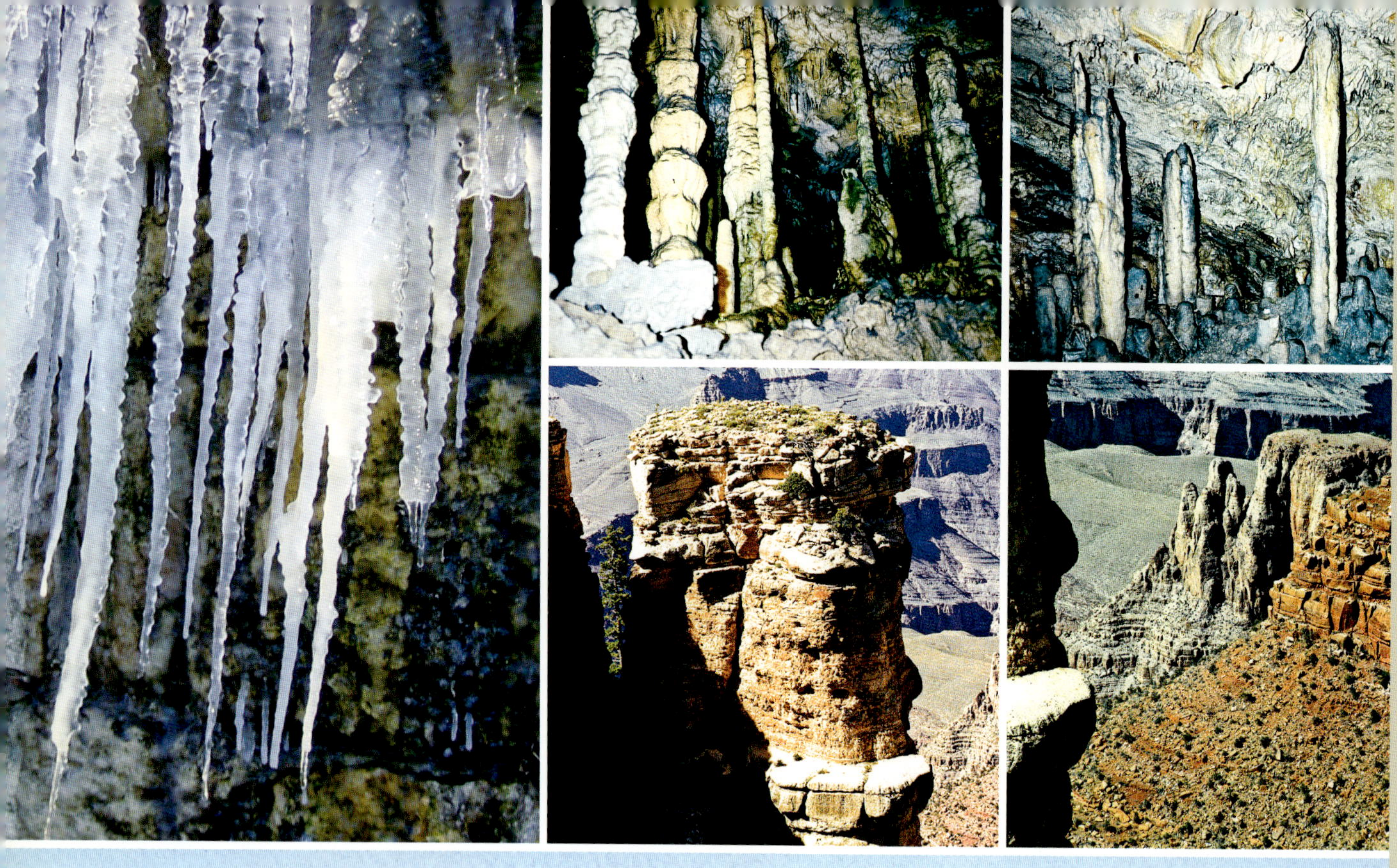

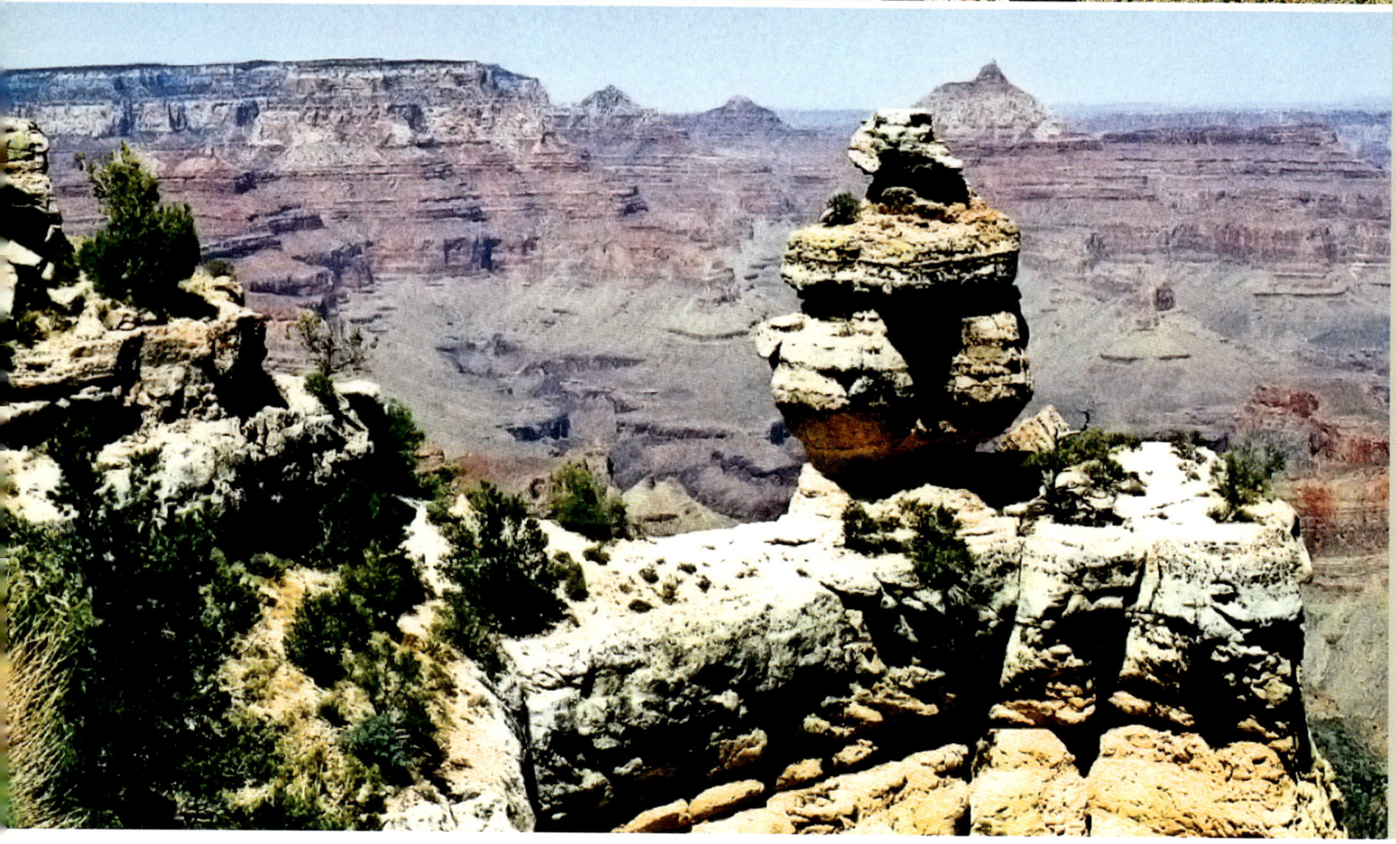

Top: Icicles in Budapest; stalagmites and stalactites in Germany; and rock formations at the Grand Canyon in Arizona

Bottom: Rock formation at the Grand Canyon in Arizona

Icicles as well as the stalagmites and stalactites in dripstone caves grow vertically and form shapes approximating cylindrical symmetry. Weather conditions often shape cylindrical formations from rocks during millennia.

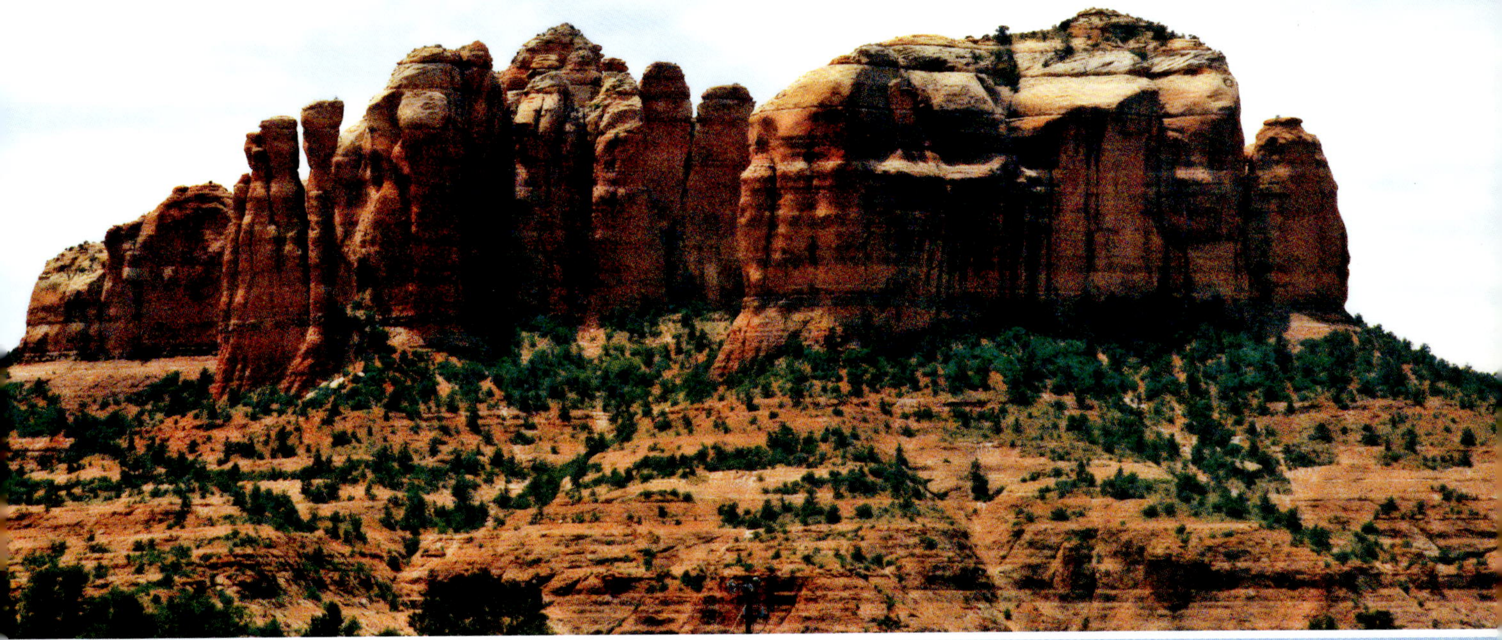

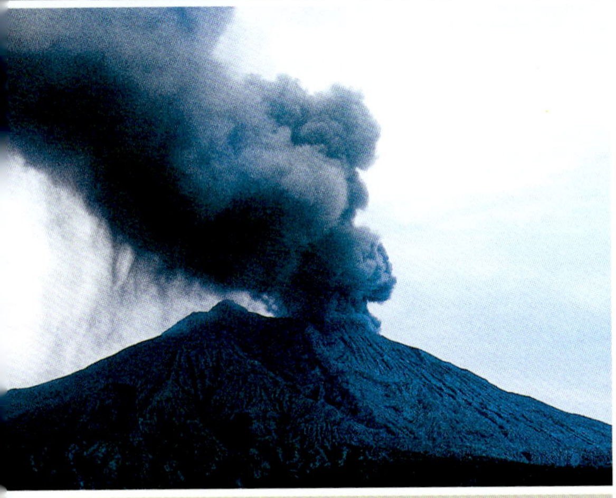

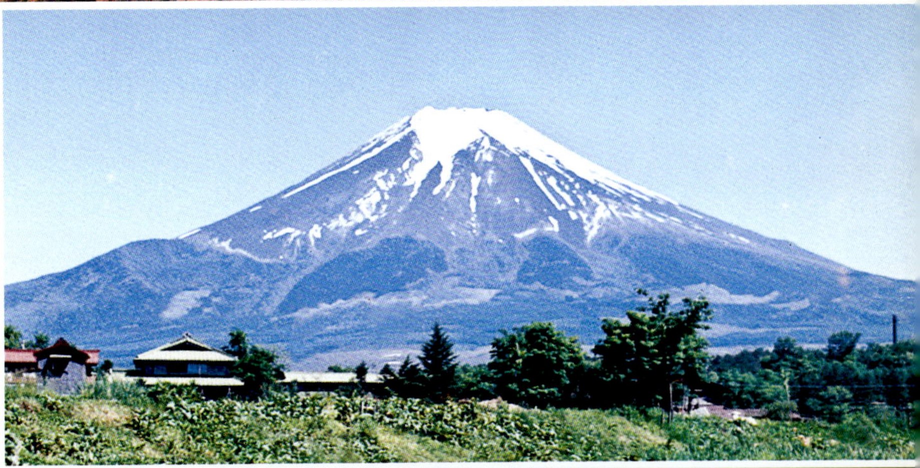

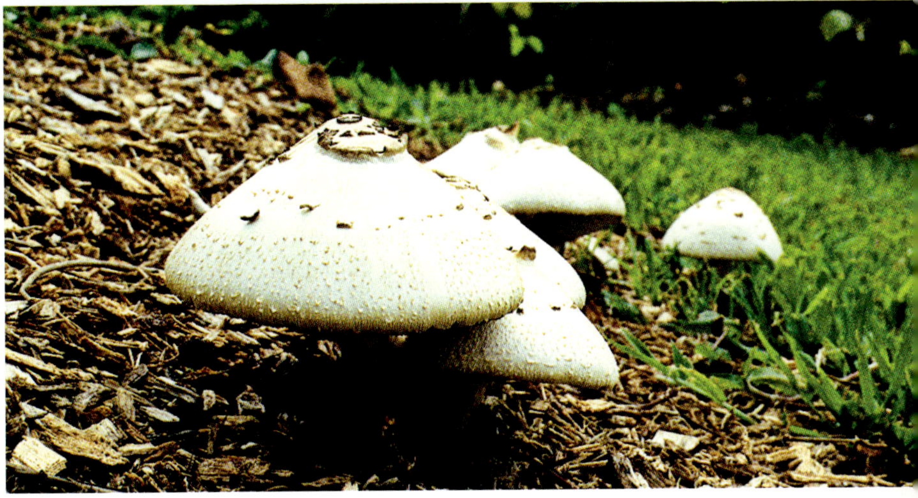

Top: Rock formations in the Arizona desert

Middle: Volcanoes in Japan

Bottom: Mushrooms

Volcanoes have usually conical shapes as conditioned by the vertical eruptions.

Mushrooms have cylindrical symmetries; their stems are of the same shape as those of other plants. Their tops are characteristically conical. The shape of the cloud formed during an atomic explosion resembles the shape of the mushroom; hence it is called mushroom cloud.

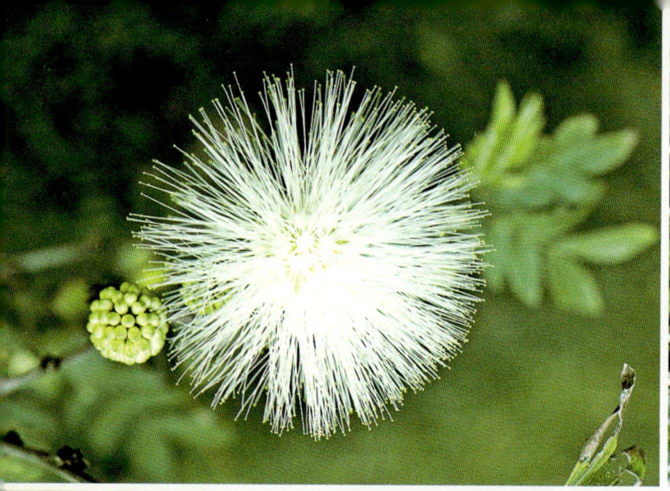
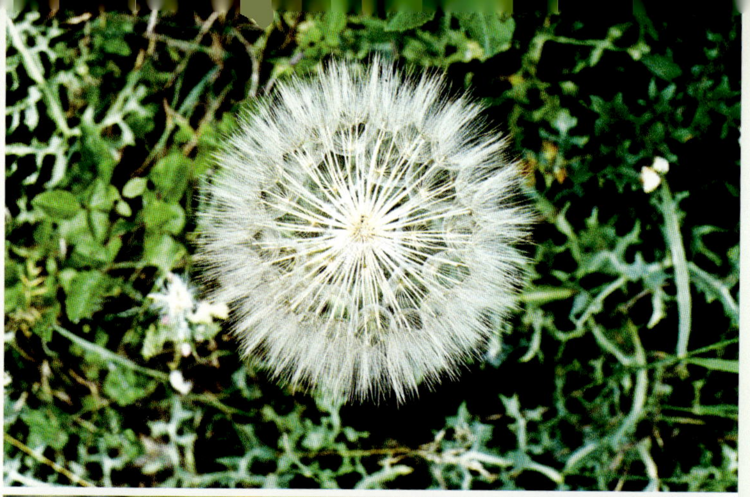
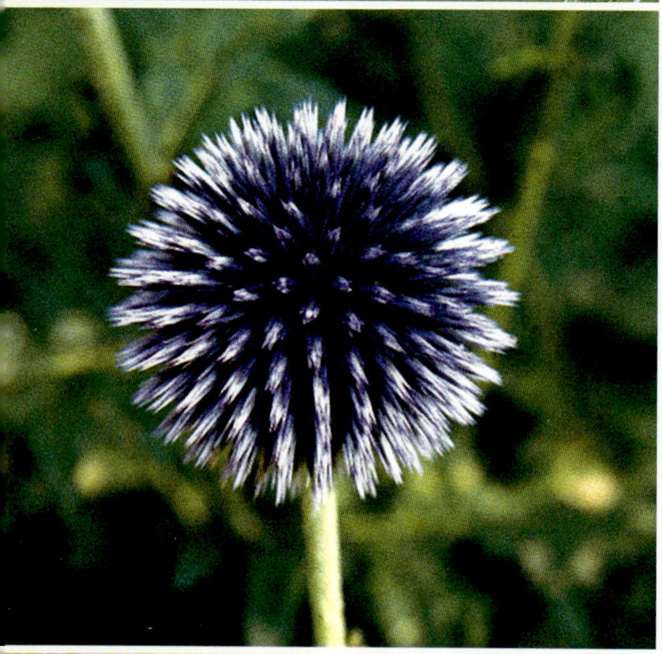
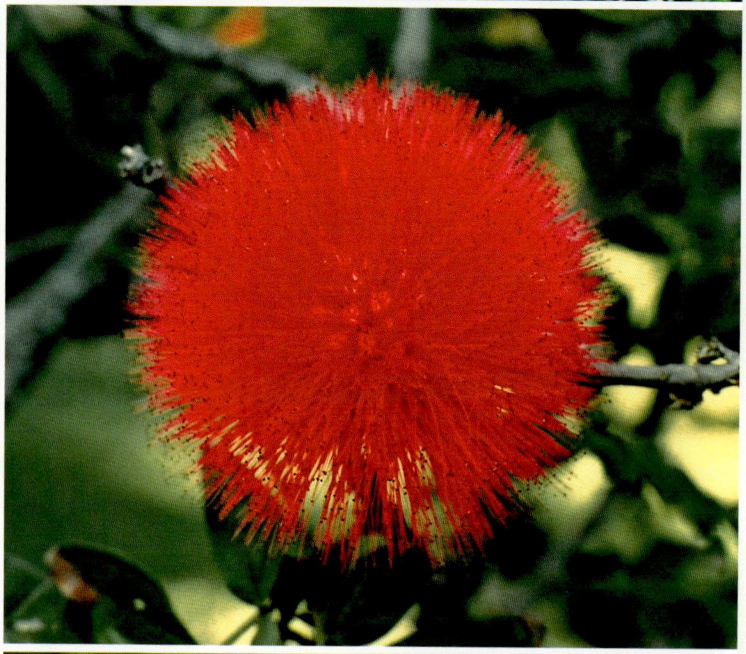
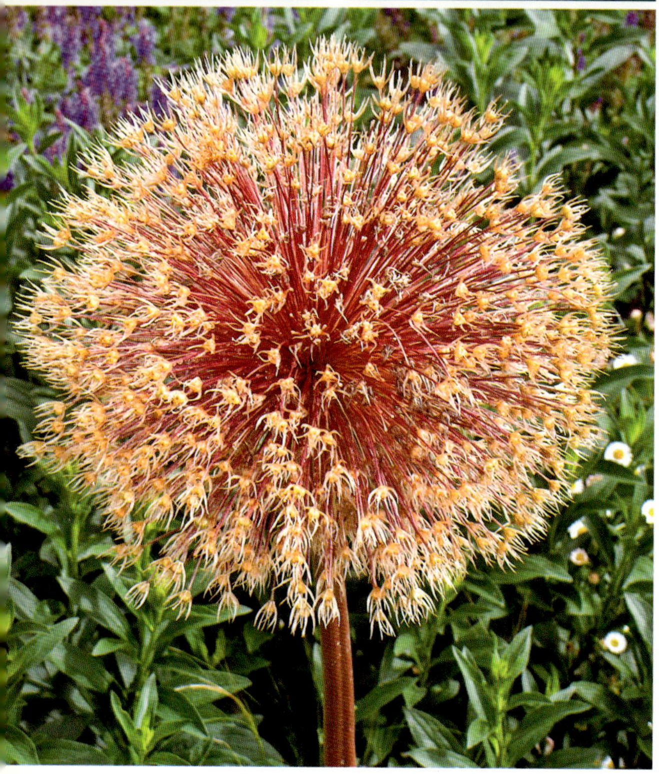
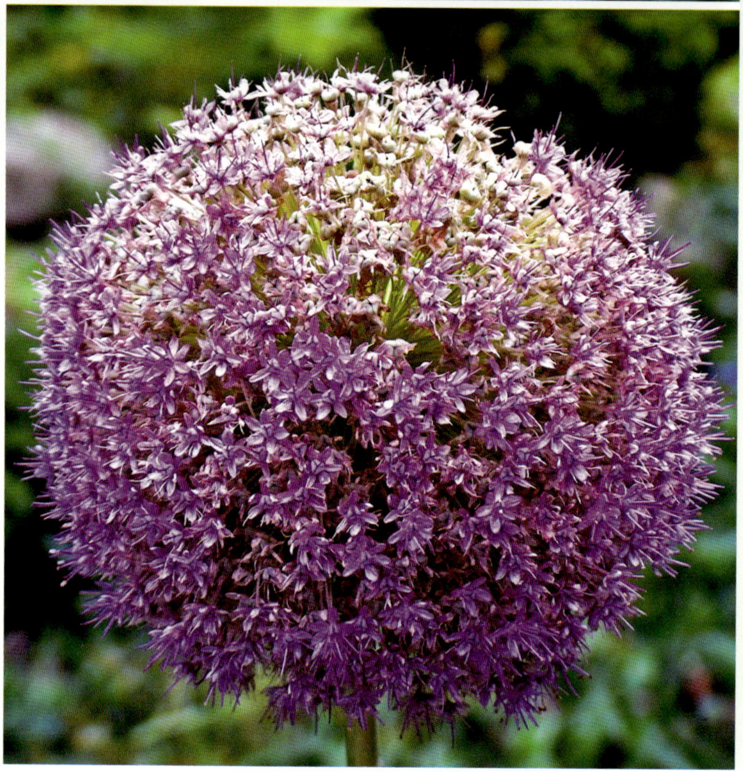

Plants do not move forward; they grow upward and outward, but stay rooted in the same place. Their stems have cylindrical symmetry and their top often has spherical symmetry. In spherical symmetry everything is the same in all directions (as on the surface of a sphere).

Characteristic examples are the dandelion and the other plants shown here.

The pollen of the hollyhock also exhibits beautiful spherical symmetry.

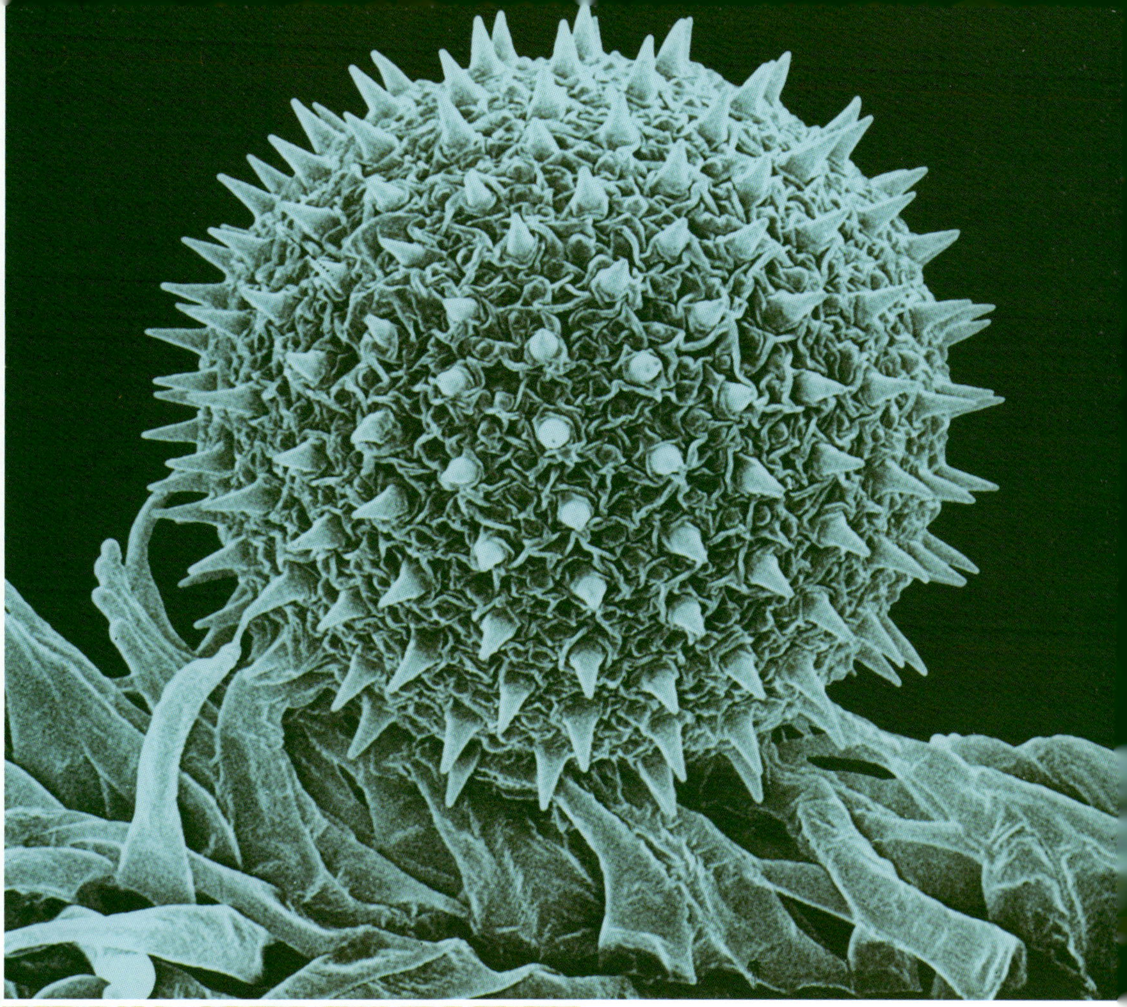

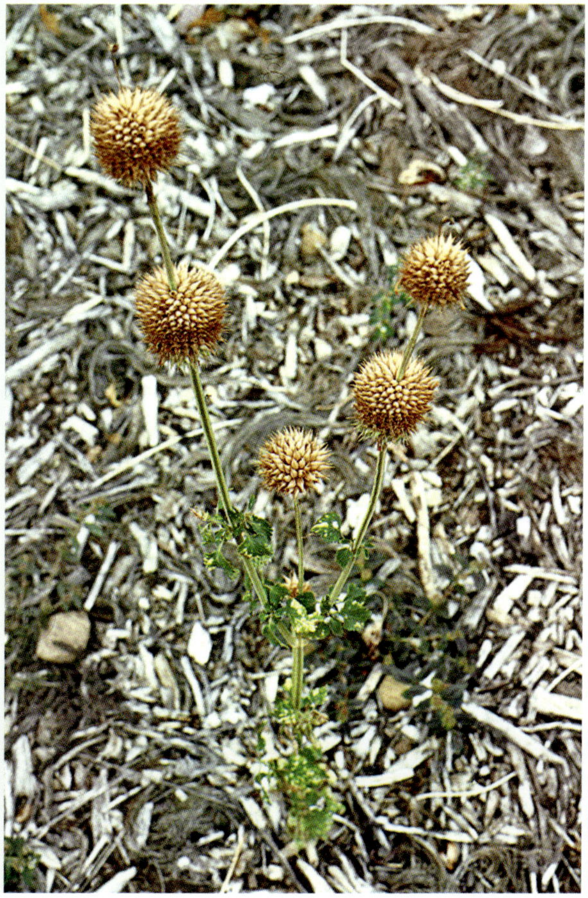

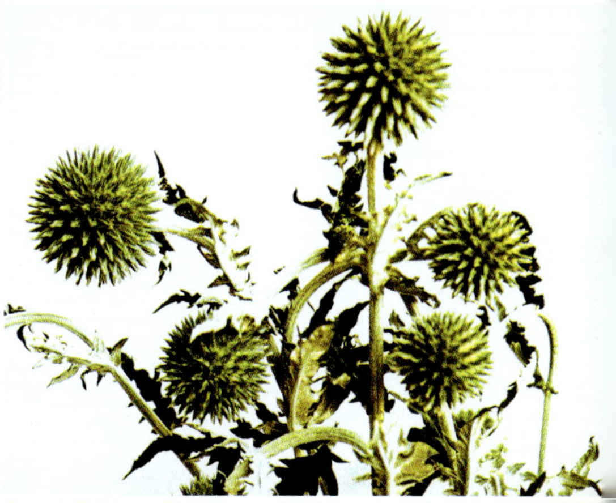

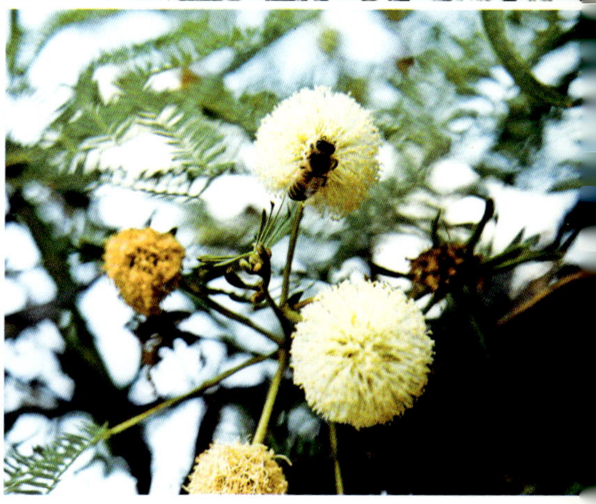

Top: The pollen of hollyhock in 100,000 times magnification; electron microscopic image (courtesy of R. Klockenkamper, Dortmund, Germany)

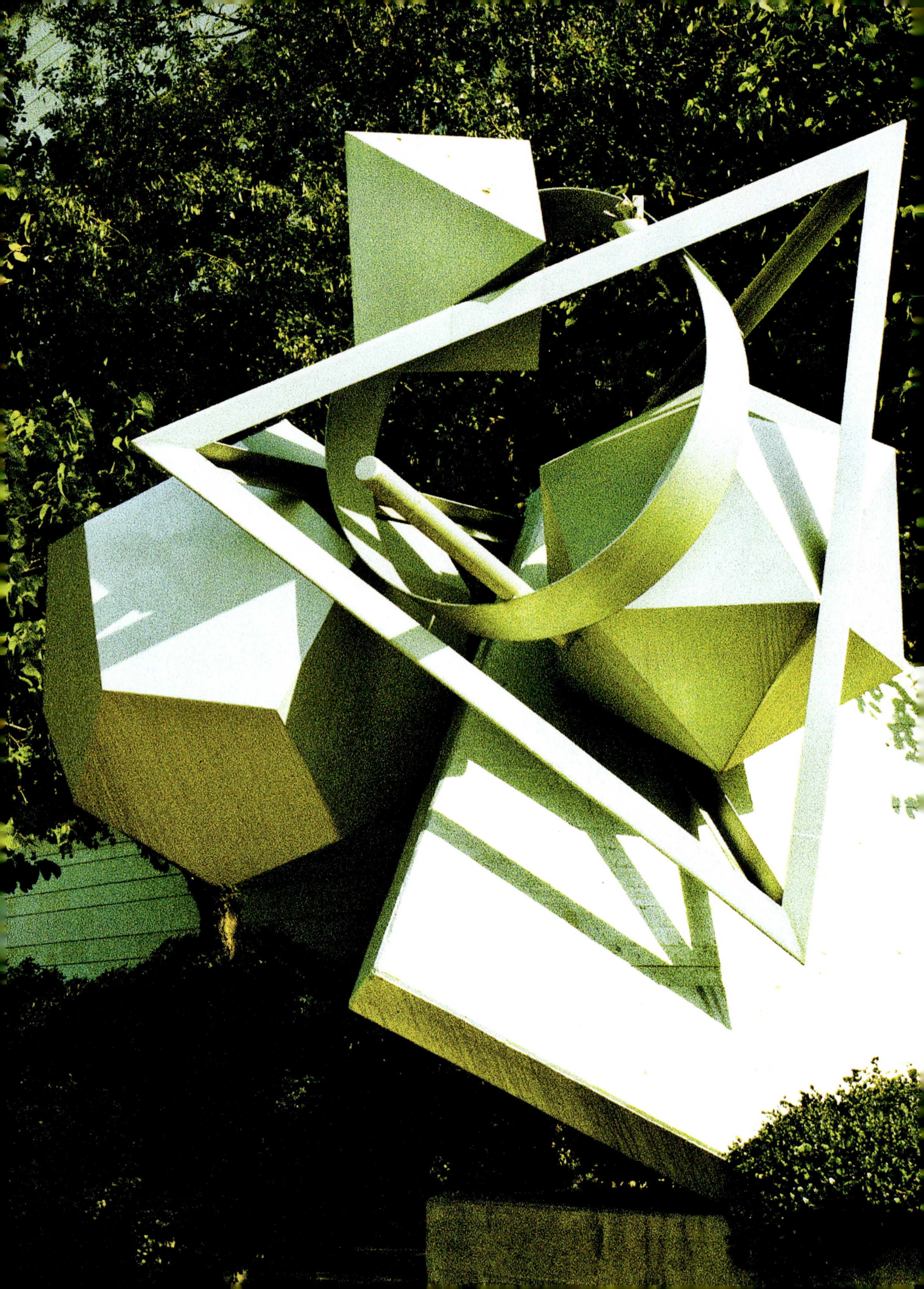

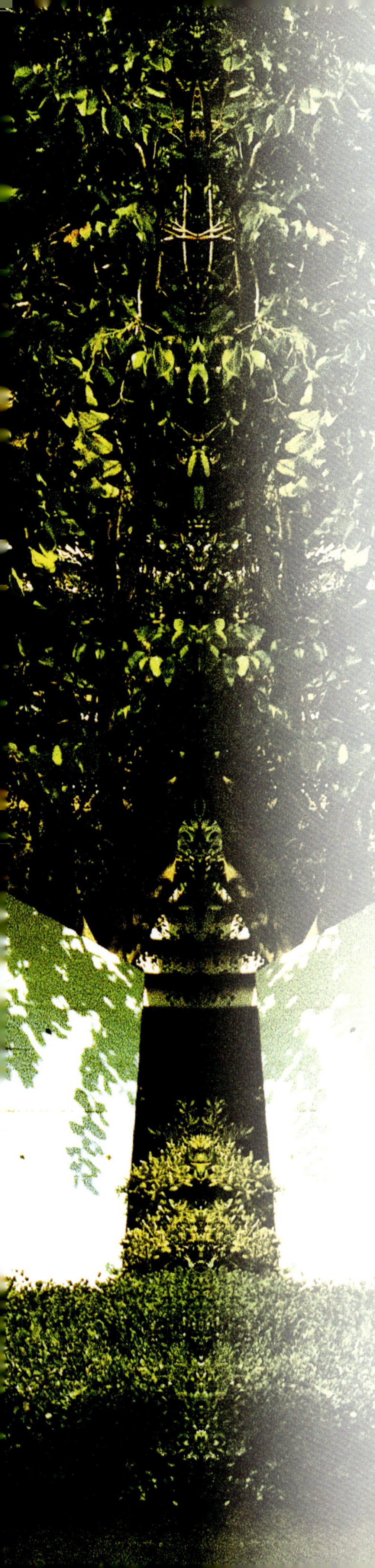

Regular Polyhedra

The faces of the regular polyhedra are all equal, regular polygons and their vertices are all alike. There are altogether five regular polyhedra, the tetrahedron, hexahedron (cube), octahedron, dodecahedron, and icosahedron. Their names indicate the number of faces.

The regular polyhedra were an important part of Plato's natural philosophy and are also called Platonic solids or Platonic bodies. The tetrahedron, cube, and octahedron are relatively simple shapes, while the discovery of the dodecahedron and icosahedron was referred to by the mathematician Hermann Weyl in his classic book *Symmetry* as "…one of the most beautiful and singular discoveries made in the whole history of mathematics."

Polyhedra

Sculpture of the regular polyhedra on the campus of Tel Aviv University

In his book *Harmonices Mundi* (The Harmony of the World), the astronomer Johannes Kepler (1571–1630) used the five regular polyhedra to represent all the elements that were known at his time. The tetrahedron represented fire, the cube the earth, the octahedron air, the icosahedron water, and, finally, the dodecahedron the Universe.

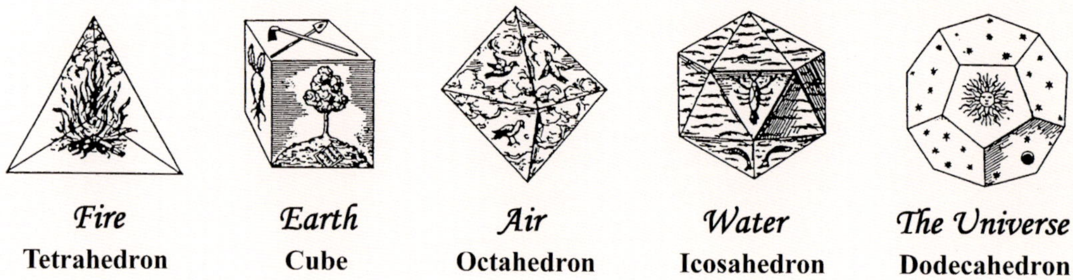

Kepler's drawings of the five regular polyhedra in his *Harmonices Mundi*, Book II, 1619

Characteristics of the Regular Polyhedra

Name	Shape of Faces	Number of Faces	Number of Vertices	Number of Edges
Tetrahedron	Triangle	4	4	6
Cube	Square	6	8	12
Octahedron	Triangle	8	6	12
Dodecahedron	Pentagon	12	20	30
Icosahedron	Triangle	20	12	30

These characteristics are related by Euler's formula: $v + f = e + 2$, where v stands for the number of vertices, f for the number of faces, and e for the number of edges.

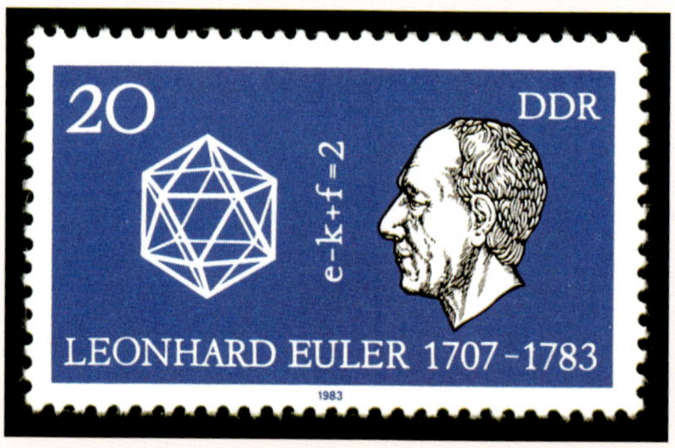

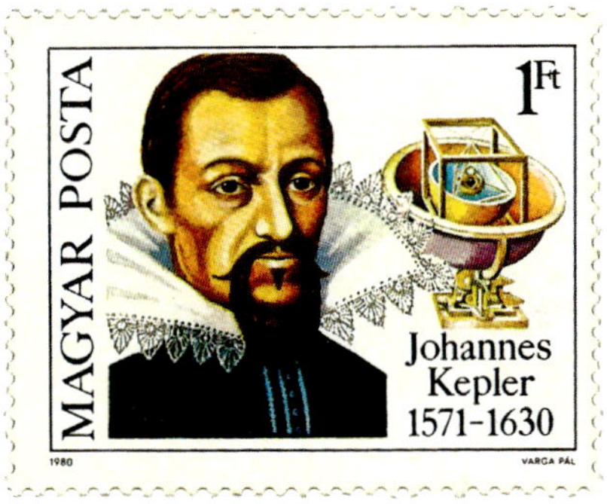

Kepler built a planetary model to describe the orbits of the six planets known at his time. He placed the five regular polyhedra around each other and suggested that the ratios of the diameters of the inscribed and circumscribed spheres would be the same as the ratios of the diameters of the planetary orbits around the Sun. The model was wrong, but attractive, and symbolized a unified approach to such diverse branches of science as what we call today astronomy and crystallography.

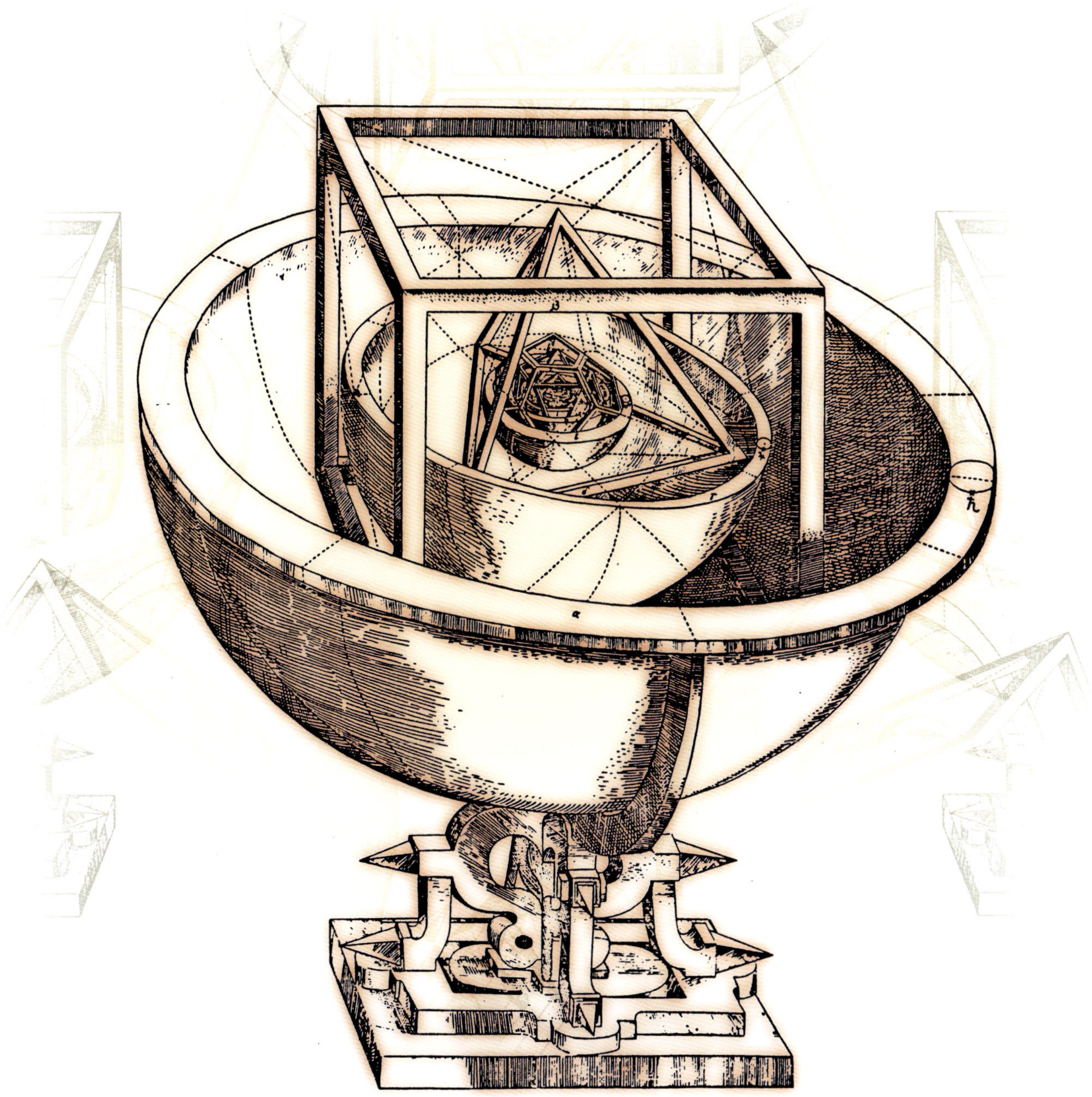

Many primitive organisms are shaped like regular polyhedra, such as radiolarians shown below from Ernst Häckel's book.

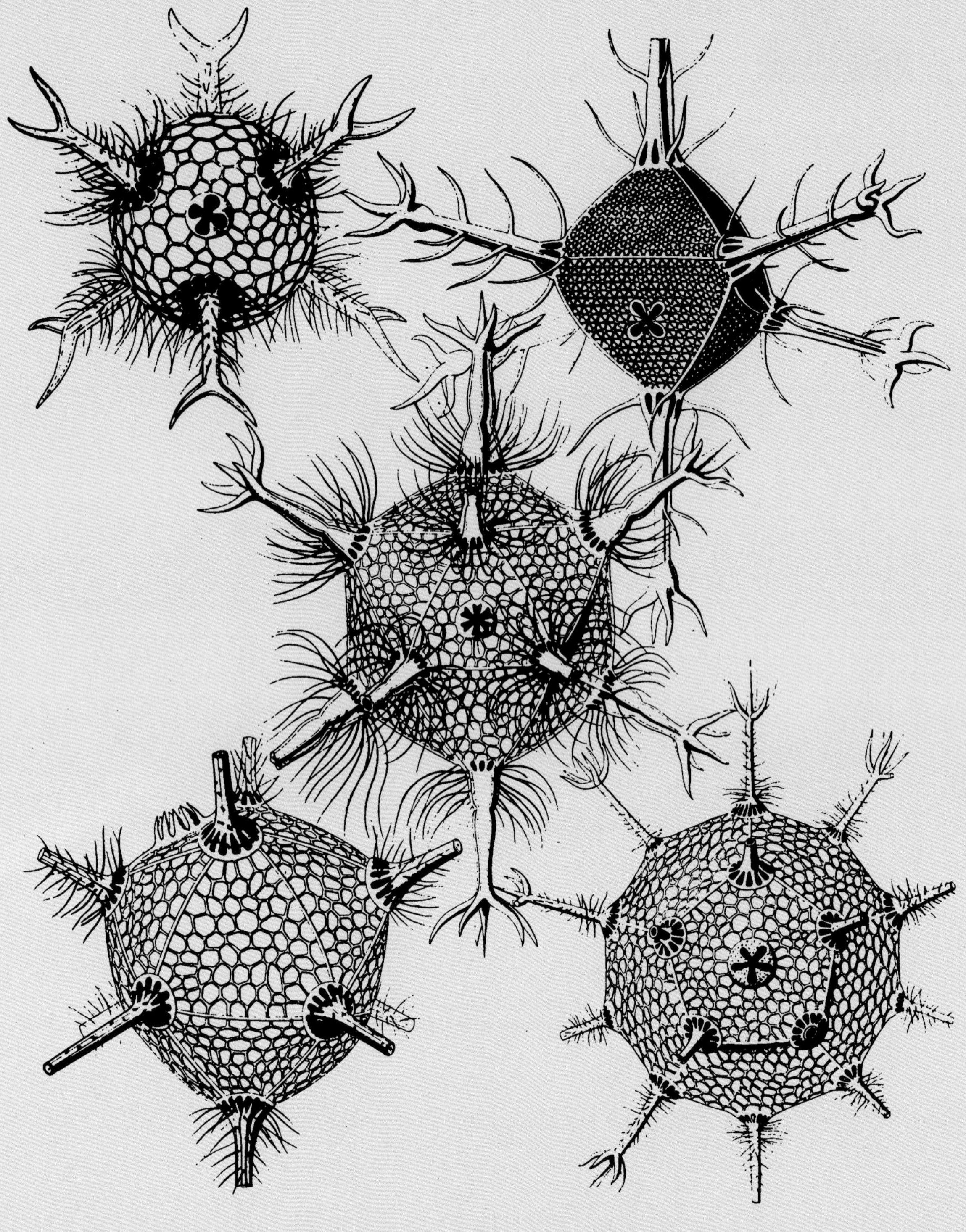

Ernst Häckel, *Kunstformen der Natur* (Verlag des Bibliografischen Instituts, Leipzig, 1899–1904)

The regular polyhedra often appear in artistic creations.

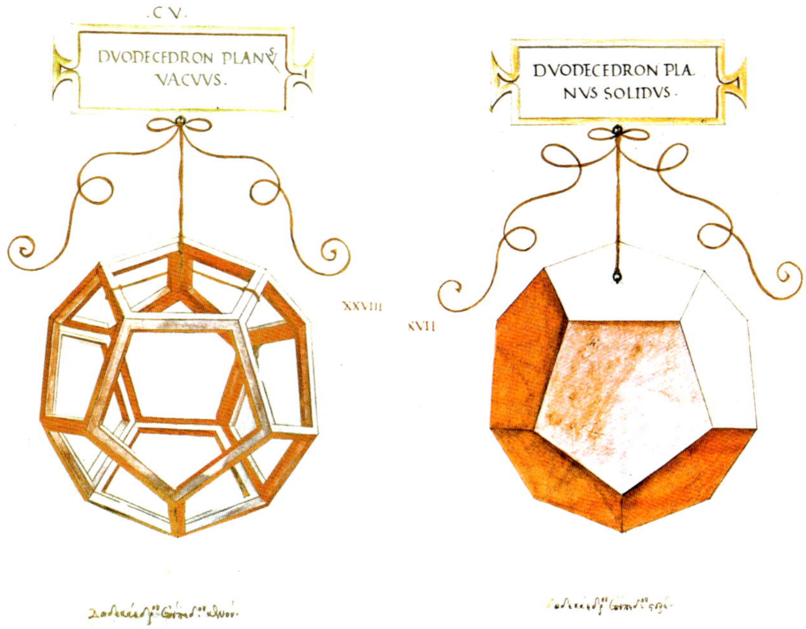

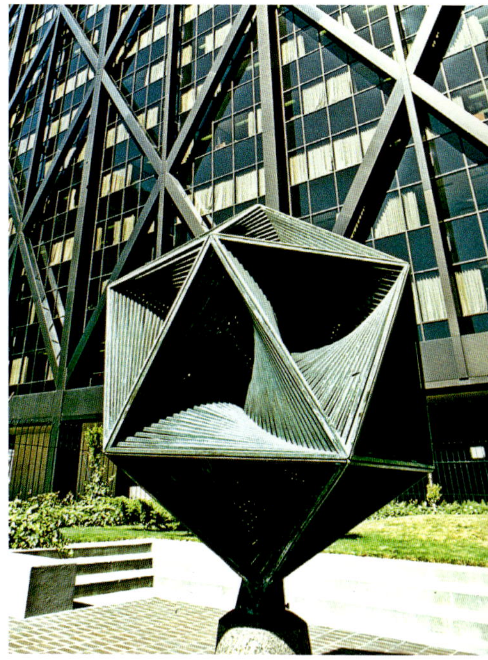

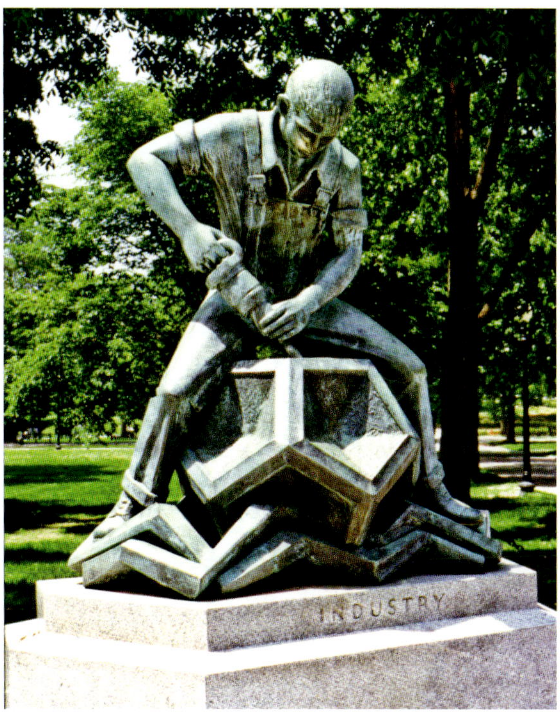

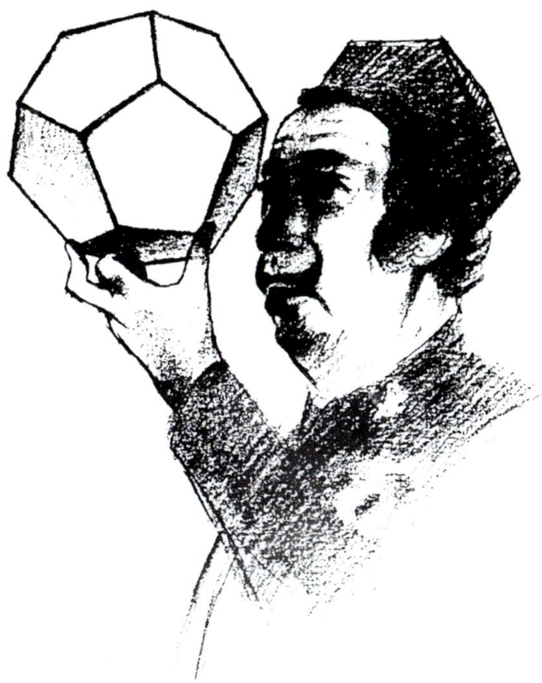

Top left: Two drawings of the dodecahedron by Leonardo da Vinci for Luca Pacioli's book *De Divina Proportione* (1509)

Top right: Charles Perry's icosahedron in San Francisco

Bottom left: Dodecahedron in the sculpture symbolizing "Industry" in Boston

Bottom right: Drawing by Ferenc Lantos, Pécs, Hungary, after one of Salvador Dali's pictures in which the artist is holding a dodecahedron

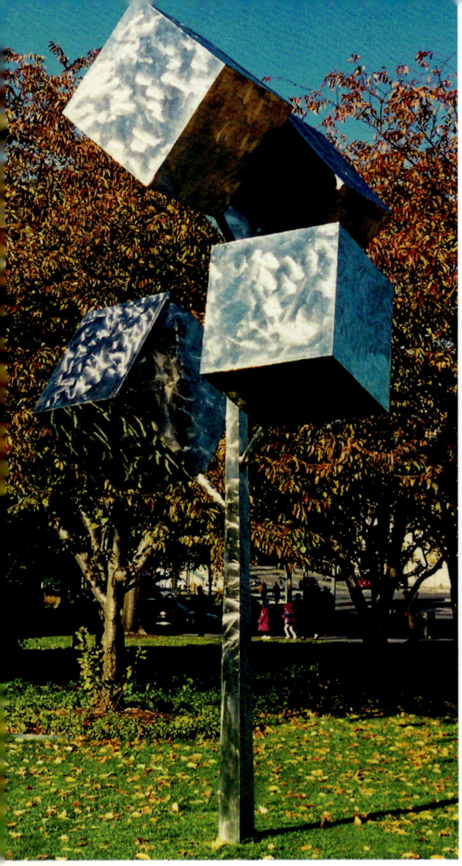

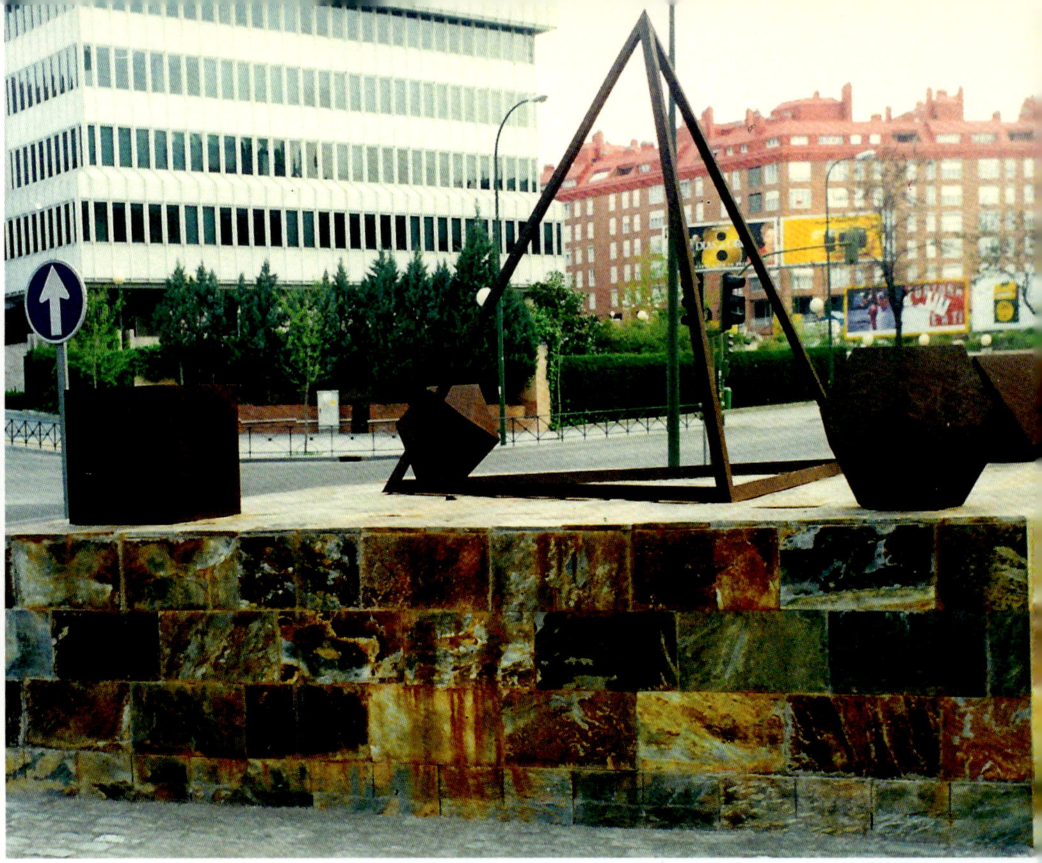

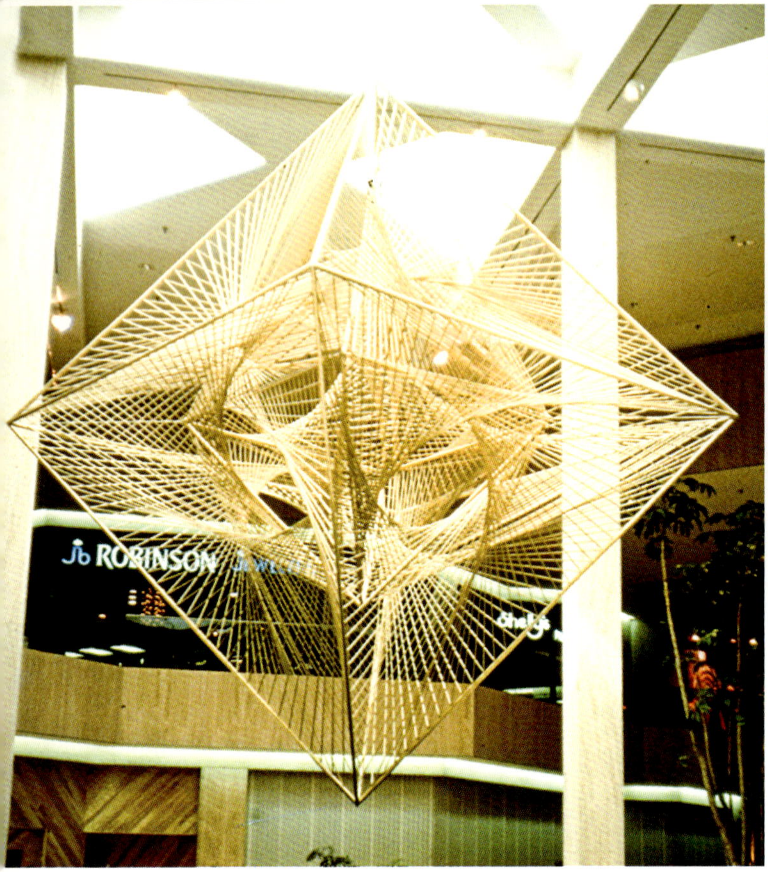

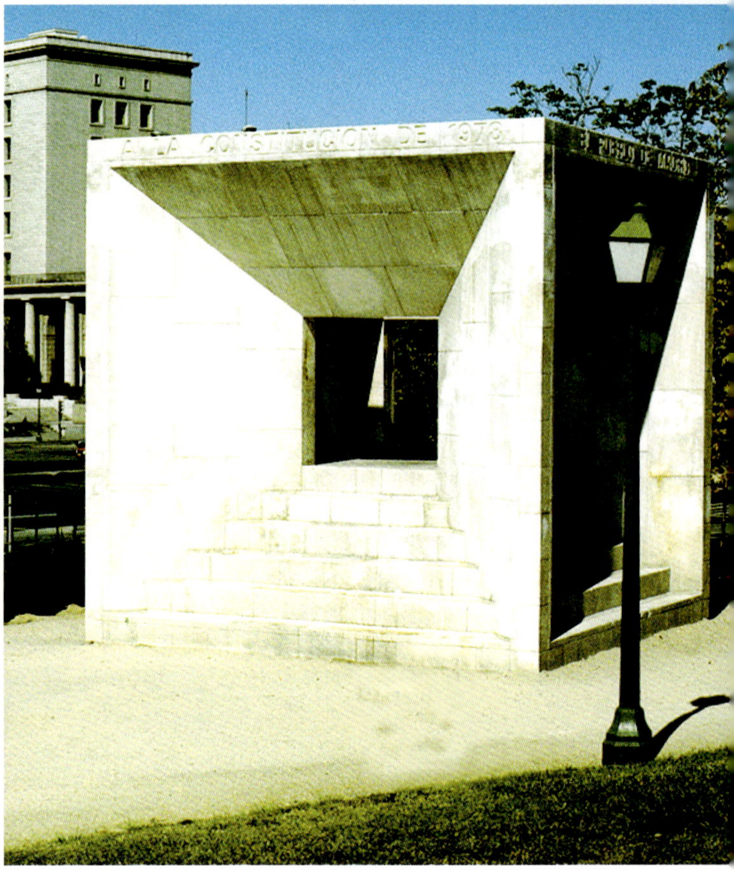

From upper right, anticlockwise:

The five Platonic polyhedra in Madrid
(courtesy of Jose Elguero, Madrid)

Cubes in Washington, D.C.

Octahedron by Charles Perry (Norwalk, Connecticut)

Memorial to the Constitution of Spain in Madrid

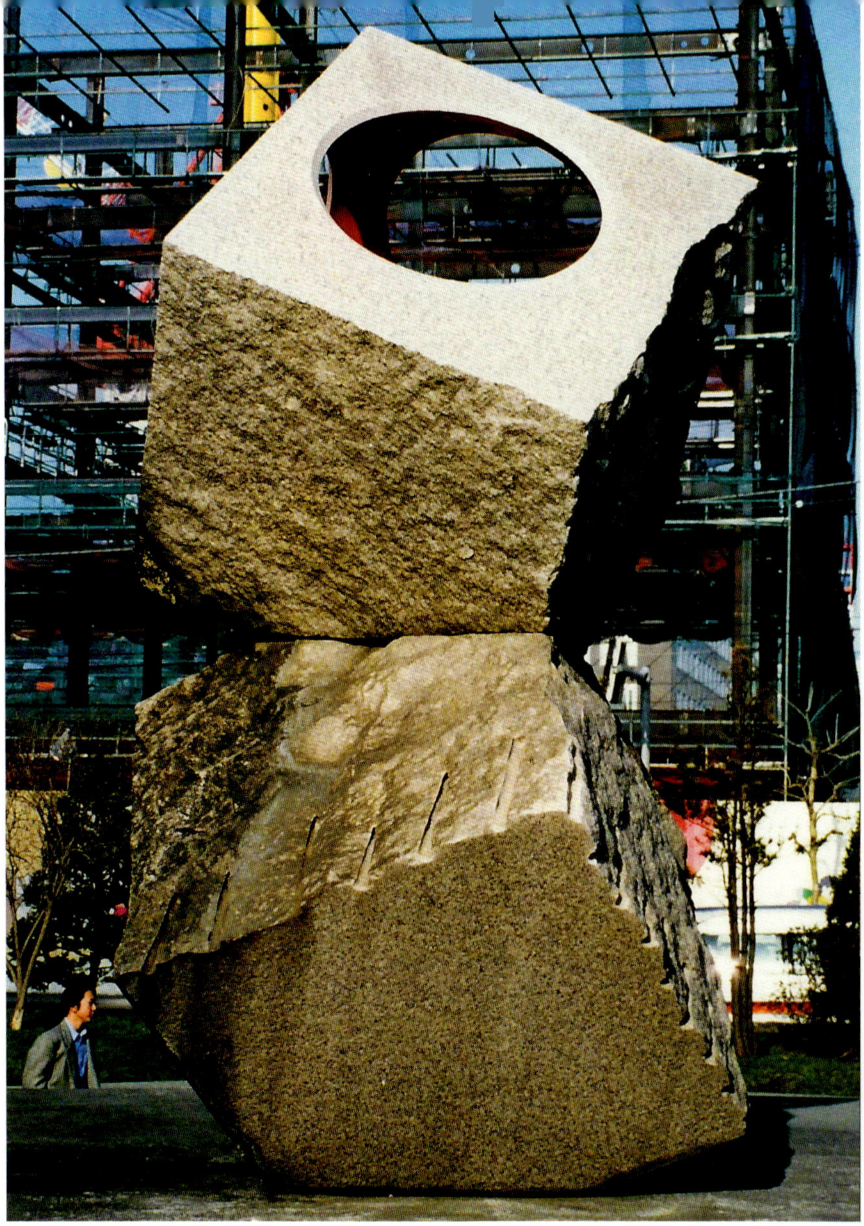

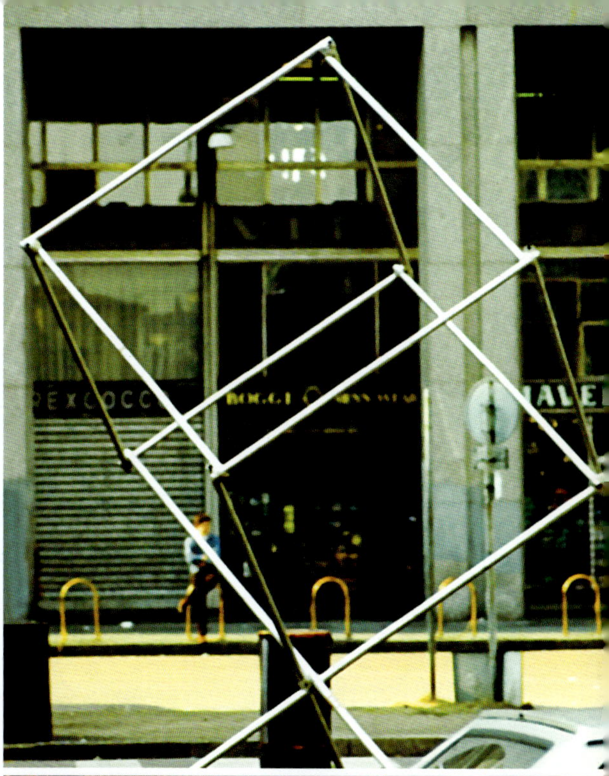

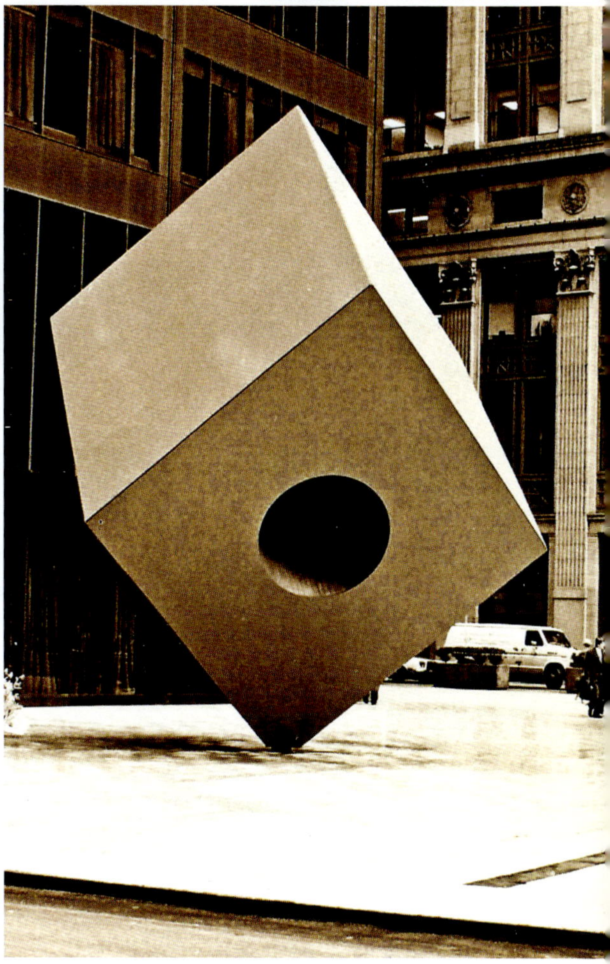

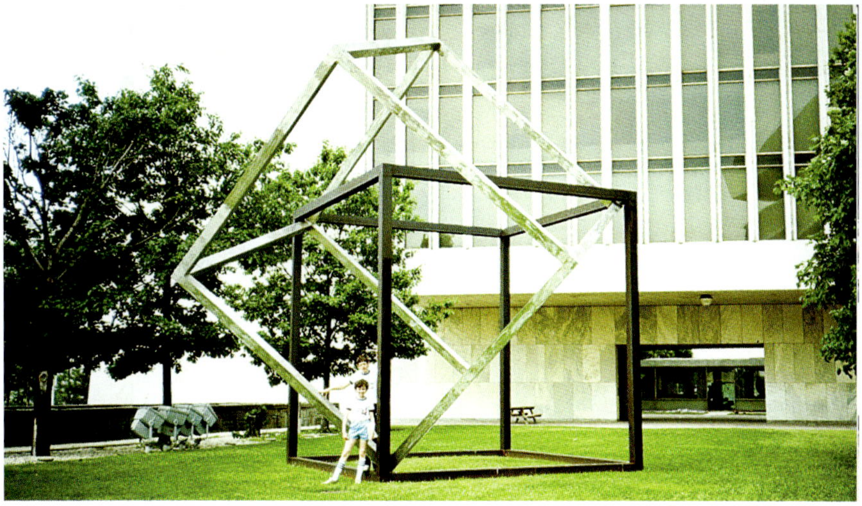

From upper left, clockwise:

Irregular cubes in Sapporo, Japan

Cube in Milan, Italy

Cube in New York City

Cubes in Albany, New York

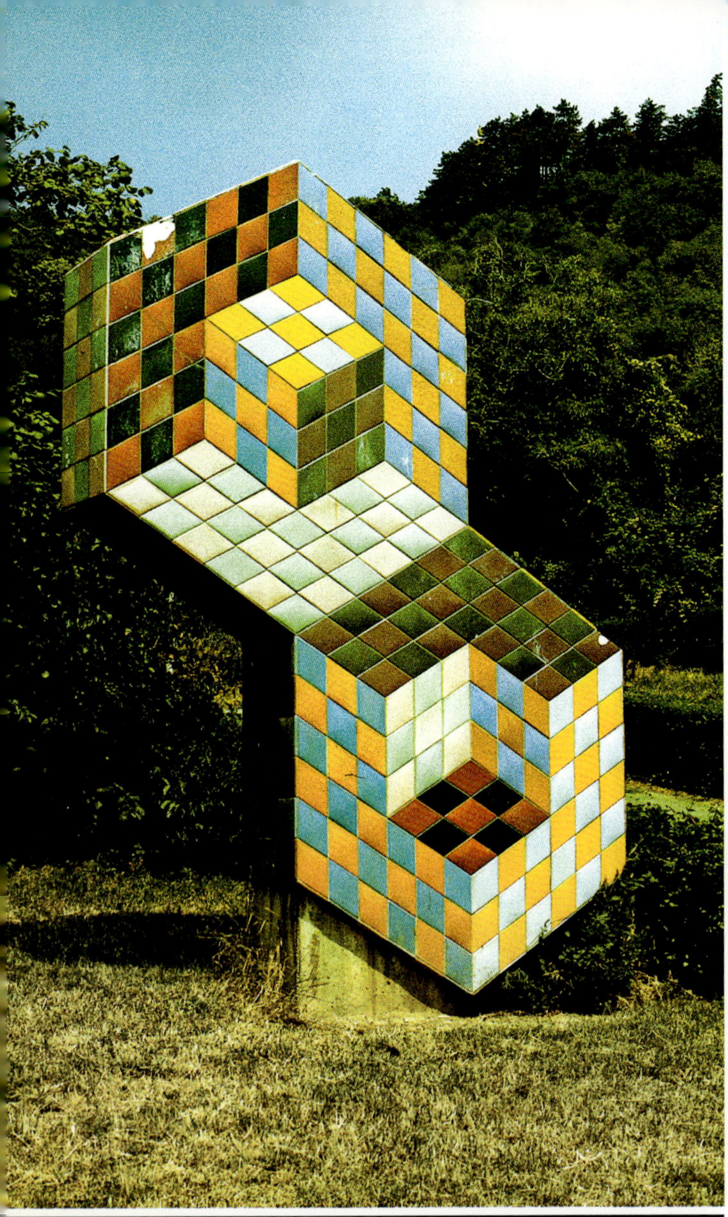

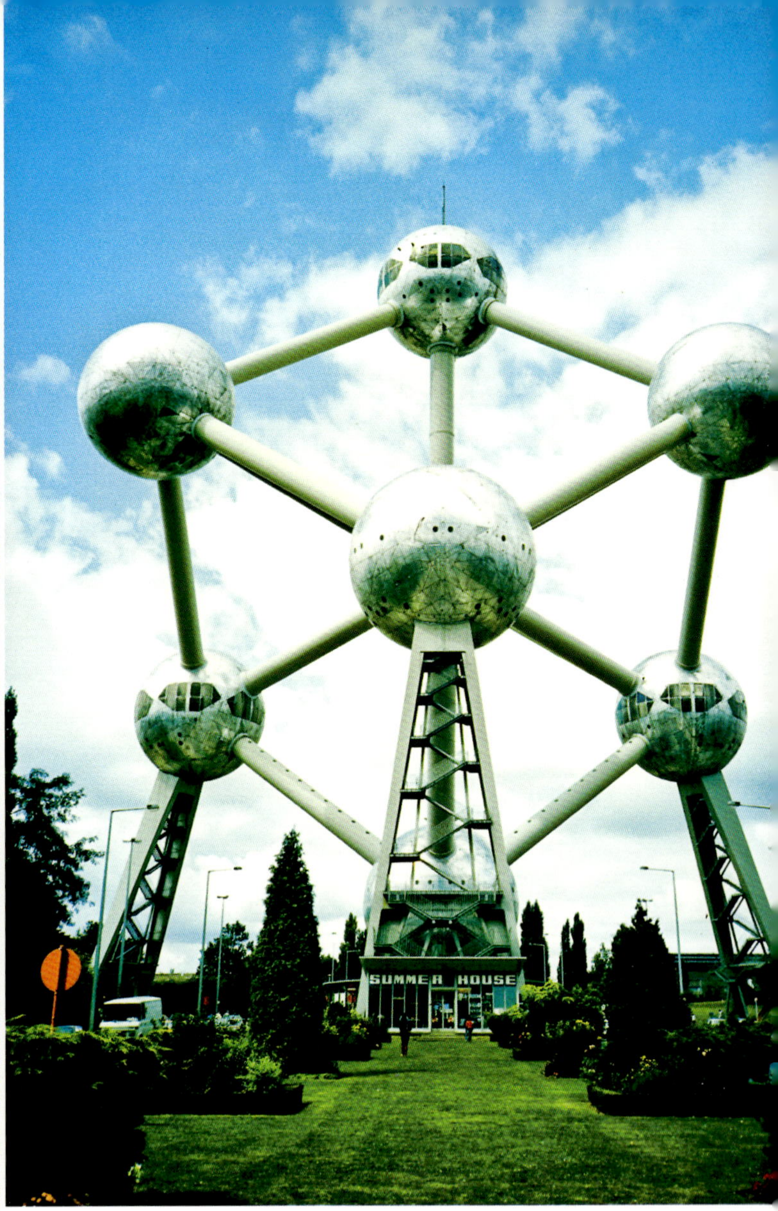

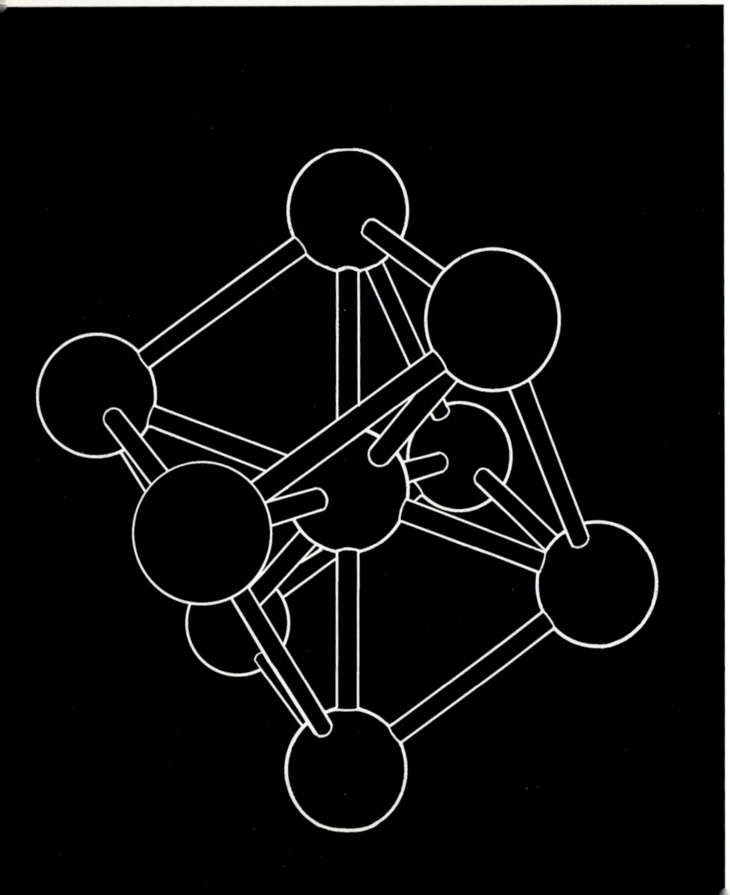

Cubes by Victor Vasarely in Pécs, Hungary

Atomium in Brussels

The drawing shows the Atomium from a different perspective, indicating its cube-shape (after a brochure advertising the Atomium).

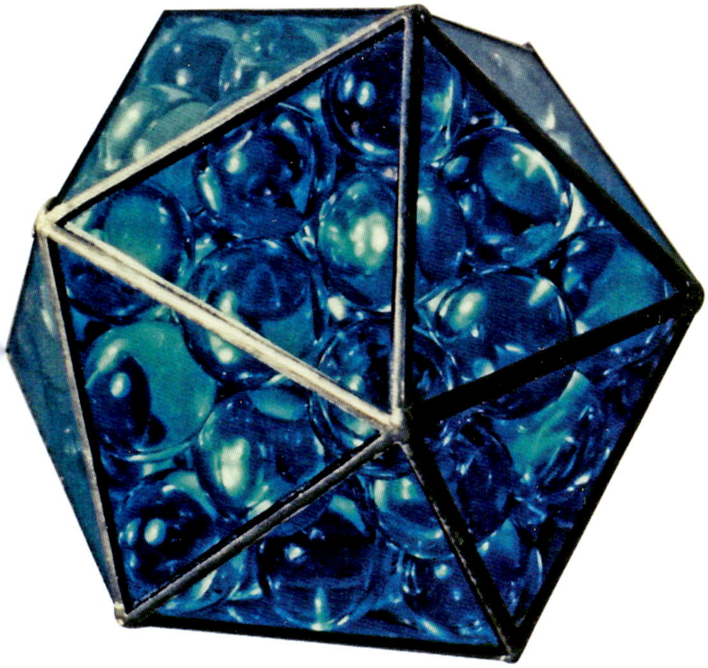

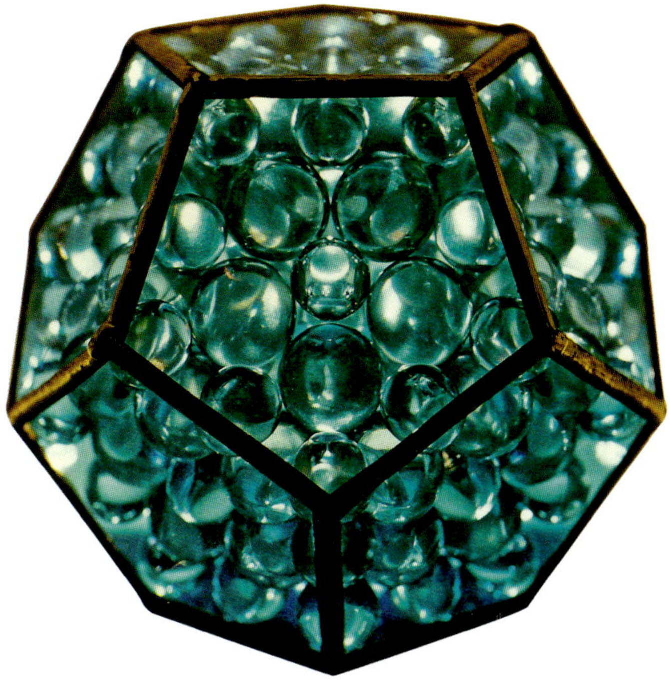

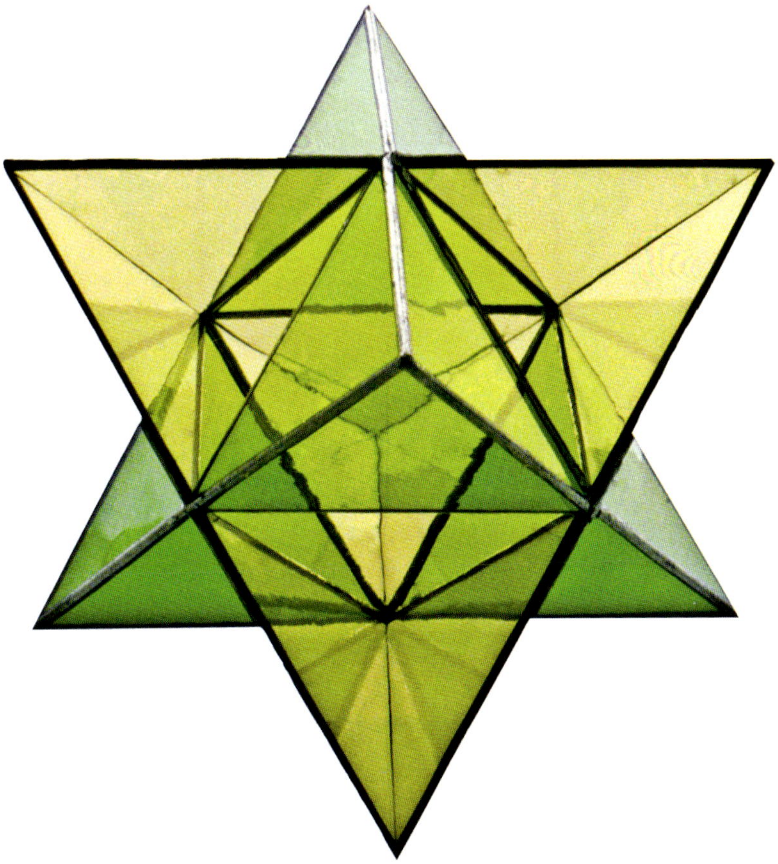

Top: Stained glass models of icosahedron and dodecahedron, filled with glass spheres

Bottom: Glass model of a star polyhedron (see next page). All three by Herbert Hauptman, Nobel laureate in Chemistry for 1985

(courtesy of Gloria J. Del Bel, Hauptman–Woodward Institute, Buffalo, New York).

Star Polyhedra

The five regular polyhedra are all convex polyhedra, meaning that they have surfaces that bulge outward. The angles formed by any two faces joined along a common edge are always smaller than 180 degrees. If we remove this restriction, four additional regular polyhedra can be constructed that are called star polyhedra.

The four star polyhedra are the great dodecahedron; great icosahedron; small stellated dodecahedron; and the great stellated dodecahedron.

We have discovered one of the star polyhedra at the top of the Sacristy of St. Peter's Basilica, just beneath the cross. It is fairly large, but too small to be easily noticeable from street level.

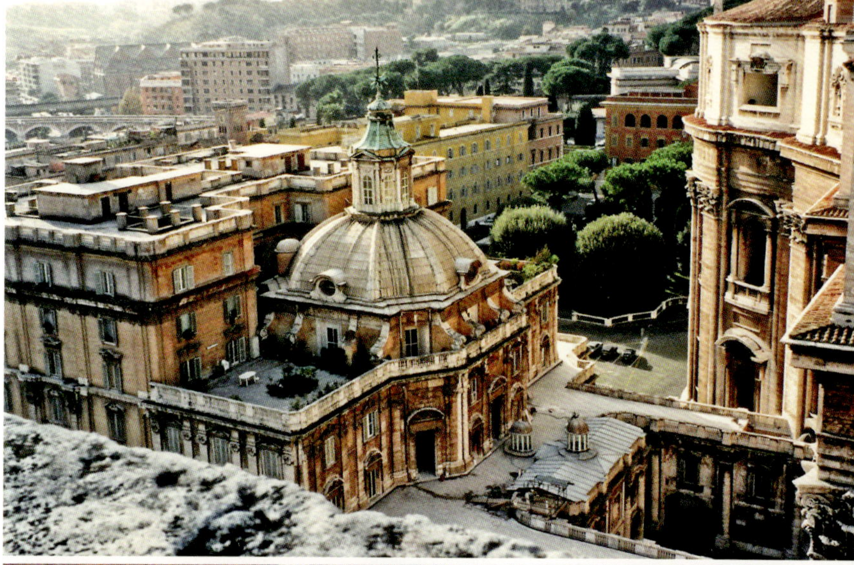

Bird's-eye view of the Sacristy of St. Peter's Basilica in Vatican City

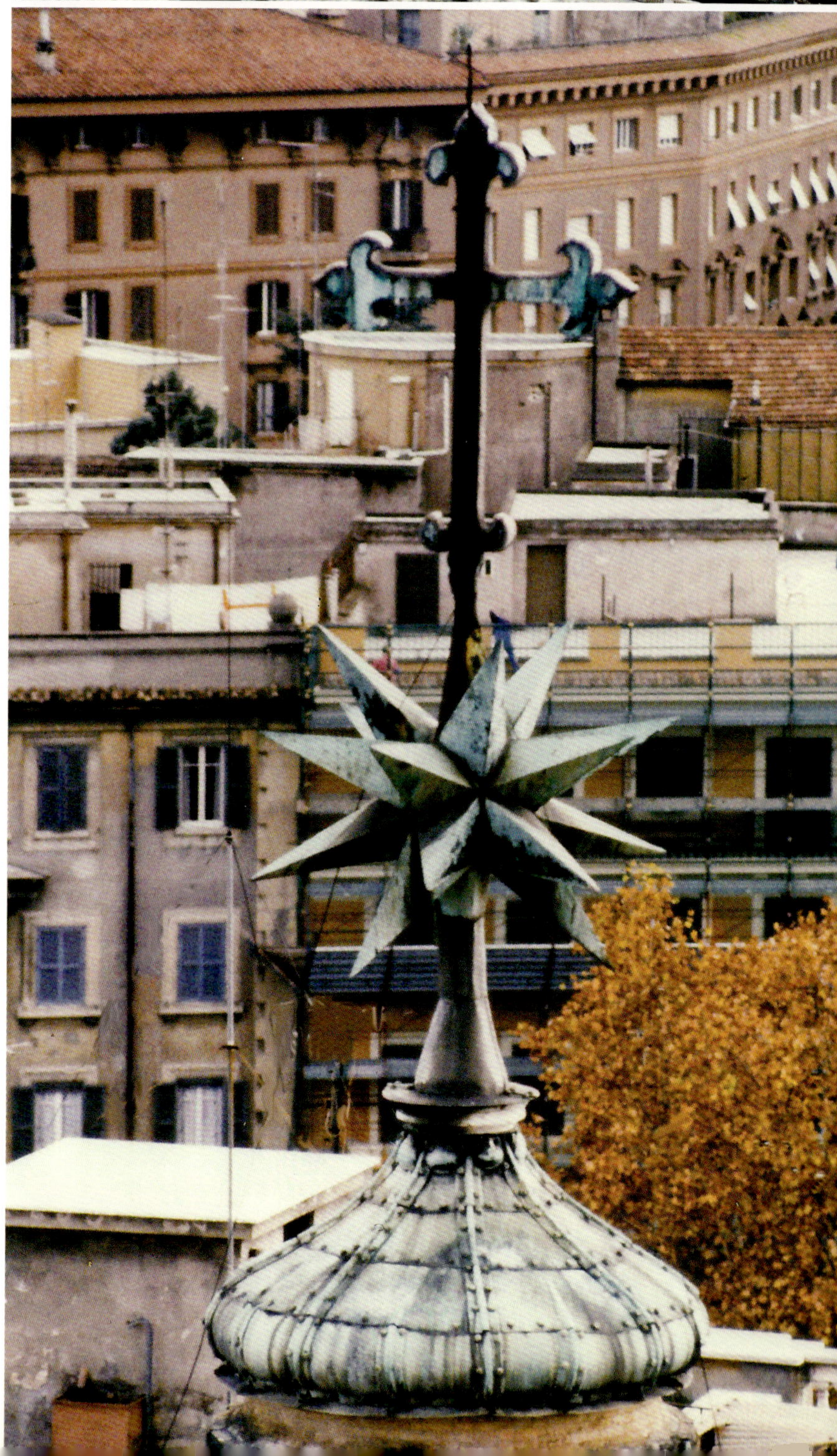

The star polyhedron beneath the cross at the top of the Sacristy of St. Peter's Basilica in Vatican City

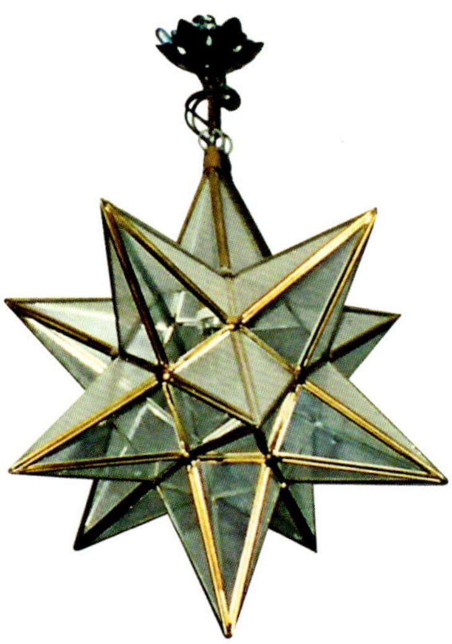

Star polyhedron as light fixture in a home in Bologna, Italy

Semiregular Polyhedra

The criteria for semiregular polyhedra are less stringent than for the regular polyhedra. The semiregular polyhedra also have regular polygons for all their faces and all their vertices are alike, but their faces are not all the same regular polygons. There are 13 polyhedra that correspond to these criteria and they are also called Archimedean polyhedra after Archimedes.

Two semiregular polyhedra are presented on this page, the snub cuboctahedron (left) and the cuboctahedron (right).

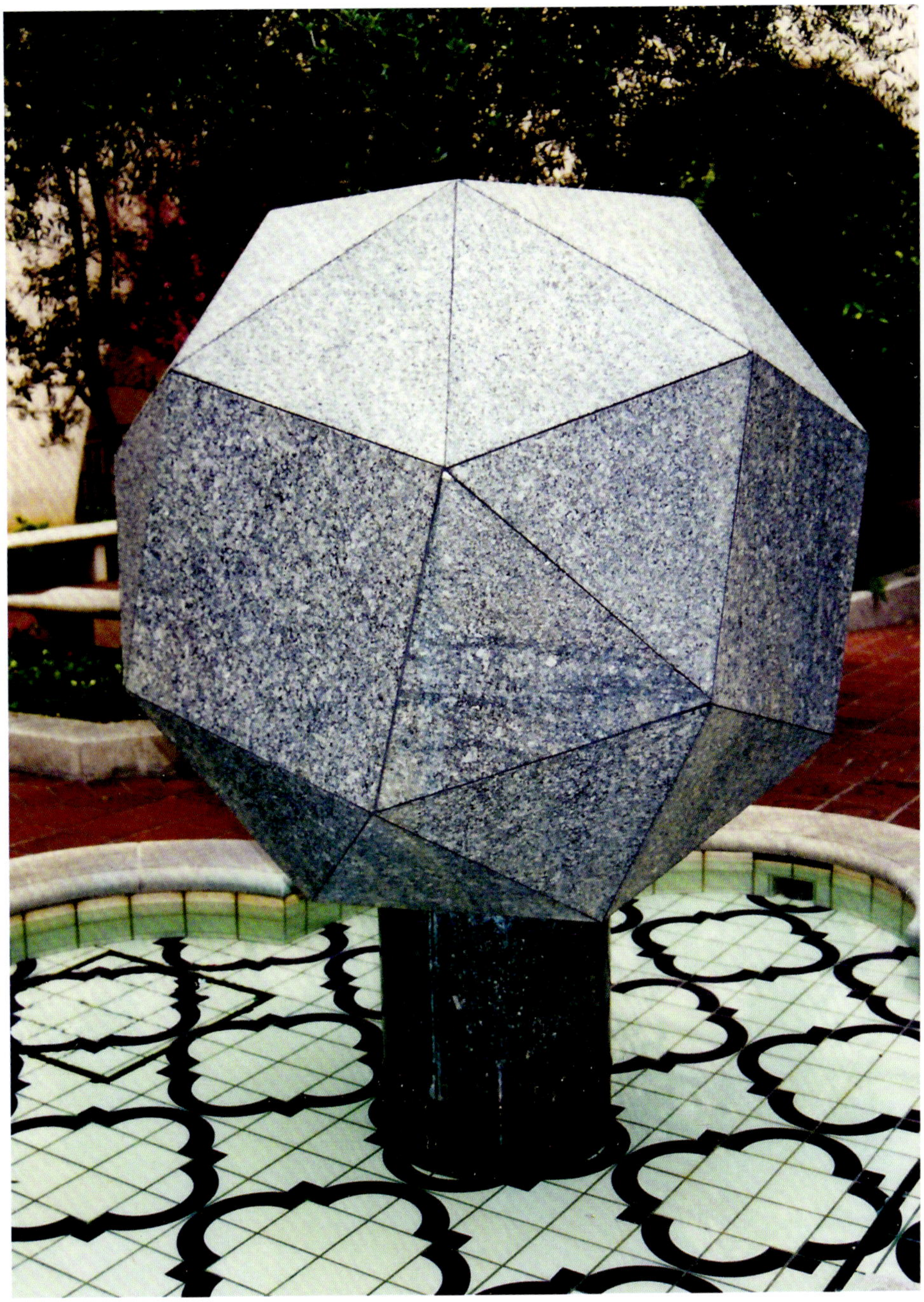

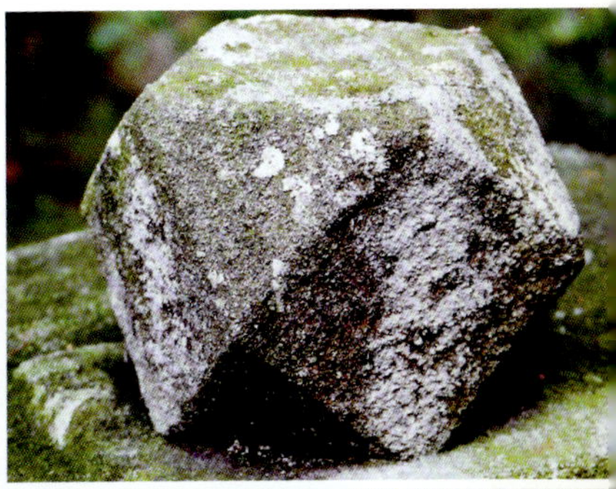

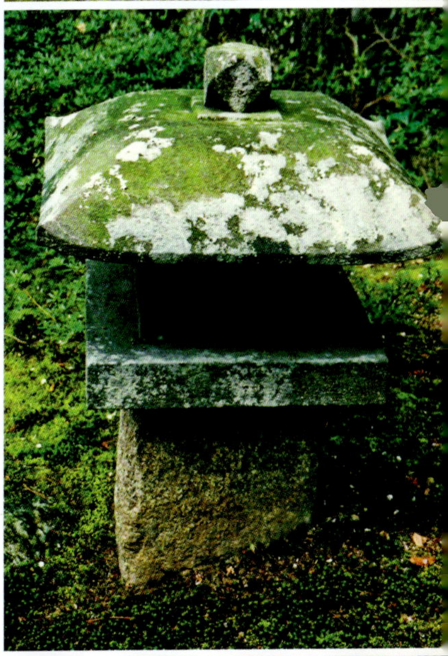

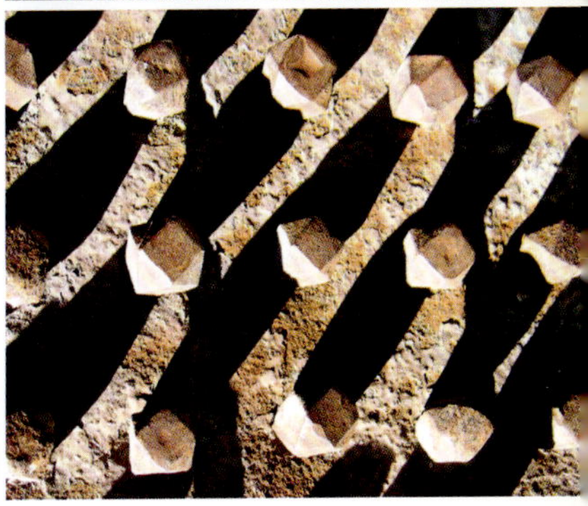

Snub cuboctahedron on the campus of the California Institute of Technology, Pasadena, California

Cuboctahedron as a top decoration of a garden lantern in the garden of the Shugakuin Imperial Villa in Kyoto, Japan and in close-up

Small cuboctahedra decorating an old iron gate in the wall of the ancient city of Acco, Israel

Buckyballs

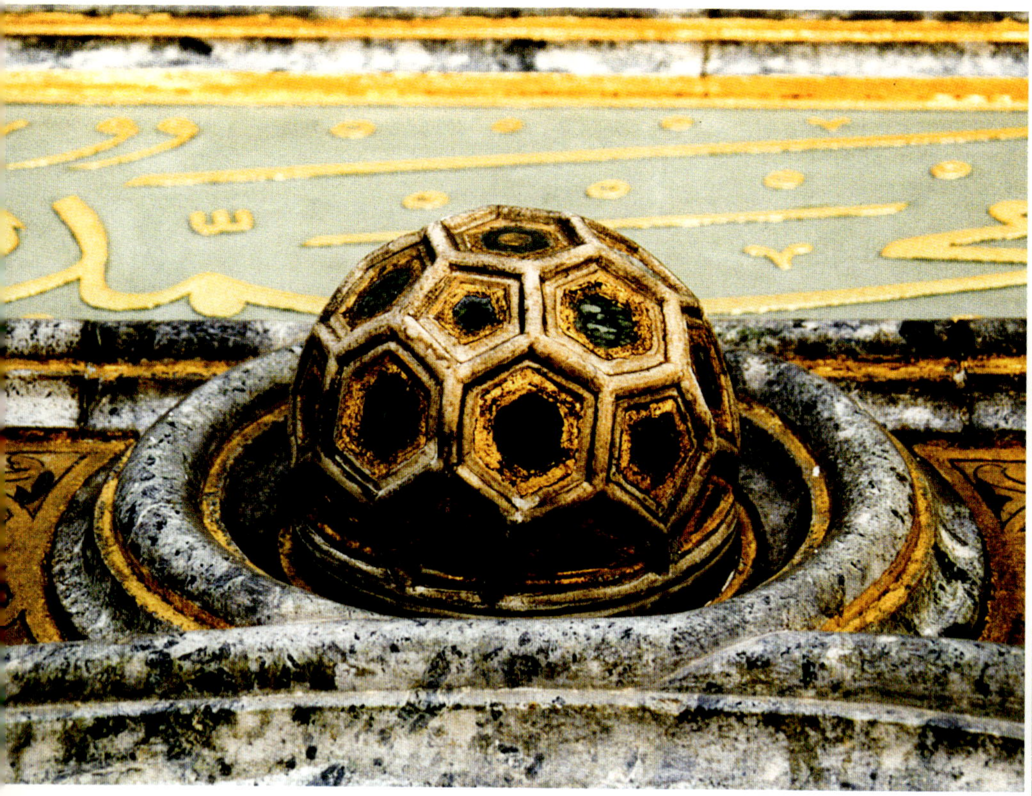

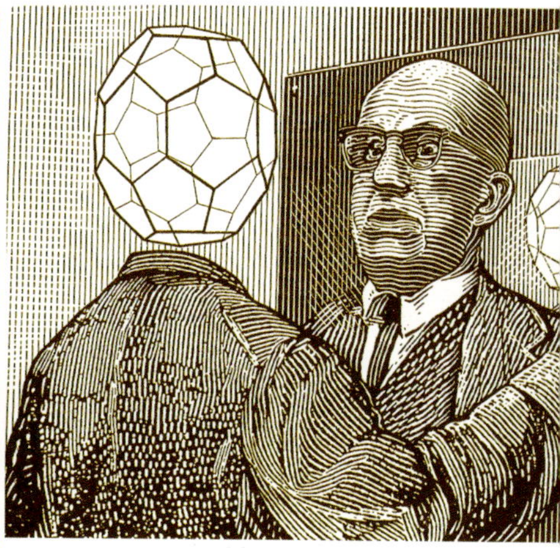

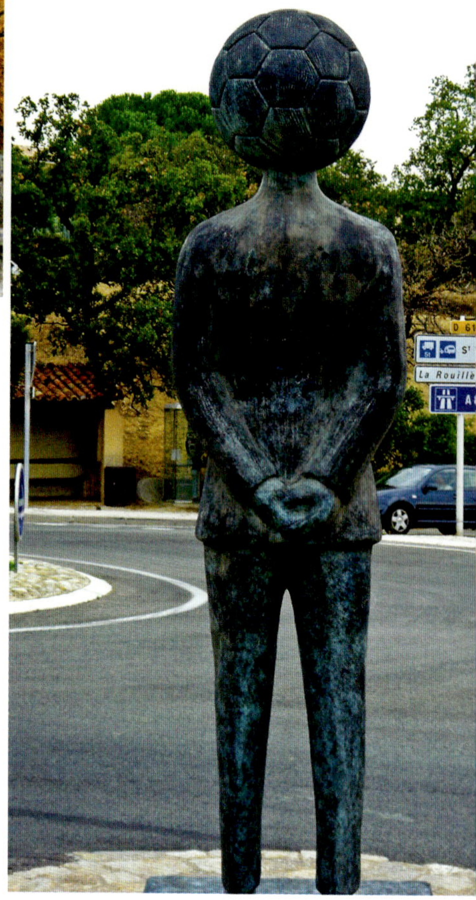

The truncated icosahedron is one of the best-known semiregular polyhedra, which is the shape of the C_{60} molecule, the third modification of carbon (after graphite and diamond). It was discovered in 1985 and named buckminsterfullerene after the American designer and inventor, R. Buckminster Fuller. It can be imagined as an icosahedron whose vertices are shaved off, hence it has 12 regular pentagons and 20 regular hexagons for its faces, and 60 vertices with a carbon atom in each vertex. The modern soccer ball is also put together from 12 pentagonal and 20 hexagonal patches.

Top left: Part of a truncated icosahedron as decoration above an entrance at the Topkapi Sarayi in Istanbul

Top right: Artist's rendering of Buckminster Fuller's portrait

(drawing by and courtesy of István Orosz, Budakeszi, Hungary)

Bottom right: Sculpture with soccer ball head in southern France

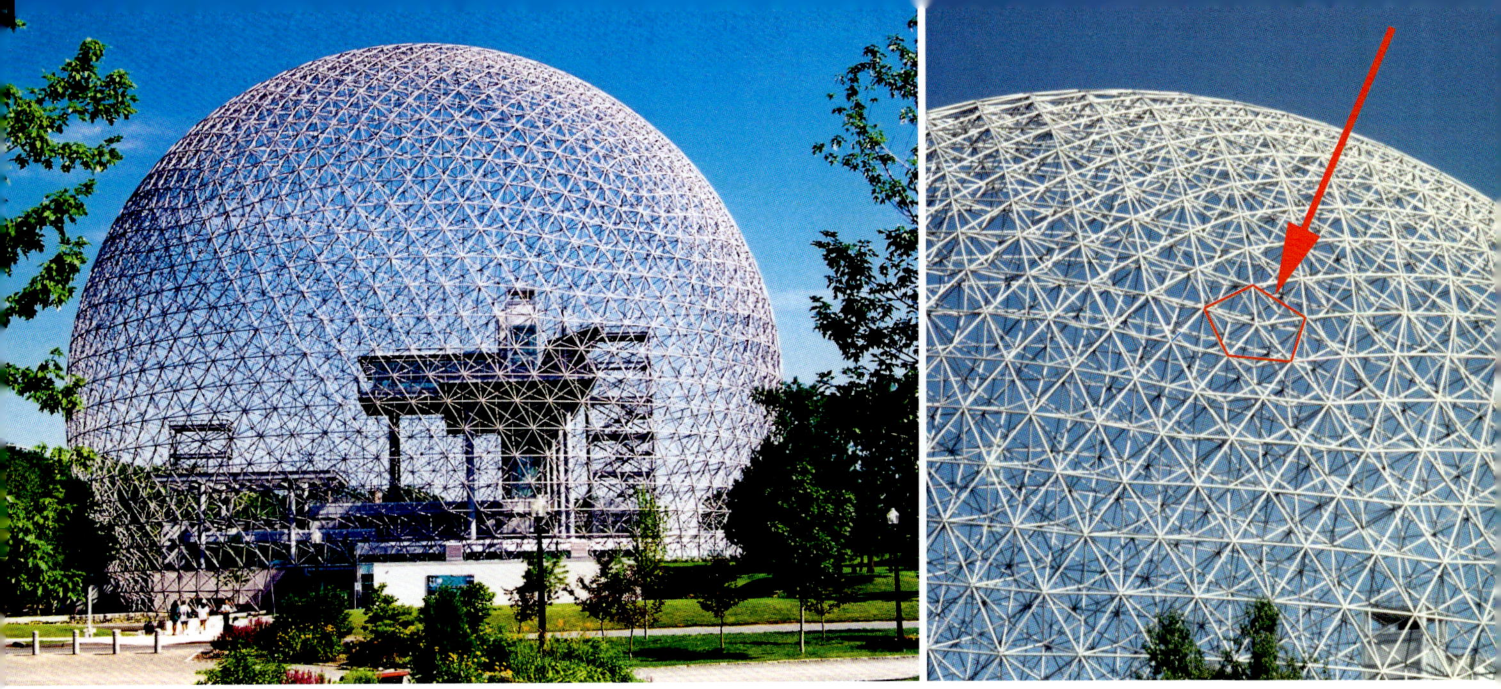

The exhibition pavilion of the United States built for the Montreal Expo in 1967. The close-up shows one of the 12 pentagons.

Buckminster Fuller constructed geodesic domes with icosahedral geometry. The most famous of his construction is in Montreal, Canada, that served as the U.S. pavilion for the Montreal Expo in 1967.

The Montreal pavilion is not a model of buckminsterfullerene because it has many more vertices than the C_{60} molecule, but it is a fullerene-type structure.

There is an early example of fullerene-type models. Dragon sculptures are common in China as guards in front of important buildings and they are often imitated in front of Chinese restaurants in the western world. In the genuine version, the female dragon has a baby dragon under her left paw and the male has a sphere under his right paw. This sphere is said to represent a ball made of silk strips that was a favorite toy in ancient China. The surface of the ball is usually decorated by a regular hexagonal pattern. Because the complete sphere cannot be covered with the regular hexagonal pattern, considerable chunks of the sphere are usually hidden by the dragon's paw and by the stand on which it stands. However, there is a dragon sculpture under whose paw the sphere is decorated by a hexagonal pattern with pentagons interspersed among the hexagons, and hardly any part of this decoration is hidden. This sculpture is in front of the Gate of Heavenly Purity in the Forbidden City in Beijing. It dates back to the reign of Qian Long (1736–1796) of the Qing dynasty.

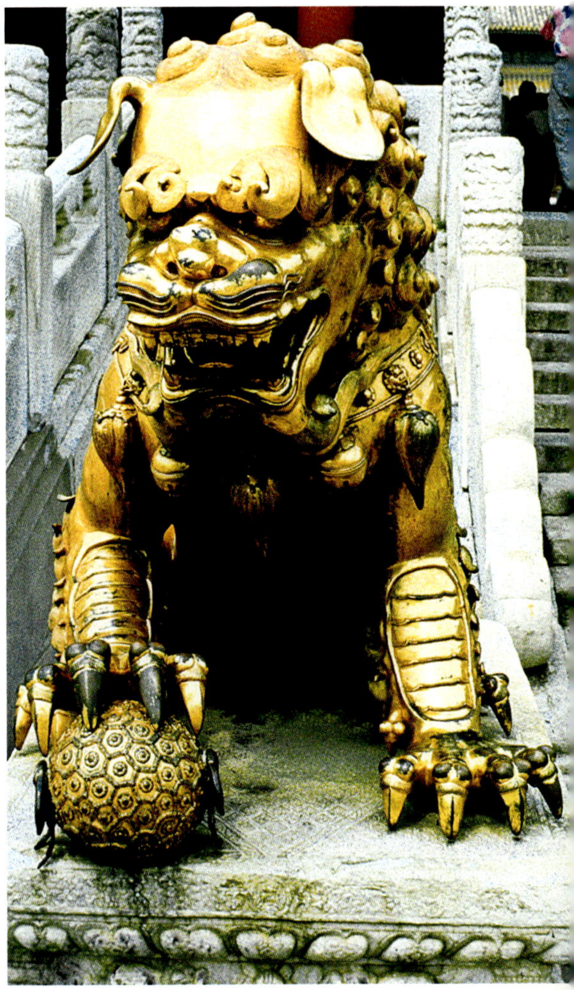

Dragon in the Forbidden City in Beijing

Molecules and Walnuts

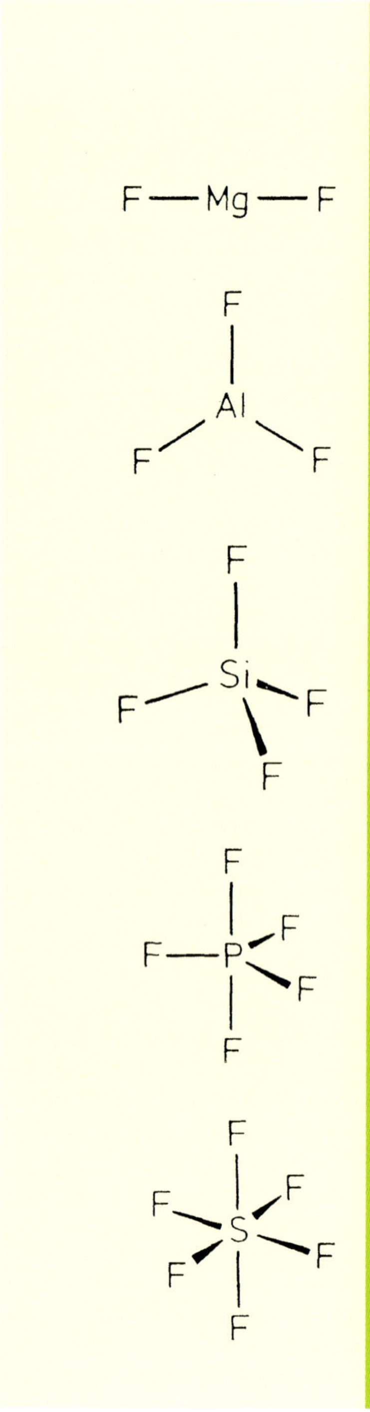

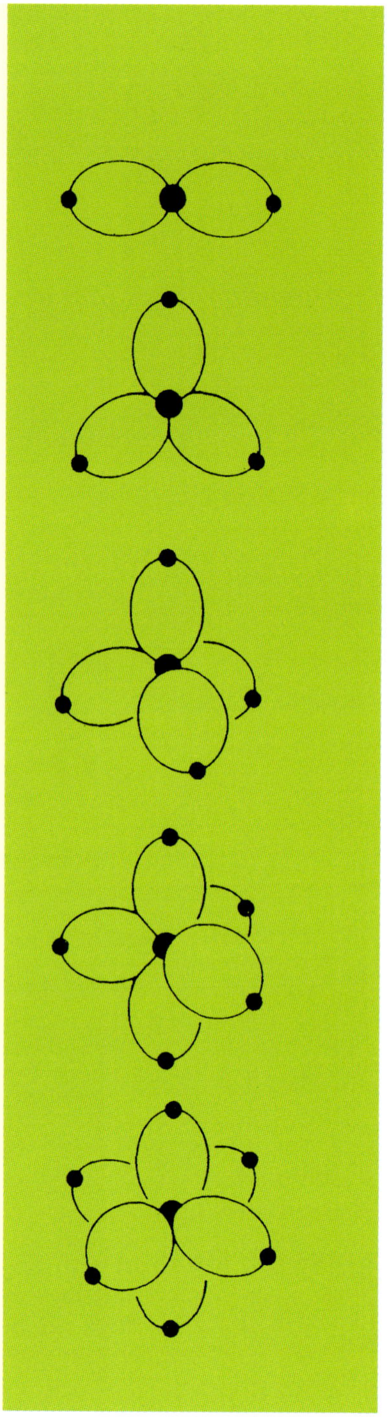

Extreme left: Shapes of molecules where the chemical bonds are represented by straight lines

Left: Shapes of molecules where the chemical bonds are represented by space domains (very much like balloons)

Simple molecules often have the shapes of polyhedra. We show here a set of molecular models with characteristic shapes. If we imagine the chemical bonds as domains of electron pairs linking two atoms, it is quite realistic to think about these domains as balloons that elbow each other for space around the central atom. In forming the appropriate balloon clusters, the balloons are fastened together at their openings. Furthermore, it has been observed that when walnuts grow together in small groups, they form polyhedral shapes very much like those of the molecules and the balloon clusters. This is how linear, trigonal planar, tetrahedral, trigonal bipyramidal, and octahedral shapes form. It turns out that the struggle for space determines the spatial arrangement of atoms in molecules, balloons, and walnuts providing a simple model for understanding some basic concepts in chemistry.

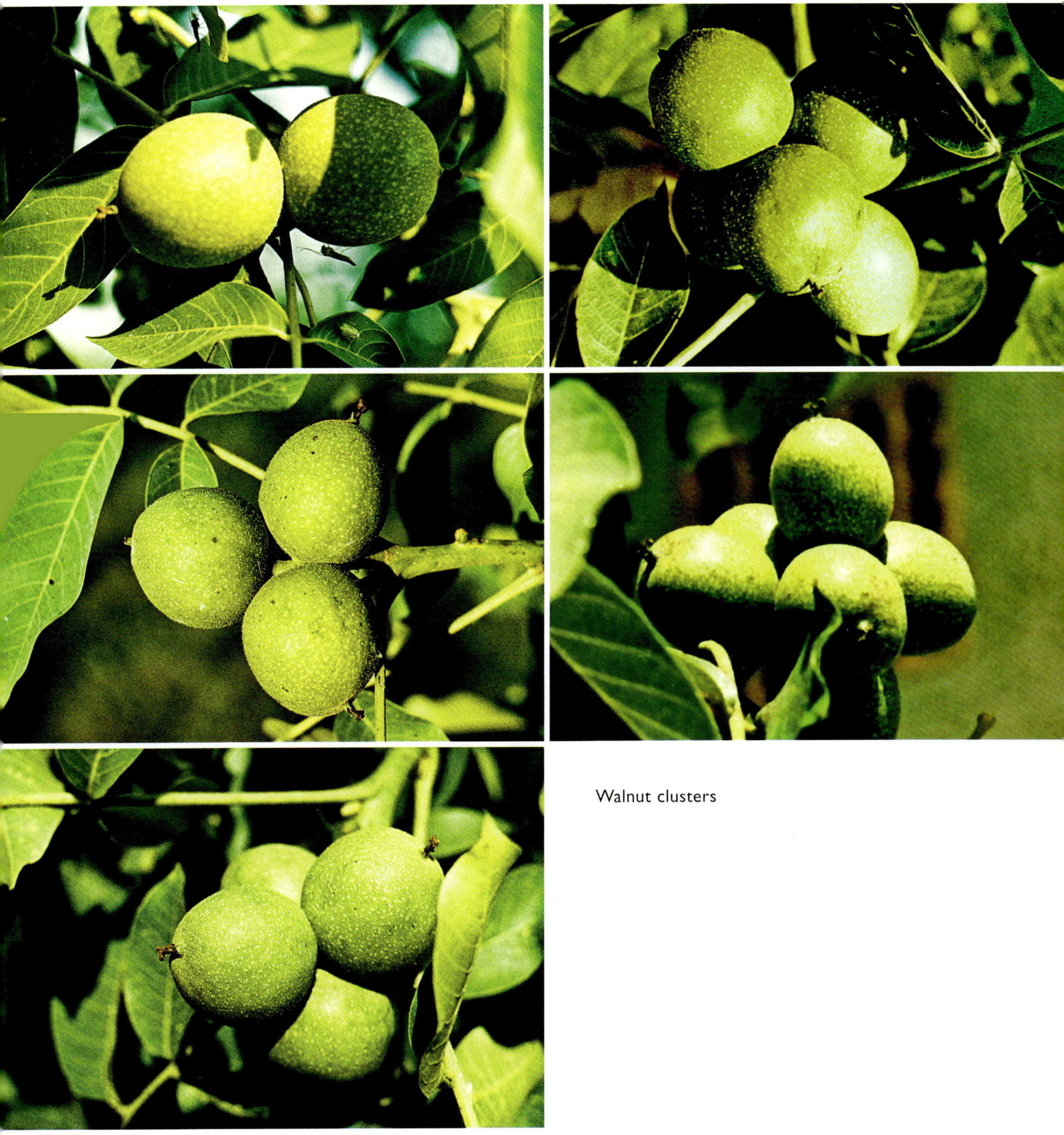

Walnut clusters

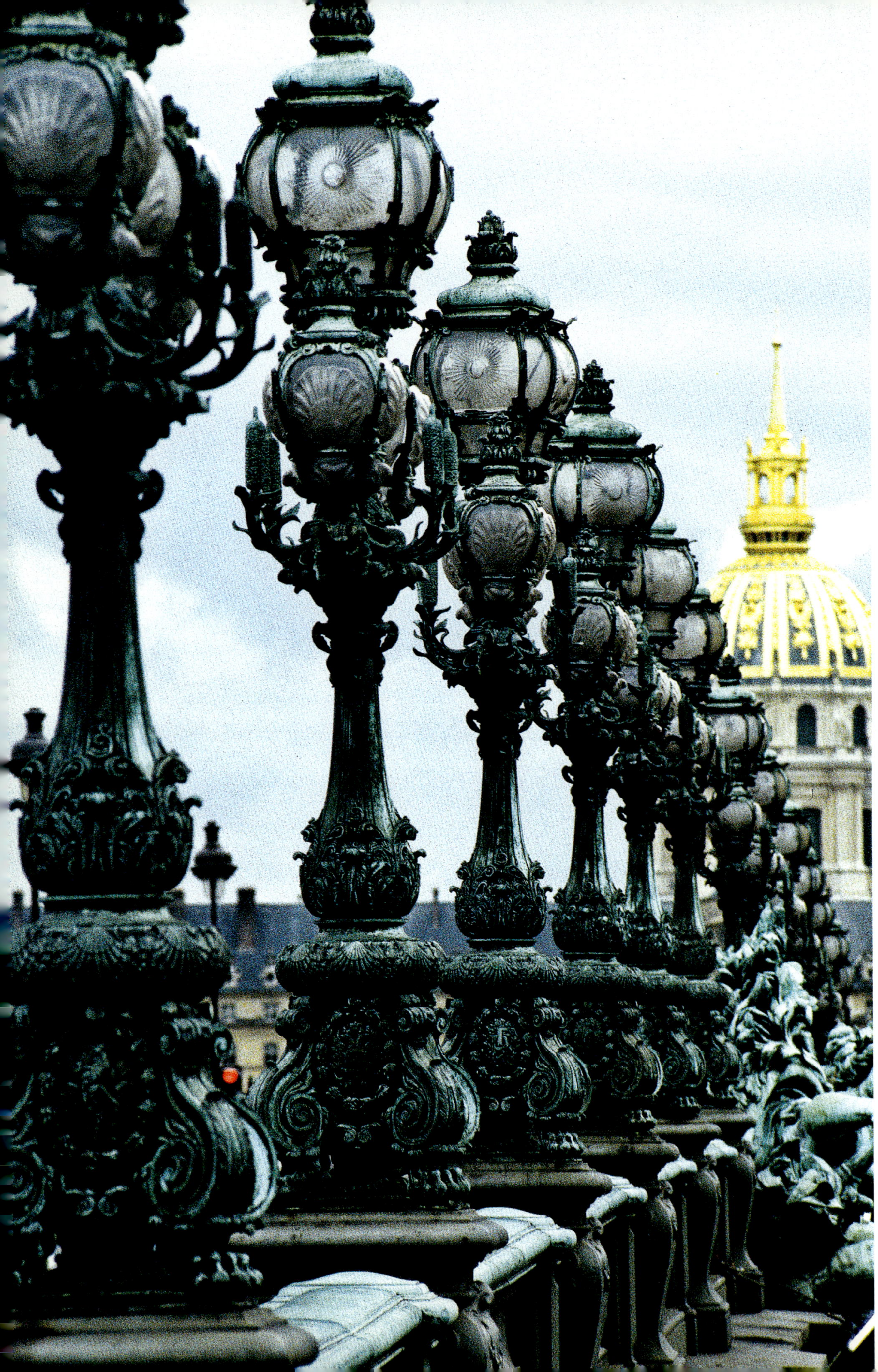

When a pattern is created by infinite repetition of a motif it has symmetry even though a different kind of symmetry from what we have seen so far. Repetition can also be looked at as translation and *translational symmetry* means shifting and repeating the motif. The pattern obtained this way is periodic. In other words, infinite repetition of the same motif leads to periodicity. In the subsequent pages we will see a great variety of translational symmetry.

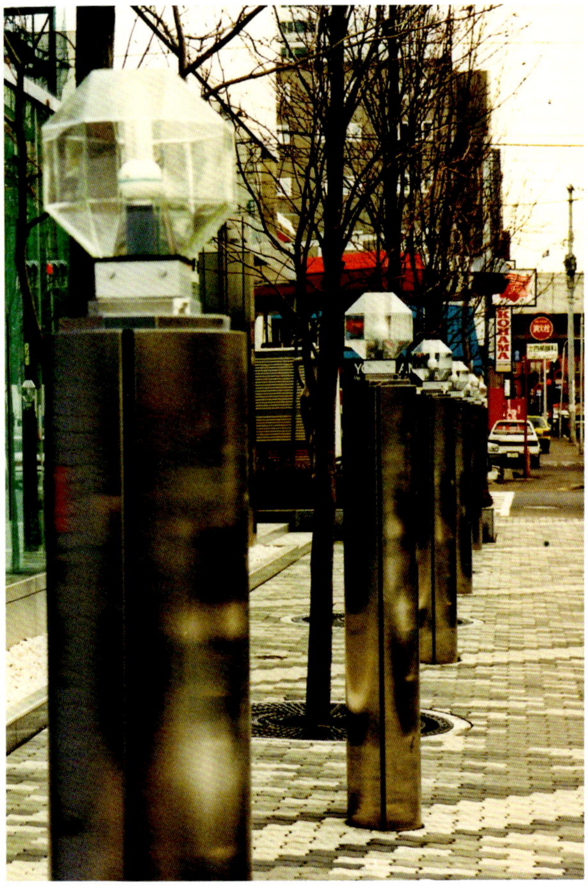

Lamps in Sapporo, Japan

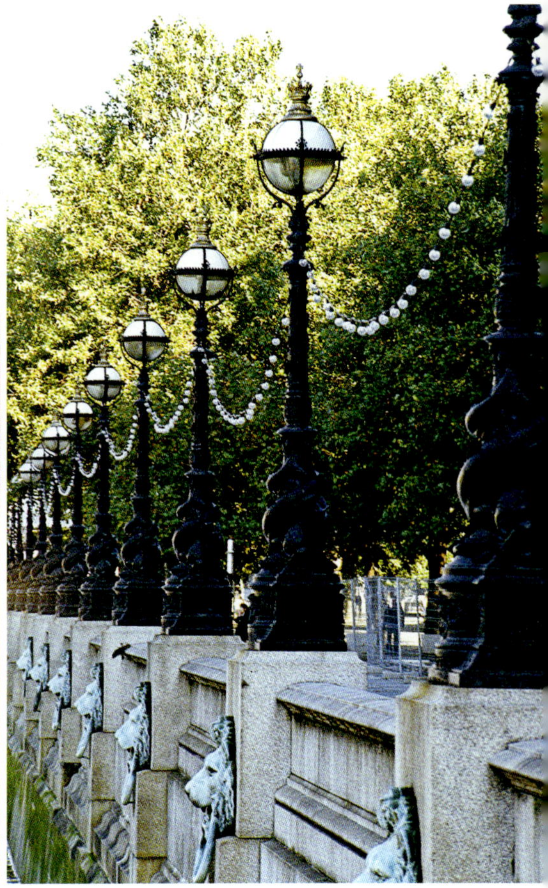

Lamps on the bank of the Thames in London

Repetitions

Lamps on the Alexander III Bridge in Paris

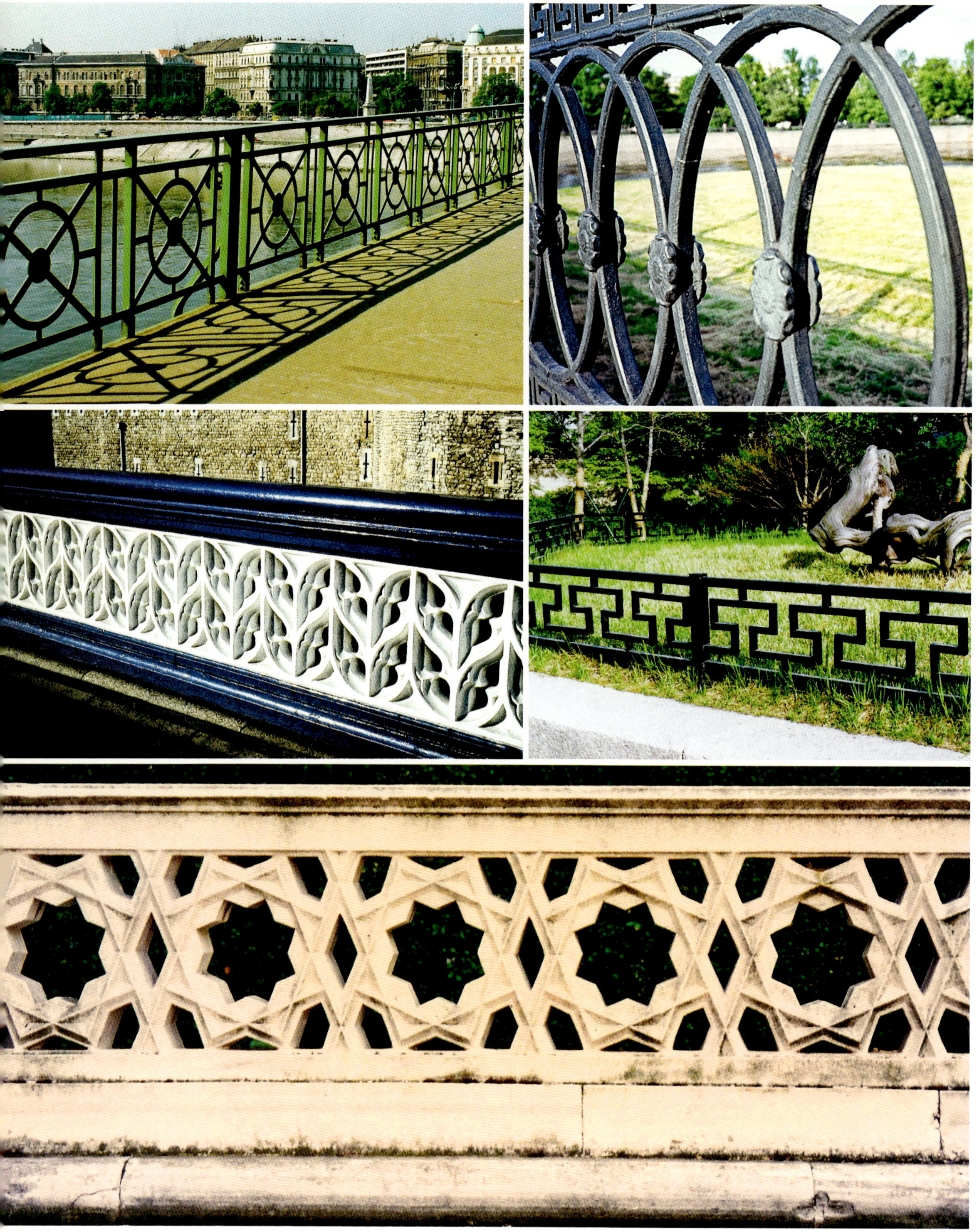

From upper left corner, clockwise:

Fences in Budapest, St. Petersburg, Seoul, Istanbul, and London

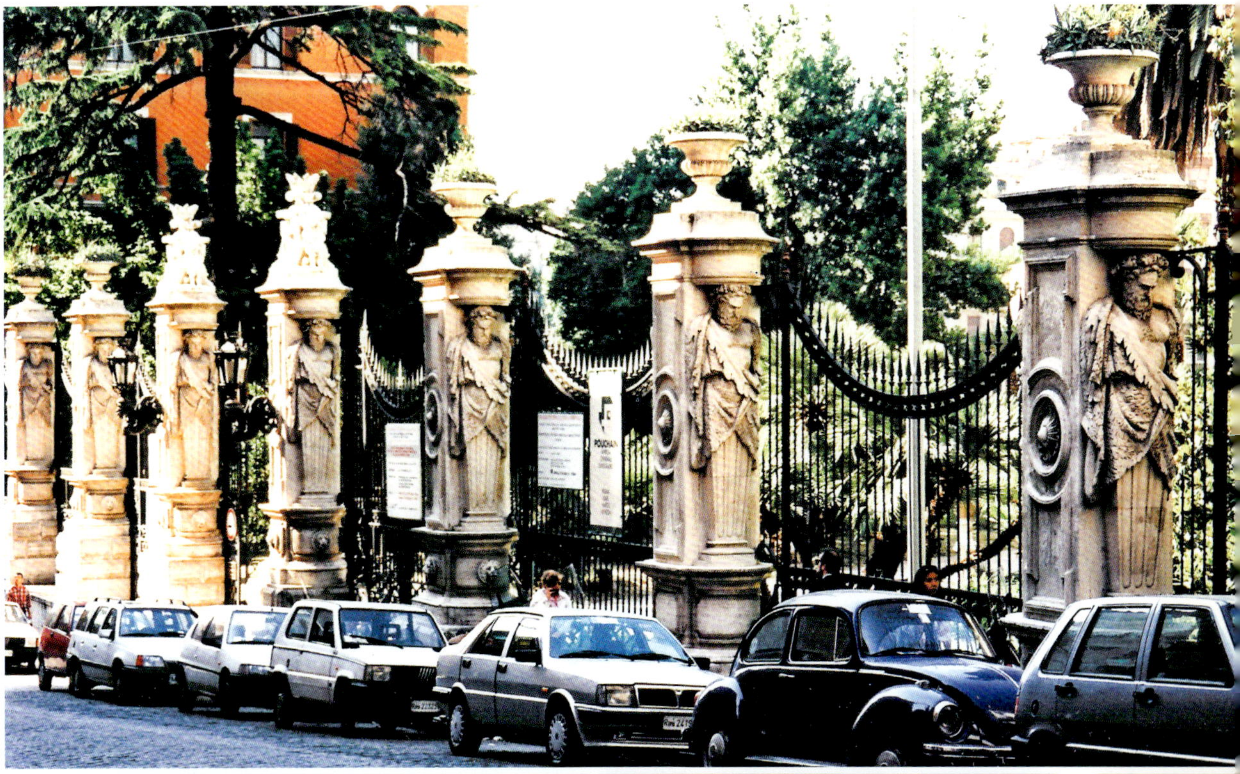

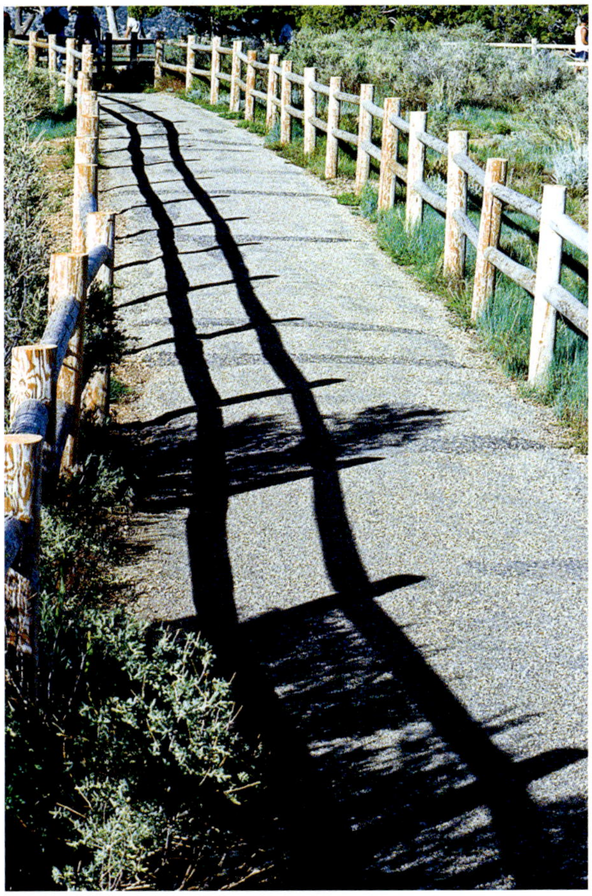

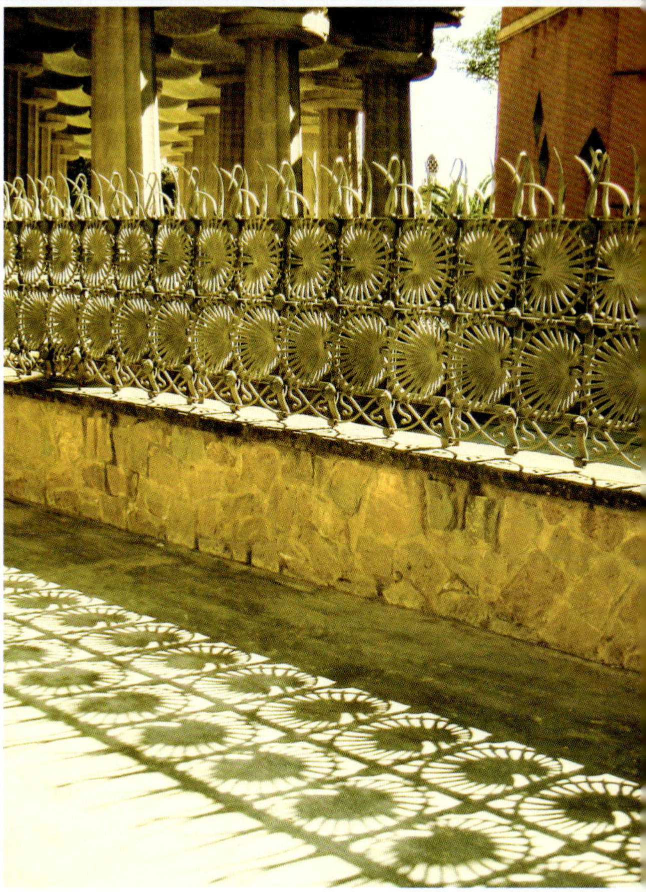

Top: Rome
Bottom left: at the Grand Canyon in Arizona
Bottom right: Barcelona

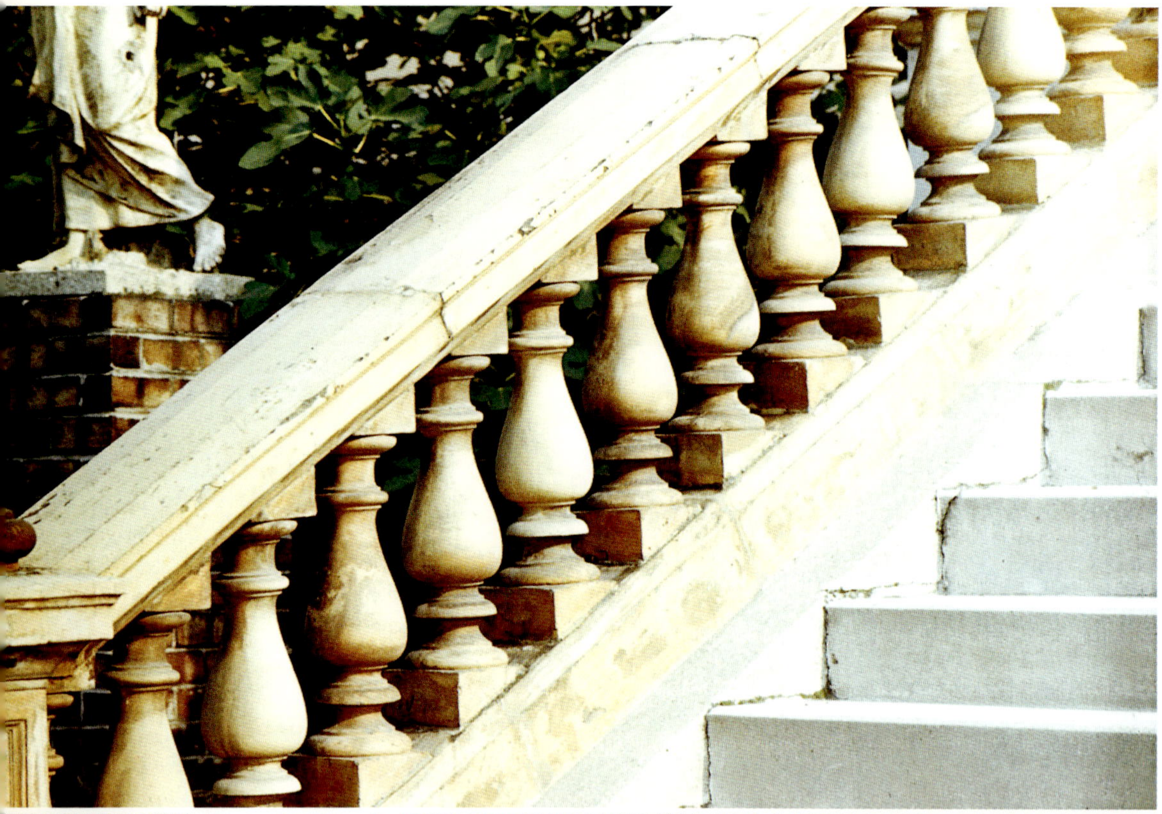

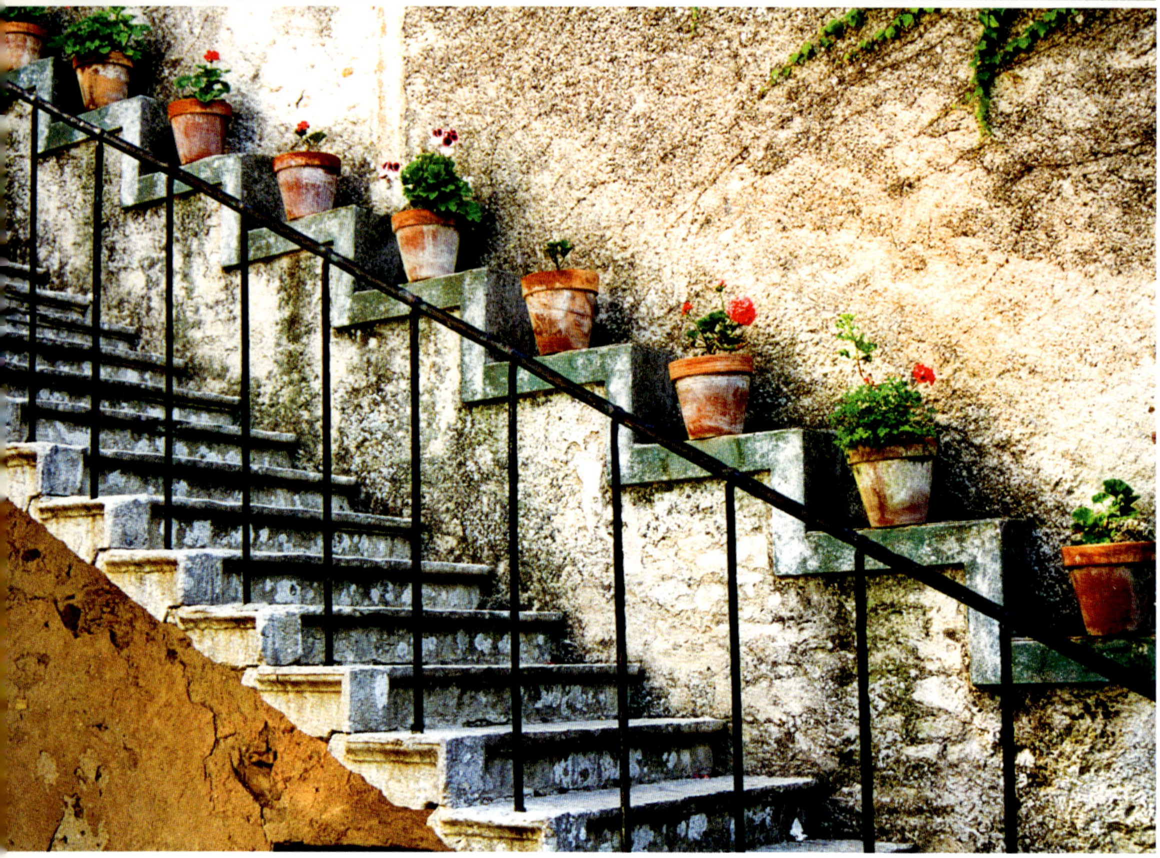

Railings and staircases in Pécs, Hungary (top), and Erice, Sicily

Colonnades (from the top left, clockwise) in Beijing; in St. Peter's Square, Vatican City; Albany, New York; and at the Palais Royal in Paris

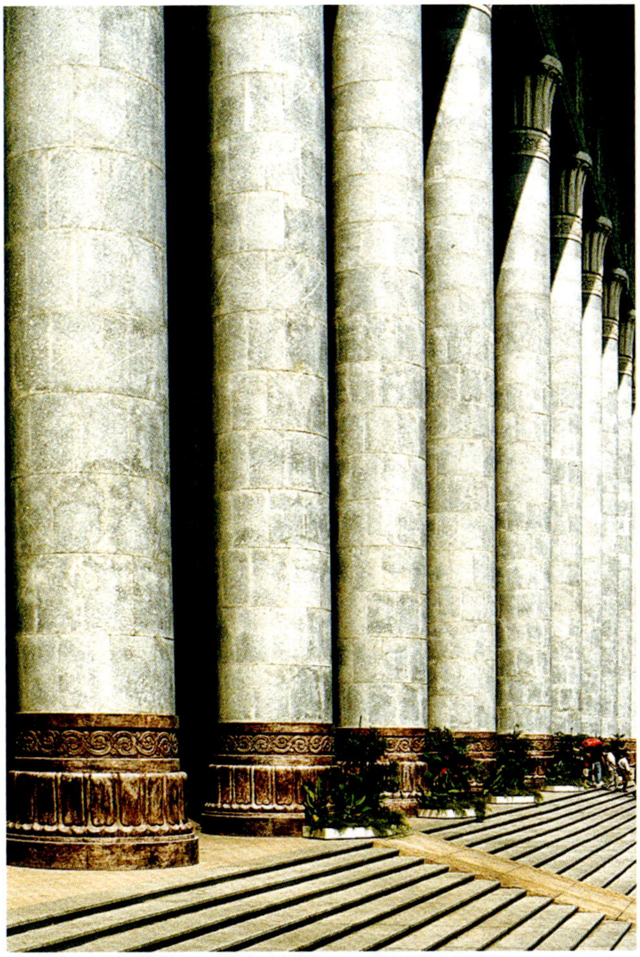
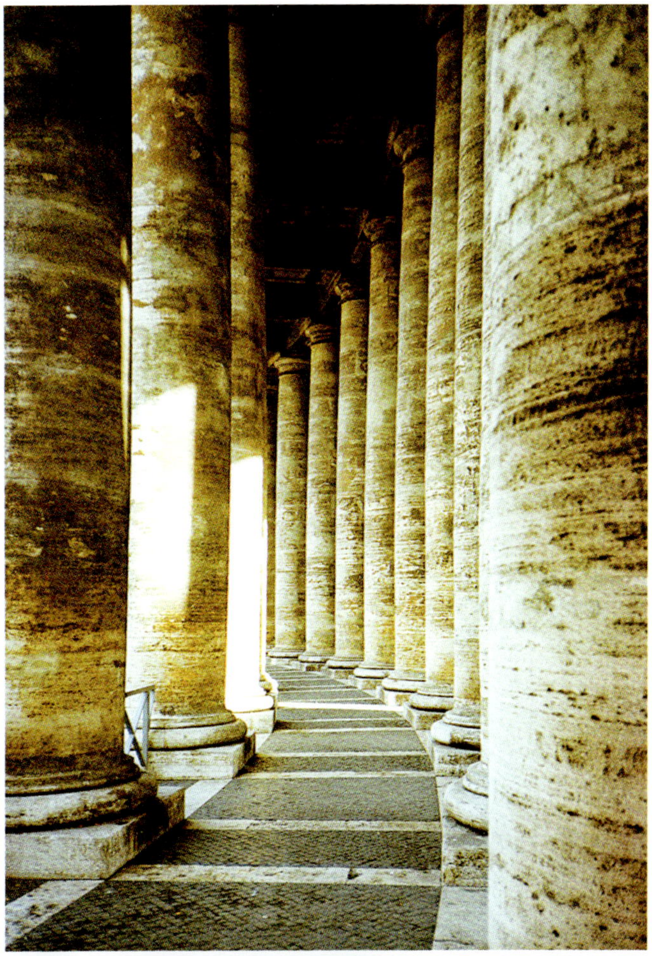
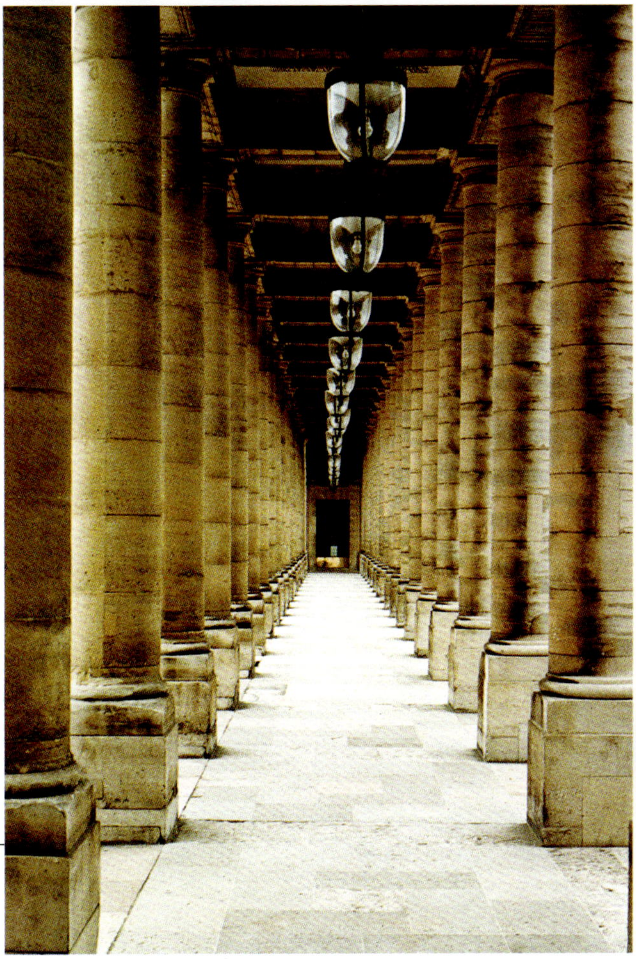
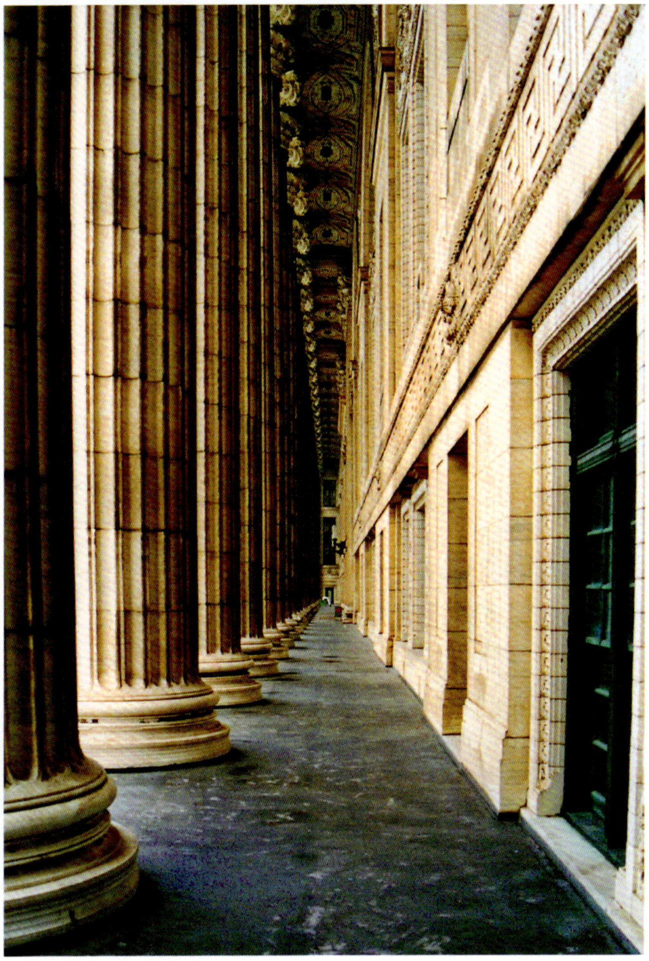

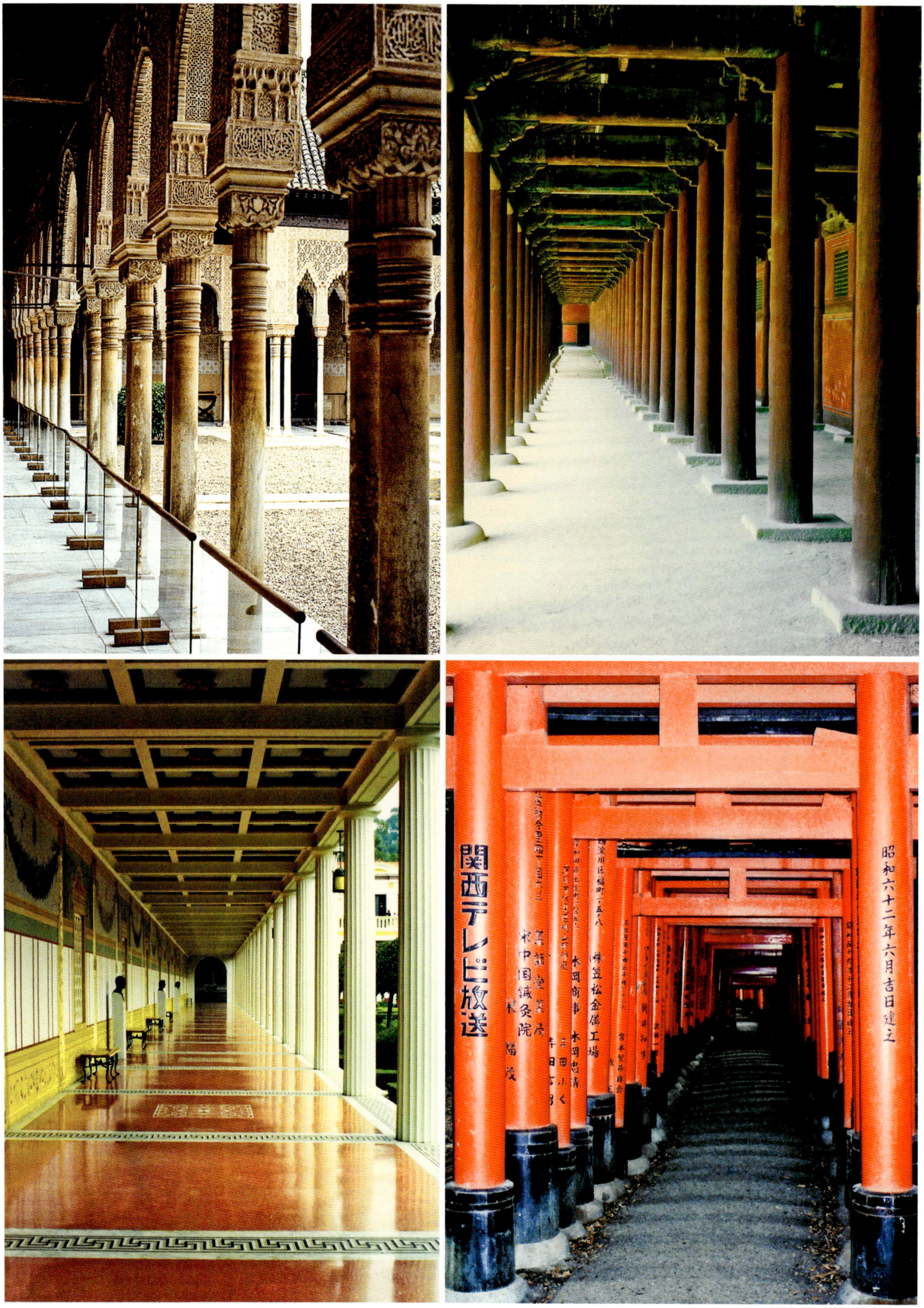

Colonnades (from the top left, clockwise) at the Alhambra, Granada, Spain; in the Republic of Korea; in Kyoto; and at the Getty Museum in Los Angeles, California

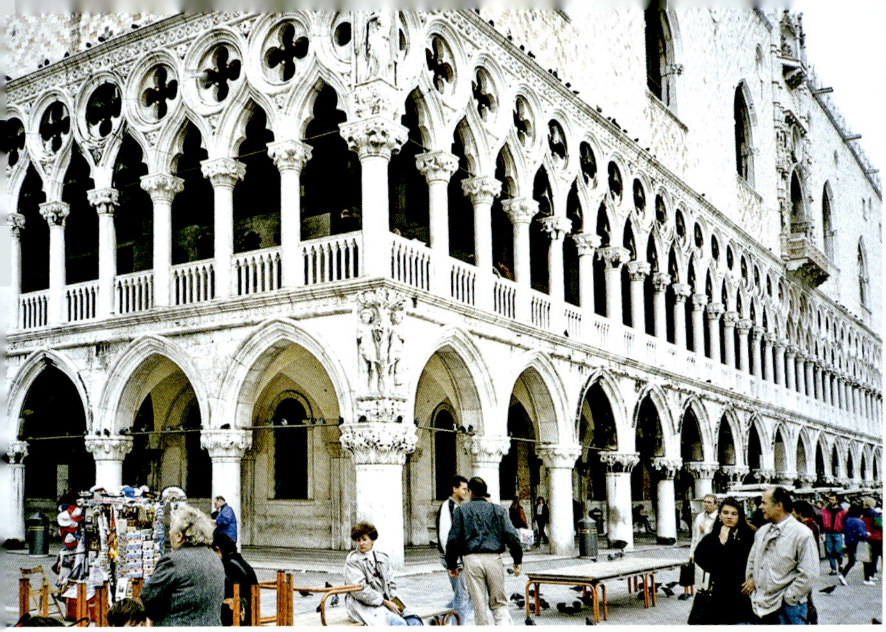

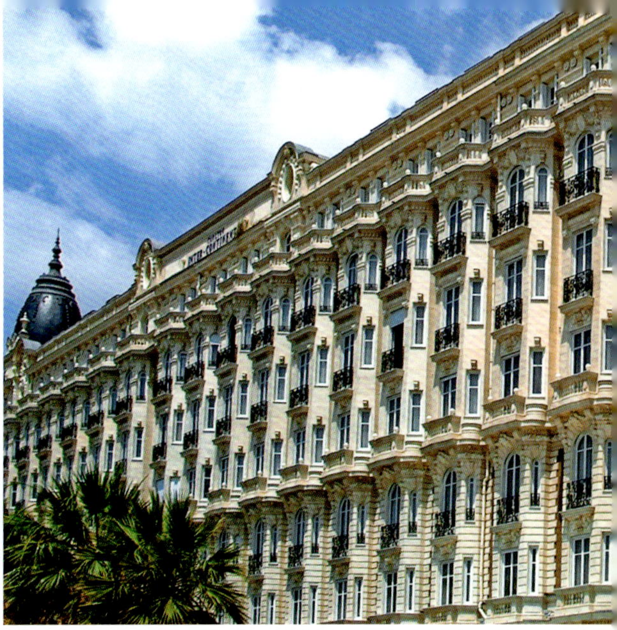

Façades (from the top left, clockwise) of the Dodge Palace in Venice; Hotel Carlton, Cannes; Termini Railway Station in Rome; and of a building in Stockholm

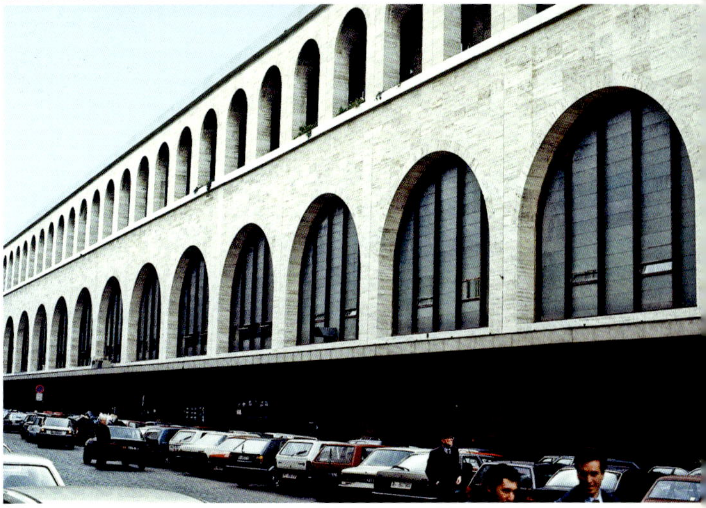

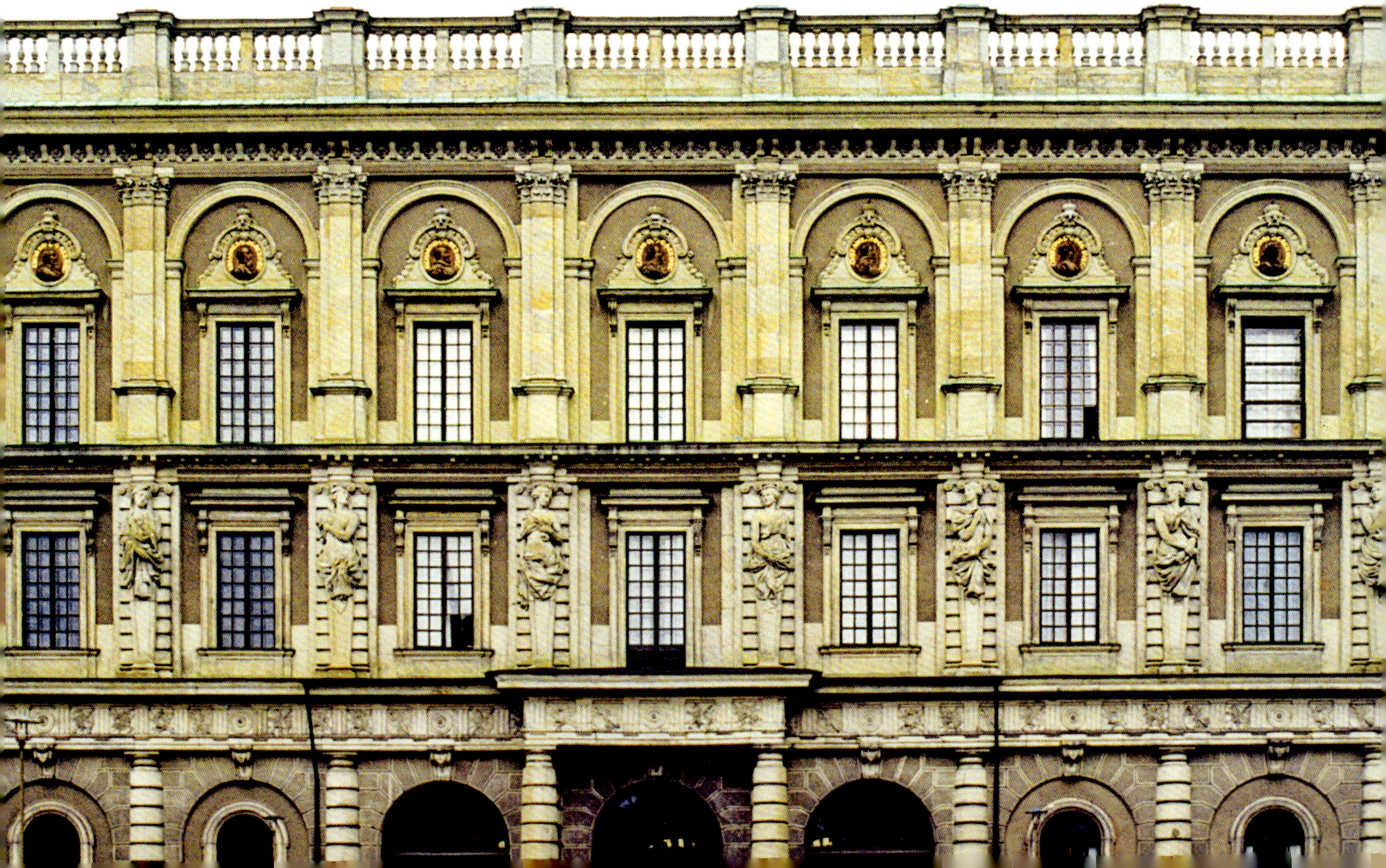

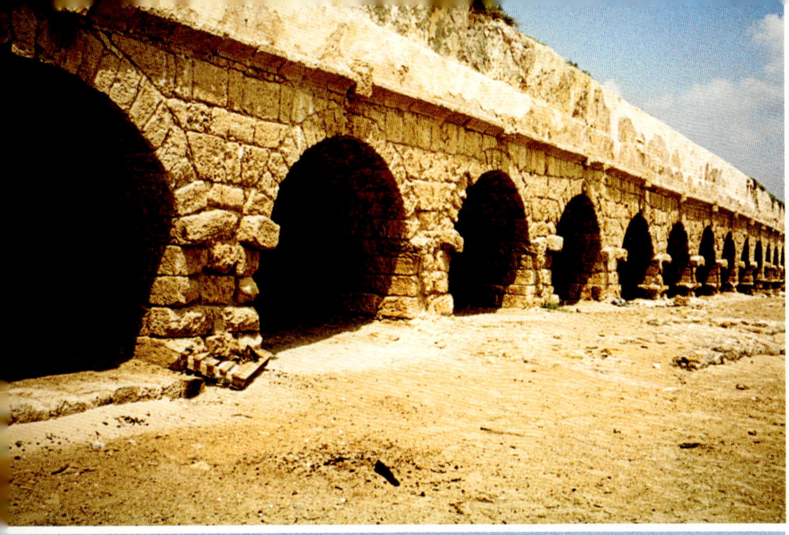

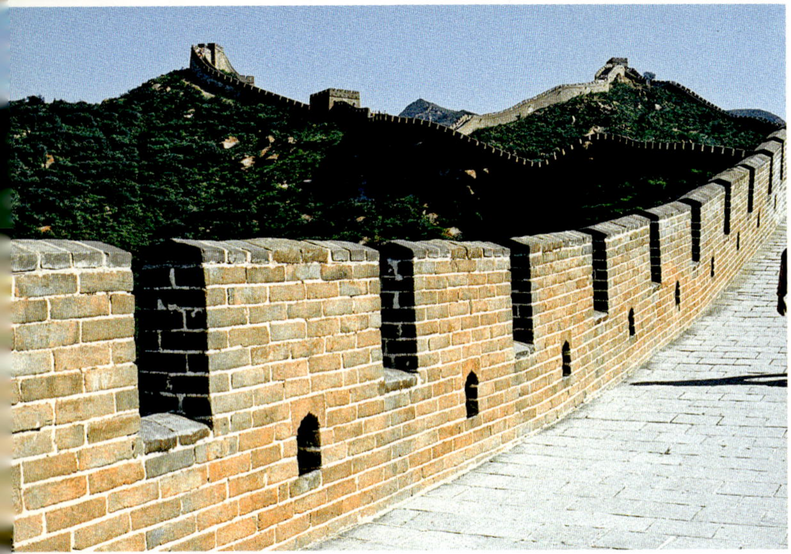

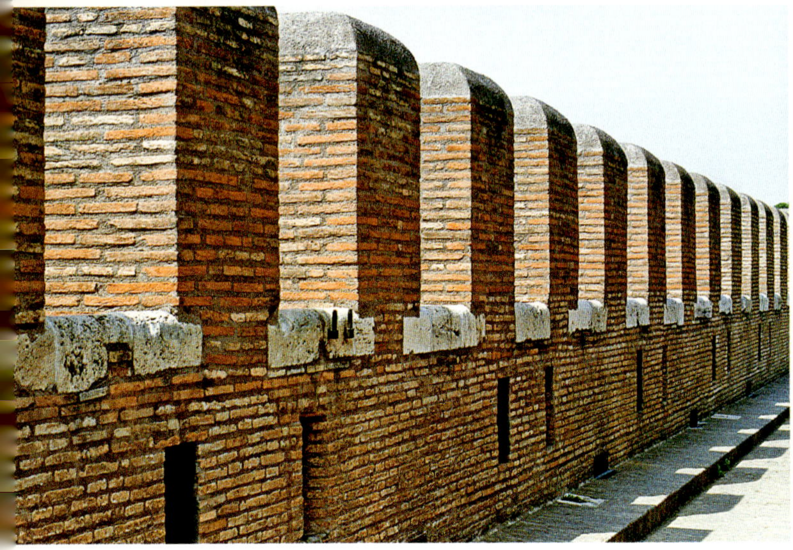

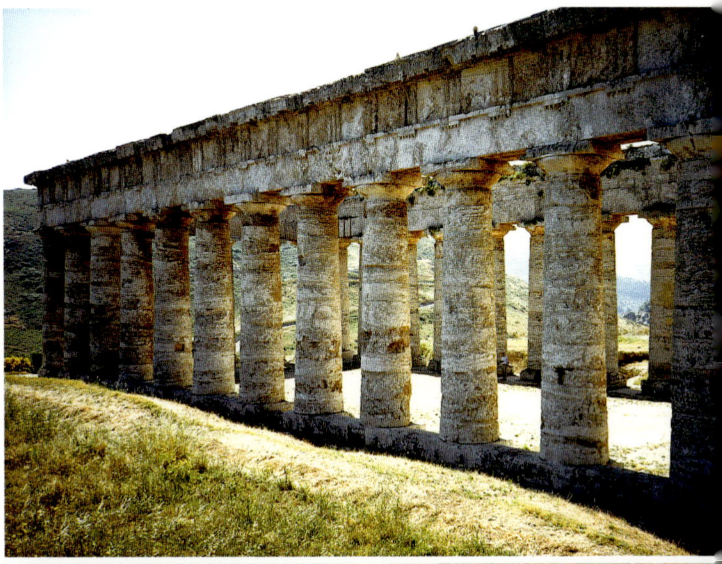

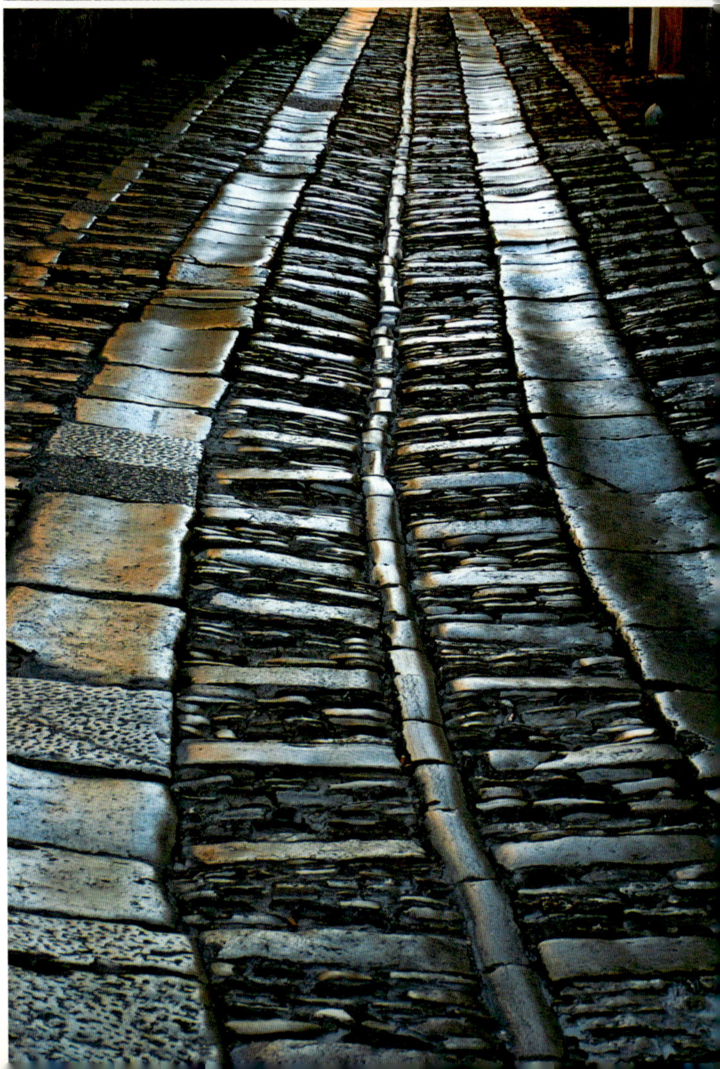

From the top left, clockwise:

Aqueduct in Cesarea, Israel; temple columns in Segesta, Sicily; pavement in Erice, Sicily; fortress wall in Rome; and the Great Wall in China

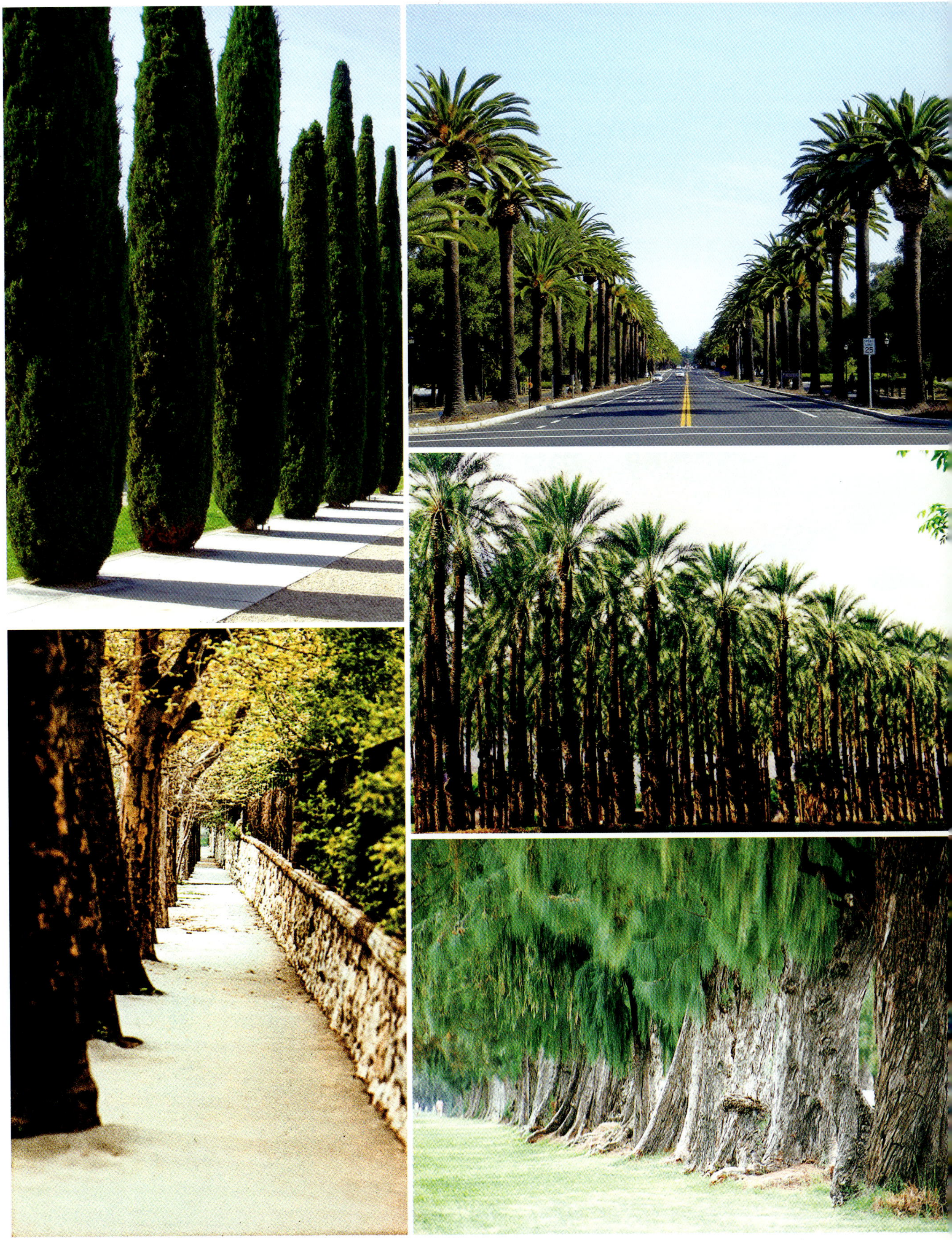

Lines of trees (from top left clockwise) in Palo Alto, California (twice); Indiana Wells, California; Honolulu, Hawaii; and Budapest

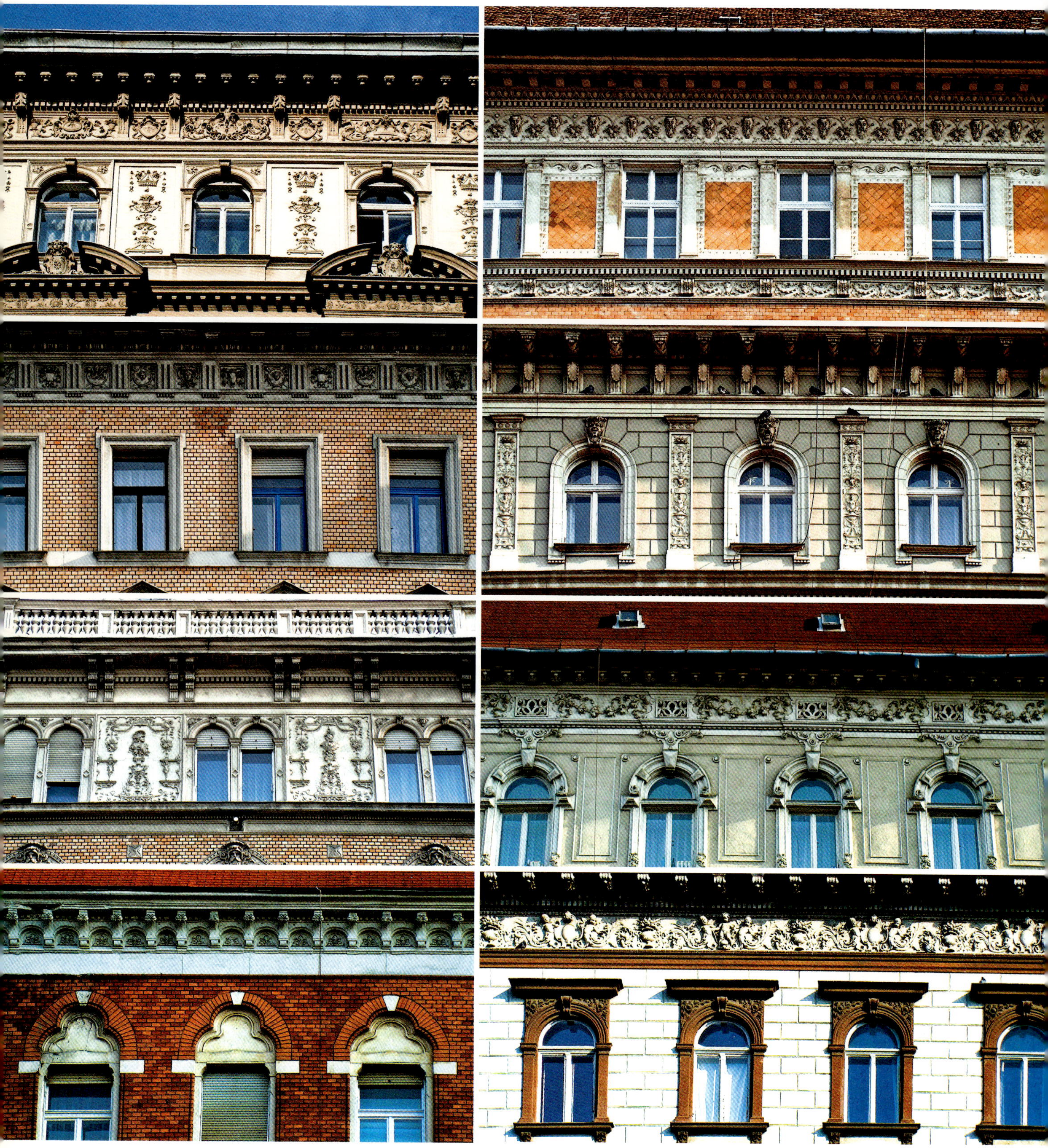

Façades of buildings on the Grand Boulevard in Budapest

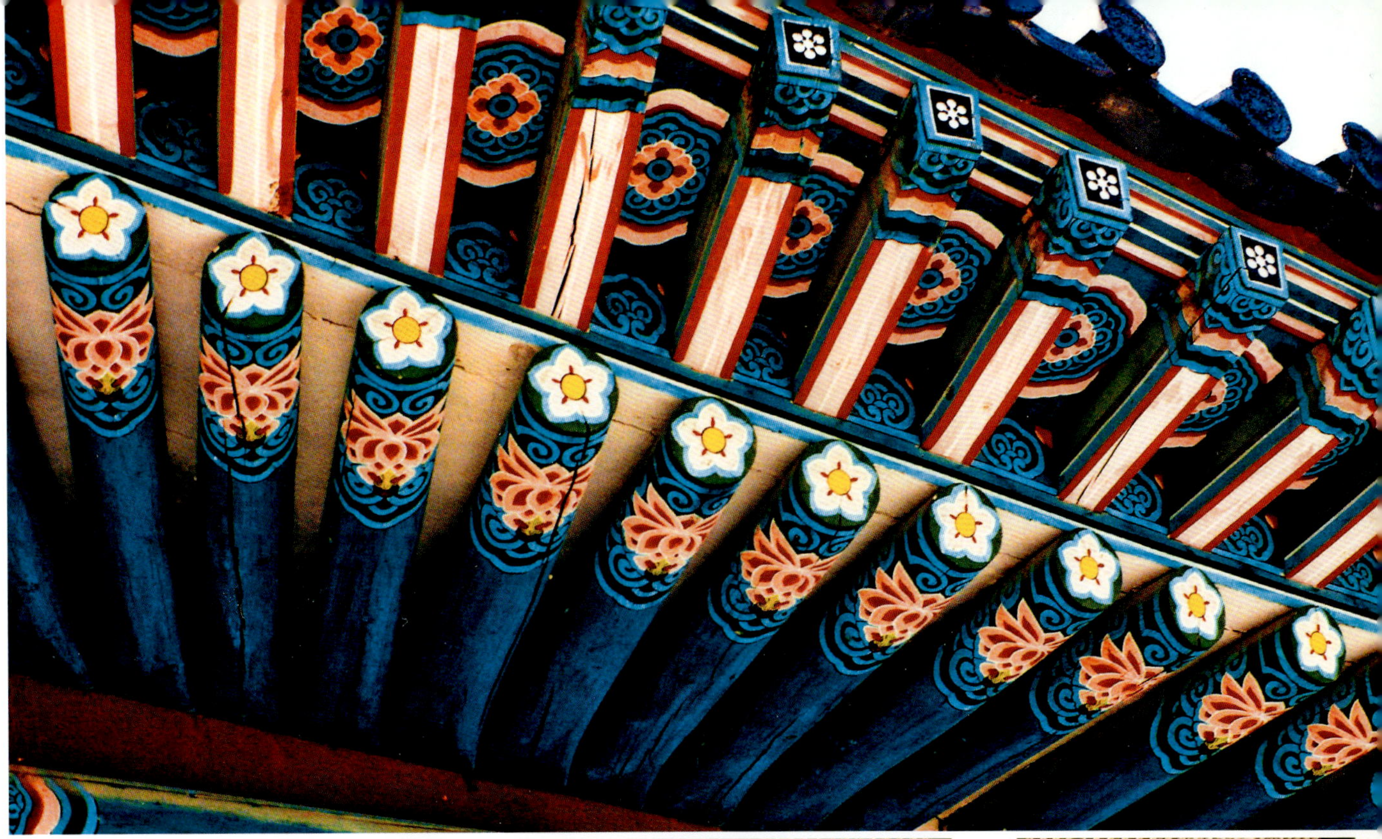

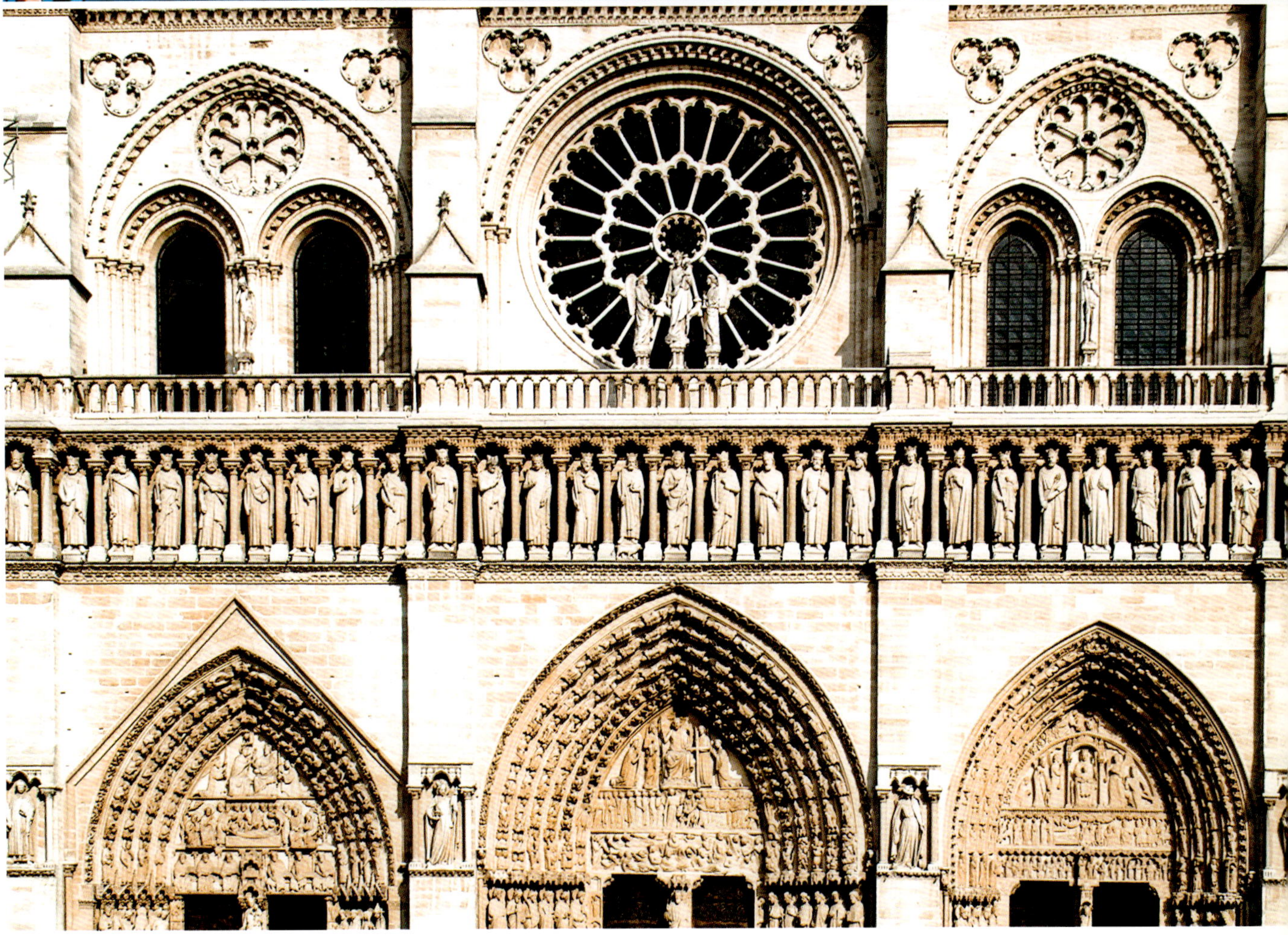

Decoration of a temple in the Republic of Korea (top) and the row of saints on the façade of the Notre Dame Cathedral, Paris

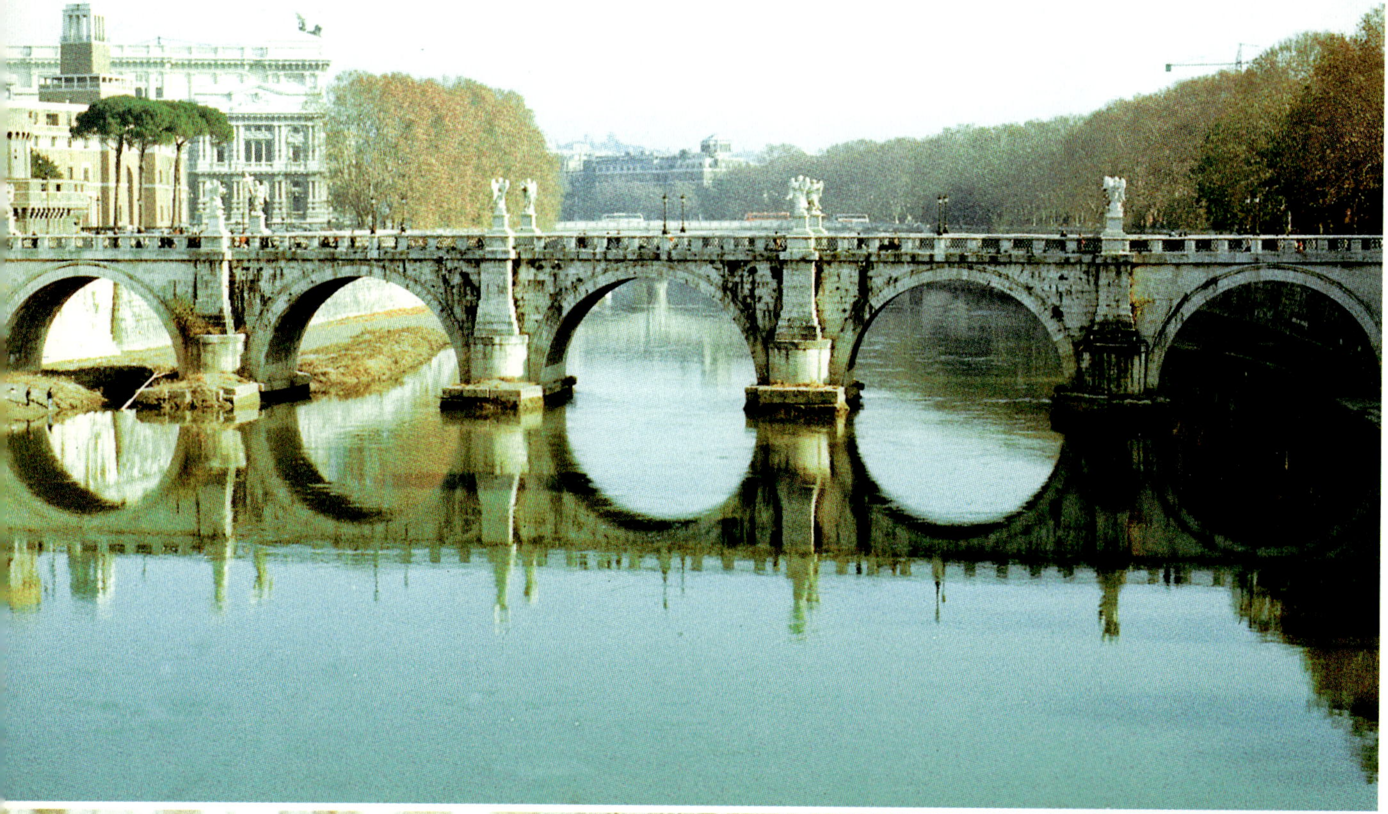

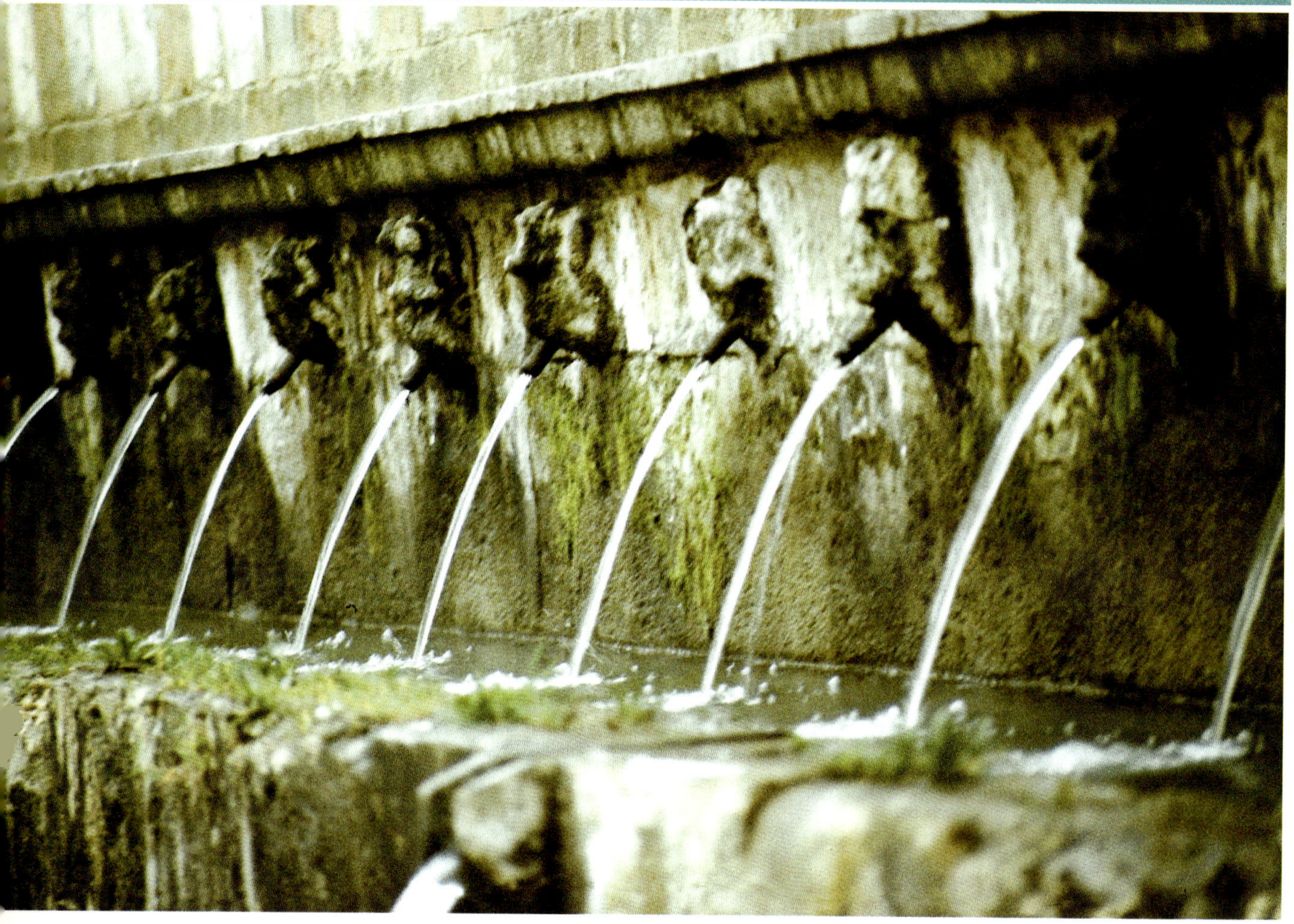

Top: San Angelo Bridge in Rome
Bottom: Fountains in L'Aquila, Italy

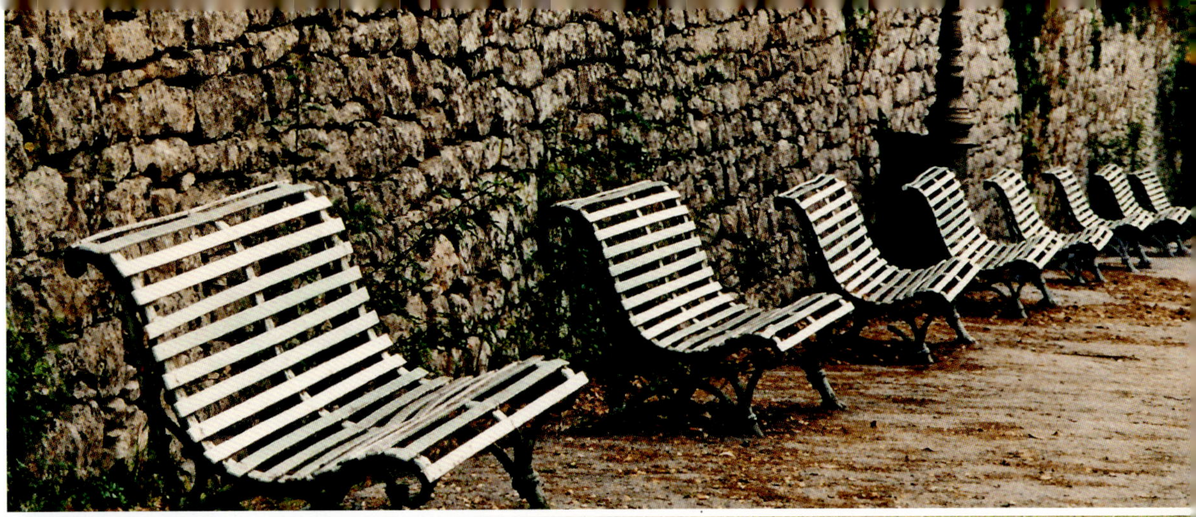

Benches in Erice, Sicily

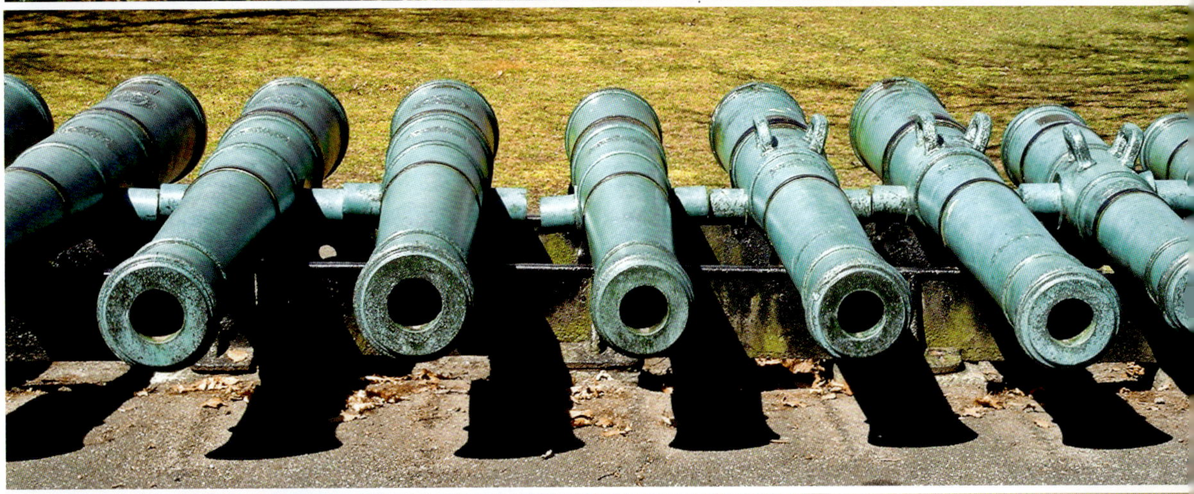

Cannon barrels at West Point, New York

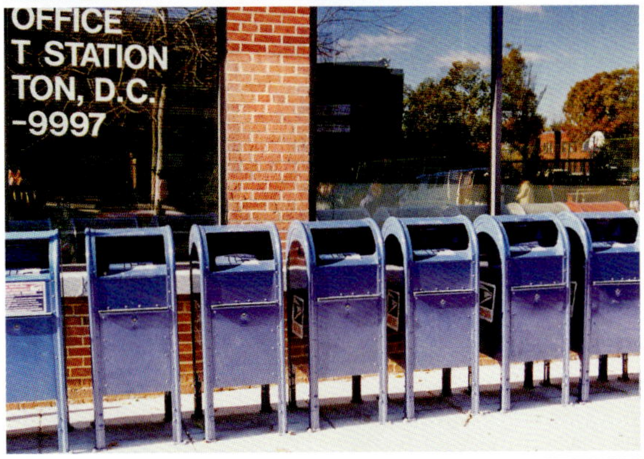

Mail boxes in Washington, DC

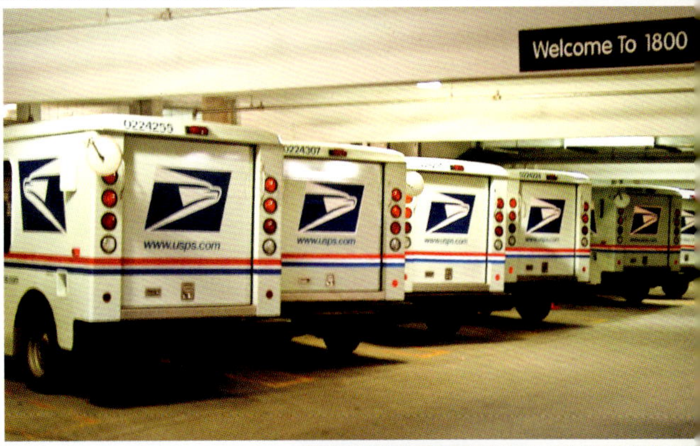

U.S. Mail trucks in Chicago

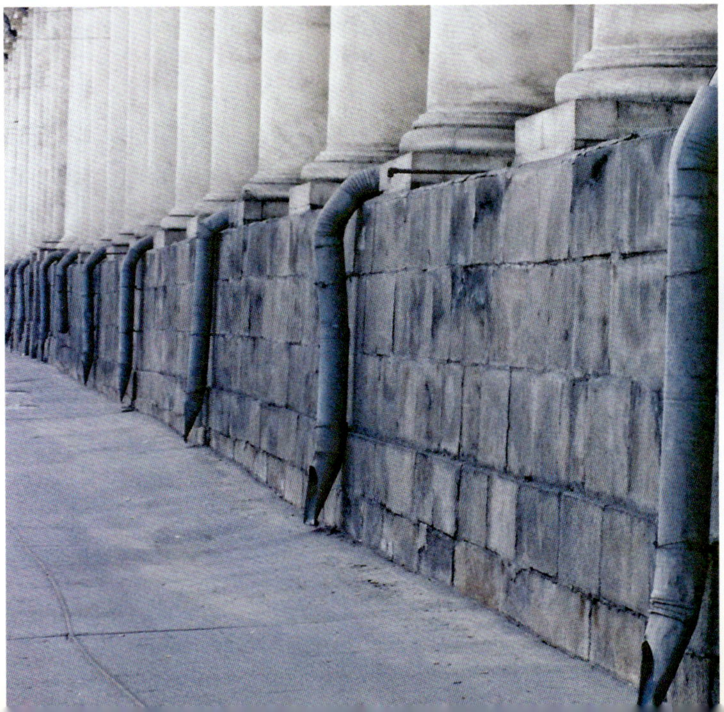

Gutters in Moscow

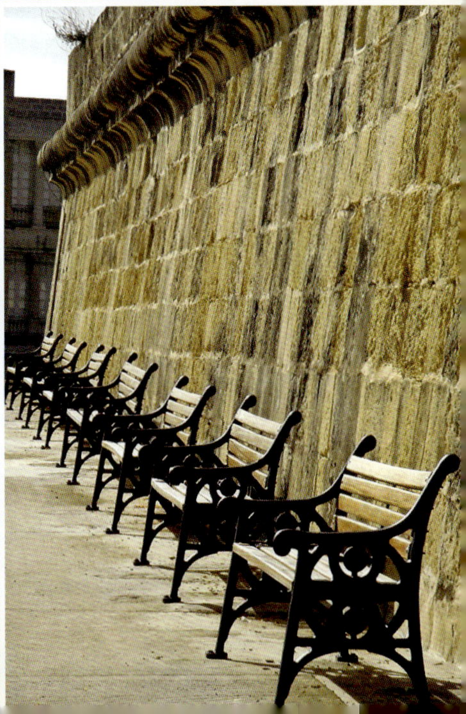

Benches in Malta

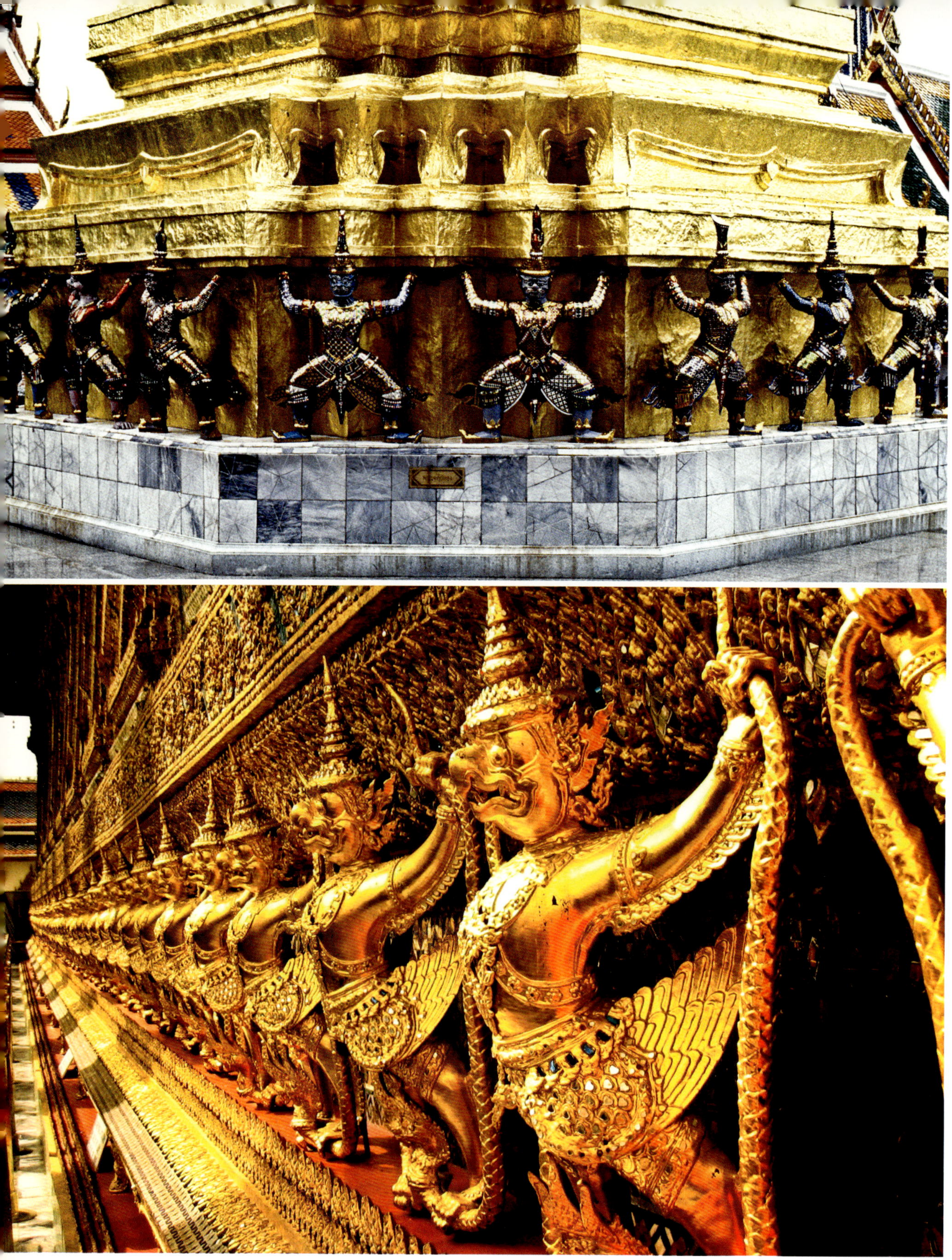

Decorations of the Royal Palace in Bangkok

With translational symmetry it is important that, in principle, the motifs should go on to infinity. Usually, as in our previous examples, this does not happen—but we can easily imagine that it might.

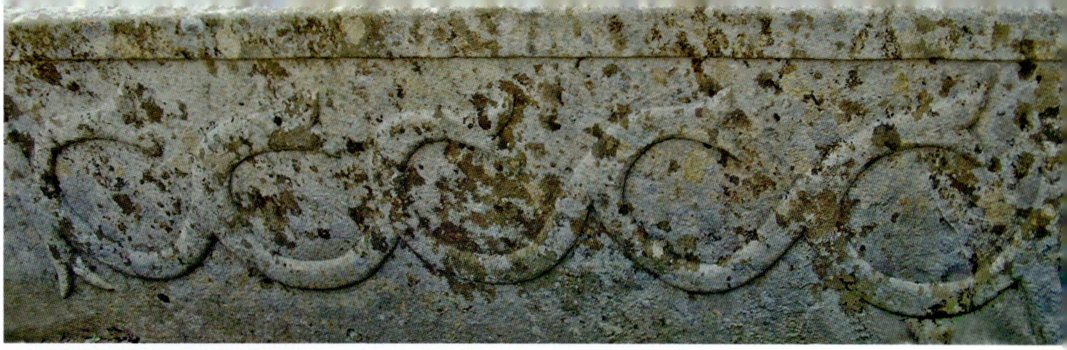

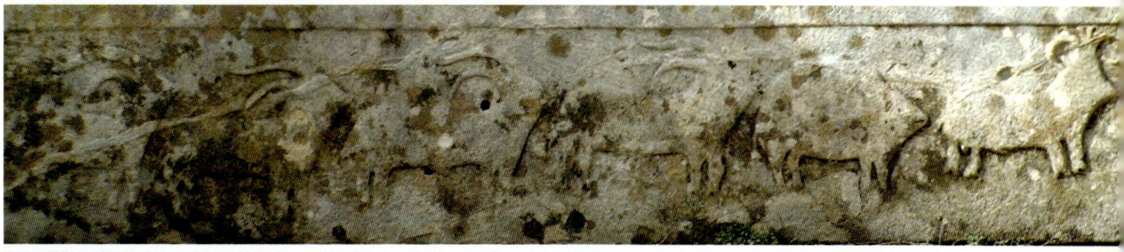

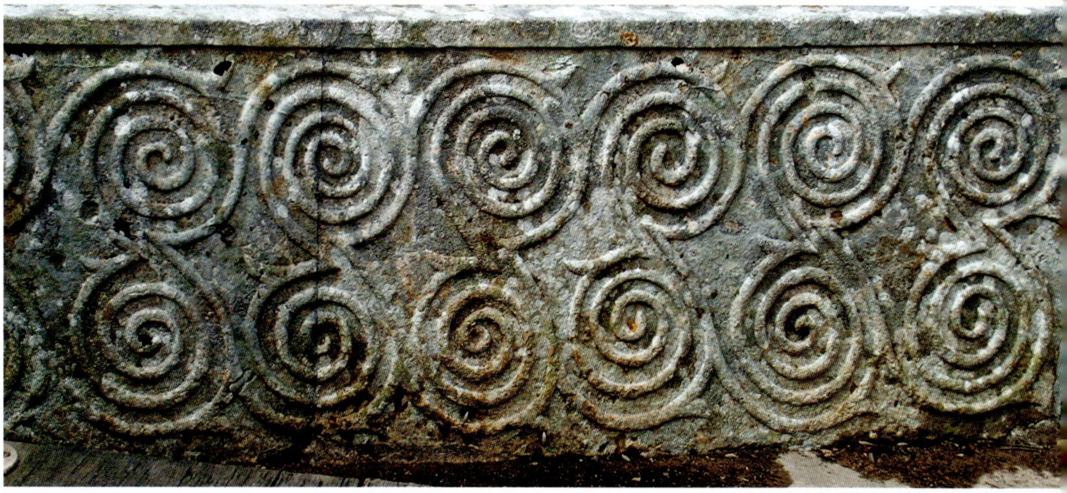

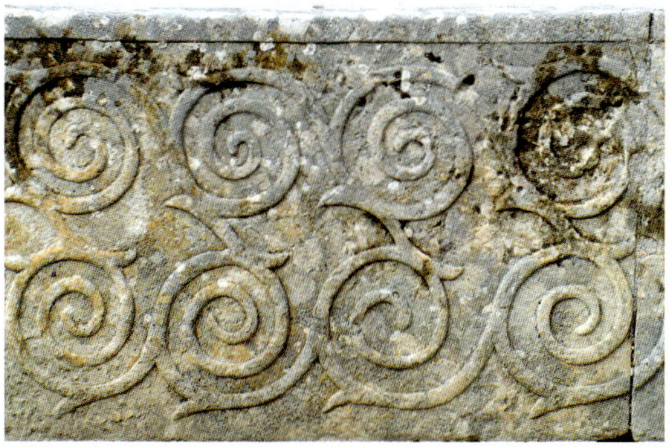

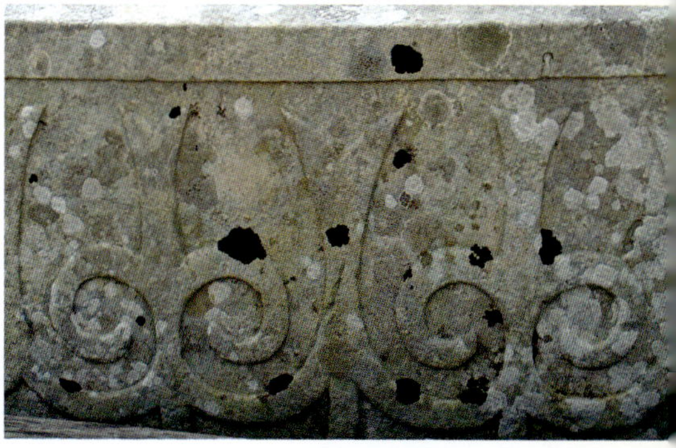

Decorations in the Tarxien temple in Malta, built c. 3000 – 2500 BCE, are early examples of translational symmetry

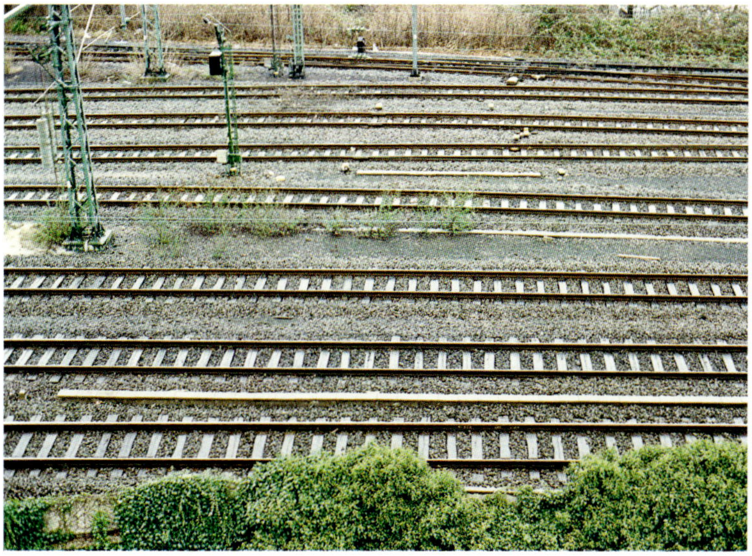

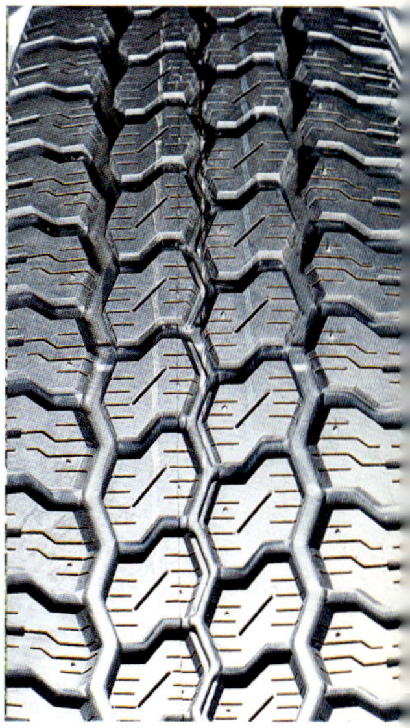

Railway tracks and tire patterns

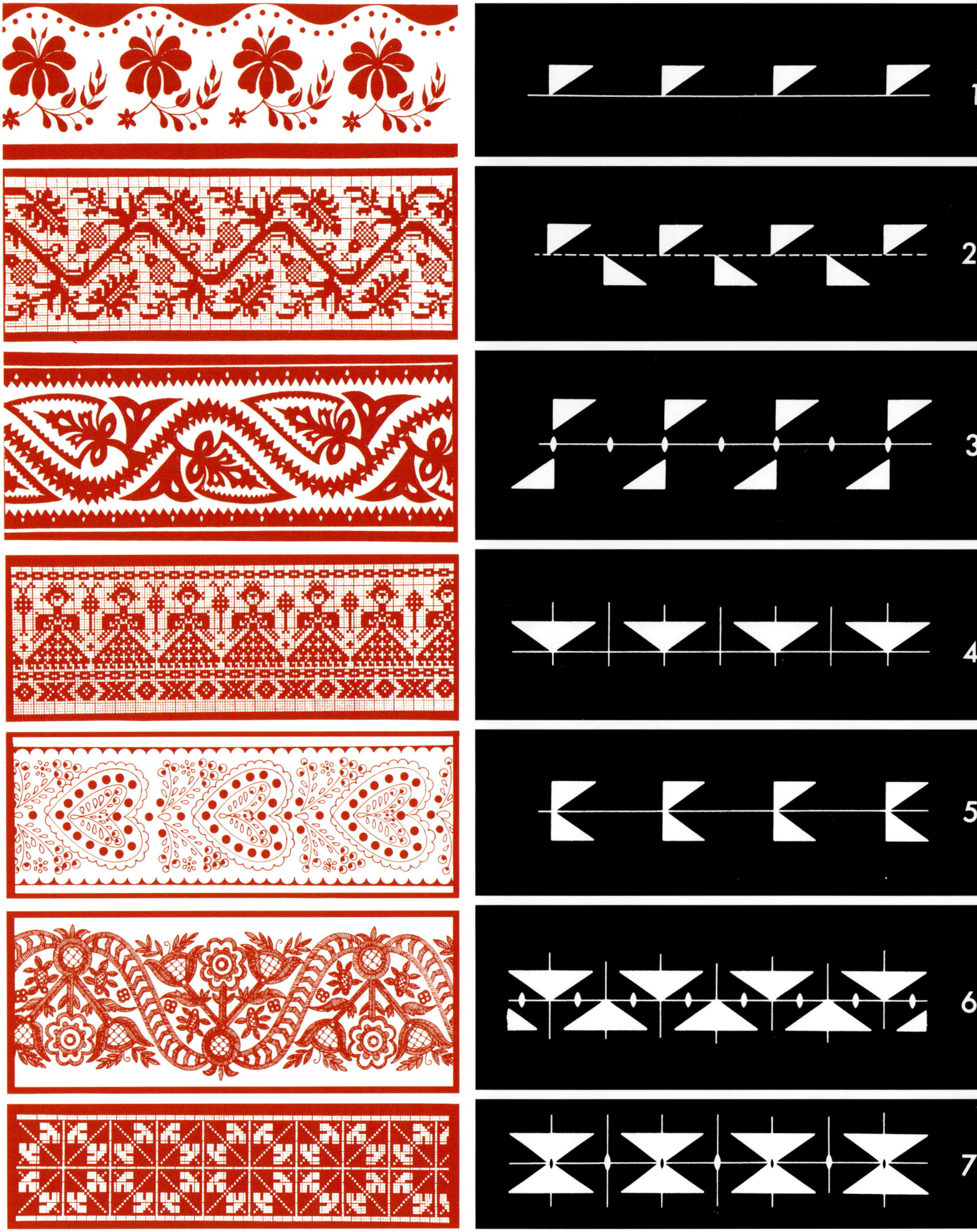
1
2
3
4
5
6
7

So far we have looked at examples of simple translational symmetry called also repetitive symmetry. There are other ways than the simple translation to achieve this kind of symmetry. Repetition may be achieved by reflection and rotation being repeated again and again.

If we have a motif and we want to repeat it in one direction, there are altogether 7 ways of doing so:

1. Repetition by simple shifting at a constant distance (we have seen many examples of this in the preceding pages).

2. Repetition by a combination of translation and horizontal reflection.

3. Repetition by consecutive twofold rotations.

4. Repetition by consecutive vertical reflection.

5. Repetition by horizontal reflection and translation.

6. Repetition by twofold rotation followed by vertical reflection.

7. Repetition by alternating vertical and horizontal reflections.

As an example of possible motifs being repeated, we offer a black triangle, and we also illustrate the patterns by a set of Hungarian needlework from the collection of the late Györgyi Lengyel (I. Hargittai, G. Lengyel, "The seven one-dimensional space-group symmetries illustrated by Hungarian folk needlework." *J. Chem. Educ.* 61 (1984) 1033–1034).

There is a new kind of symmetry operation among those enumerated above, called *glide reflection*. The simplest example is in No. 2 where the pattern is obtained by shifting the triangle and reflecting it and so on. Such symmetry may also occur as a consequence of combining two symmetry operations that we had already been familiar with. In No. 6, twofold rotation is followed by vertical reflection, which can also be visualized as gliding the double triangle and then reflecting it. Our footprints in wet sand show a nice example of glide reflection.

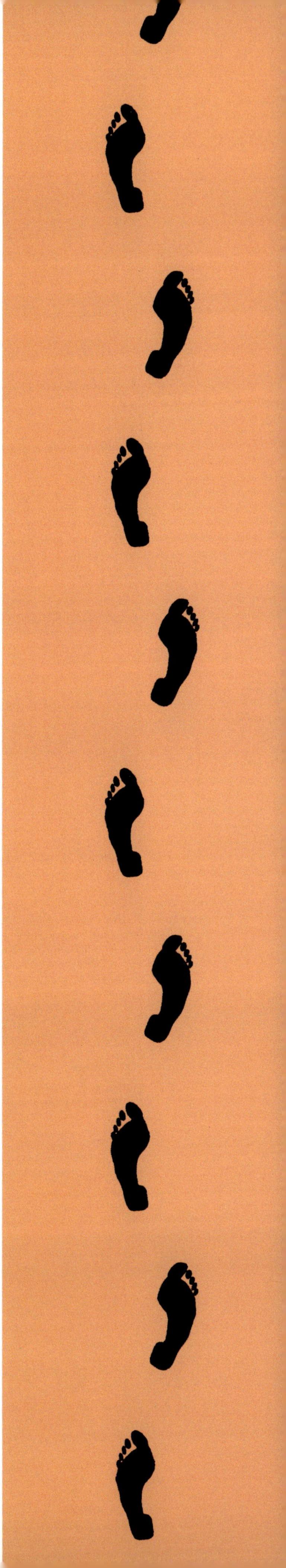

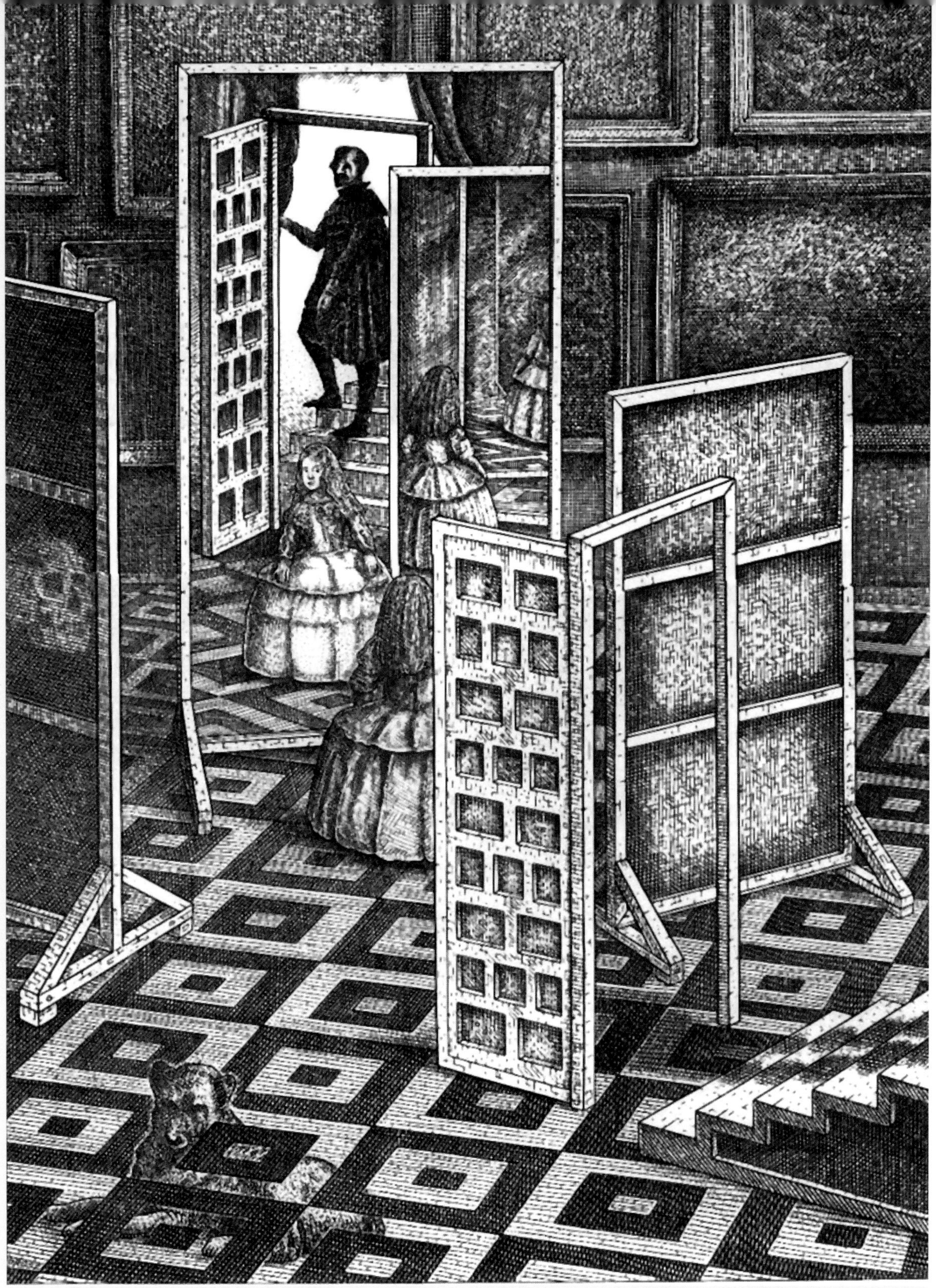

István Orosz, ***Velázquez spectaculum***

(courtesy of the artist)

An interesting artistic example of repetition by reflection is the picture titled *Velázquez spectaculum* by István Orosz (Budakeszi, Hungary). The back of the little princess is seen in a series of mirrors in the picture inspired by the famous painting *Las Meninas* by Velázquez.

Similarity Symmetry

Similarity symmetry is a kind of repetitive symmetry, but it is a special kind in that simultaneously with repetition, there is a gradual change in size.

In the painting by Lima de Freitas (Portugal), the size of the picture diminishes. Gradual but more-or-less regular changes of other properties may also accompany the repetition.

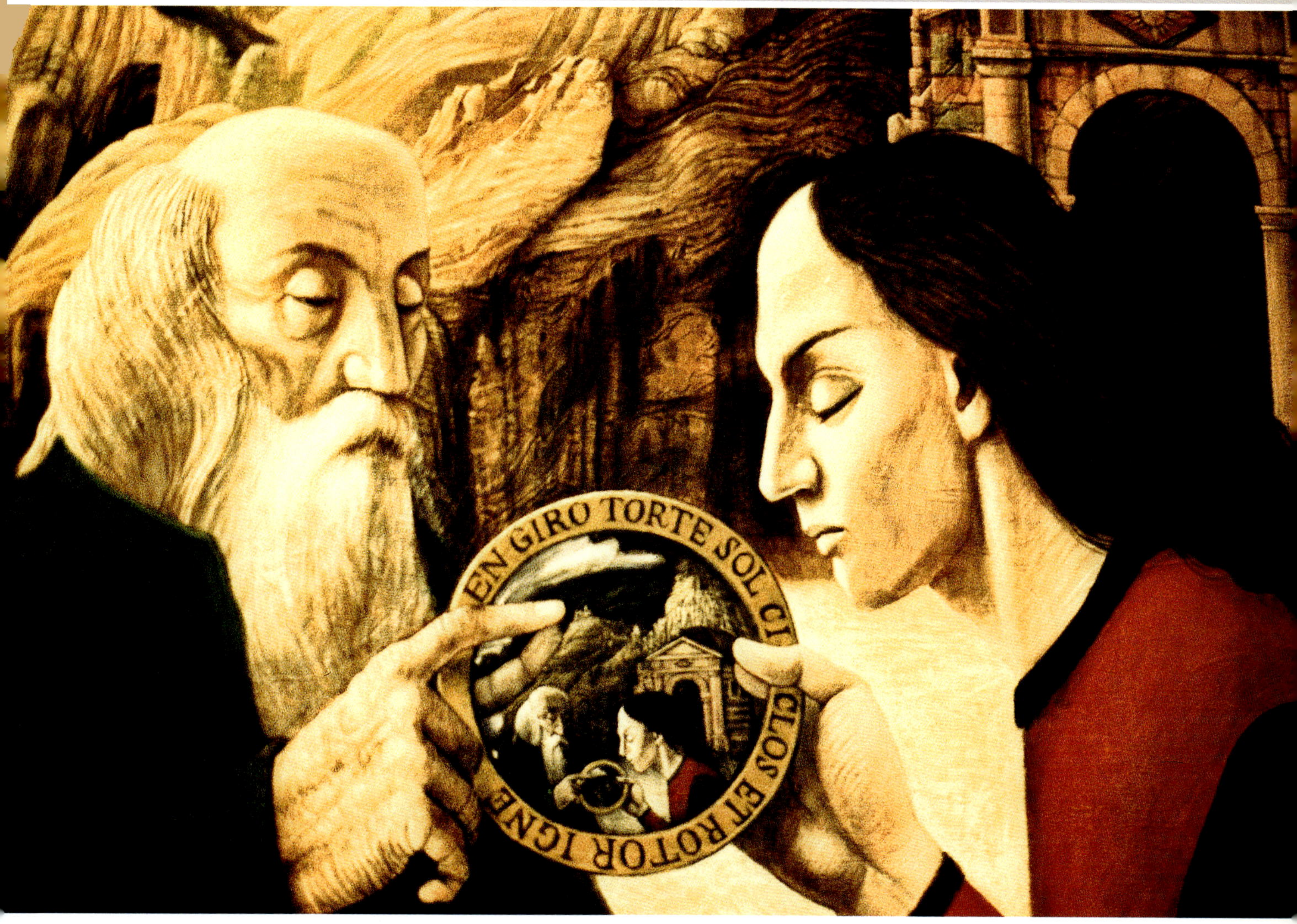

Painting by Lima de Freitas, Portugal

(courtesy of the artist)

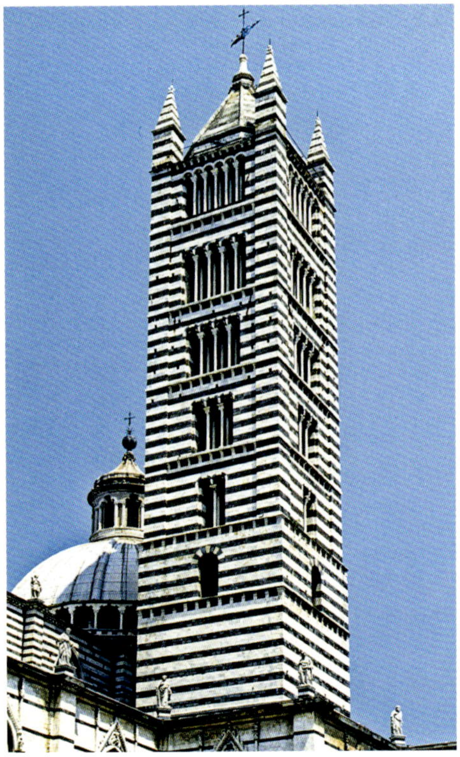

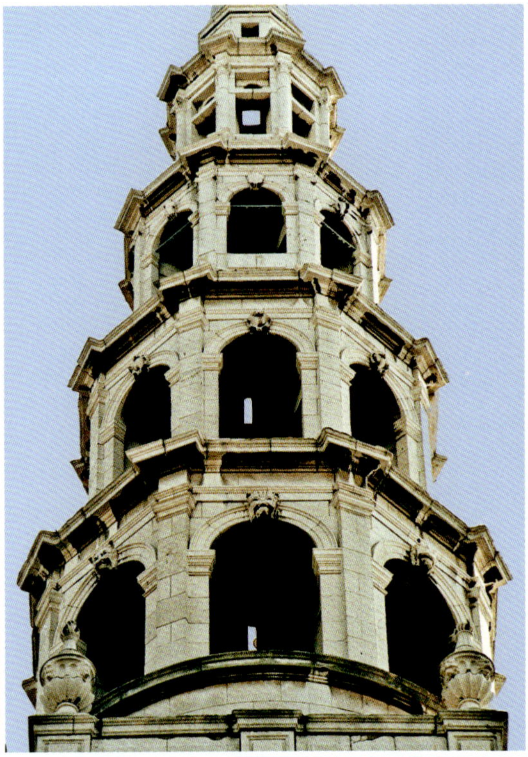

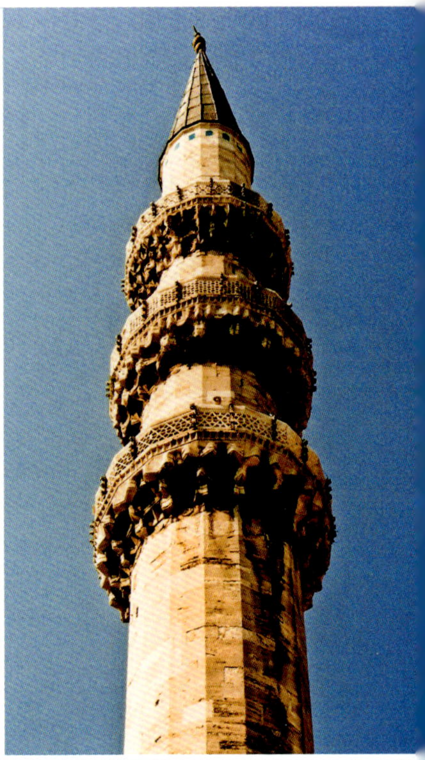

Towers (from upper left, clockwise) in Florence, Italy; London; Istanbul; and Bangkok (bottom two)

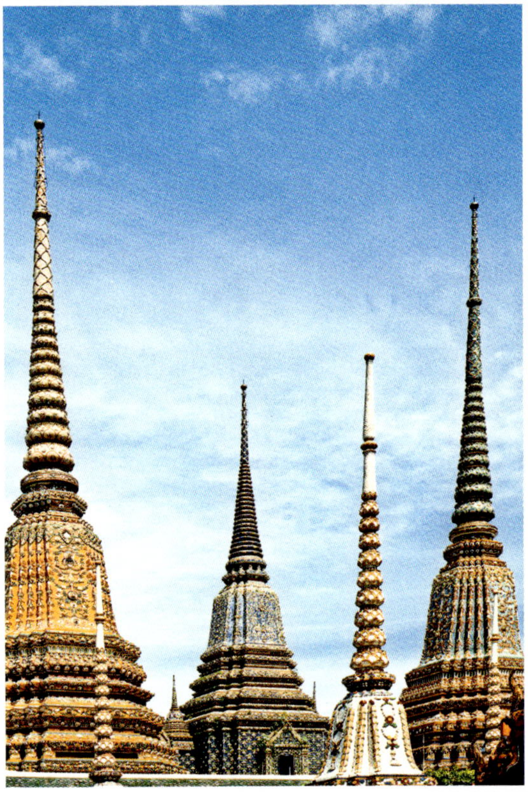

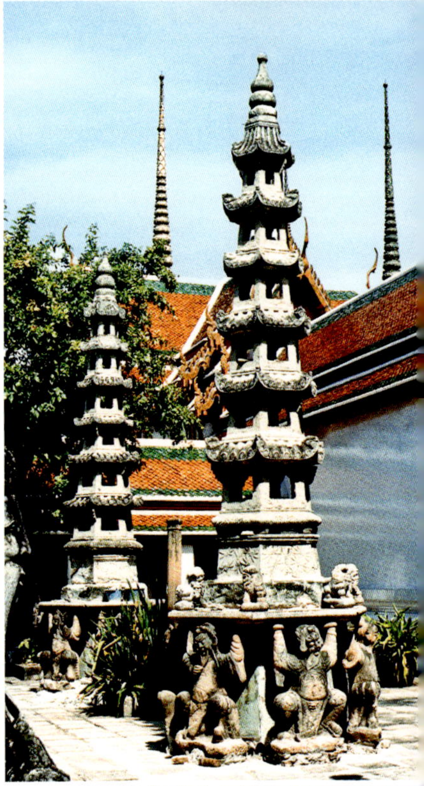

The Russian *Matrioshka* dolls are another example of similarity symmetry

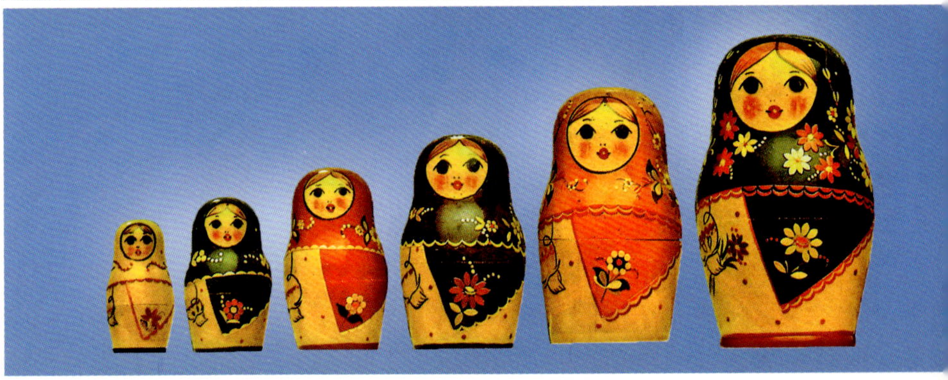

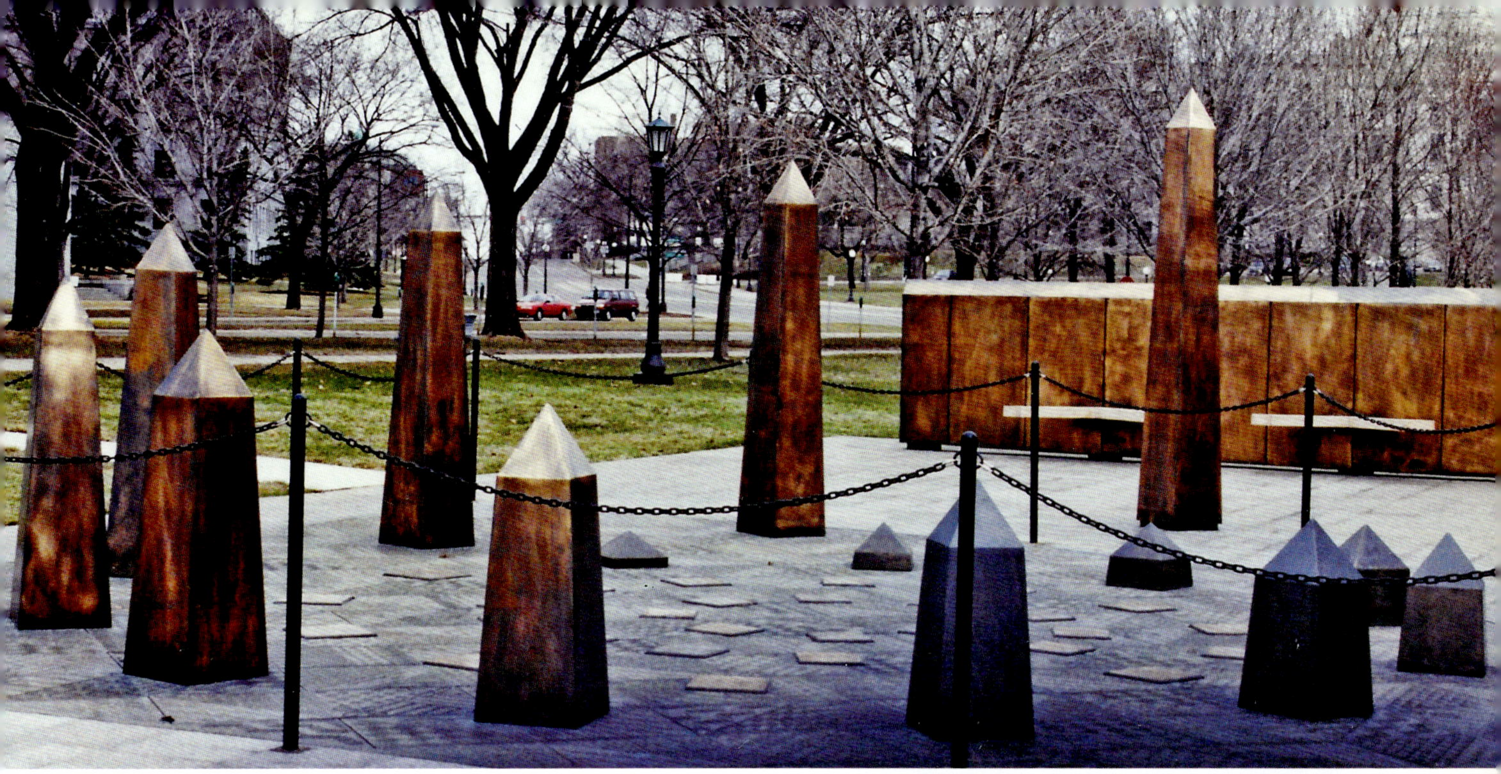

Open-air sculpture in St. Paul, Minnesota

The set of columns with tetragonal pyramid tops in St. Paul, Minnesota, provide yet another example of similarity symmetry.

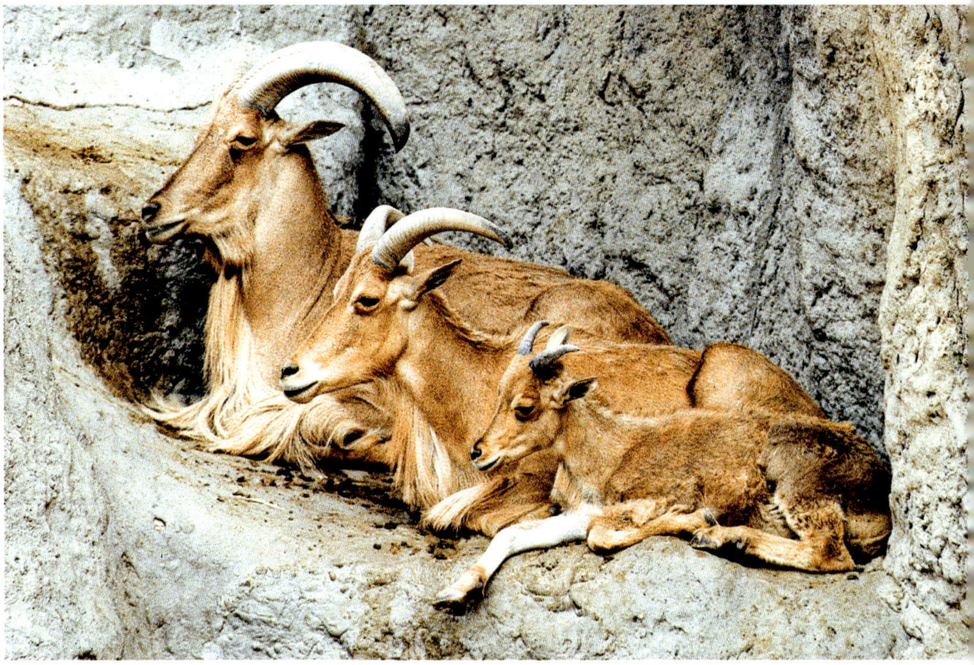

Mountain goats in the Budapest Zoo

The mountain goats in the picture show size changes (even though the changes are hardly regular). Apart from the size, age could also be considered as the property changing gradually among these animals. In order to apply symmetry considerations to the group of mountain goats, we can extend their set to infinity in our thoughts, in either direction.

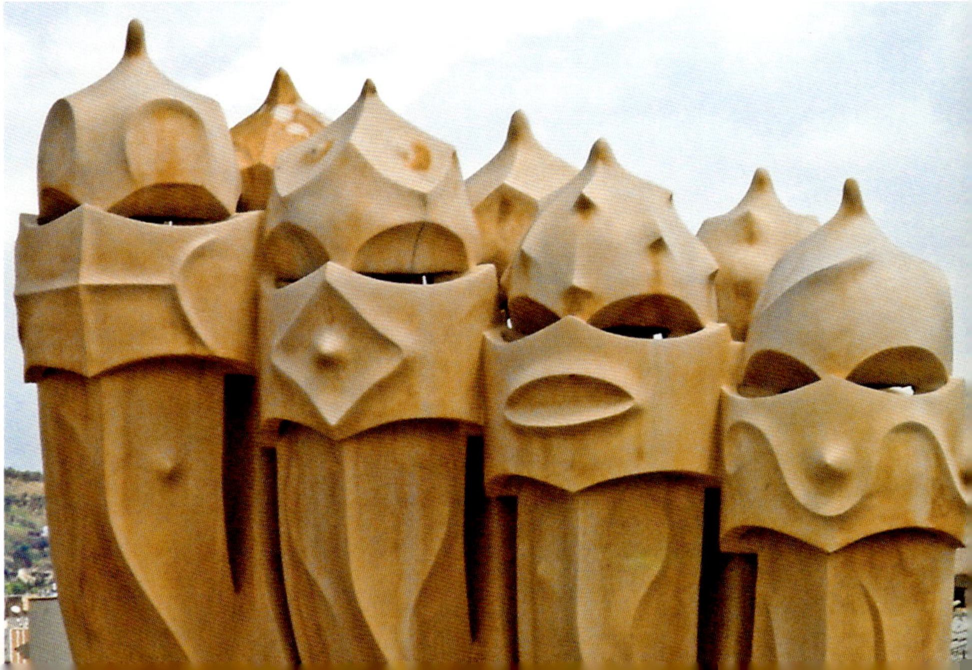

Sculptures by Antoni Gaudi in Barcelona

The Gaudi sculptures in Barcelona also exhibit similarity symmetry; not only in their size but also in the variation of the faces and heads of the figures.

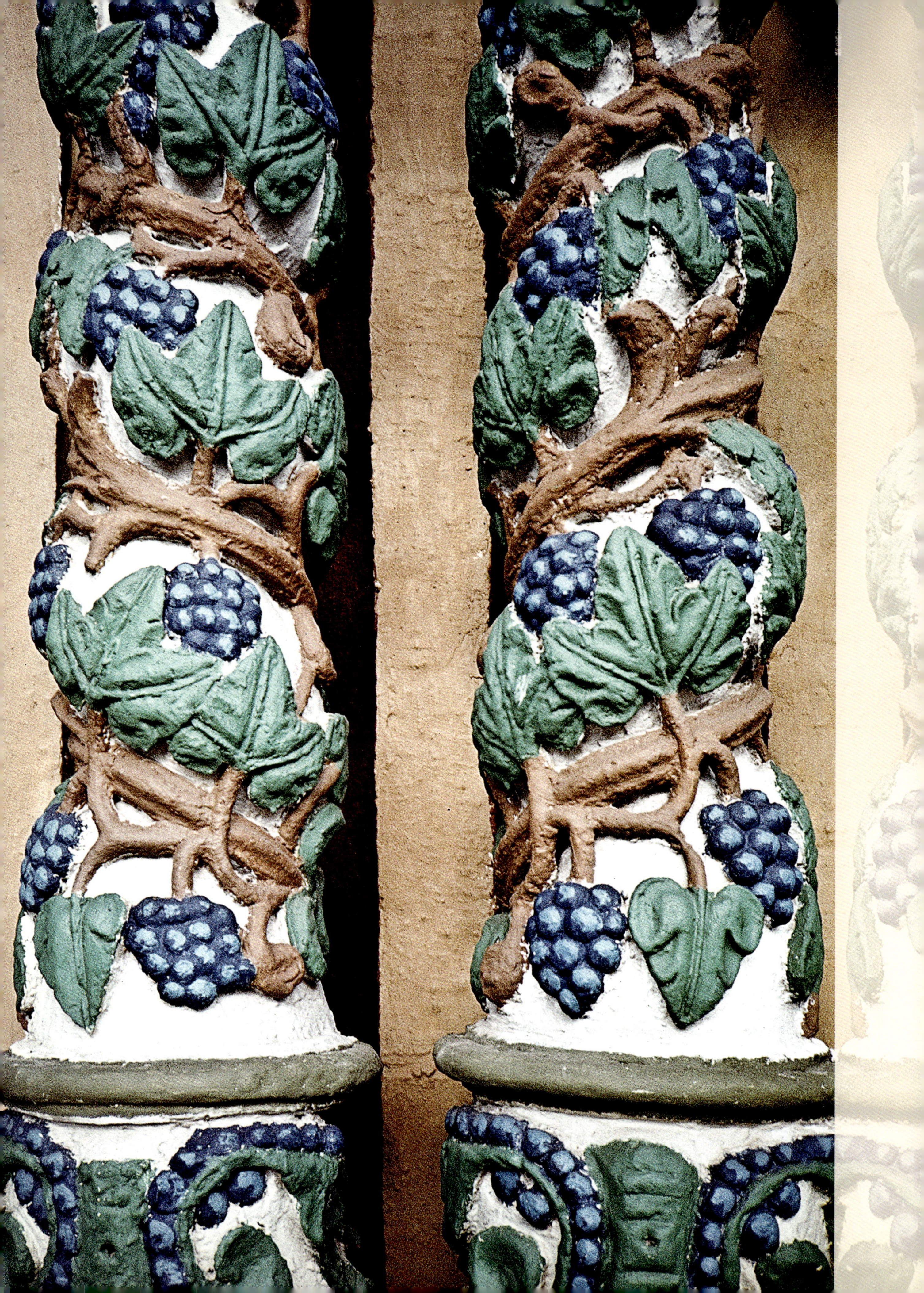

Helical symmetry is also repetitive symmetry, but in this case translation is combined with rotation. Helical symmetry implies regular amounts of translation accompanied by regular amounts of rotation. The name helical also implies a constant diameter of the object. The term spiral staircase demonstrates some inaccuracy in language. We will see that in contrast to the helix, the spiral has a gradually changing diameter. The spiral staircase, however, has the attributes of a helix.

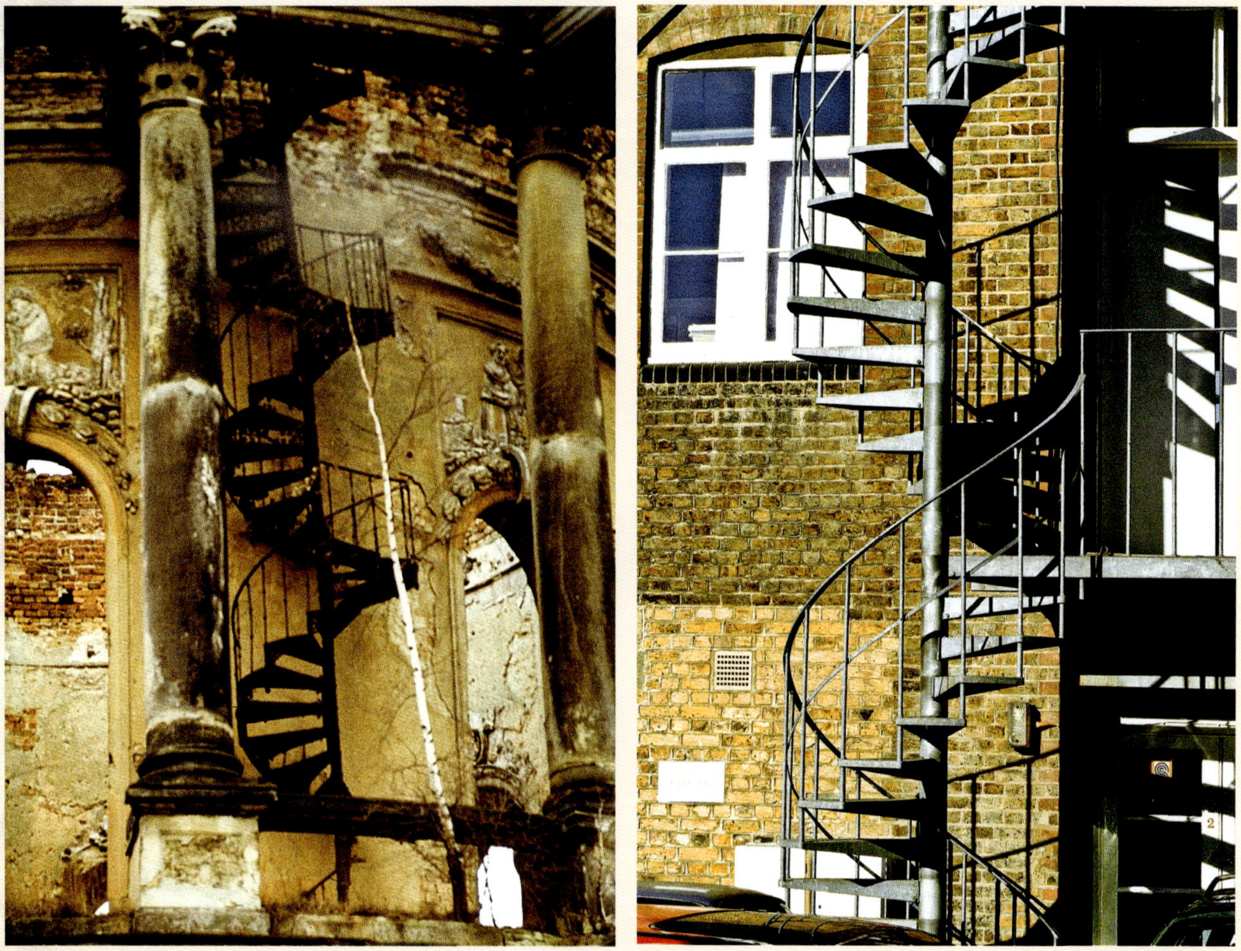

Spiral staircases in Potsdam, Germany (left) and in London

(The Potsdam example came from the inside of a palace that was bombed out during World War II; it remained in ruins for decades after the war, but has probably been restored since.)

Helical Symmetry

Columns at an Eastern Orthodox monastery in Zagorsk, Russia

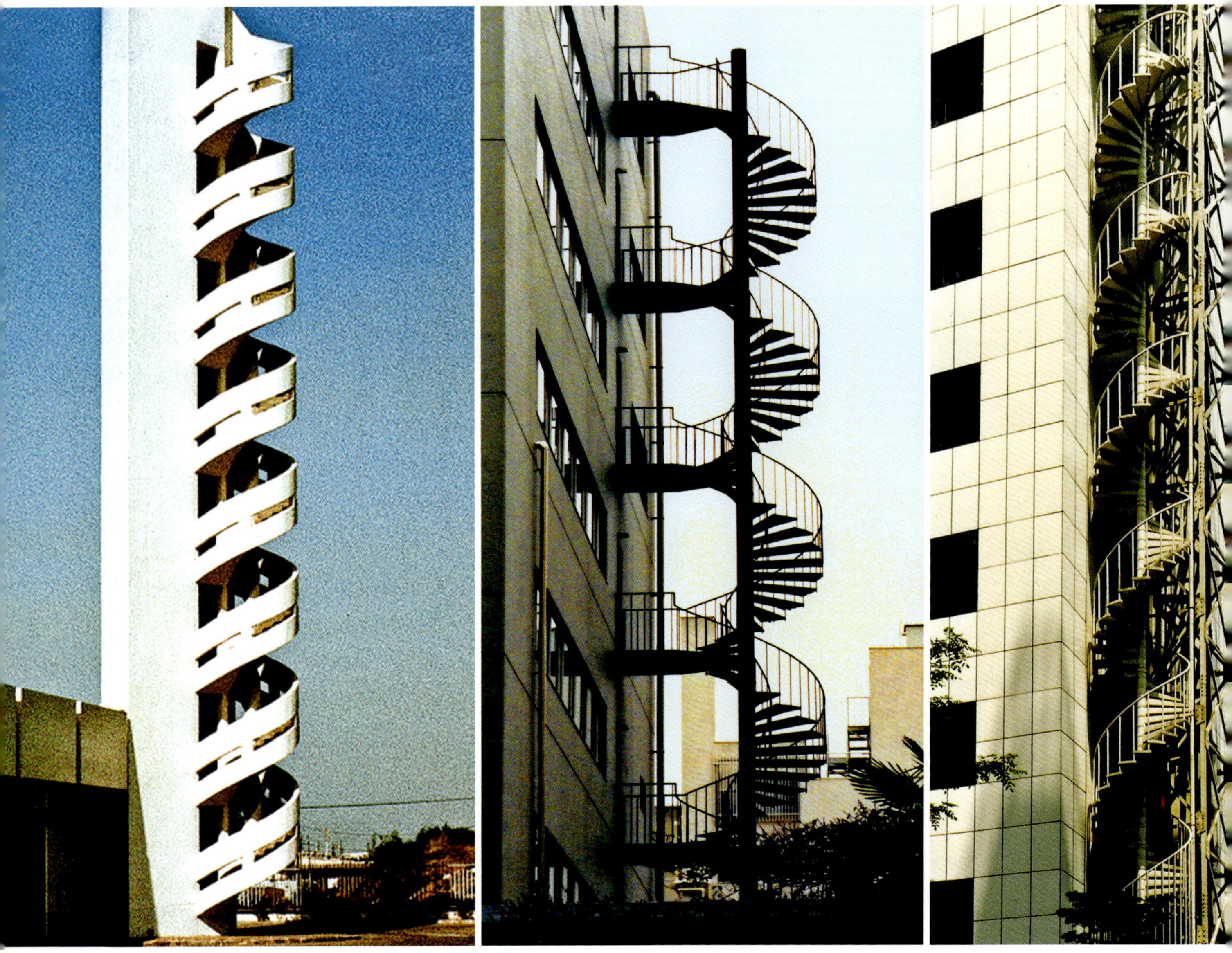

Spiral staircases (from left to right) in Tel Aviv; Kyoto; and Seoul

Spiral staircase in Cambridge, UK

Its shadow also provides a beautiful image of a spiral staircase. This is why we can say that this picture shows a doubled helix (though not a double helix). Incidentally, the building at whose side this staircase is, houses the Laboratory of Molecular Biology. The double helix structure of deoxyribonucleic acid, DNA, was discovered in its predecessor research unit.

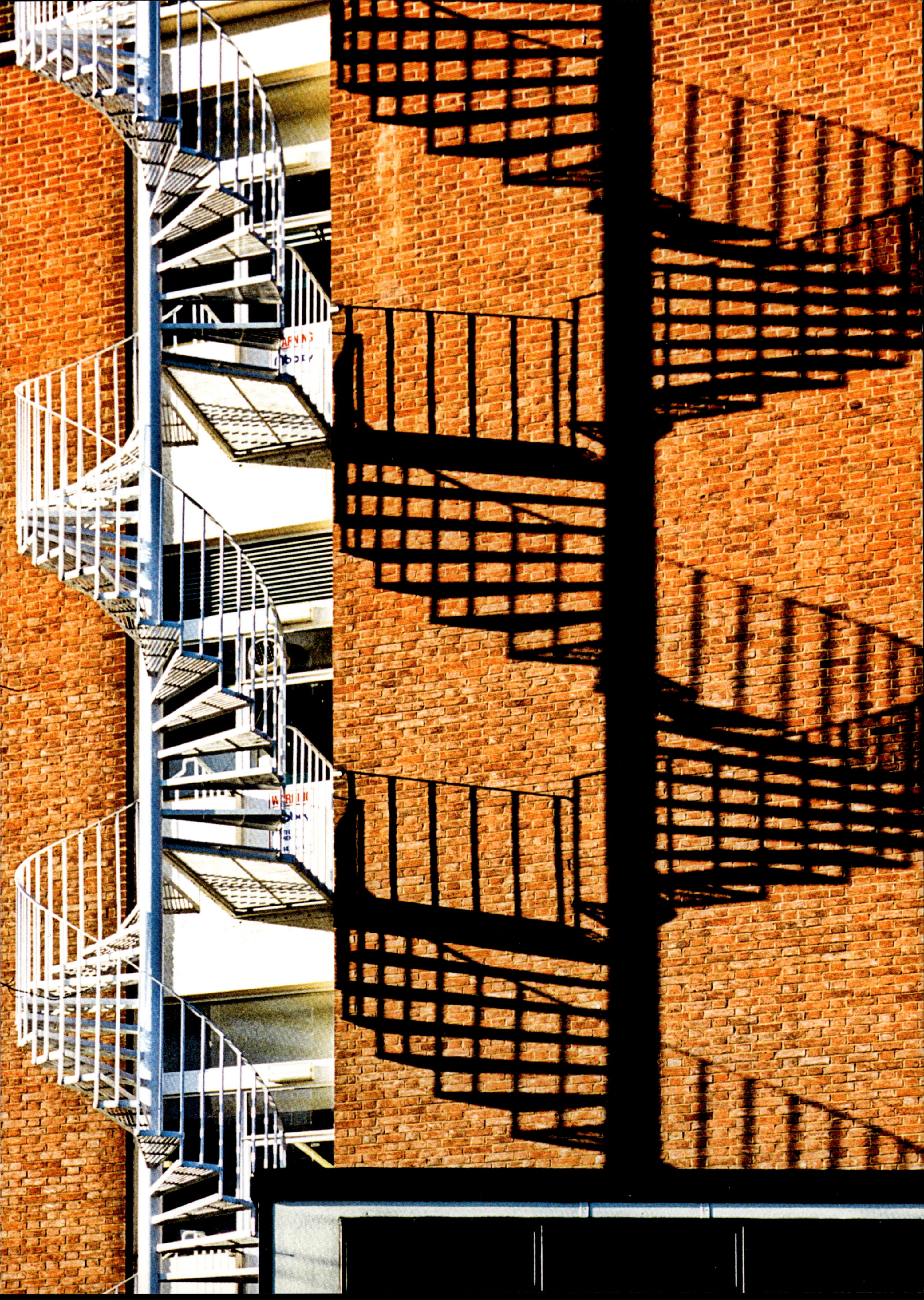

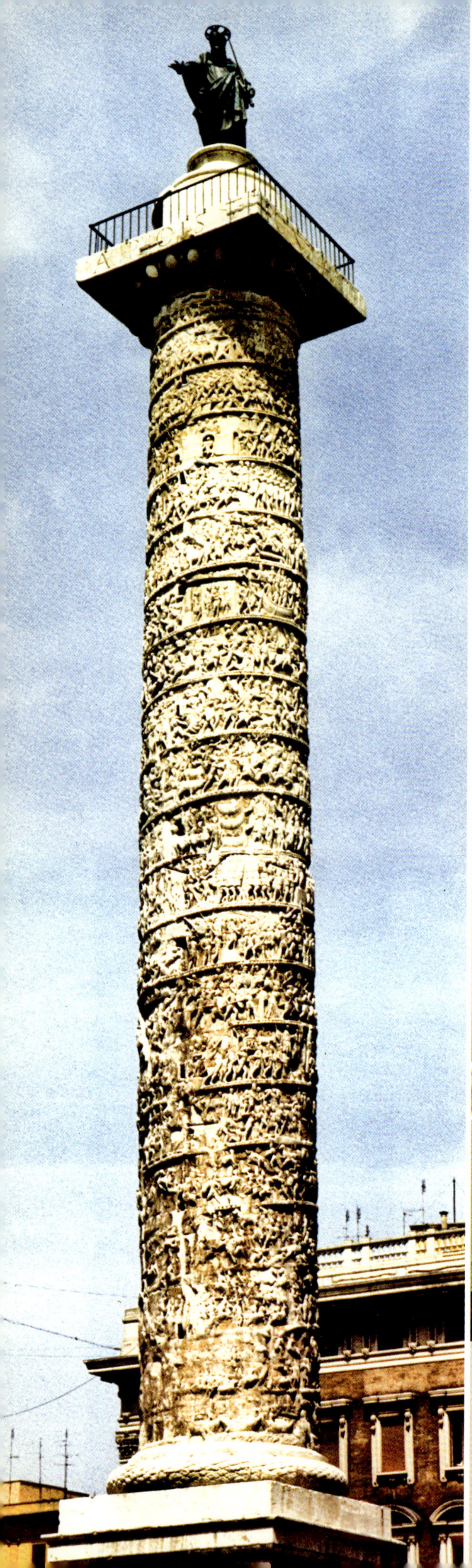

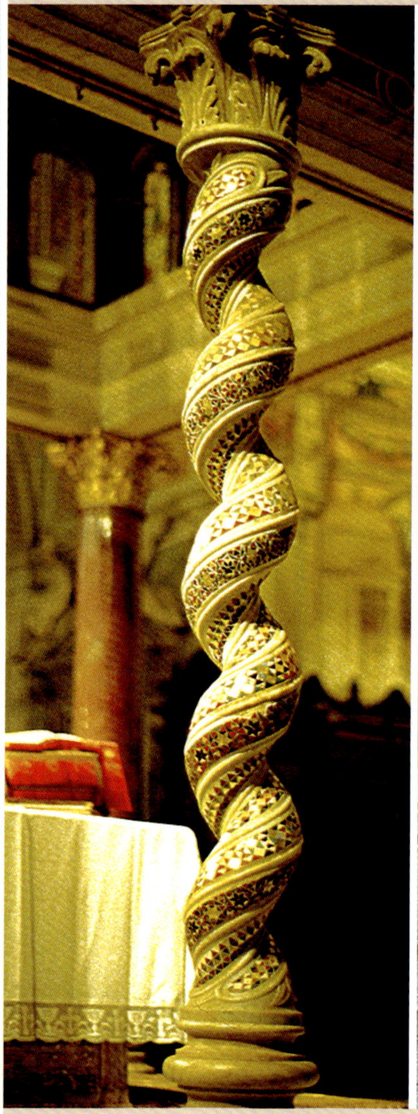

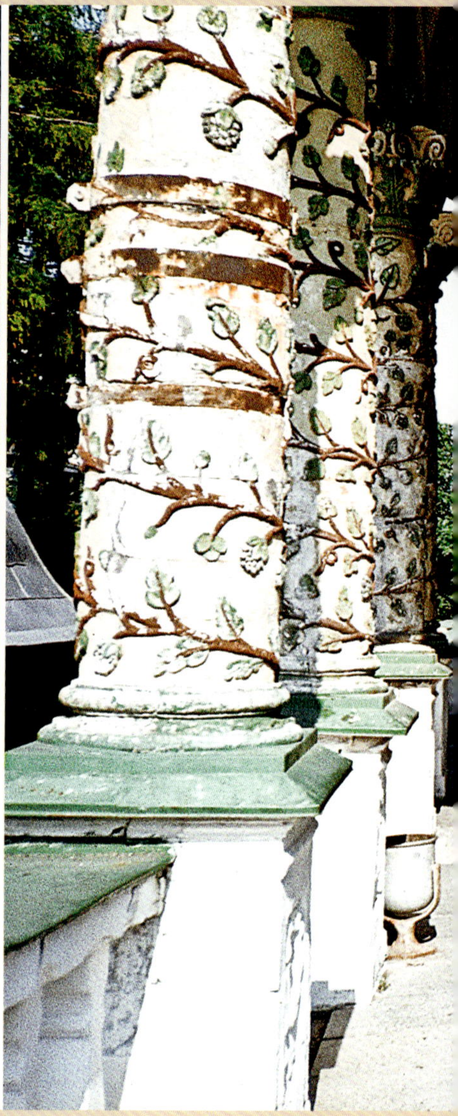

Helical column for interior decoration in a church in Rome

Trajan's Column in Rome

Helices may have directionality. They may be "right-handed" or "left-handed." The chapter-opening image of the Zagorsk columns provide a beautiful illustration for the handedness of helices. This property is "vital" for biologically important macromolecules.

Right top and bottom:

Columns in a Rome monastery

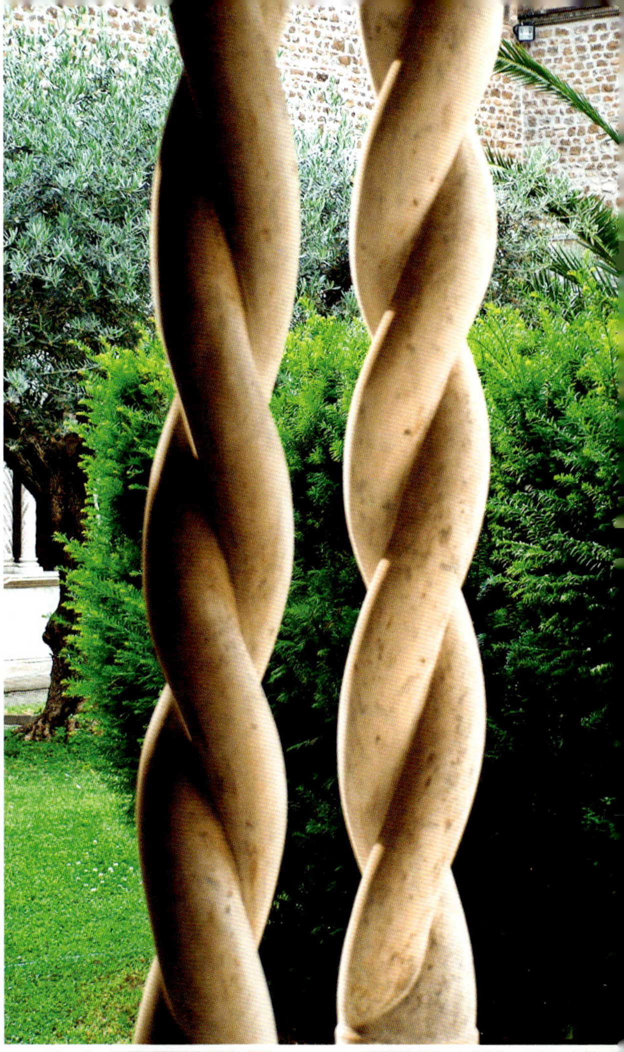

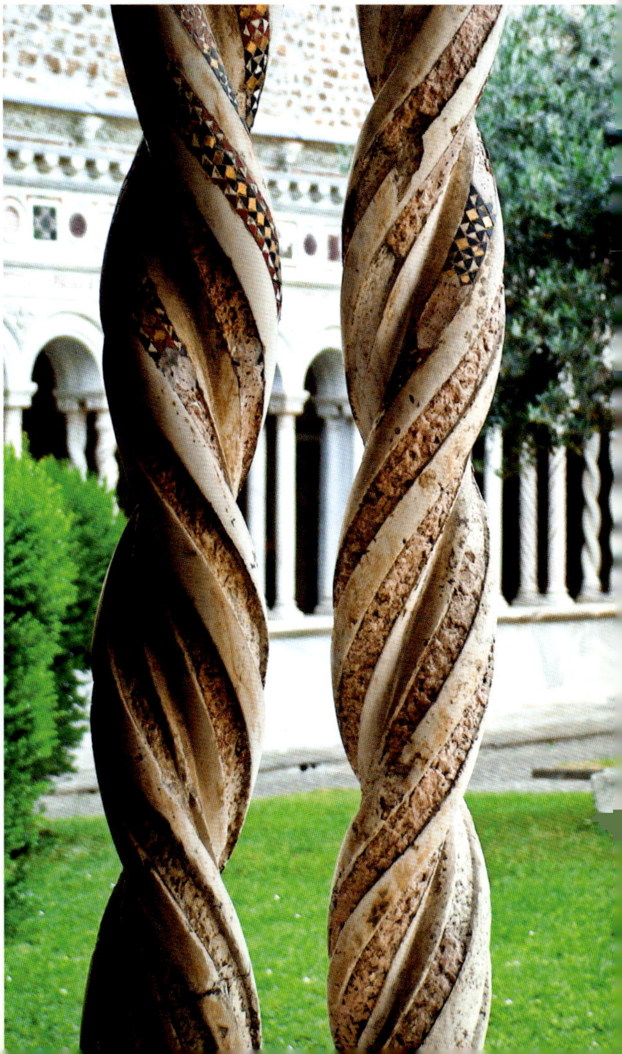

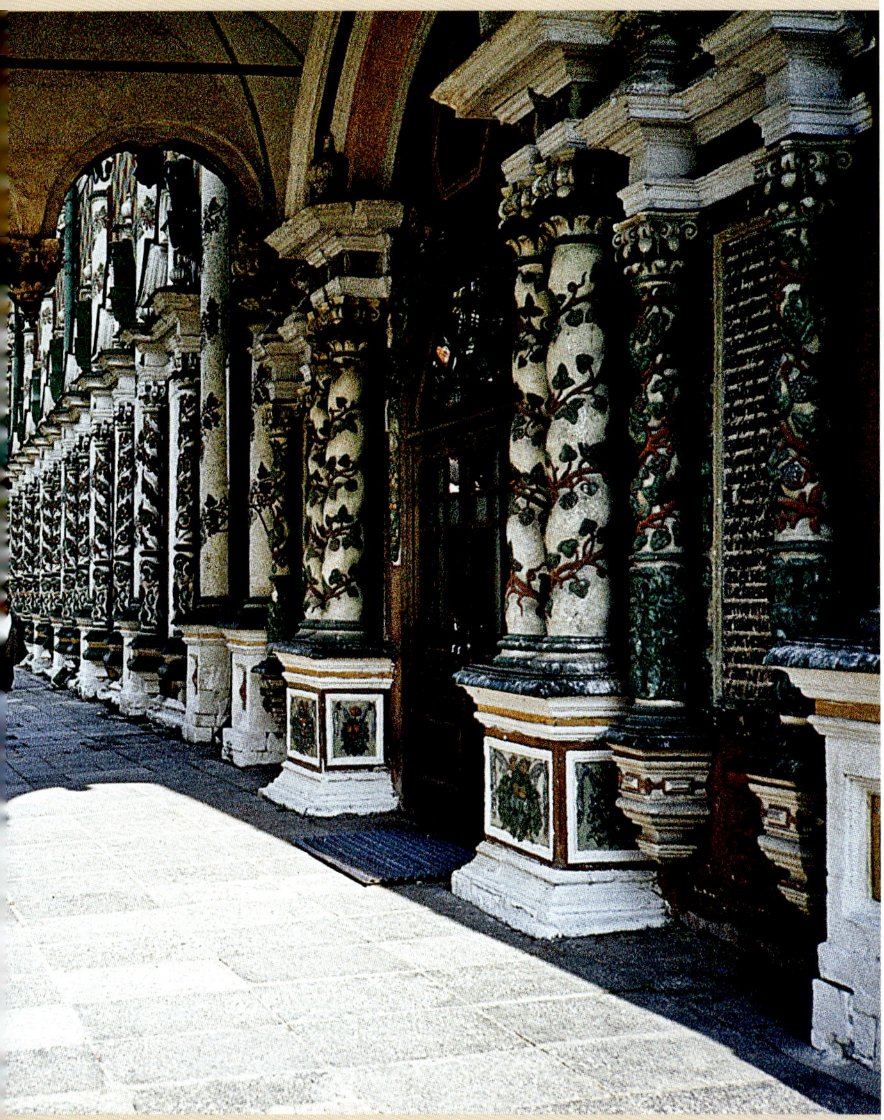

Columns of the Eastern Orthodox monastery in Zagorsk, Russia

Proteins and Nucleic Acids

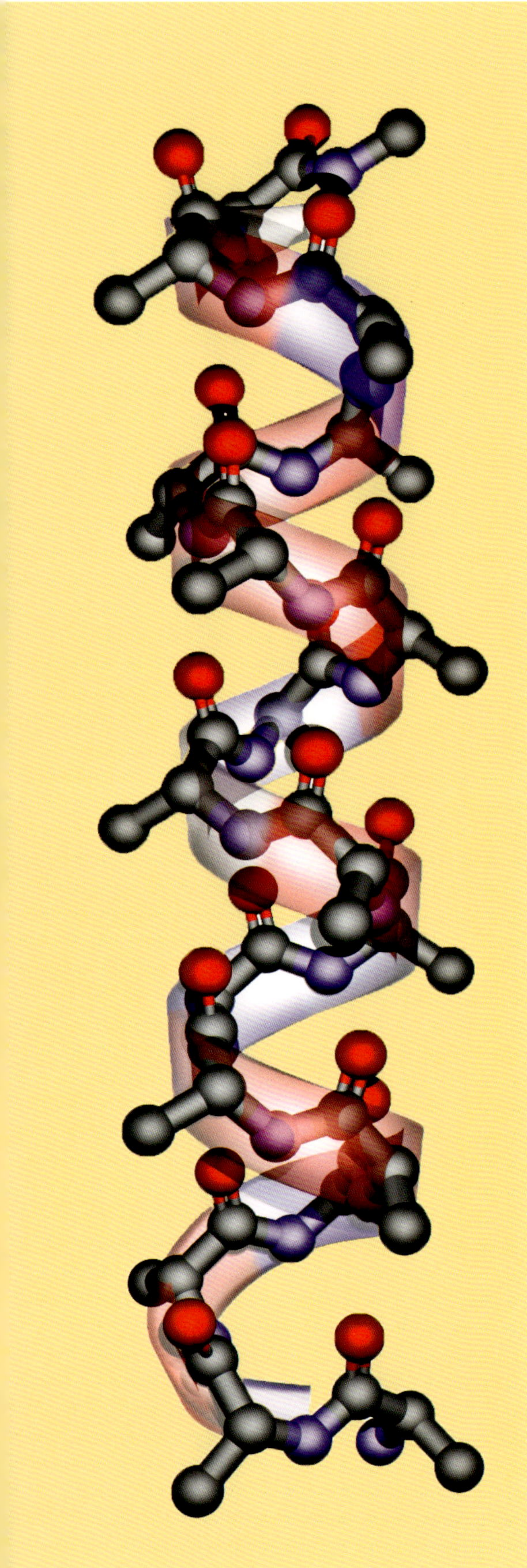

The biologically important macromolecules have helical structures. Linus Pauling made a ground-breaking discovery when he proposed a model, called the alpha-helix, for protein structures. It was soon followed by James Watson and Francis Crick's suggestion for a double helix structure of DNA.

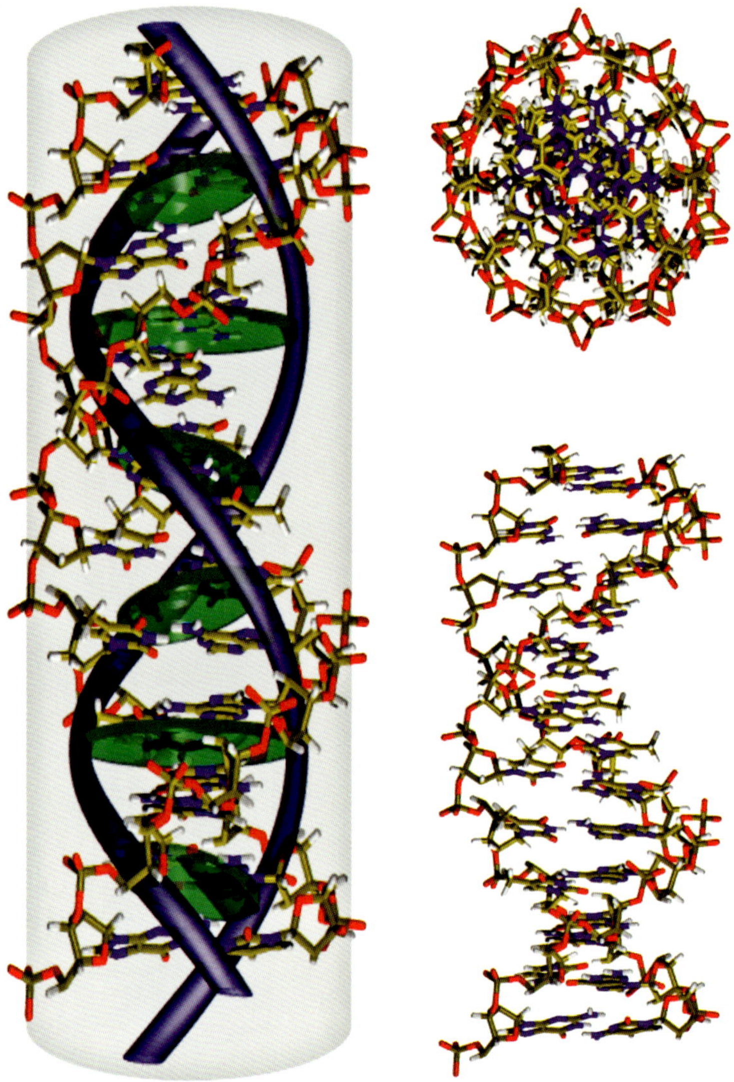

Left: Alpha-helix model of proteins

(courtesy of Ilya Yanov, Jackson, Mississippi)

Right: Models of DNA in three variations represented by computer graphics

(courtesy of Ilya Yanov, Jackson, Mississippi)

Due to the importance of DNA, as the substance of heredity, and the aesthetic appeal of its structure, the double helix has become a popular subject for sculptors.

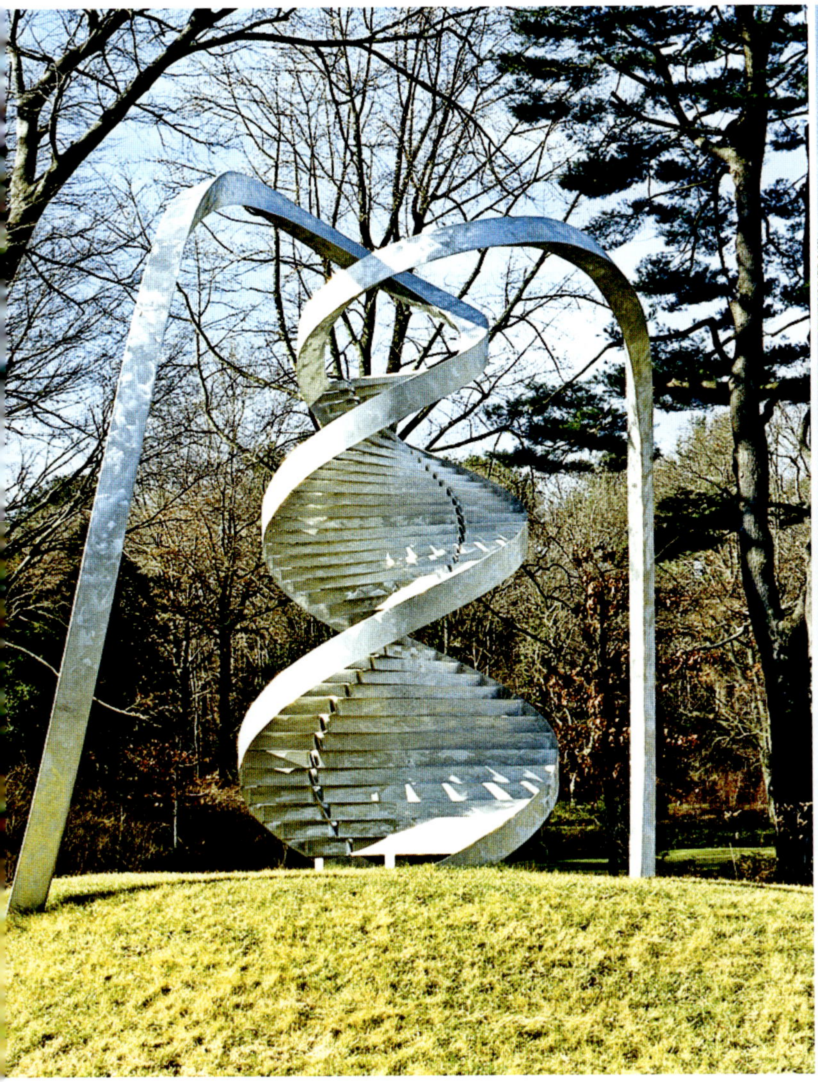

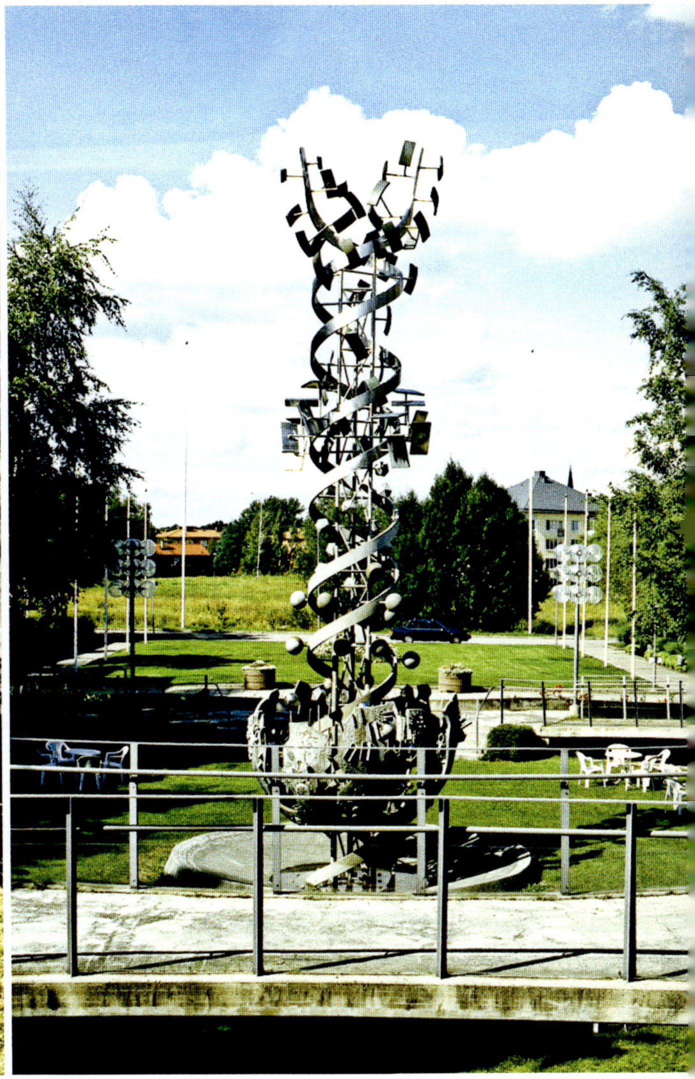

Left: *Spirals Time – Time Spirals* by Charles A. Jencks at Cold Spring Harbor Laboratory in Cold Spring Harbor, New York

Right: *The Double Helix* by Bror Marklund at Uppsala University in Uppsala, Sweden

Top: Interior spiral staircase in an office building in Budapest

Bottom: Double-helical spiral staircase at the Vatican Museum in Vatican City

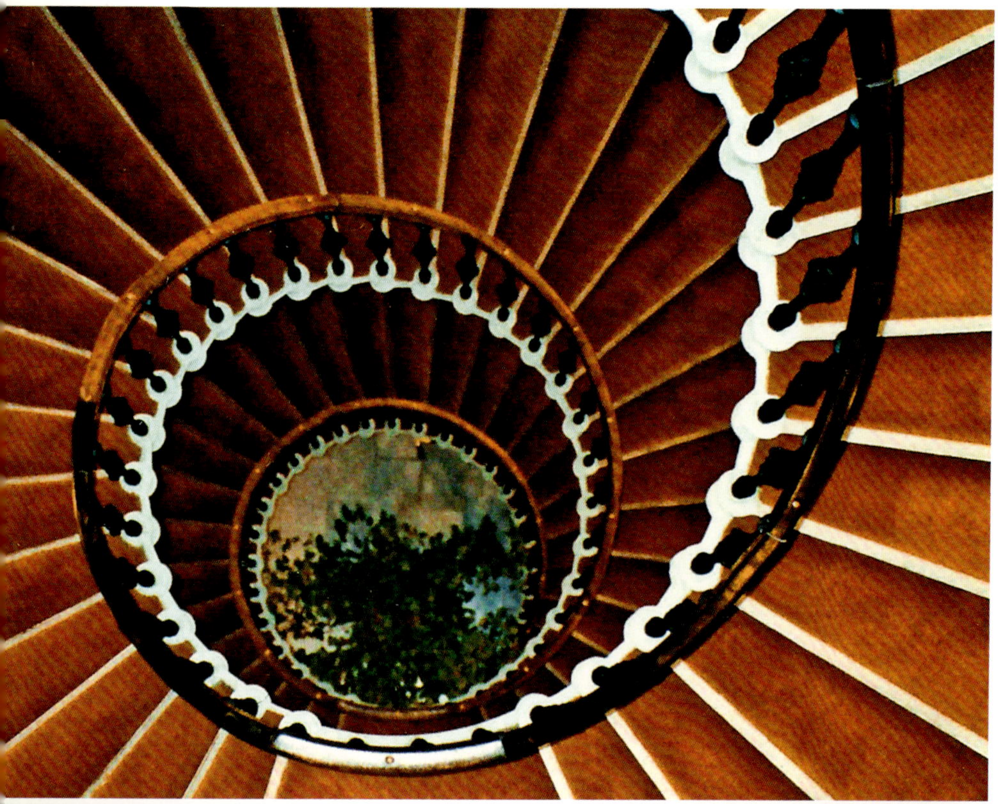

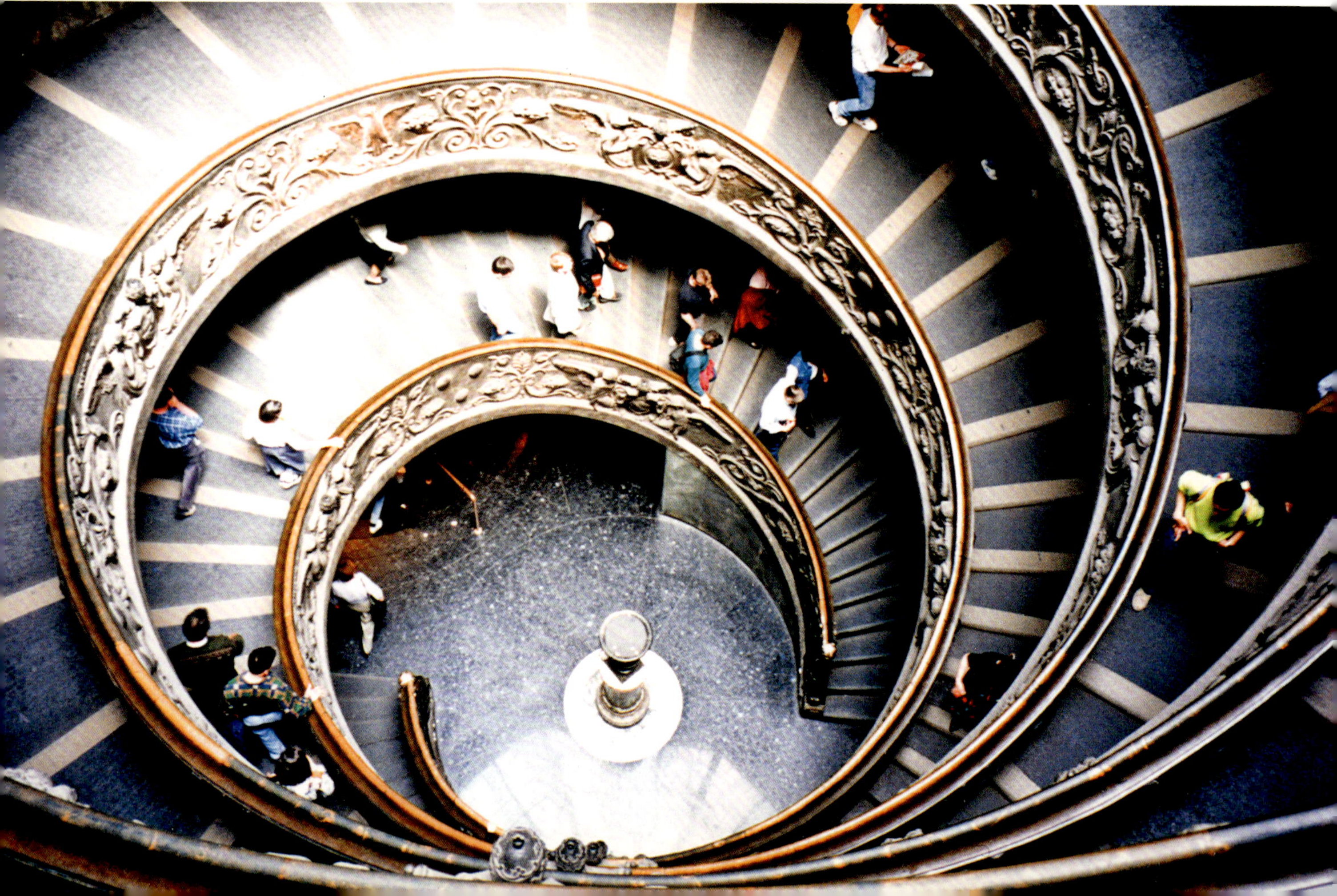

As we have mentioned, the spiral staircases are not spirals in the strict geometrical meaning of this term; rather, they are helices even though when looking at spiral staircases in perspective, they appear indeed as spirals. There is a true double spiral as a sculpture in the garden of the Weizmann Institute in Rehovot, Israel.

Spiral staircase at the Norwegian University of Science and Technology in Trondheim

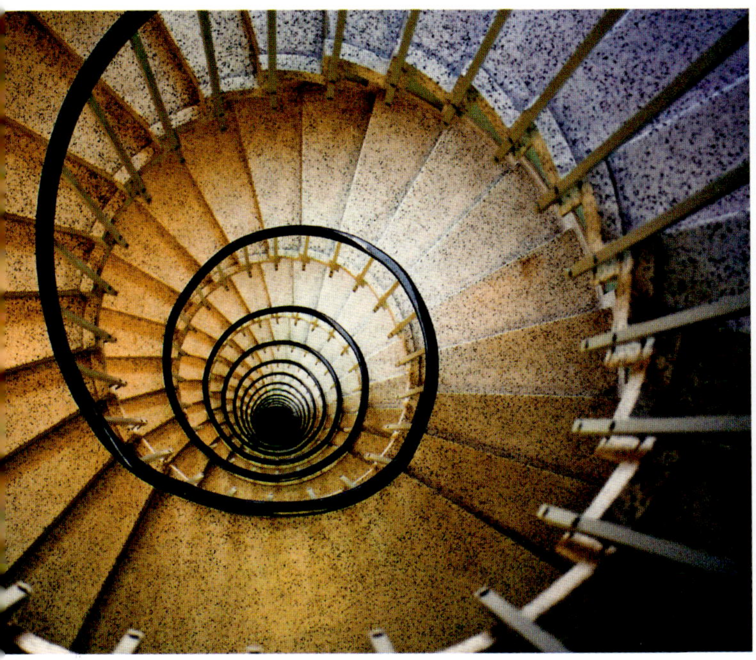

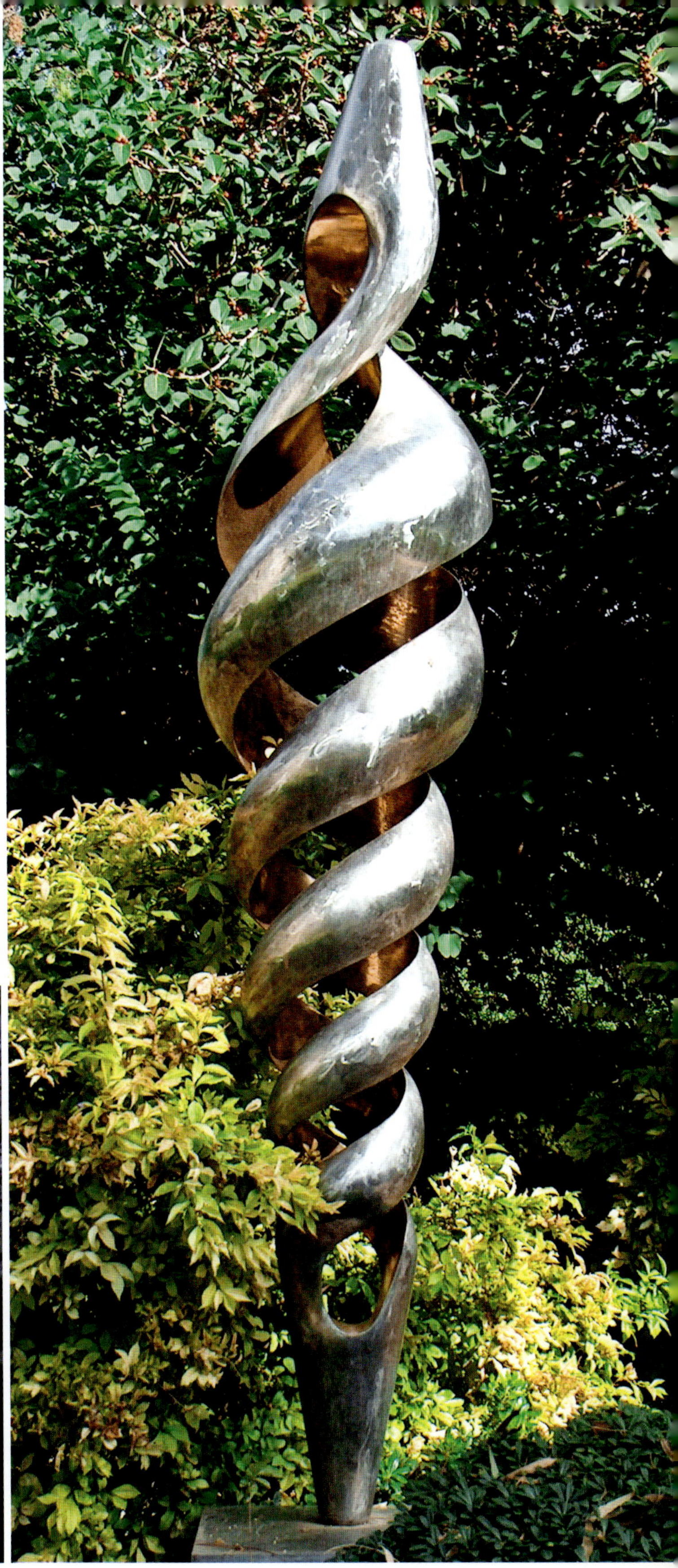

Double spiral sculpture at the Weizmann Institute in Rehovot, Israel

Two-dimensional Spirals

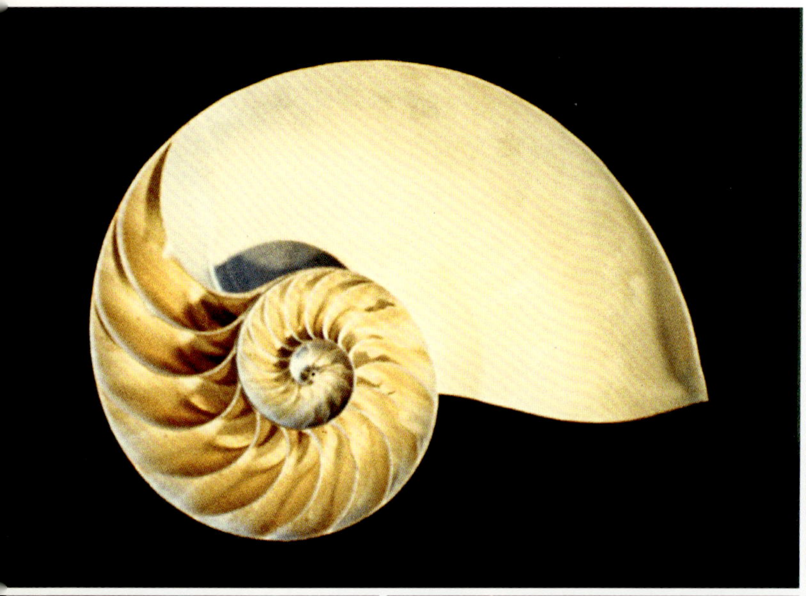

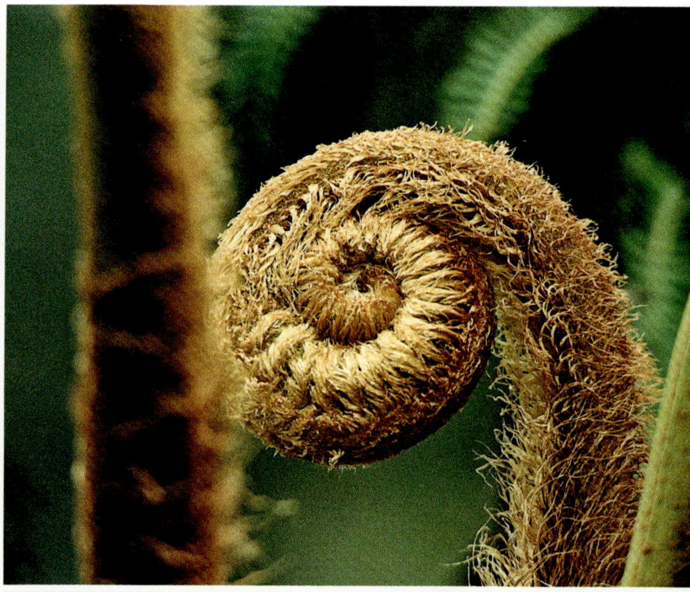

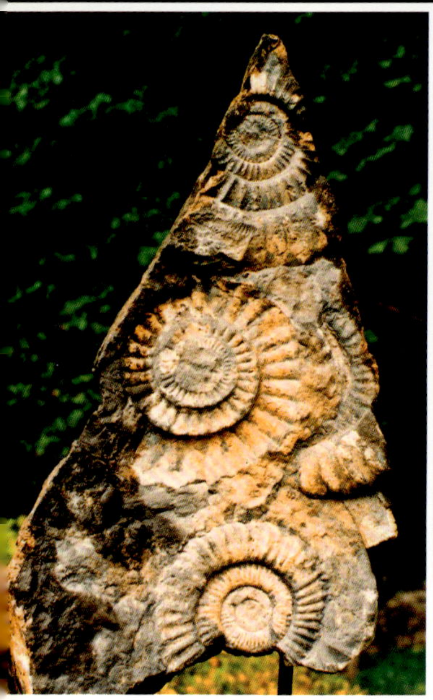

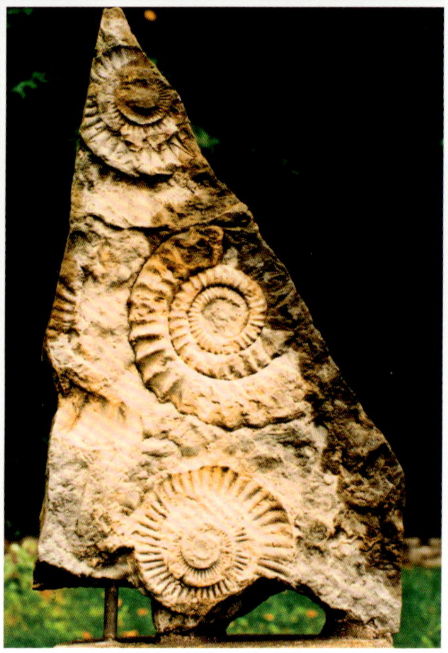

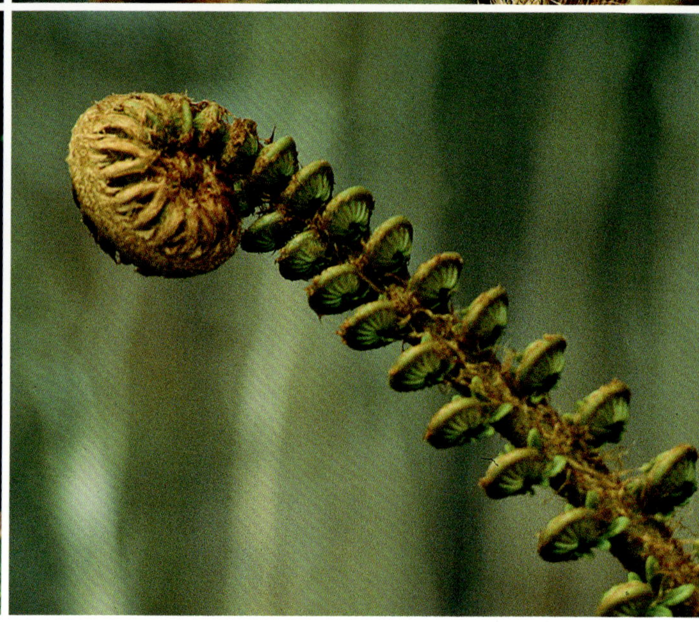

There are two-dimensional spirals whereas a helix in two dimensions would merely appear as a circle. The shapes of snails are three-dimensional spirals but projected onto a plane they represent two-dimensional spirals.

Top left: Nautilus snail

Bottom left: Fossil snails in a park in Vaduz, Lichtenstein

Top/bottom right: Spiraling ferns in Puerto Rico

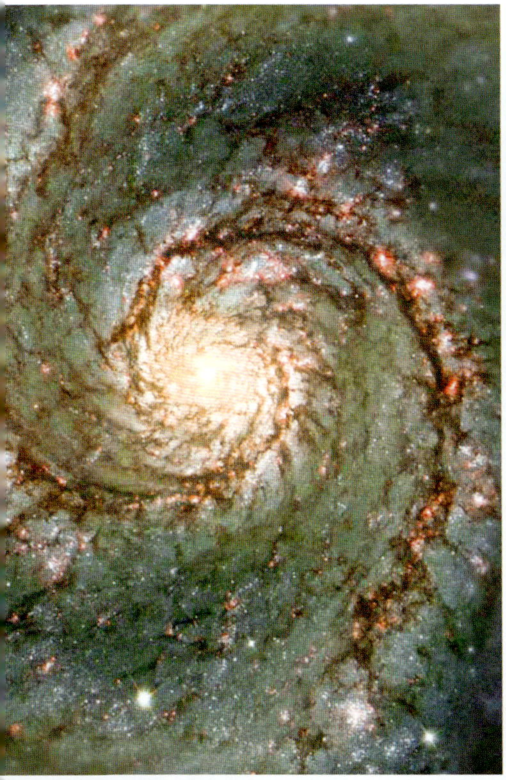

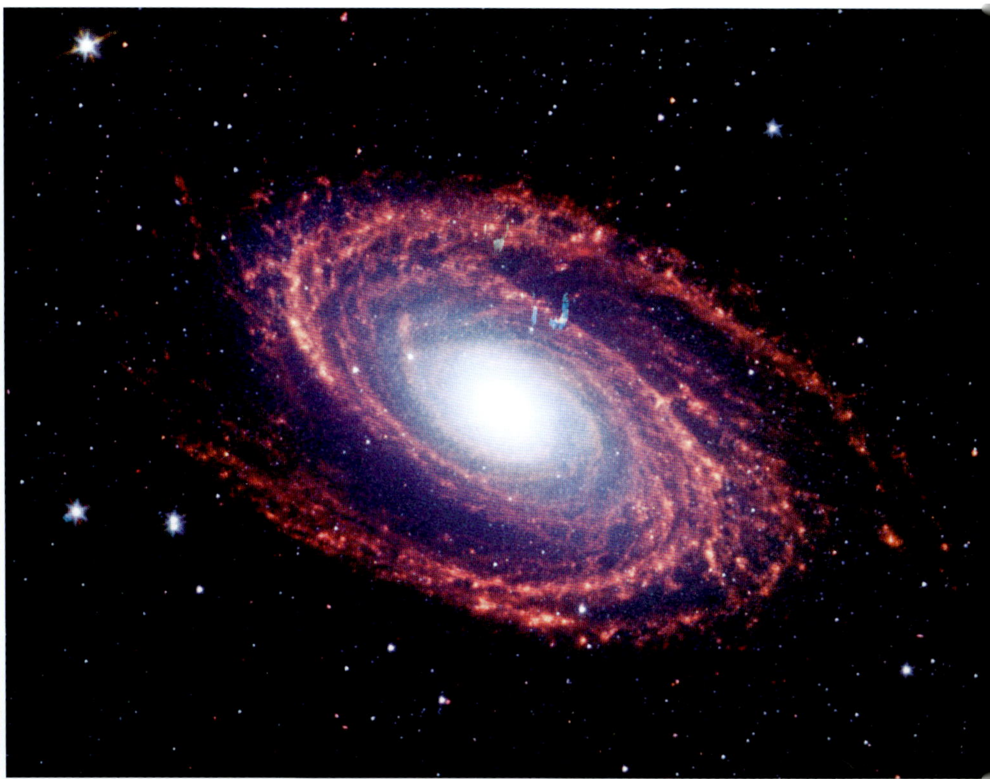

Top left:
Whirlpool Galaxy with well visible lights of the stars of the galaxy
(NASA and the Hubble Heritage Team (STScI/AURA), photograph by N. Scoville (Caltech) and T. Rector (NOAO, reproduced with permission)

Top right:
Messier-81 Galaxy at 12 million light years from the Earth
(Photograph by NASA/JPL/Caltech/Harvard–Smithsonian Center for Astrophysics, used with permission)

Spiral formations are frequent among natural phenomena. Examples include galaxies and hurricanes.

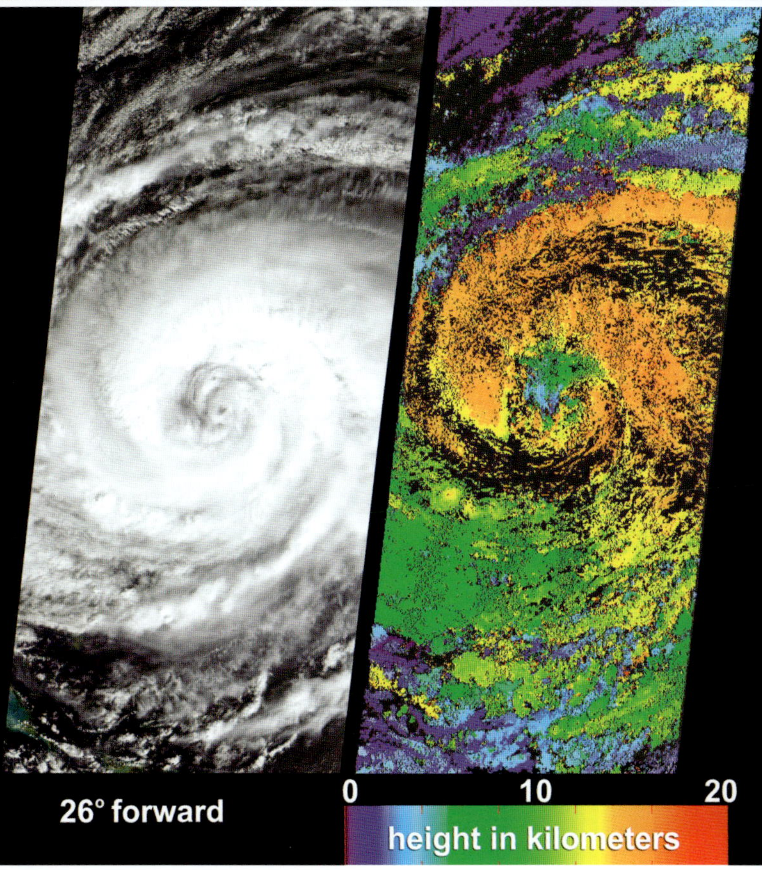

Hurricane Jeanne over Florida in 2004
(Photograph by NASA/GSFC/LaRC/JPL, MISR Team, used by permission)

Further Three-dimensional Spirals

The spiraling towers are, of course, man-made spatial spirals.

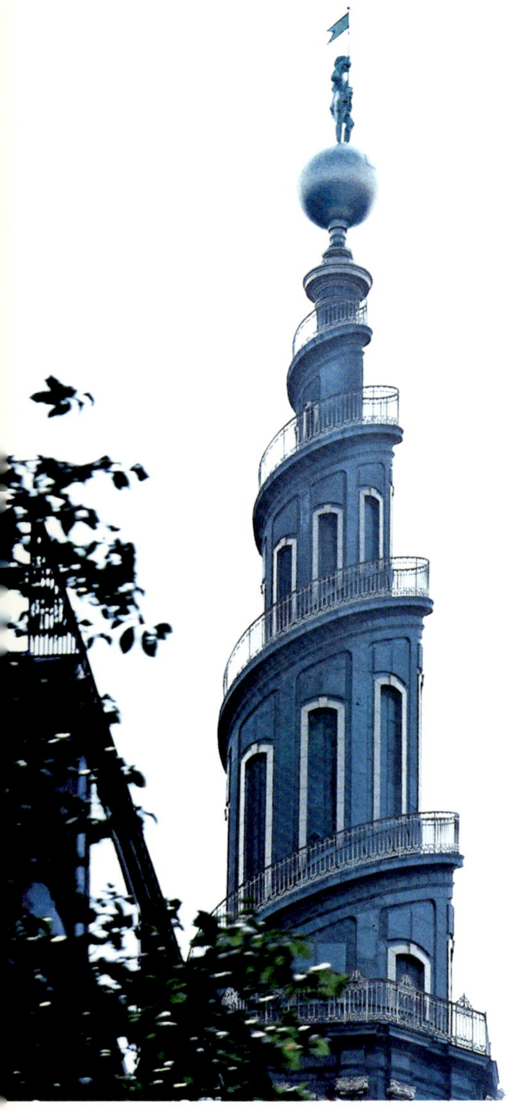

Copenhagen

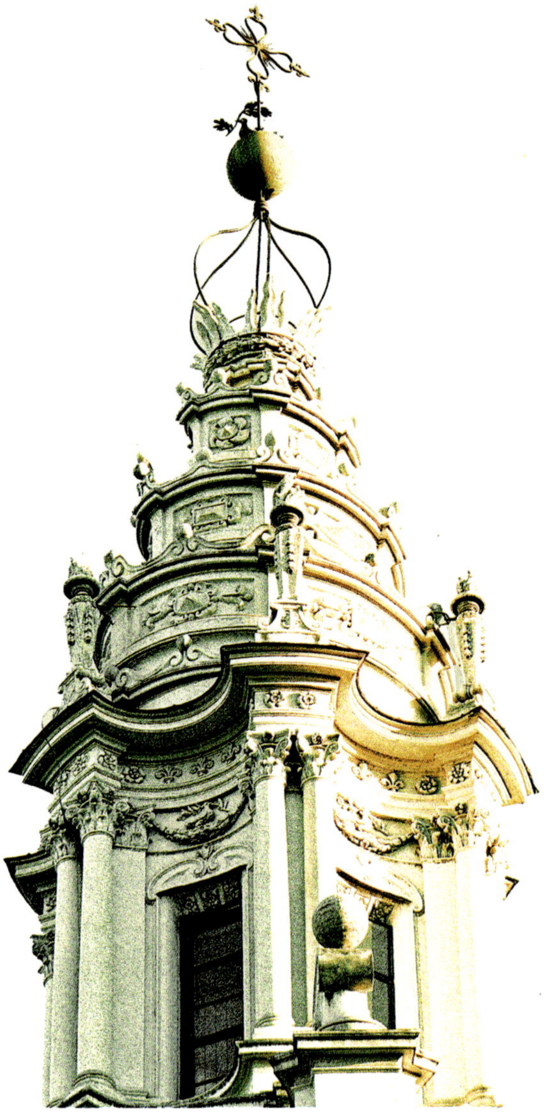

Rome

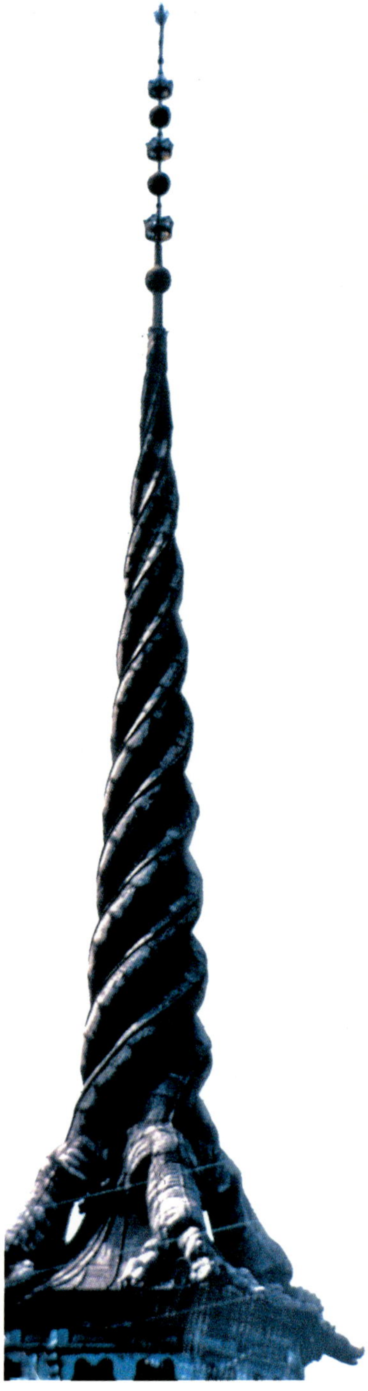

Copenhagen

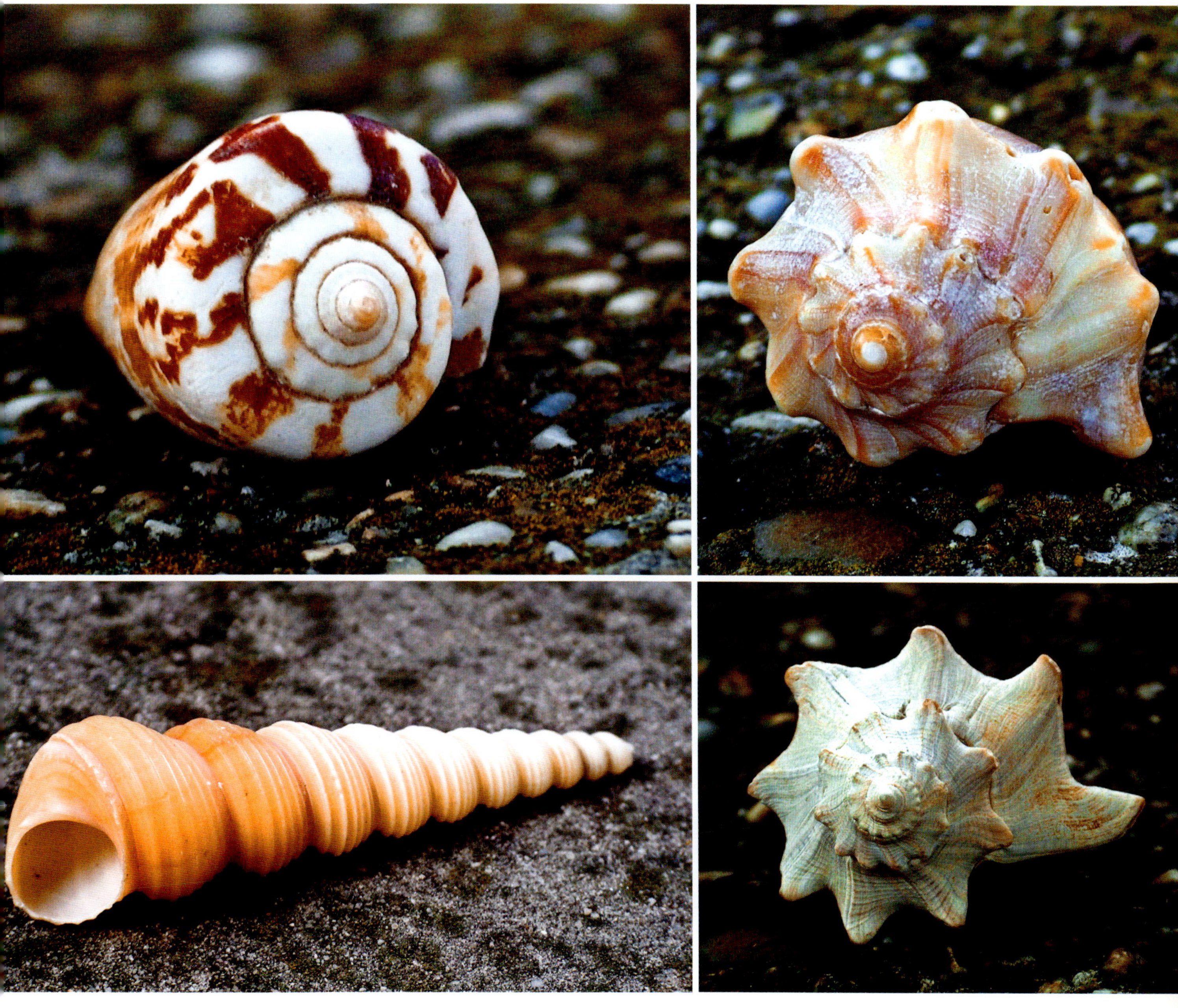

The shapes of snails are three-dimensional spirals; some of them are so flat that they approximate two-dimensionality.

Phyllotaxy

(arrangement of leaves on the stem)

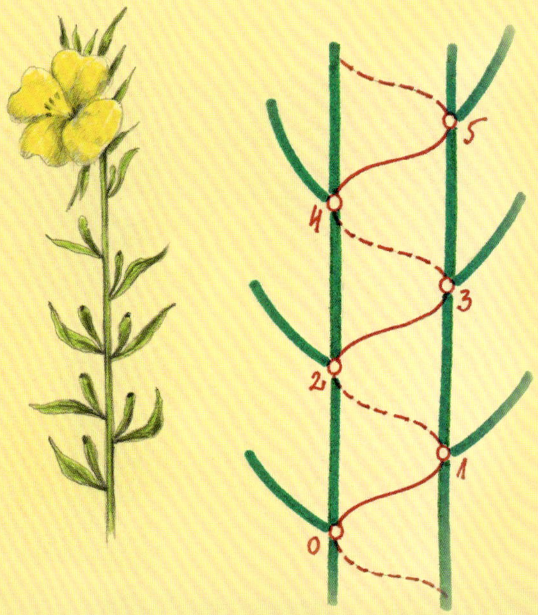

The scattered leaf arrangements around the stems of some plants provide intriguing examples of helical and spiral symmetries. The phenomenon is called *phyllotaxy* by botanists. The simplest case is when the leaves occur on opposite sides of the stem as we move along it. An example is the leaf arrangement of the yellow flower *Oenothera biennis*.

We start with a leaf labeled 0, and circle the stem until we find another leaf exactly eclipsing the leaf 0. This will be leaf 2, after one complete circle around the stem. The ratio of the two numbers is ½, it tells us that a new leaf is always found at half of the circumference of the stem.

Oenothera biennis

(drawing courtesy of Ferenc Lantos, Pécs, Hungary)

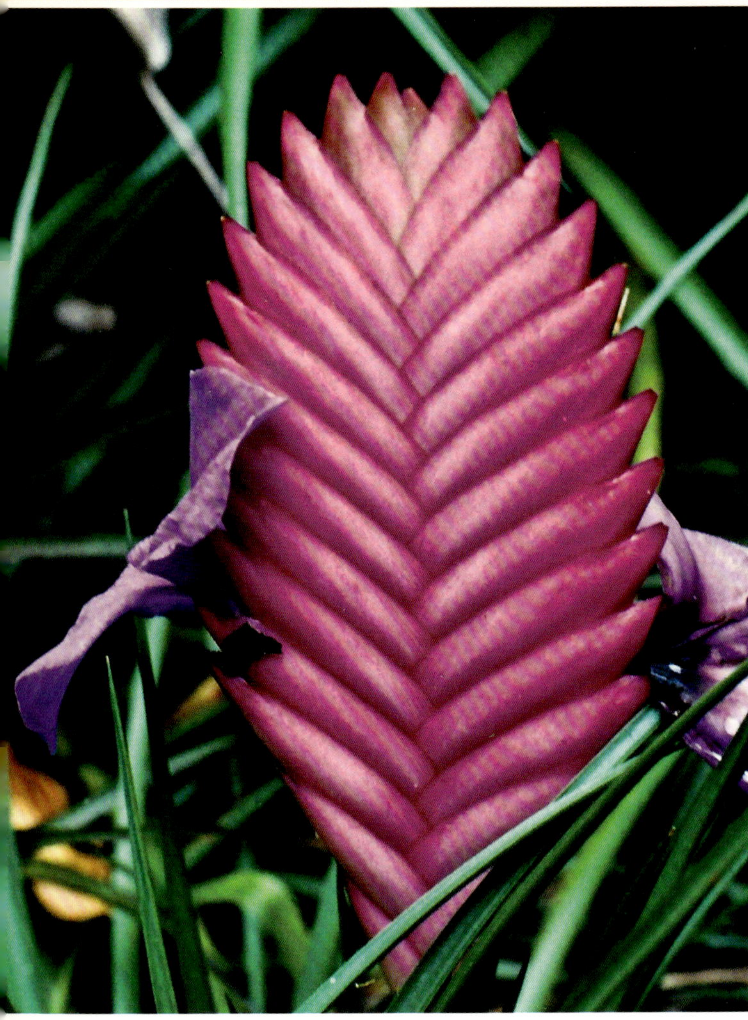

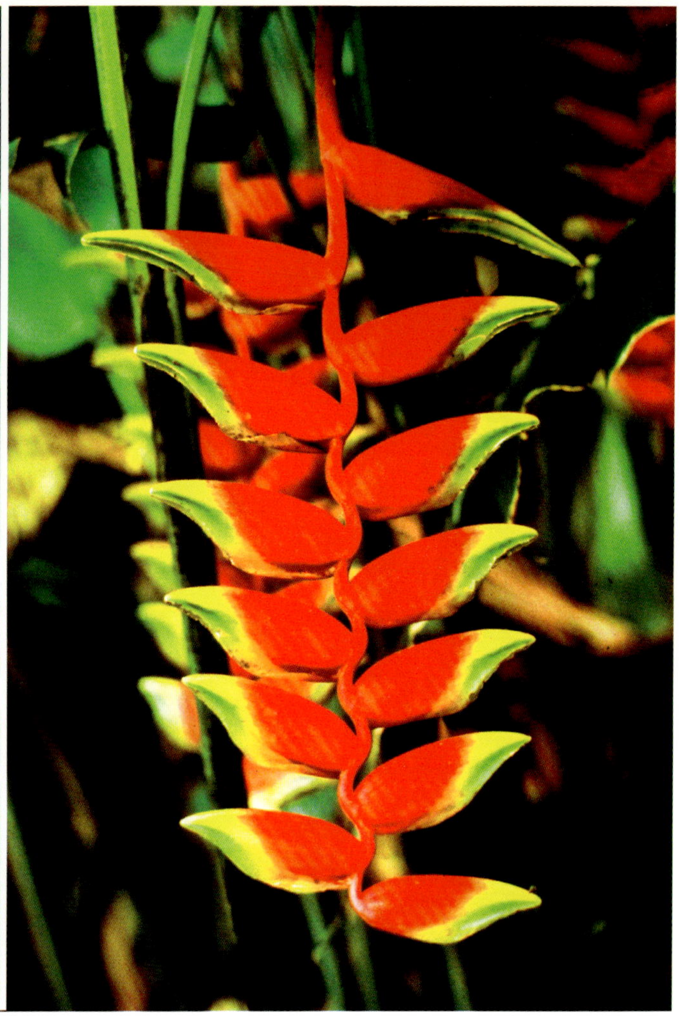

Two flowers from Hawaii displaying an arrangement similar to the leaves of *Oenothera biennis*

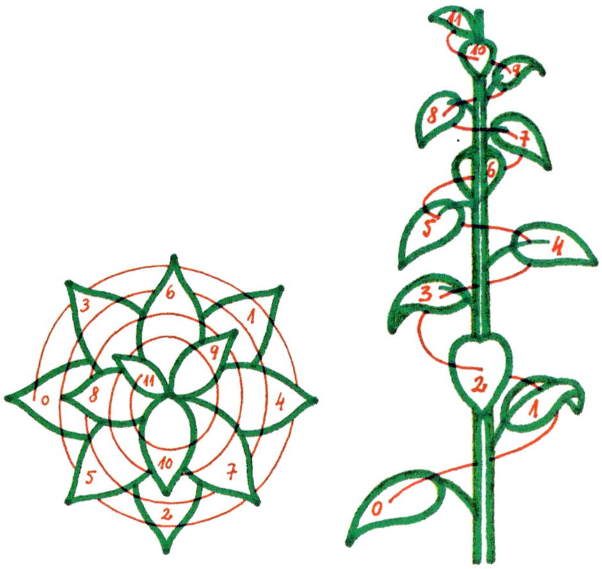

Plantago media

(drawing courtesy of Ferenc Lantos, Pécs, Hungary)

For *Plantago media*, starting with leaf 0, we have to circle the stem three times before reaching leaf 8 that is exactly above the initial leaf 0. The ratio of the two numbers is 3/8, and this tells us that a new leaf occurs at each 3/8 fraction of the circumference of the stem.

In order to consider these successions of leaves, we have to extend them to infinity, which we can do in our imagination.

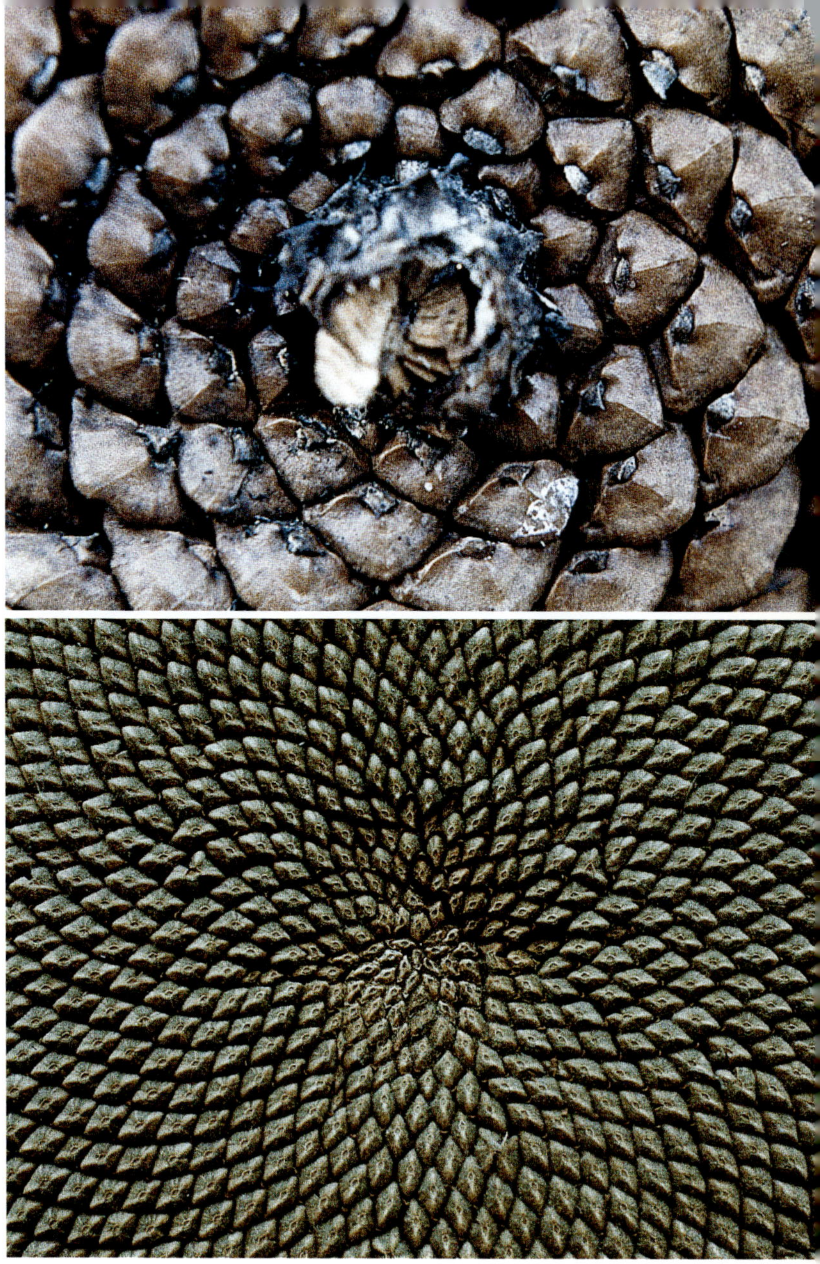

Top: Pine cone

Bottom: Sunflower

(courtesy of Sándor Kabai, Budapest)

There are known plants with scattered leaf arrangements that are described by 1/2, 1/3, 2/5, 3/8, 5/13, etc. Each of these series, that is, the series of the numerators, 1, 1, 2, 3, and so on, and that of the denominators, 2, 3, 5, and so on, is a so-called Fibonacci series named after its discoverer, Leonardo of Pisa (Fibonacci), an Italian mathematician who lived in the 13th century. In a Fibonacci series each number is the sum of the previous two.

The Fibonacci numbers also occur in the numbers of spirals of the scales of pine cones and seeds of sunflowers. There are invariably 13 left-bound and 8 right-bound spirals of scales in the pine cones, and both are Fibonacci numbers. Much larger Fibonacci numbers can be observed by counting the left-bound and right-bound spirals of the seed arrangements of sunflowers. The spirals of pine cone scales and sunflower seeds can be considered as if they were compressed around their respective stems.

It is easy to count the scales of pine cones and the seeds of sunflowers because they are in a plane. It is more difficult to distinguish the spirals on flowers and, for example, around the pineapple. These two pages display examples from Hawaii.

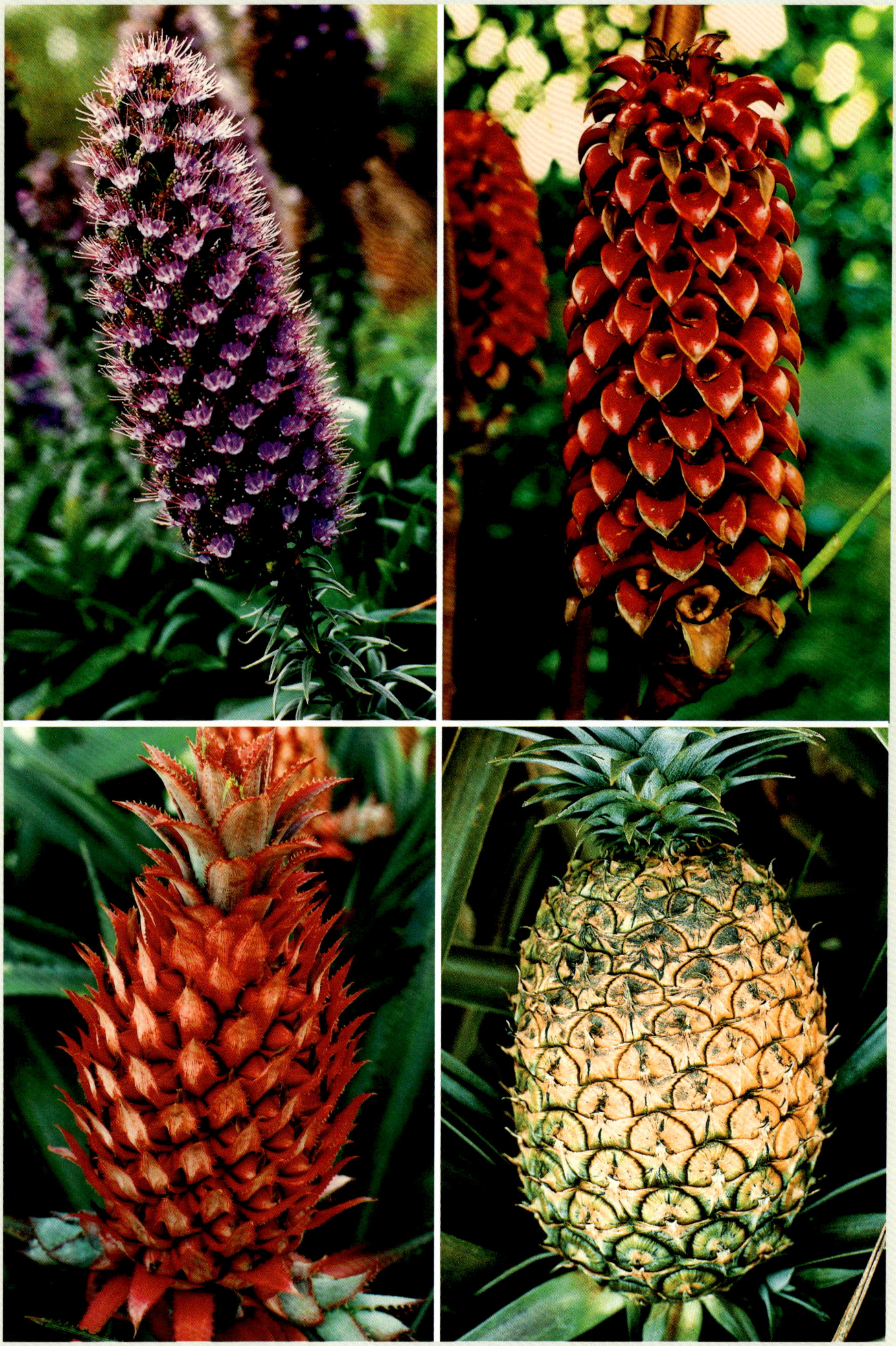

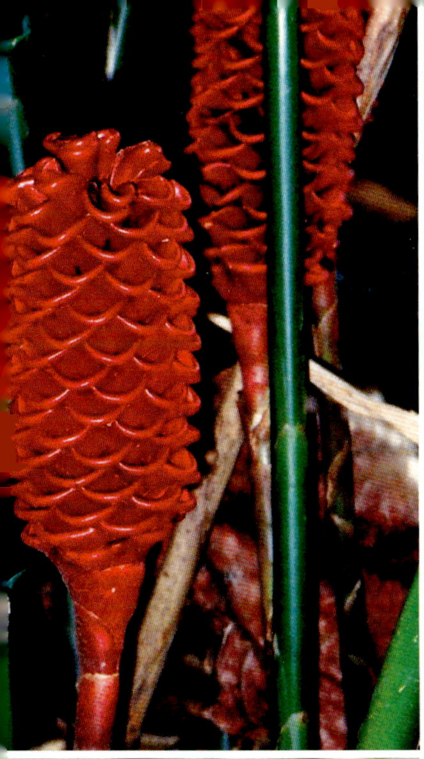
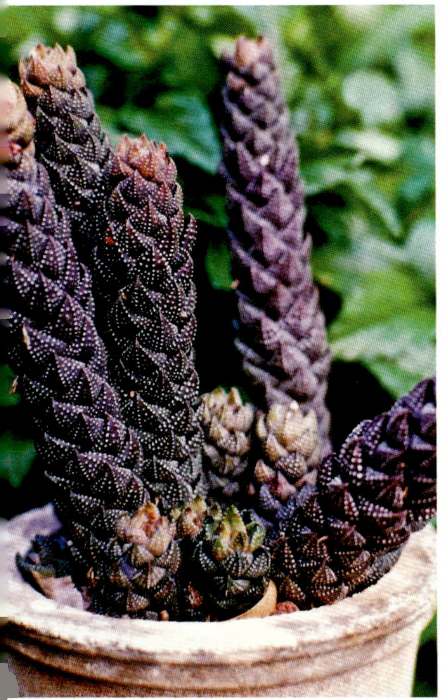
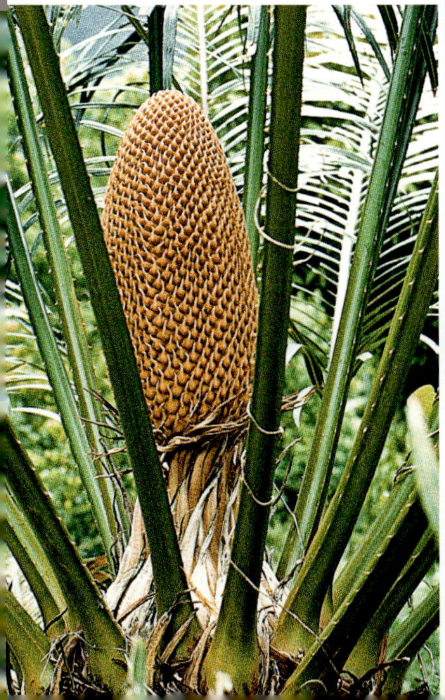
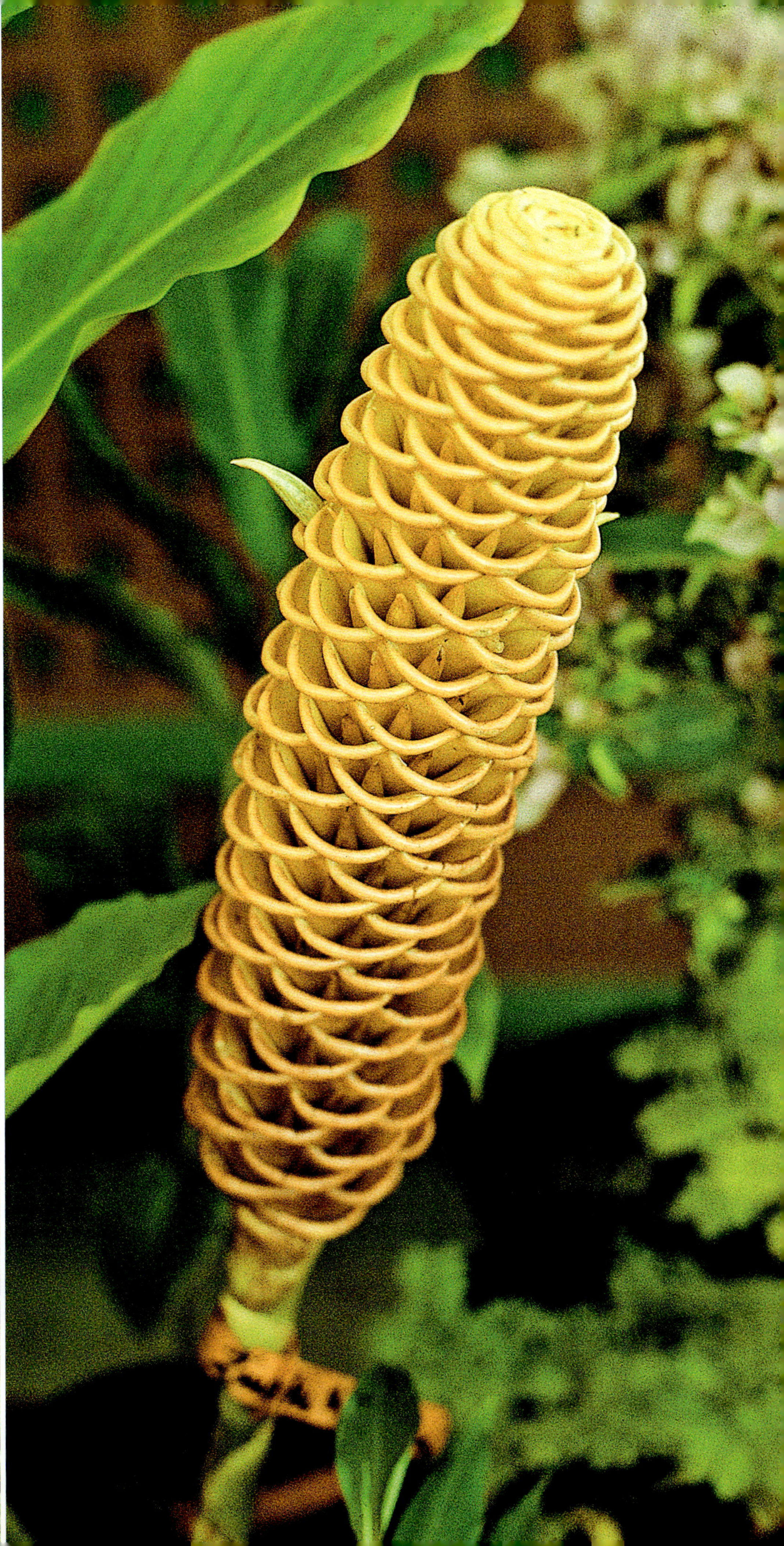

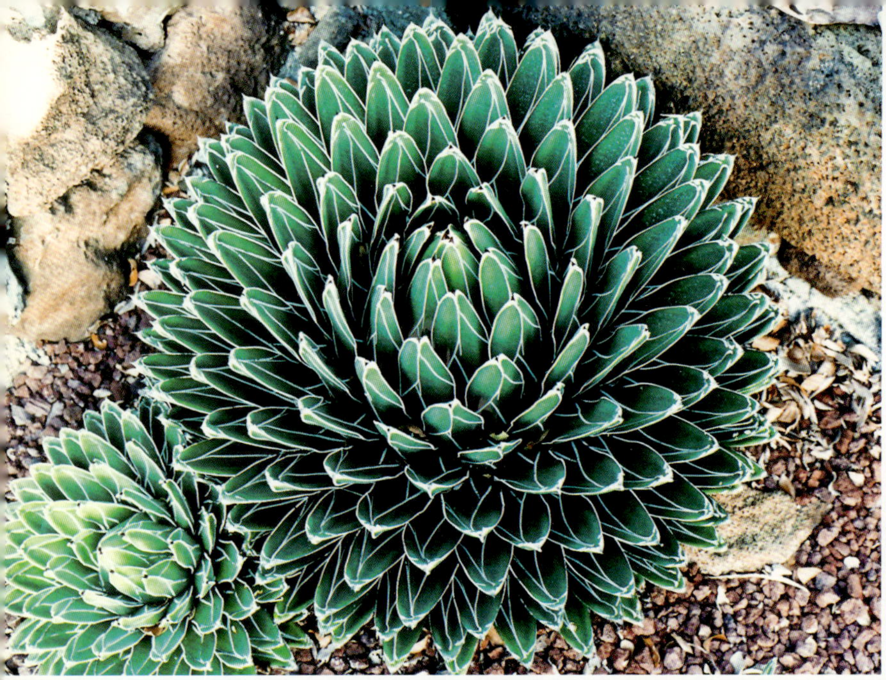 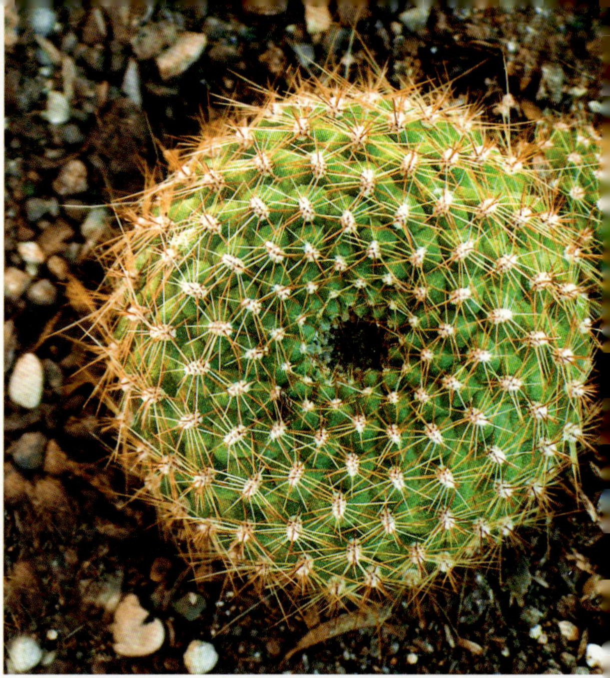

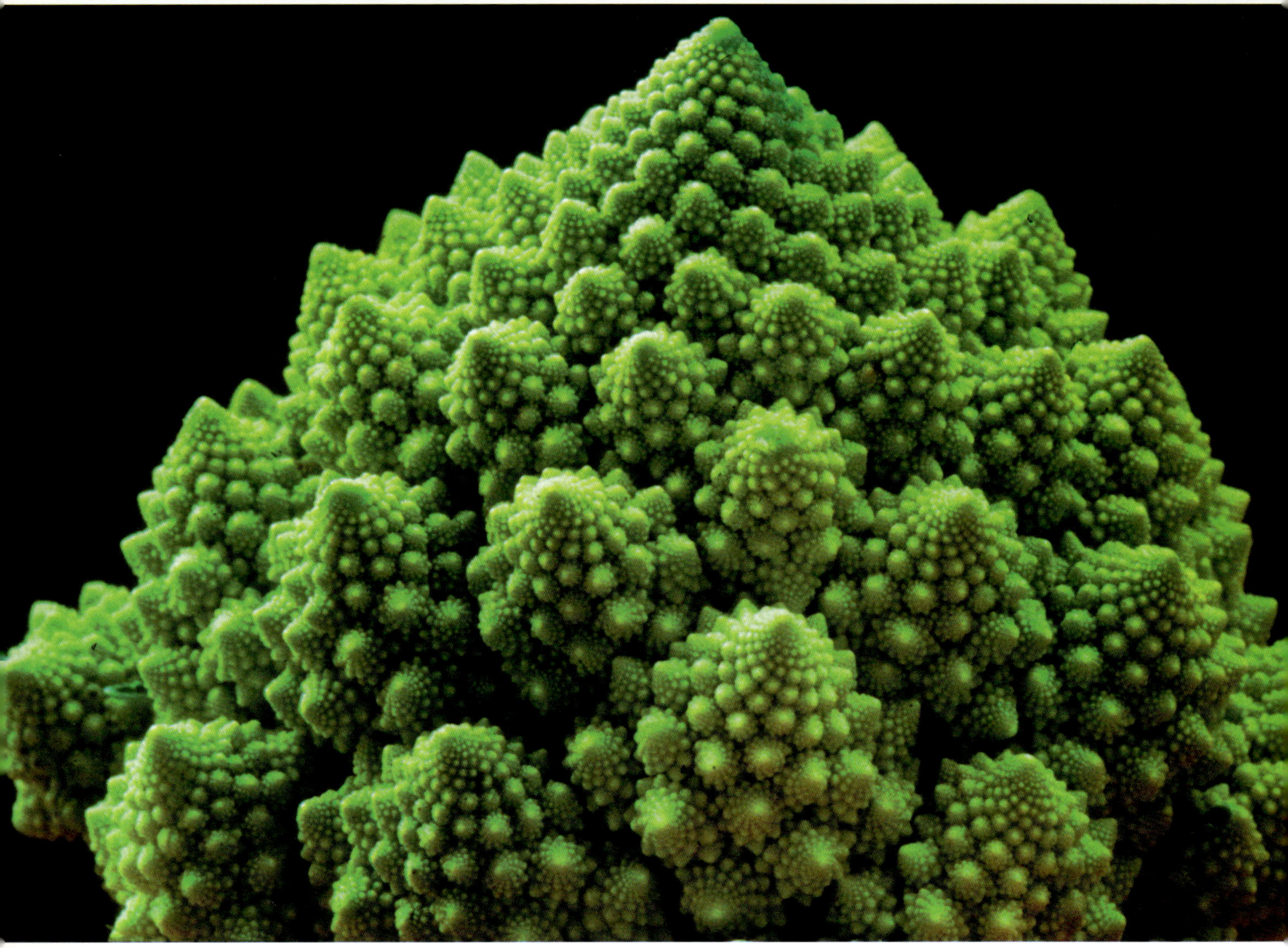

Cacti and cauliflowers provide further examples of spiral symmetries. The image of cauliflower here (courtesy of Benoit Mandelbrot, Yorktown Heights) is an intriguing case as it shows countless numbers of spiraling arrangements. It is also a good example of the gradual changes in the sizes that has been present in most other examples as well. So the cases of spiral symmetries are also often examples of similarity symmetries.

If we continue the ratios characteristic for phyllotaxy, the actual values lead to the golden proportion: ½ = 0.500, 1/3 = 0.333, ..., 5/13 = 0.385, 8/21 = 0.381, ..., 13/34 = 0.382, ..., 0.38196...

The golden proportion characterizes the golden section, also called the *Divine Proportion*. The golden section means that a certain length (any length) is divided in such a way that the ratio of the longer part to the whole is the same as the ratio of the shorter part to the longer part. If the whole is 1.000, the longer part will be 0.618 and the shorter part, 0.382.

The golden section often occurs in architecture, musical compositions and elsewhere in human creations.

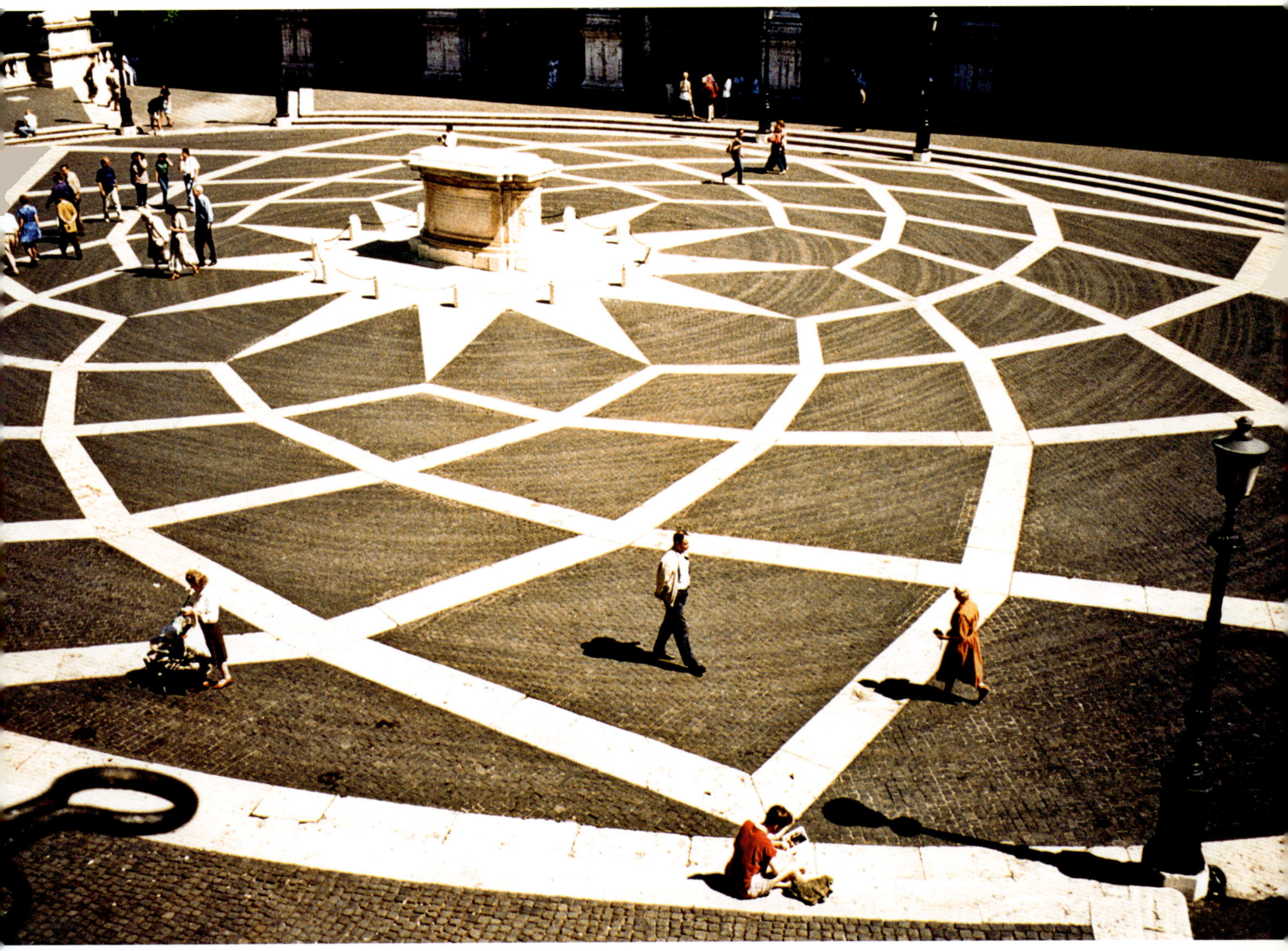

Michelangelo's design in the quadrangle of the *Capitolium* in Rome provides a large-scale example of spirals in the plane.

When the repeated patterns extend "to infinity" in two directions, they cover the plane.

The honeycomb is a network of equal-size regular hexagons. It is periodic because there is an endless repetition of its basic motif and its periodicity extends in two directions, thus covering the whole plane (here we ignore the curvature of the honeycomb in the picture).

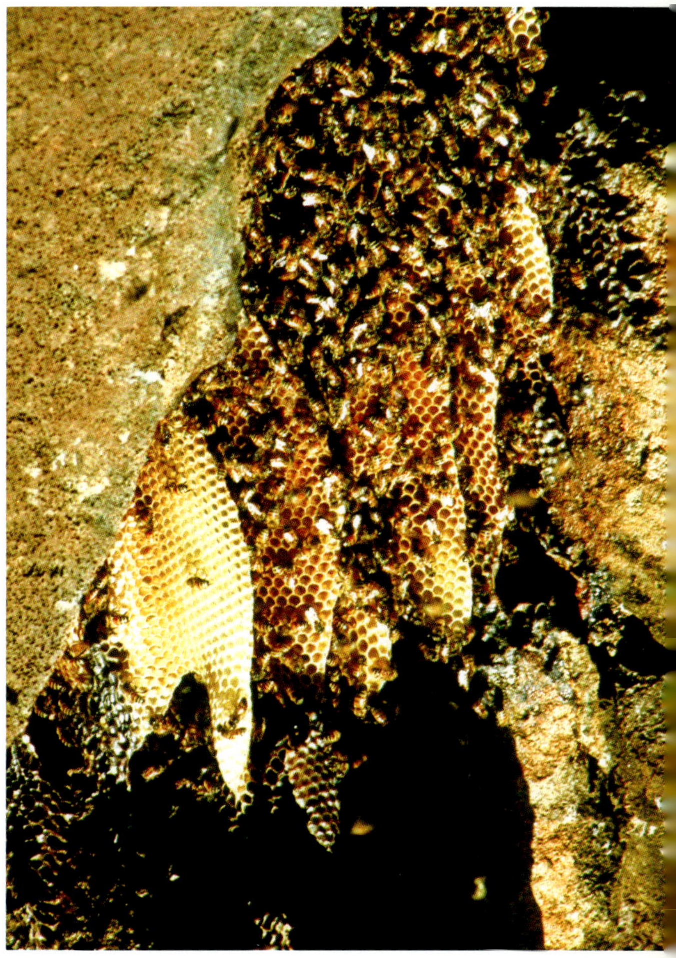

Planar Patterns

Indigo-dyed textile decoration from Pápa, Hungary

To build a honeycomb from wax, the bees first form a network of closely packed circles. The bees are of nearly equal size and move around in circles. Although the circles they create are as closely packed as possible, they do not cover the available surface completely. The liquid wax flows into the spaces between the circles and forms hexagons. The hexagons then cover the entire available surface without gaps.

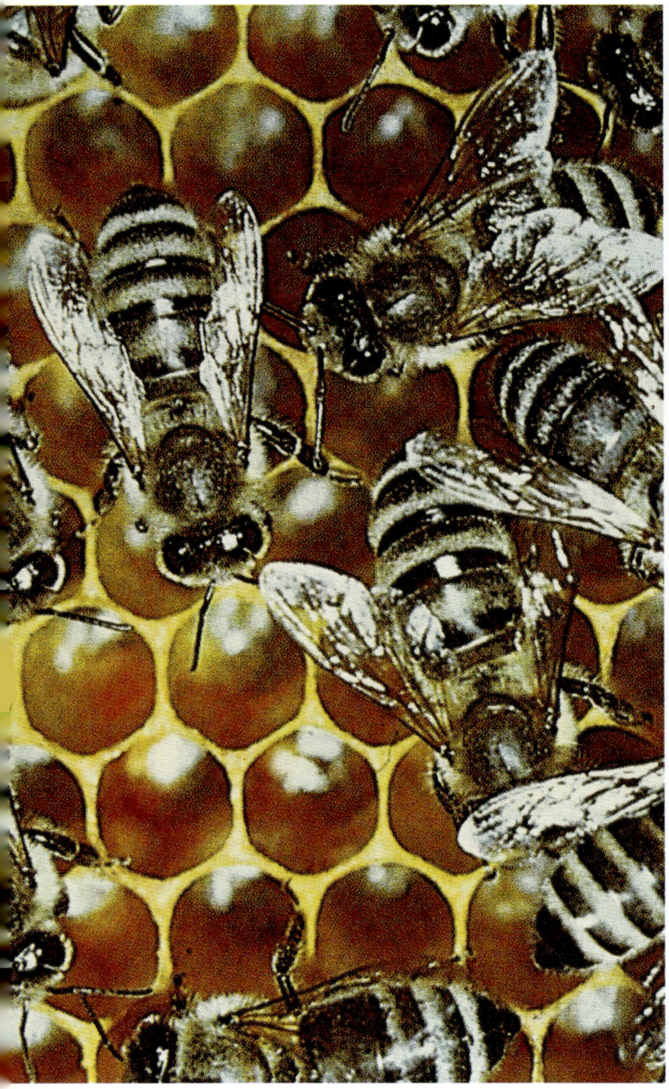

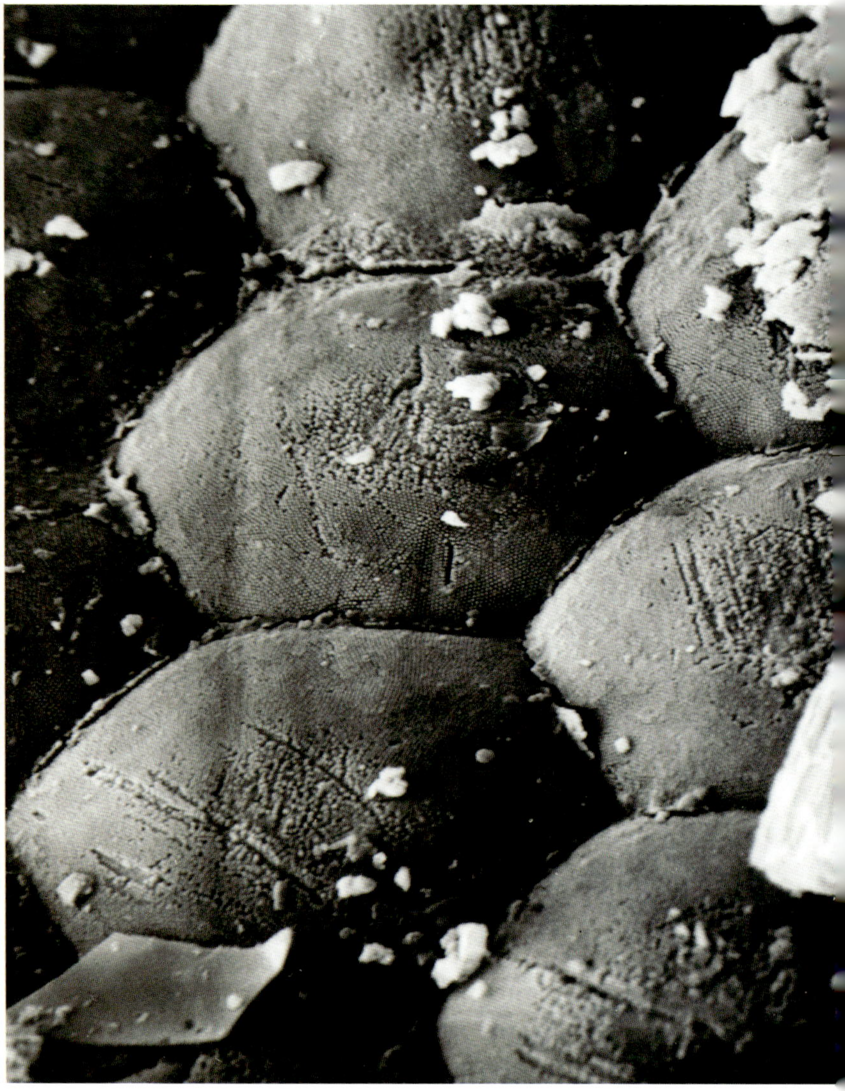

Left: Honeycomb

(photograph by Donath Waltenberger, Mindelheim, Germany, from J. Rudnai, L. Beliczay, *Mézkönyv* {in Hungarian, Honeybook}, Corvina, Budapest, 1987).

Right: The moth's compound eye

(magnification ×2000, courtesy of J. Morral, Storrs, Connecticut)

The concrete base under construction for a Norwegian offshore oil platform in the North Sea (the picture is from over two decades ago) looks like a honeycomb. The base consists of a network of regular hexagonal shapes. The symmetries of the honeycomb and the base of the oil platform come from repetition of the hexagons in two directions. The hexagons of carbon atoms in a graphite layer form a similar pattern with the same symmetry. The iron fence in front of a window at the Topkapi Sarayi in Istanbul may be looked at as a model of the hexagonal graphite pattern.

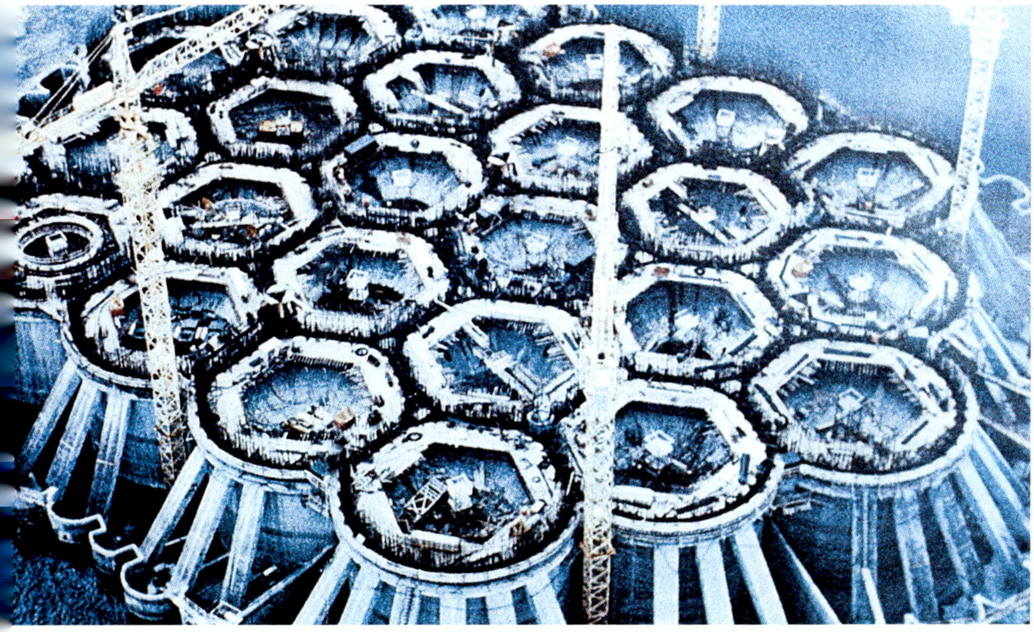

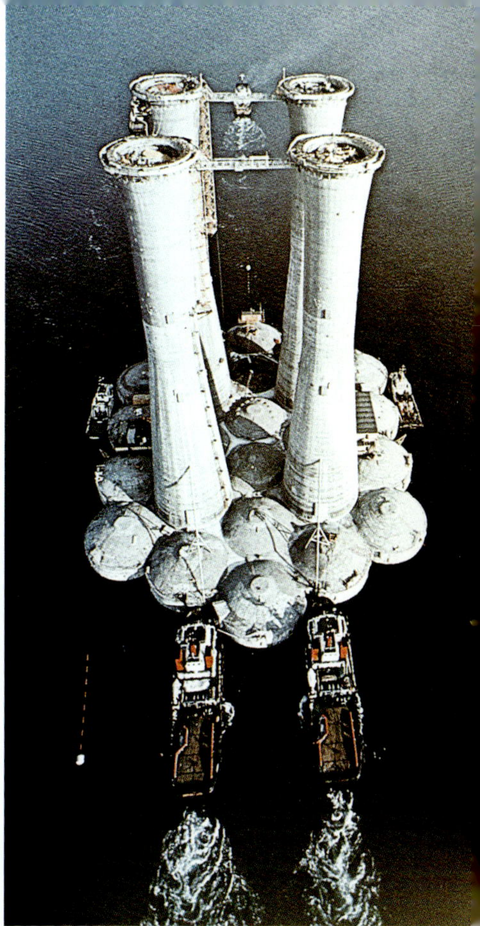

The only regular polygons (of equal size) that can cover the surface without gaps or overlaps are the equilateral triangle, the square, and the regular hexagon. If we try to cover a flat surface with, for example, equal-size regular pentagons, there will always be gaps, no matter how we arrange these pentagons. The same is true for equal-size regular octagons. There are no such restrictions for polygons of arbitrary shapes.

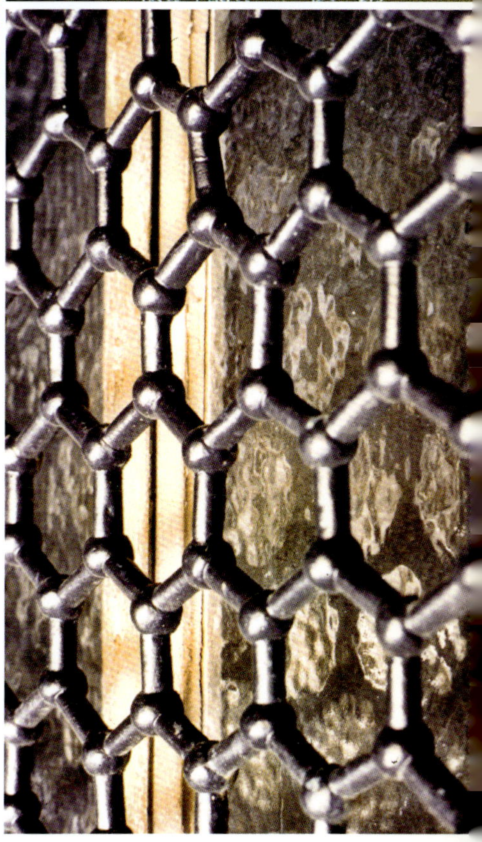

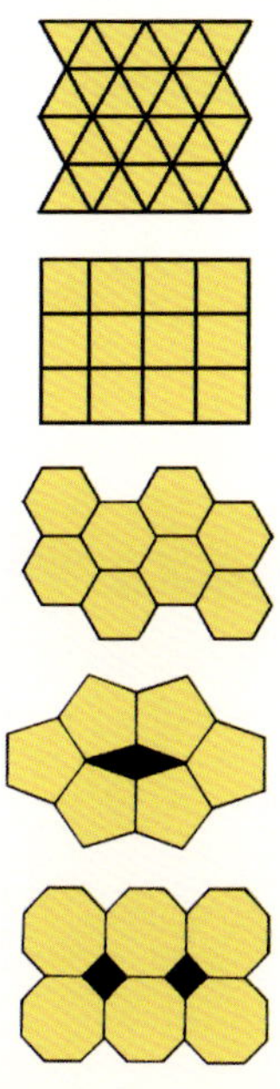

Top left: Base of offshore oil platform after it is turned upside down

(from a Statoil brochure)

Top right: Oil platform being towed to destination before being turned upside down

(from a Statoil brochure)

Bottom right: Iron fence in front of a window at the Topkapi Sarayi in Istanbul, Turkey

Bottom left: Patterns of regular polygons for covering the plane: for equilateral triangles, squares and regular hexagons the plane is covered without gaps; in both the regular pentagonal and regular octagonal patterns, gaps occur

Just as there are 7 possibilities for repeating the black triangle in one direction, there are exactly 17 symmetry variations for repeating the black triangle in two directions in covering the surface. Here the 17 possibilities are shown by patterns of the black triangles and by Hungarian patterns from the collection of the late Györgyi Lengyel (I. Hargittai, G. Lengyel, "The seventeen two-dimensional space-group symmetries in Hungarian needlework." *J. Chem. Educ.* 62 (1985) 35-36). For specialists, we indicated the international designation of the symmetries of the patterns and drew some of the characteristic symmetry elements into the patterns of the black triangles.

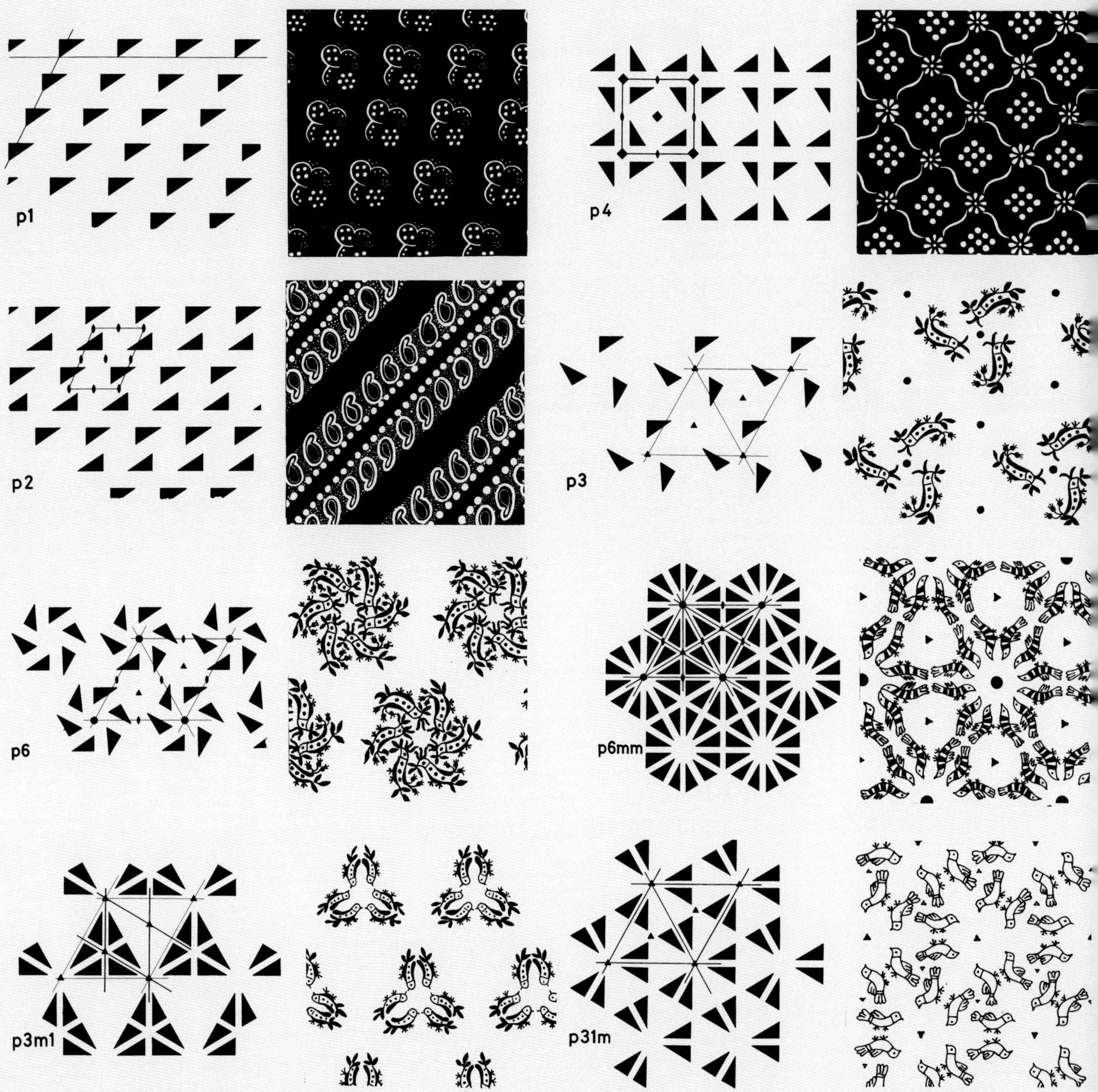

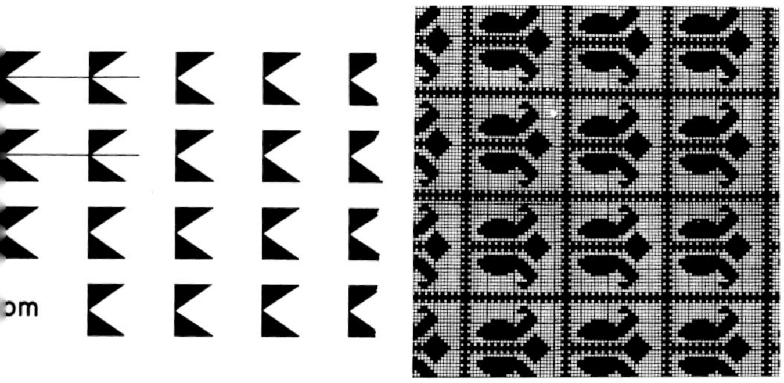

pm

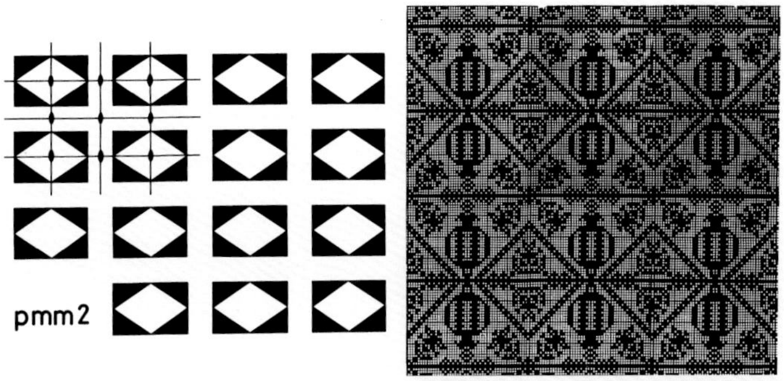

pmm2

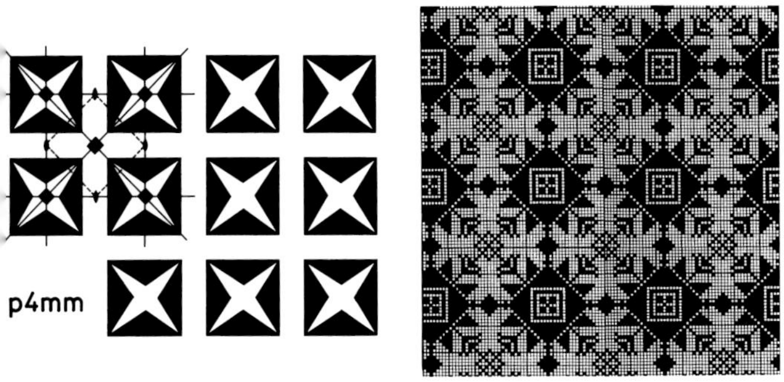

p4mm

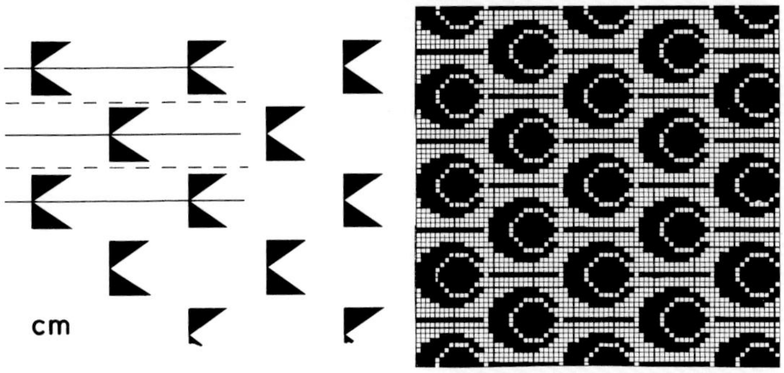

cm

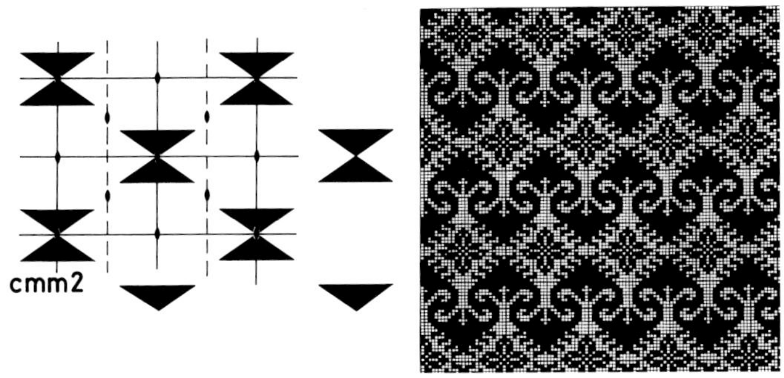

cmm2

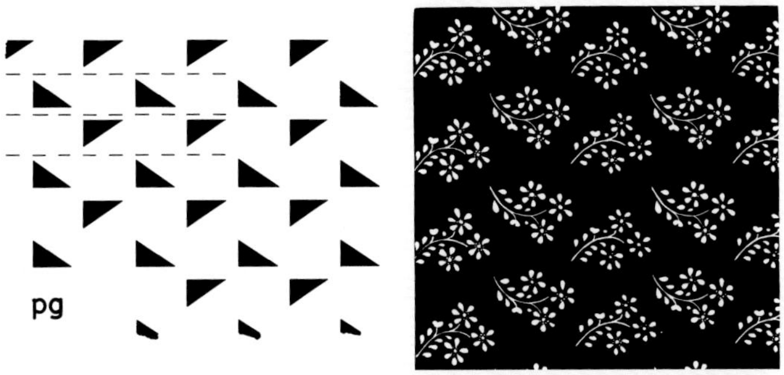

pg

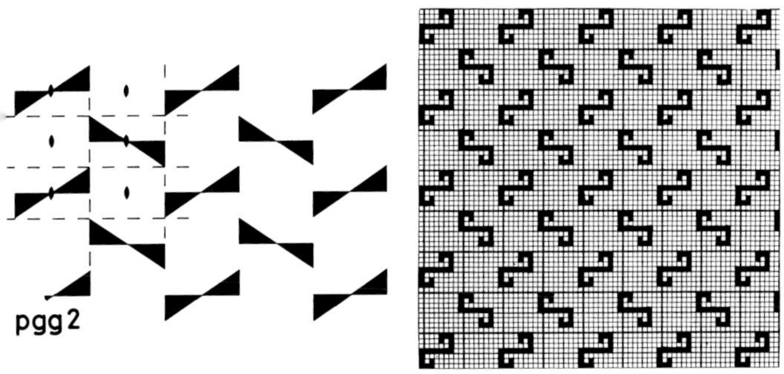

pgg2

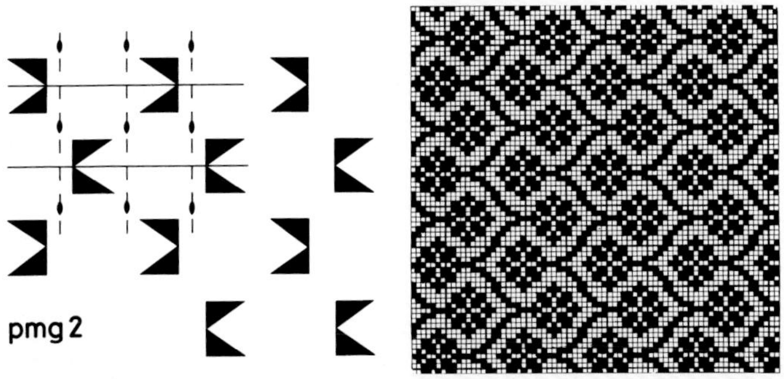

pmg2

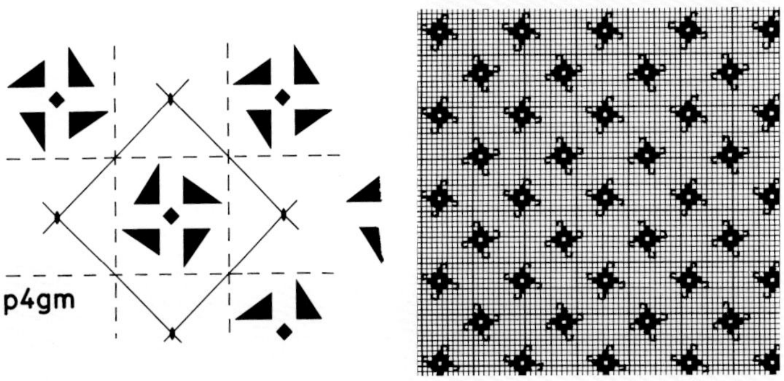

p4gm

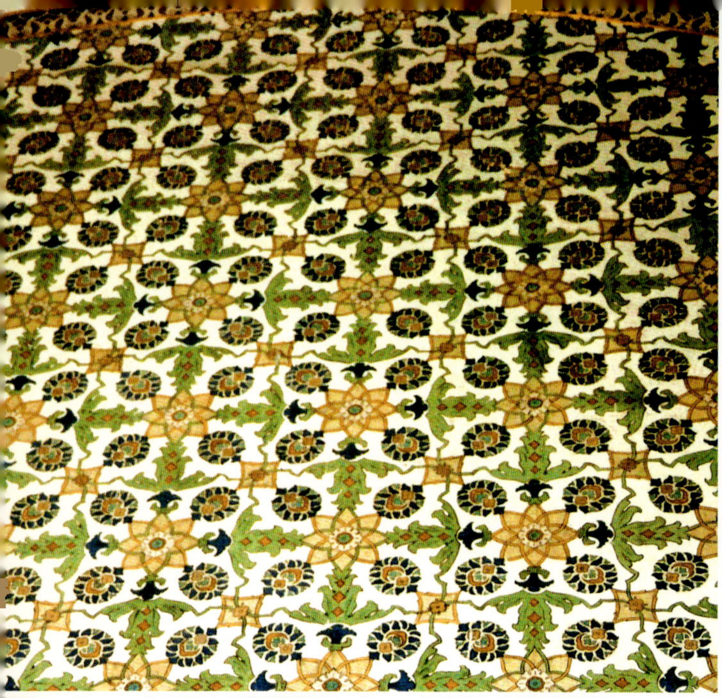

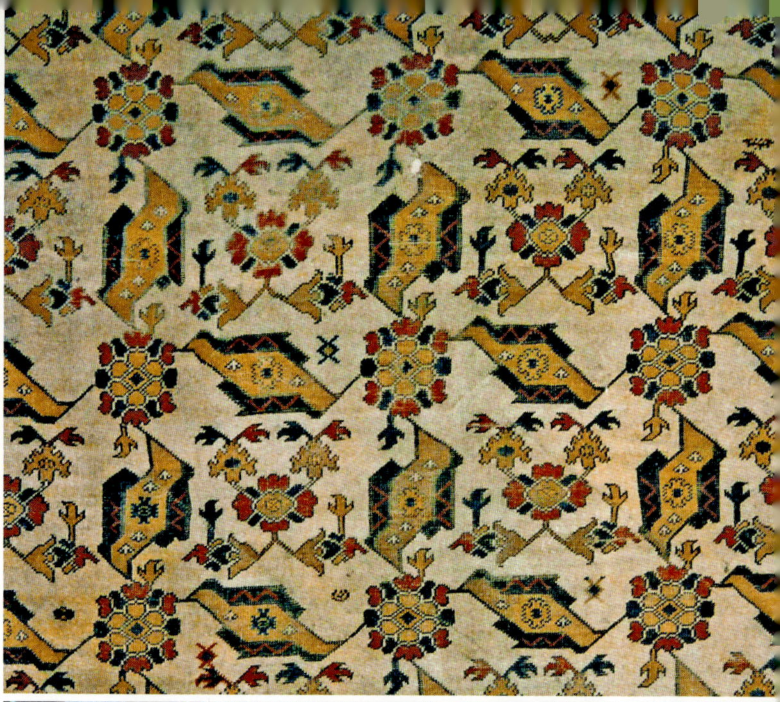

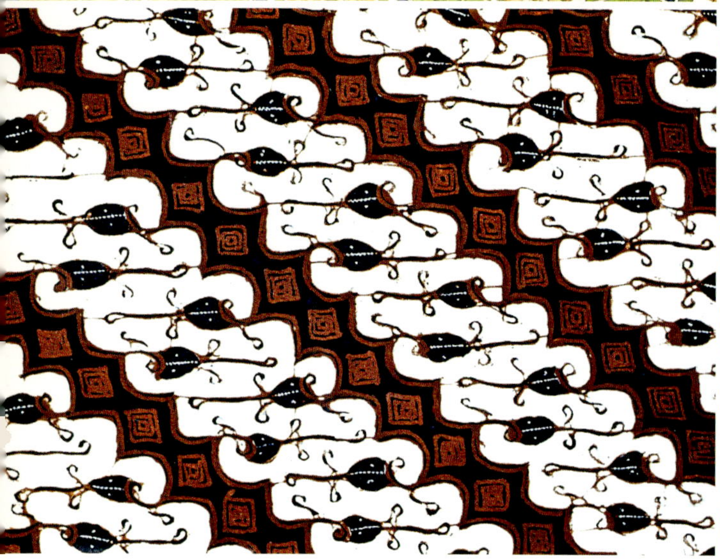

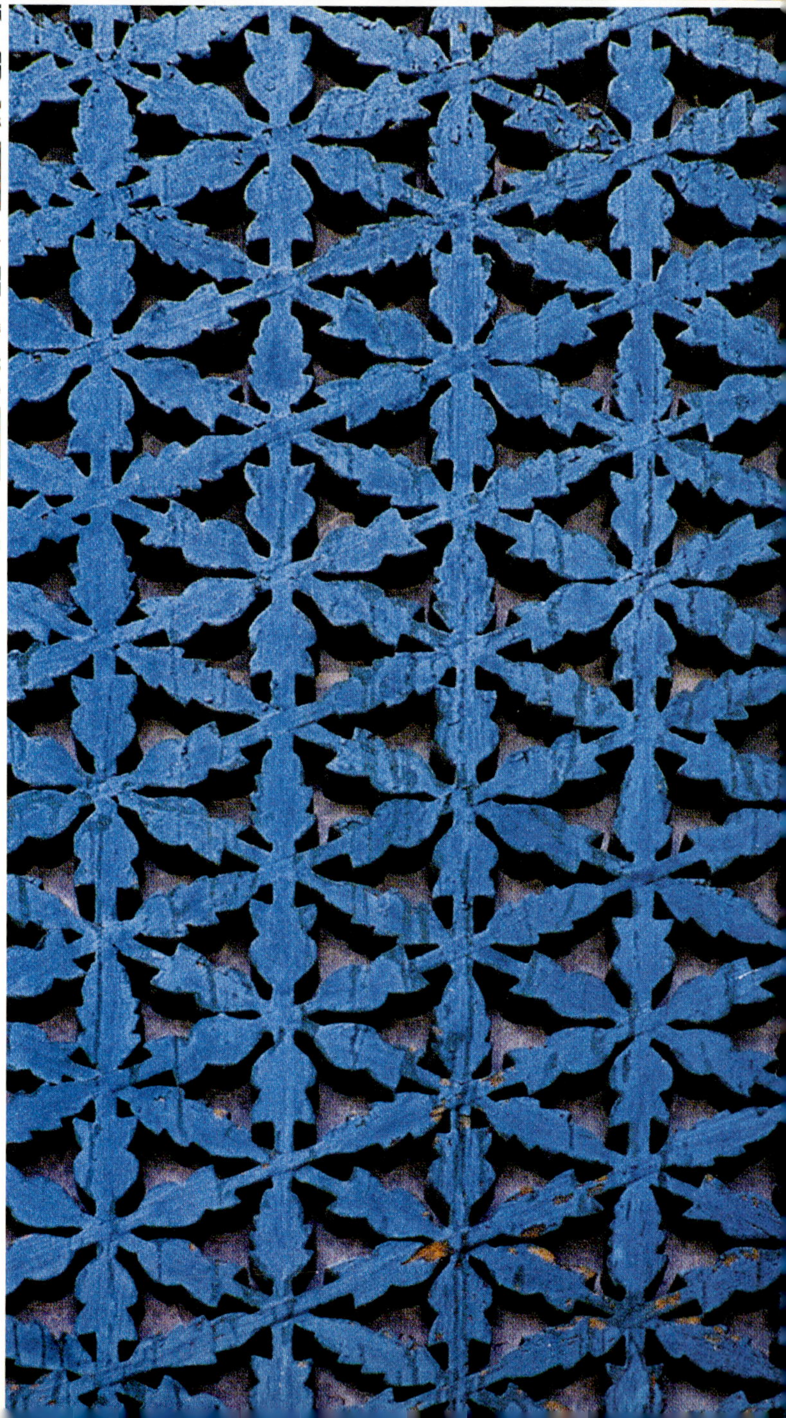

There are countless examples demonstrating planar patterns. These and the following pages display a sample to illustrate the wealth of their variations.

This page: Wall decorations and a batik pattern

Opposite page: Mosaics from the villa of the Roman Emperor Hadrian in Tivoli, near Rome and a batik pattern

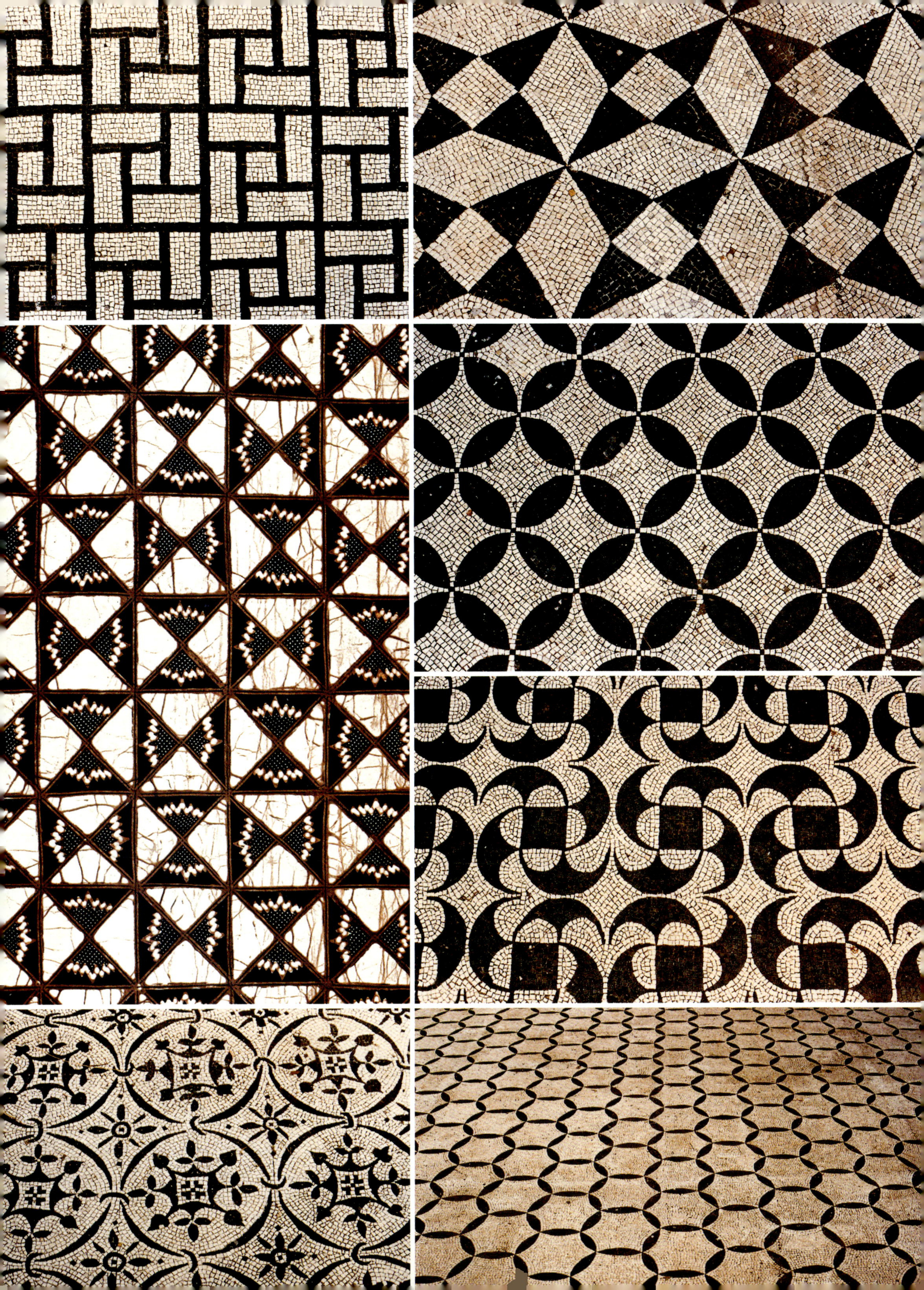

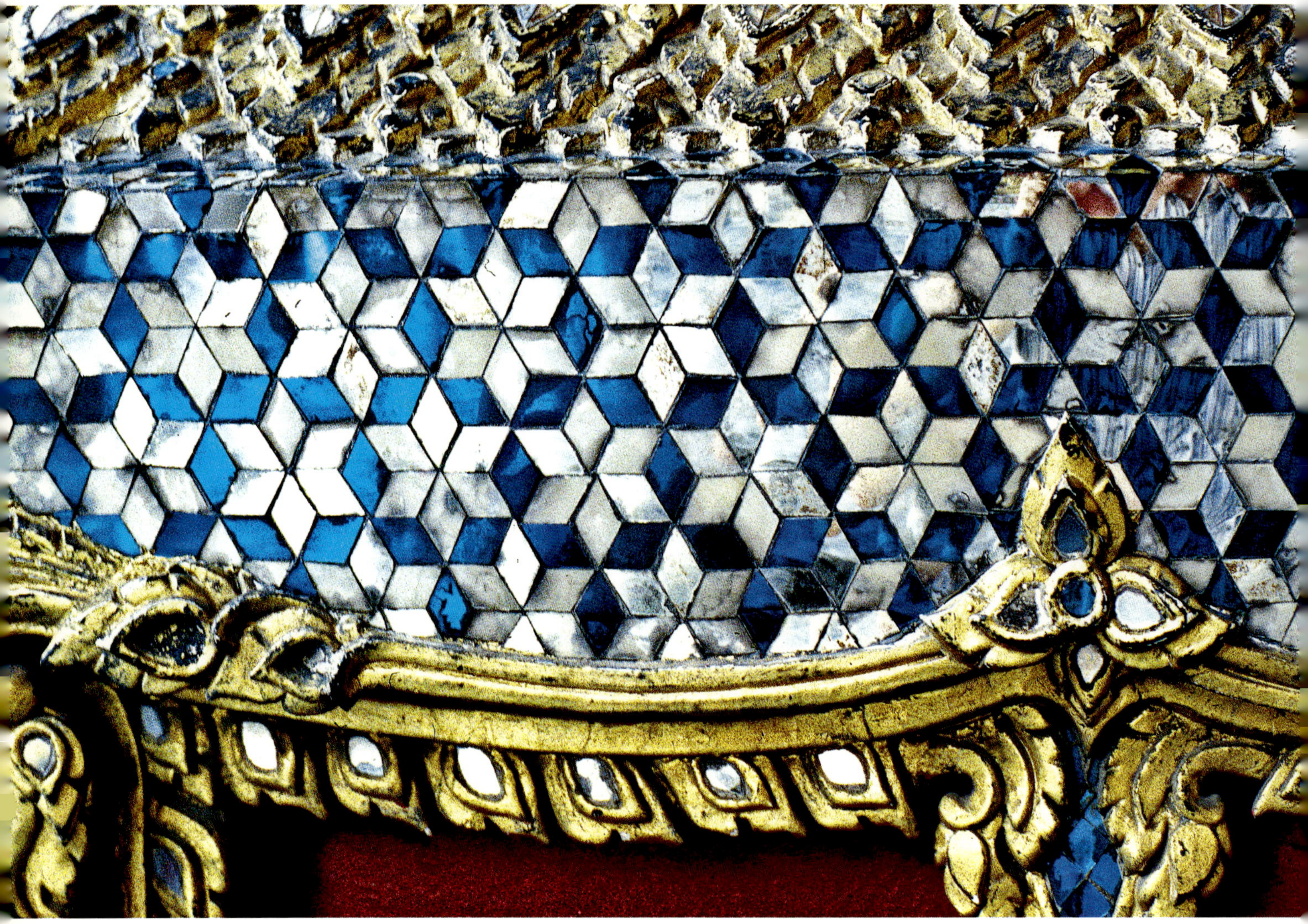

Decoration in the Royal Palace in Bangkok

Top left and bottom right: Decorations in the Royal Palace in Bangkok

Top right: Wall decoration in the Alhambra Palace in Granada, Spain

Bottom left: Wall decoration in Córdoba, Spain

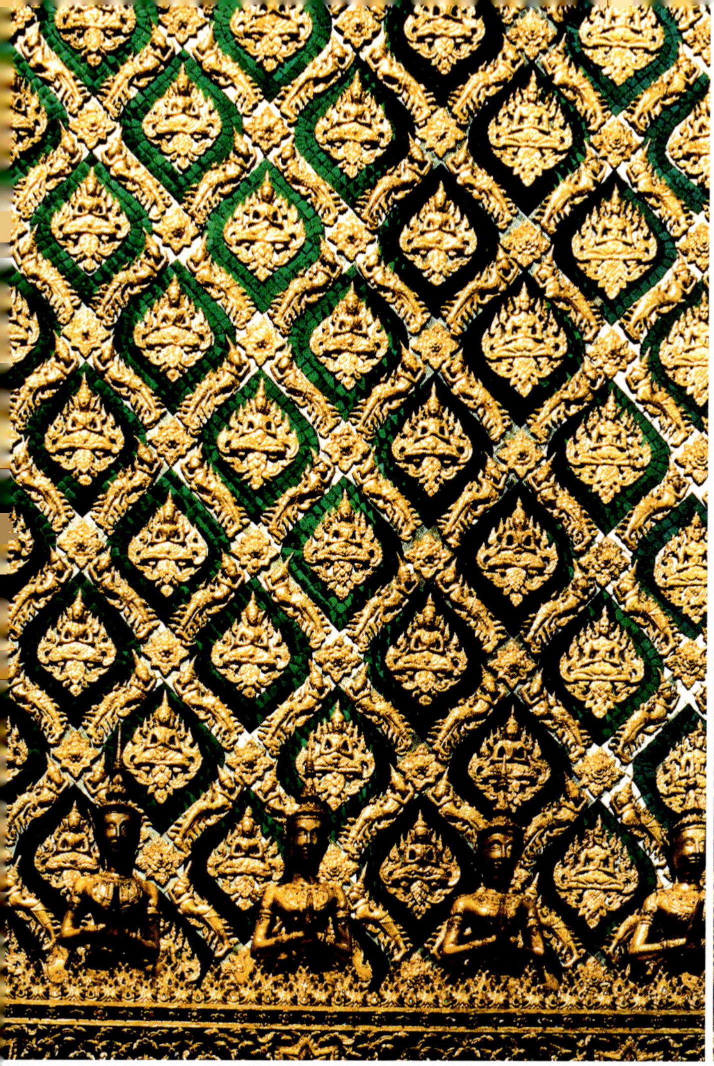

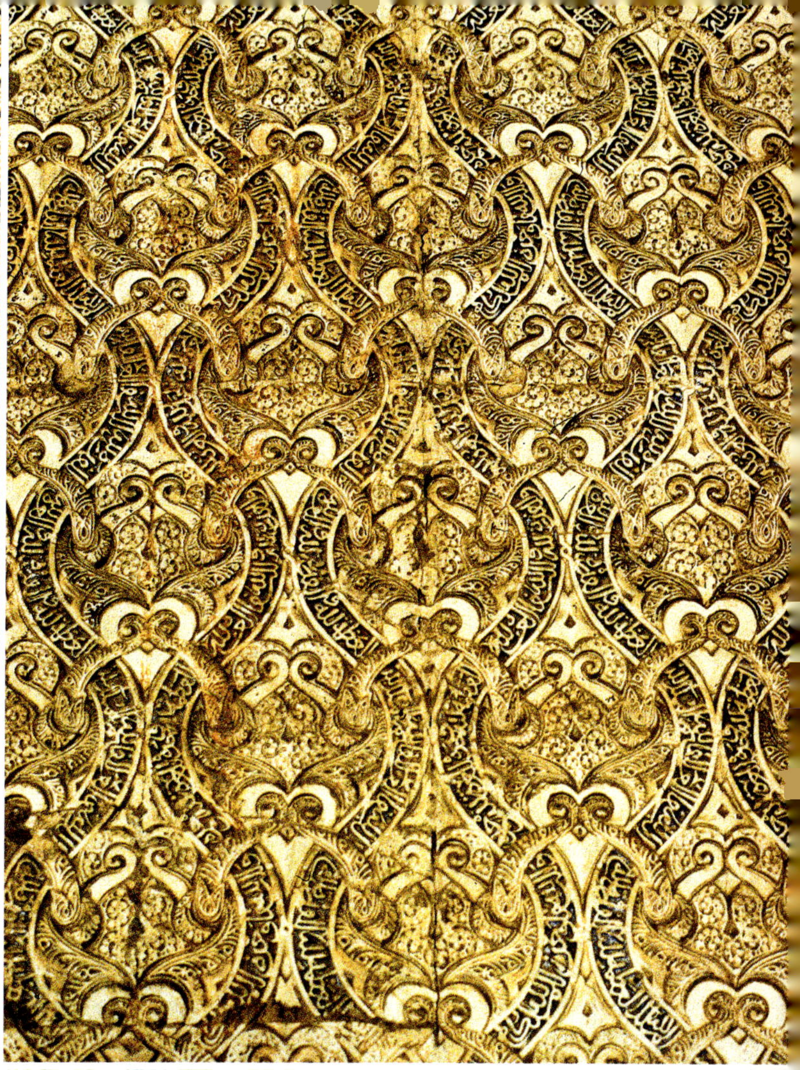

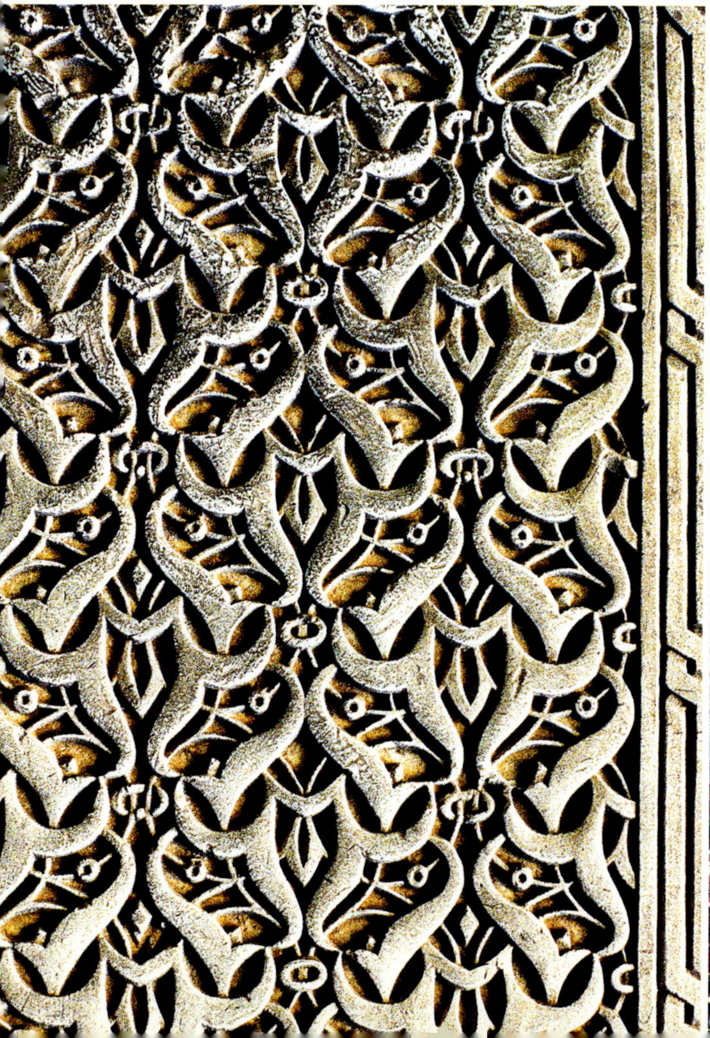

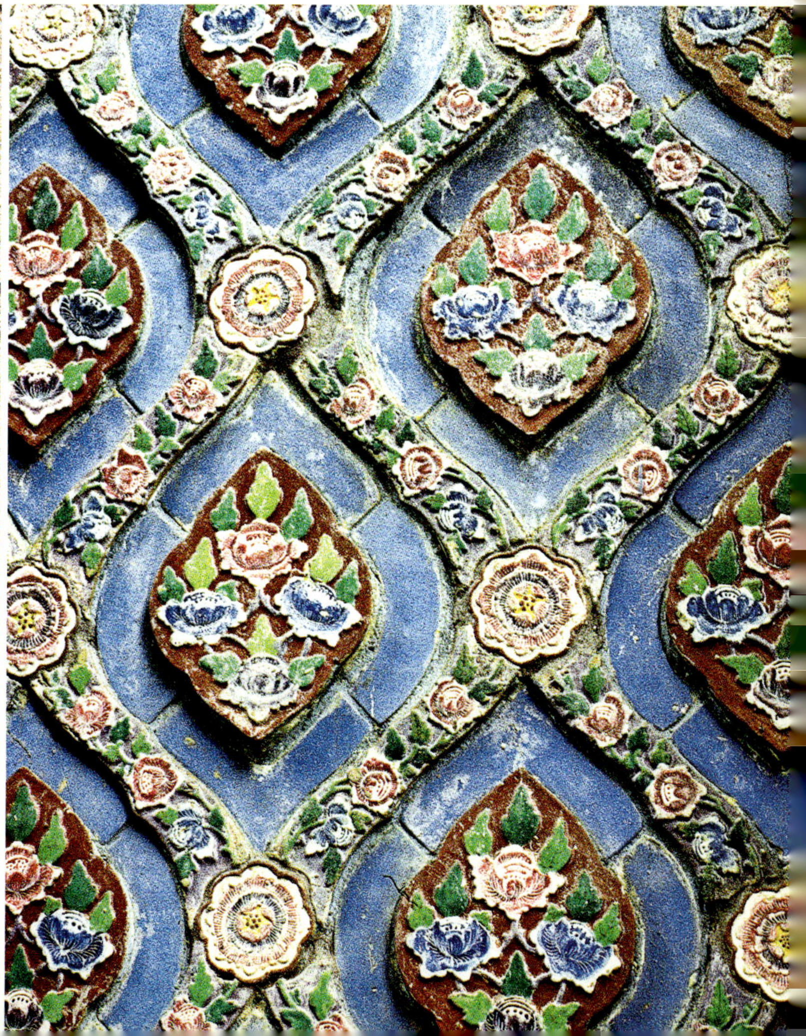

Top left: Pavement of the Main Square in Baja, Hungary

Top right: Pavement of the Main Square in Erice, Sicily

Bottom left: Pavement at the Alhambra Palace in Granada

Bottom right: Pavement in an Italian church

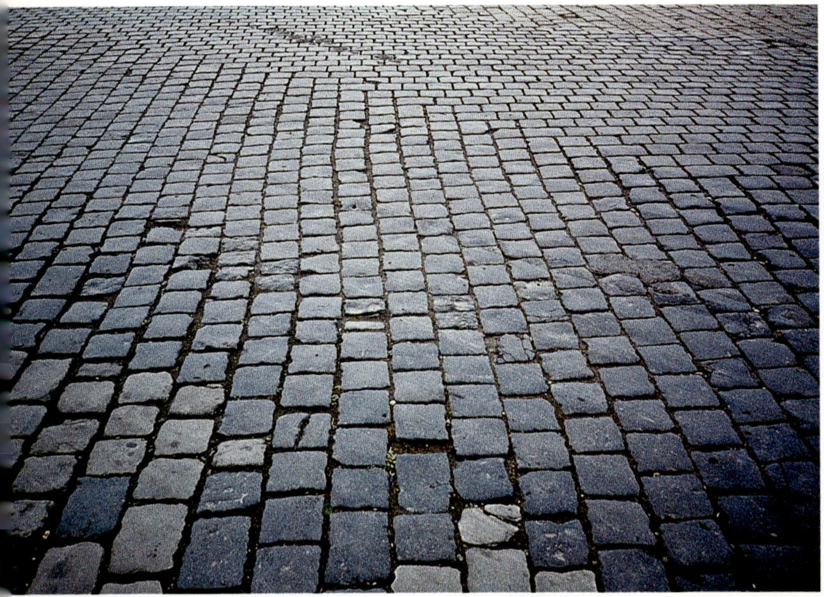

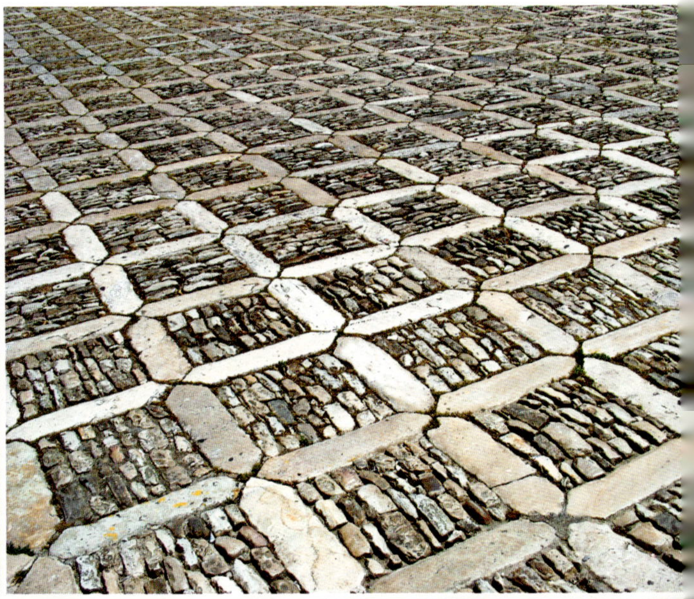

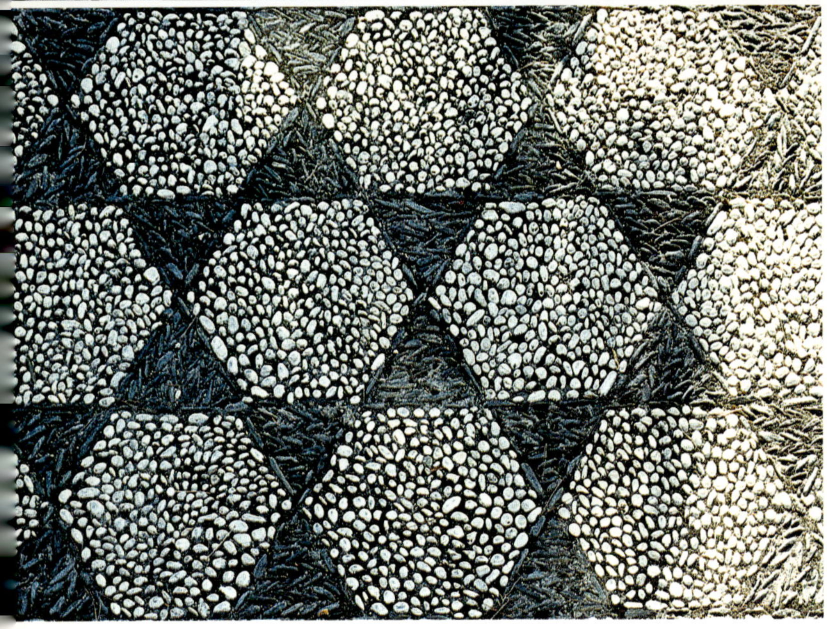

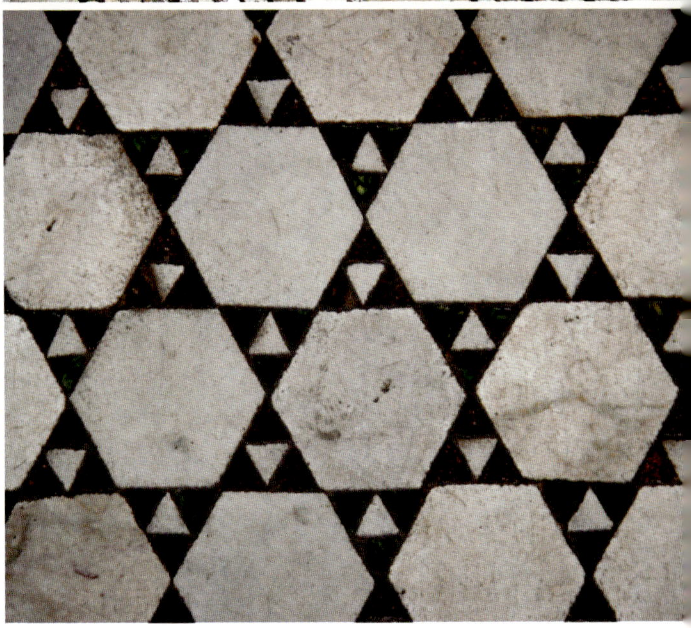

Top left: Wall mosaic at the Alhambra Palace in Granada, Spain

Top right: Brick pattern of an industrial chimney

Bottom left: The wall of the birthplace of Leonardo da Vinci in Vinci, Italy

Bottom right: Brick wall in Moscow

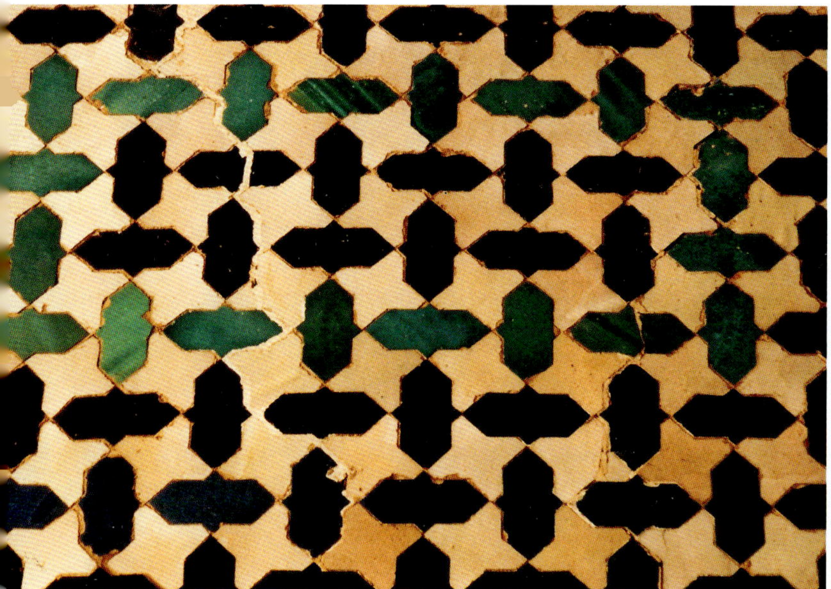

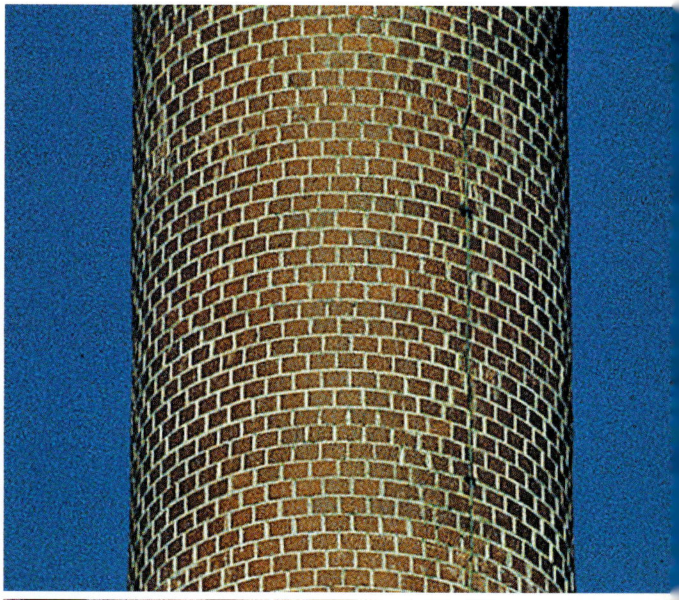

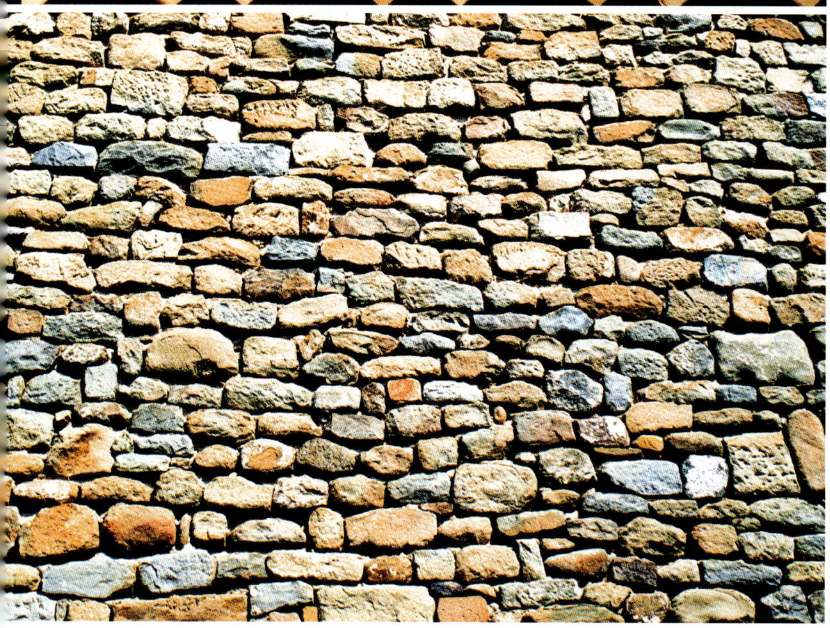

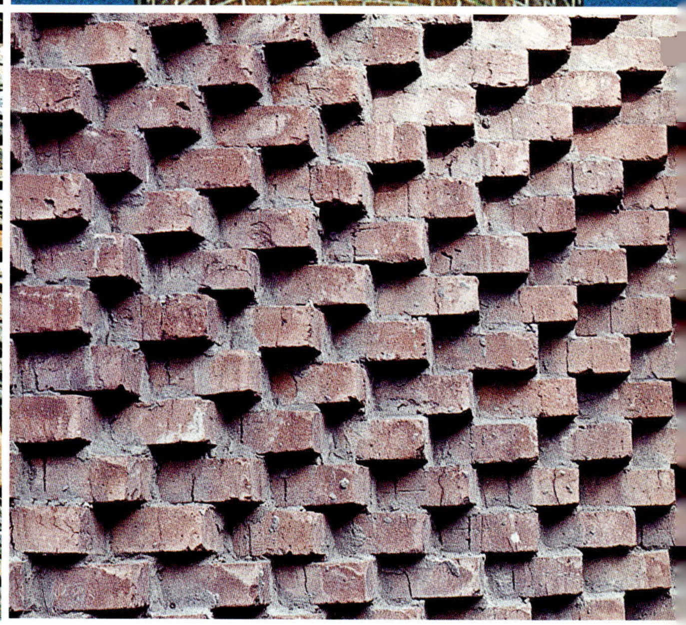

Top left: Detail of the roof of the Matthias Church in Budapest and [top right] of St. Stephen's Cathedral in Vienna, Austria

Bottom: Roof in Japan

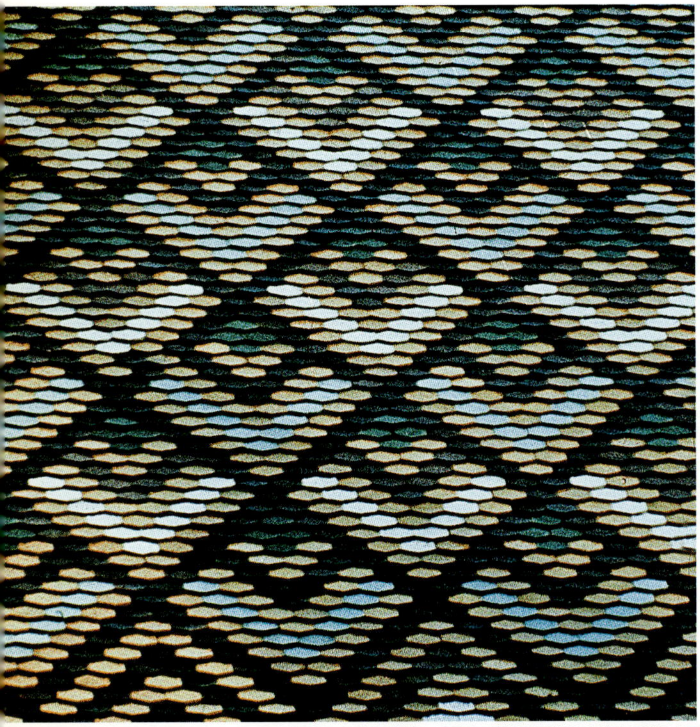

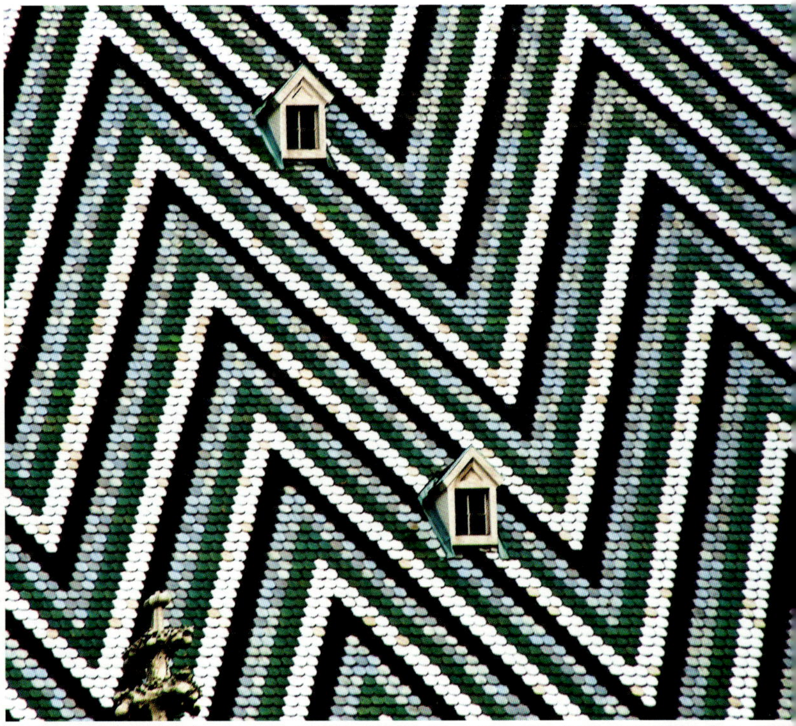

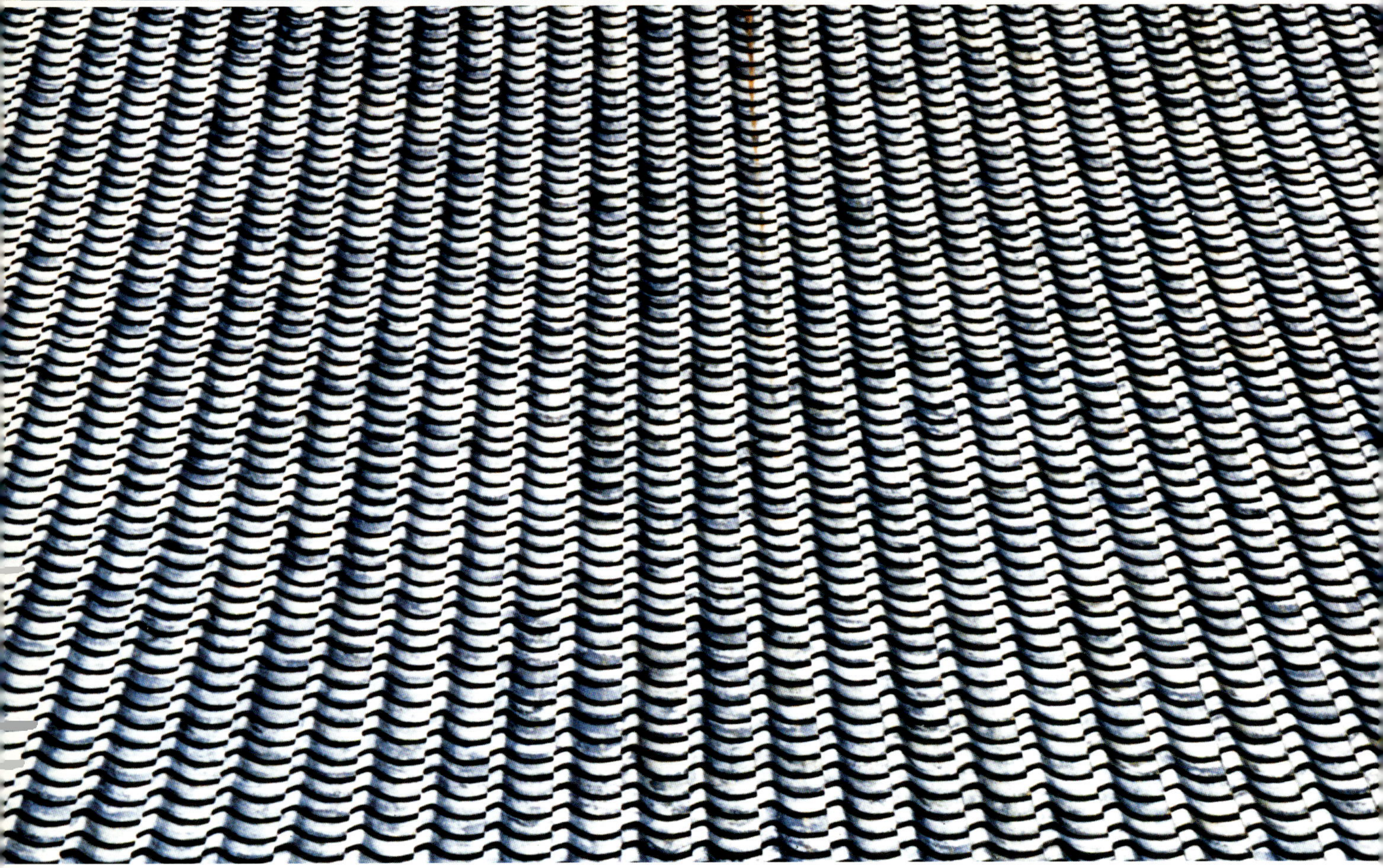

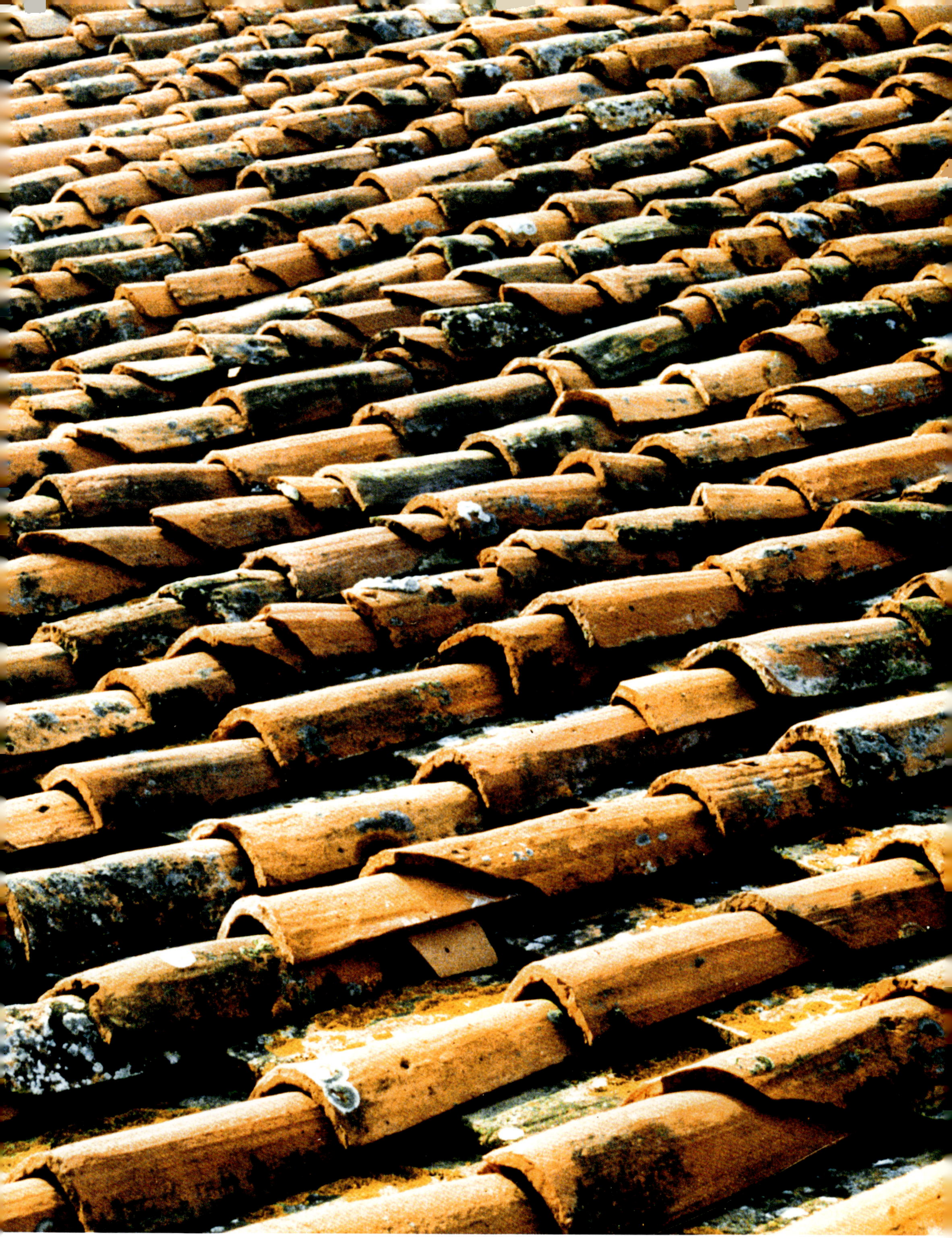

Roof in Vinci, Italy

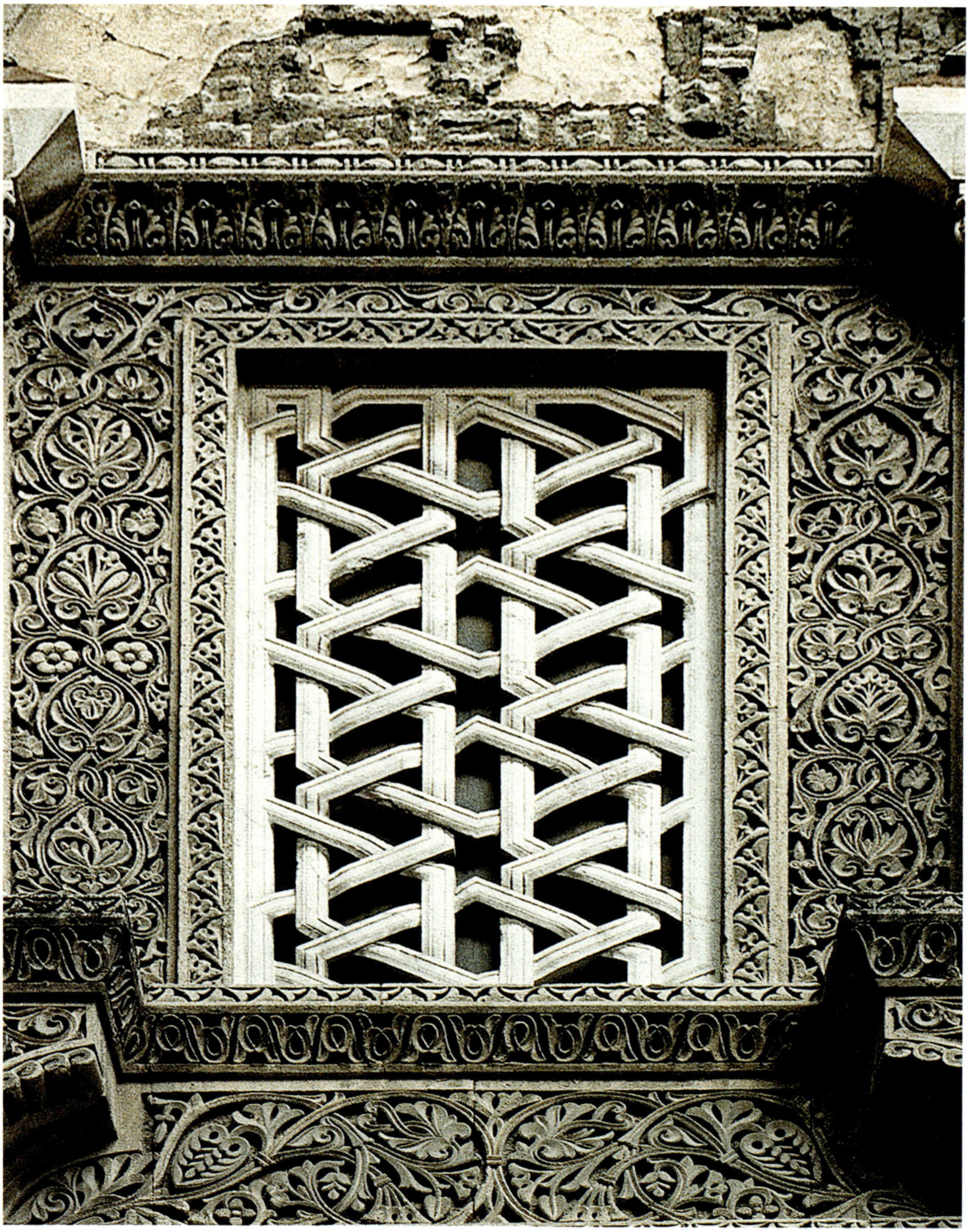

Window fences in Córdoba, Spain

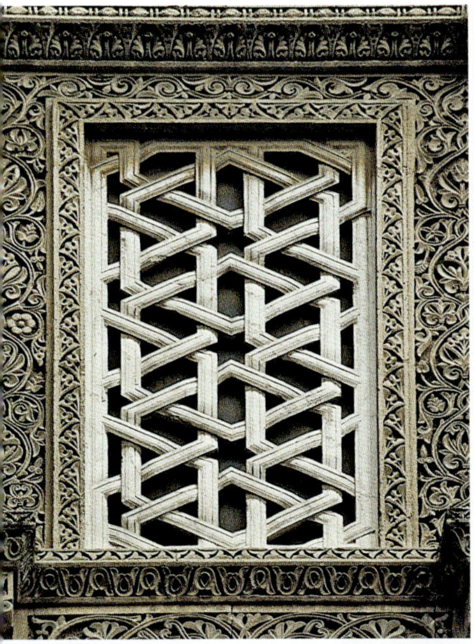
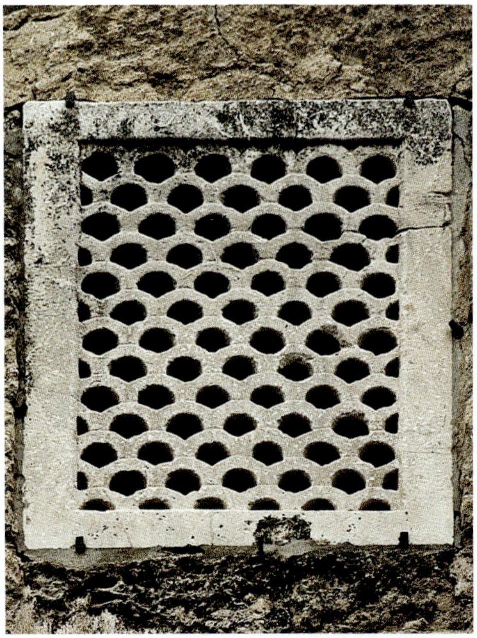
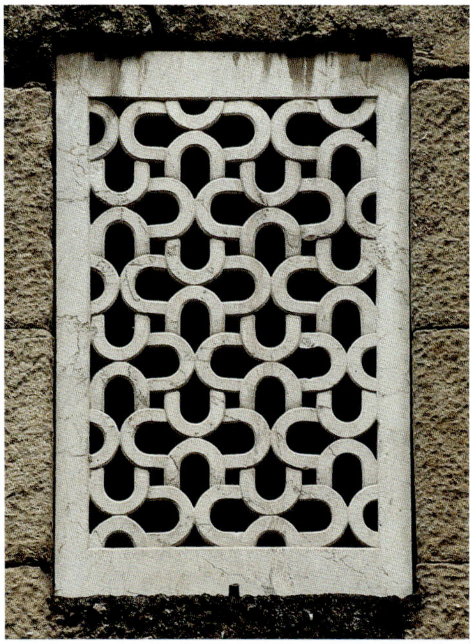
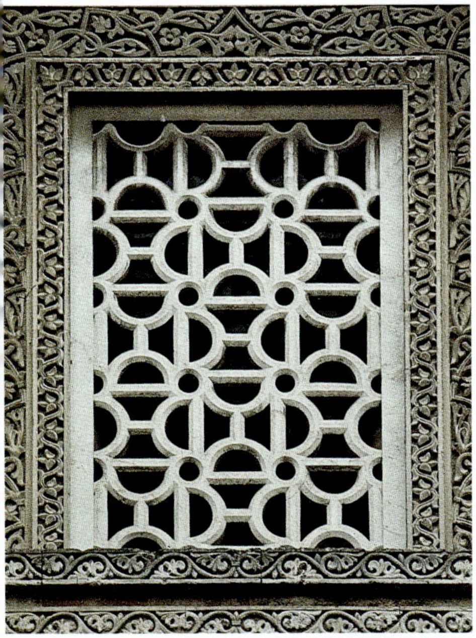
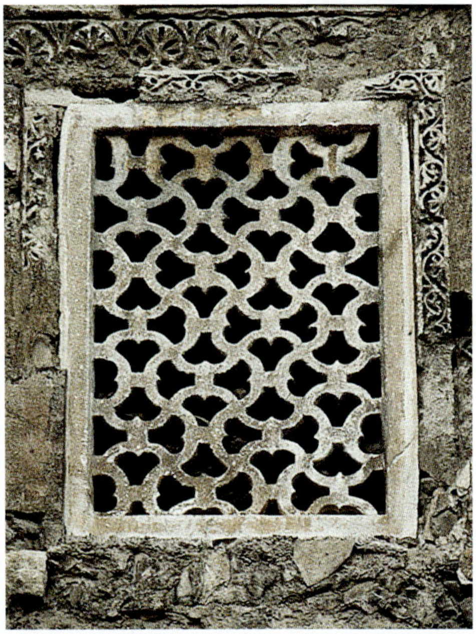
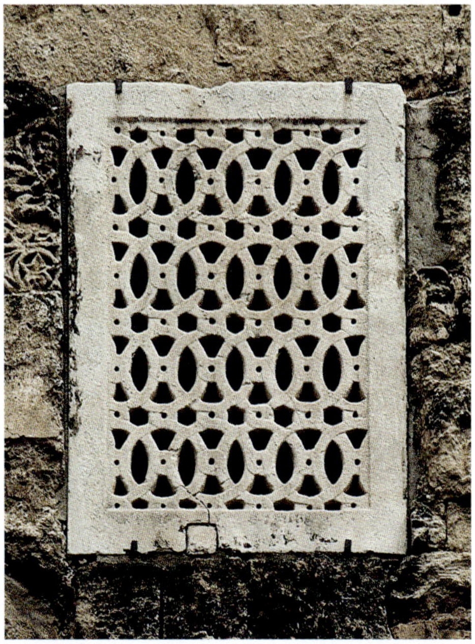
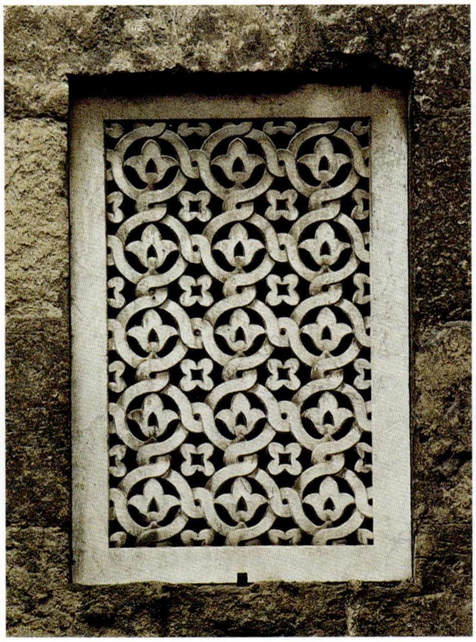
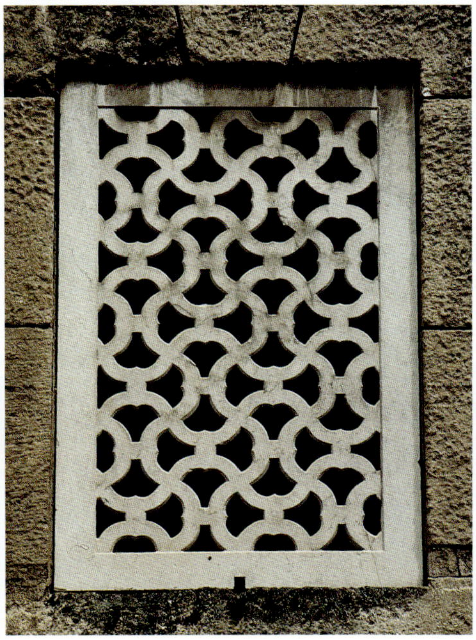
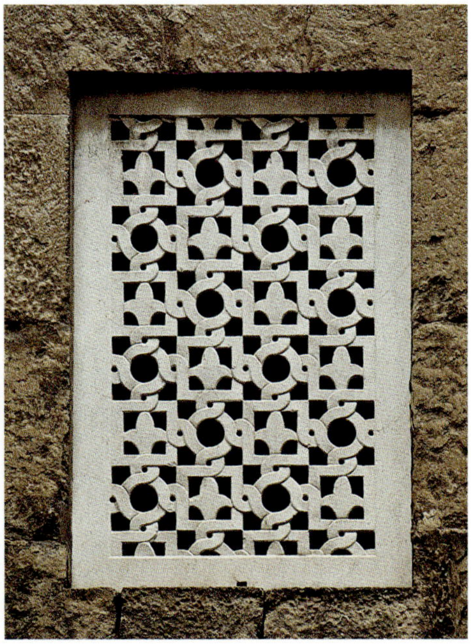

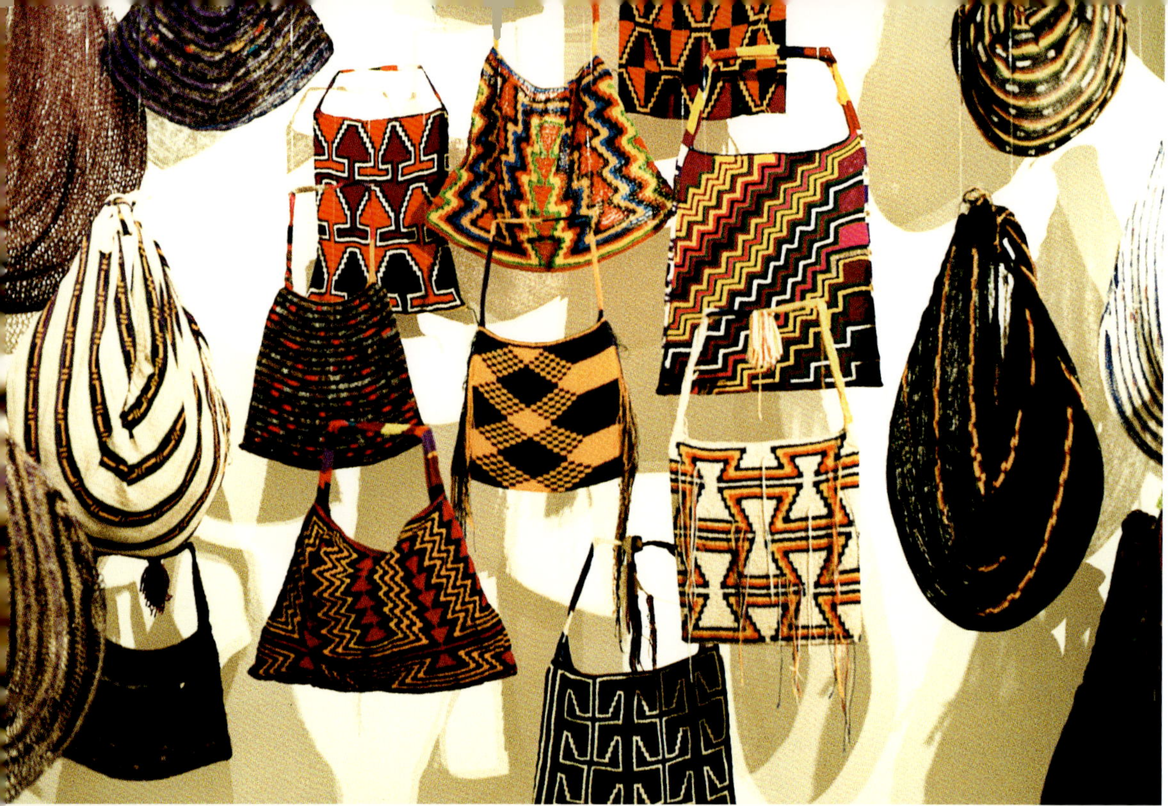

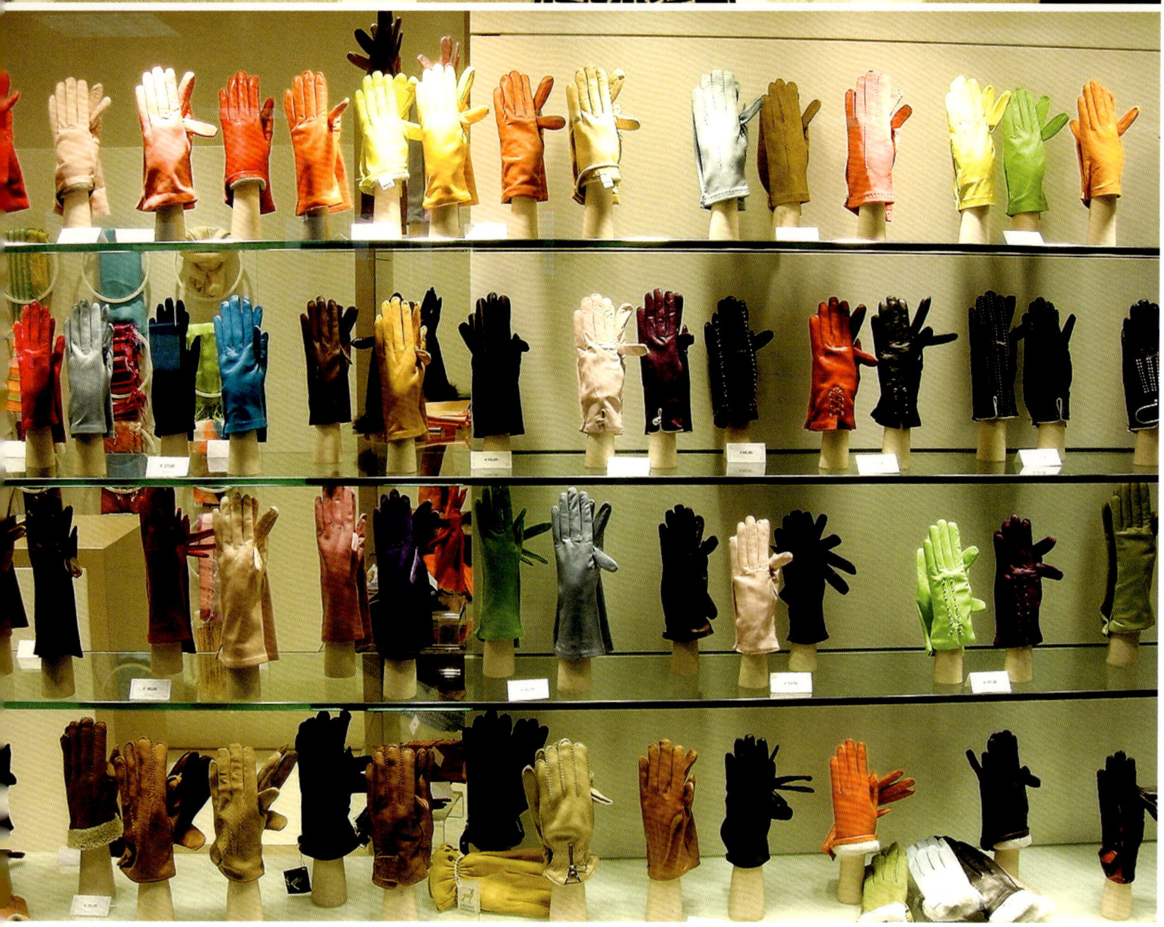

Top: Decorations on bags at an ethnographic exhibition

Bottom: Gloves in a shop window

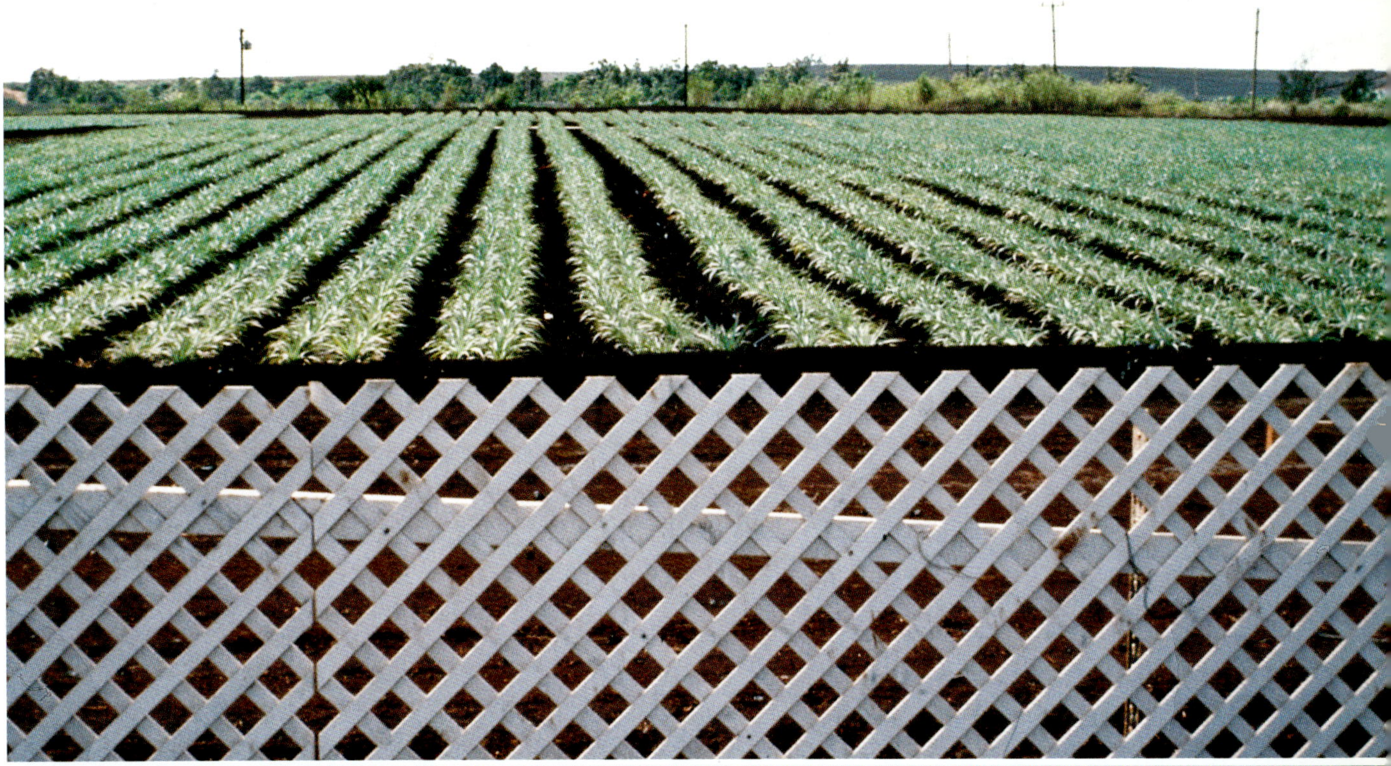

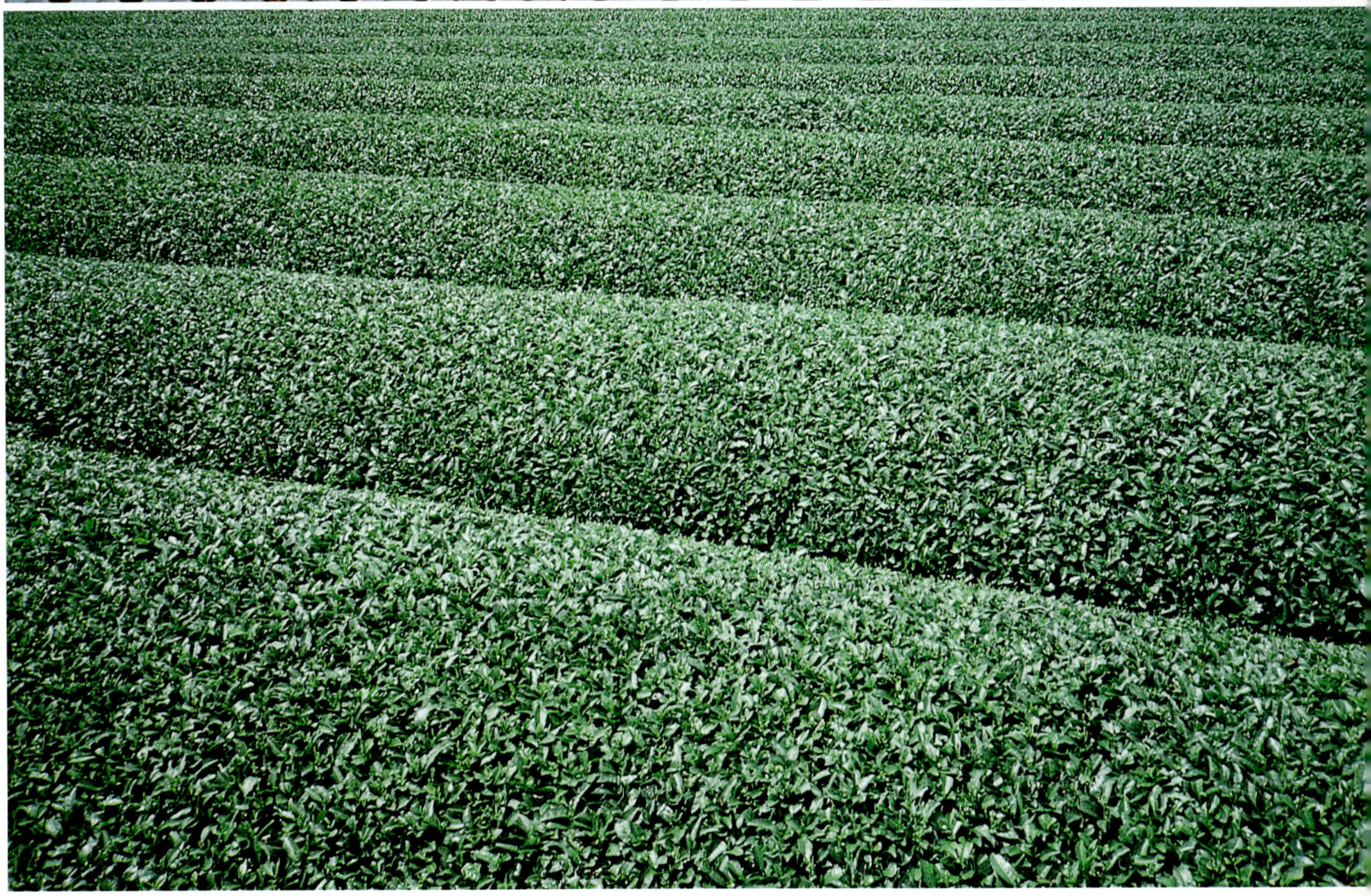

Top: Pineapple plantation behind a grated fence in Oahu, Hawaii

Bottom: Tea plantation in Fukuroi, Shizuoka, Japan

Façades

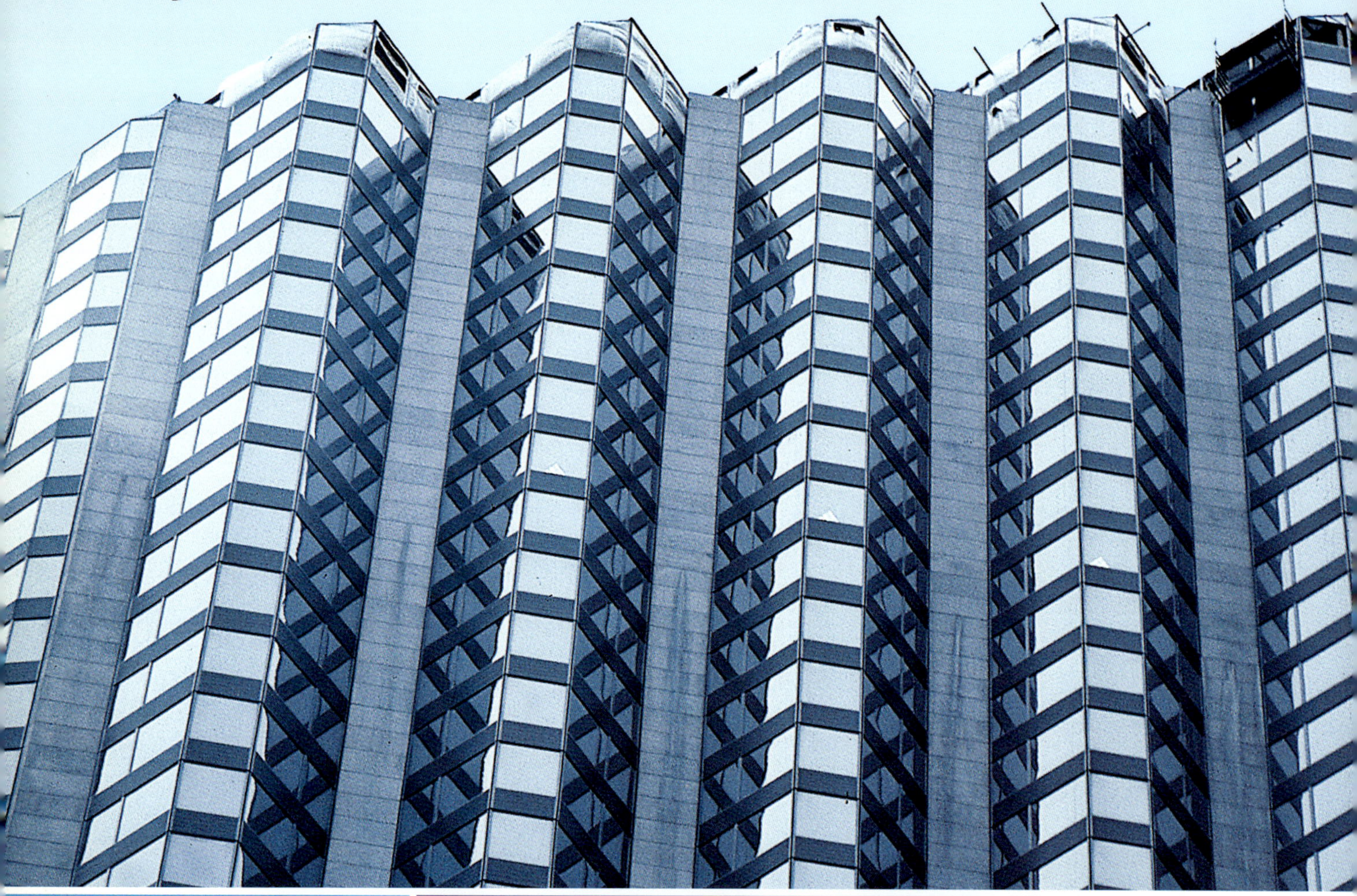

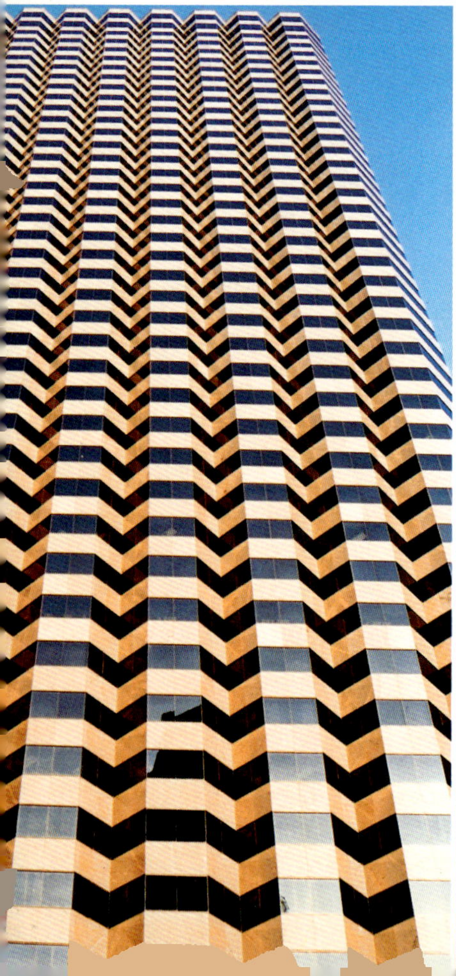

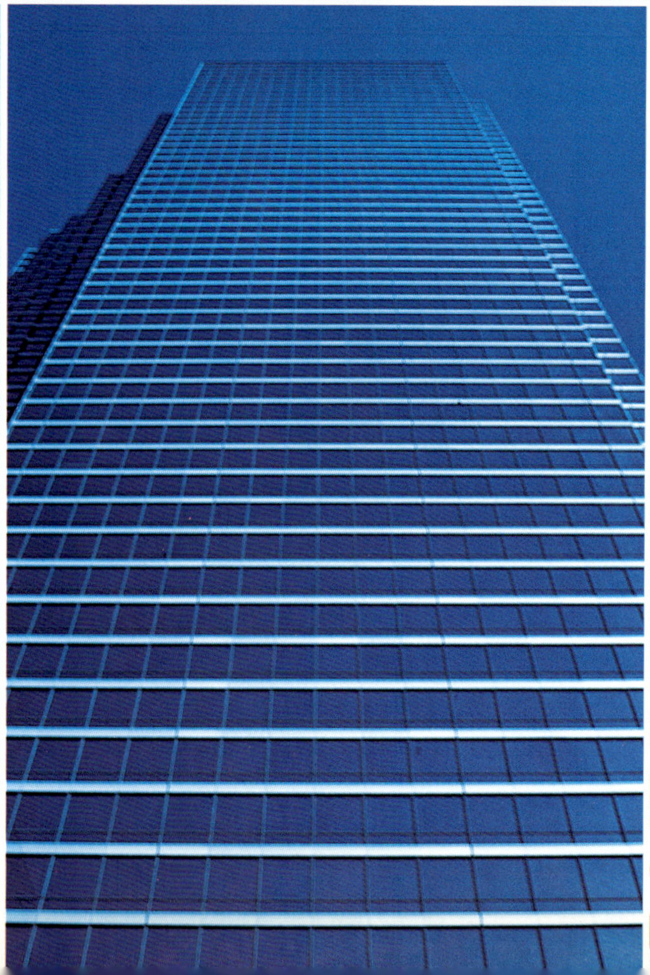

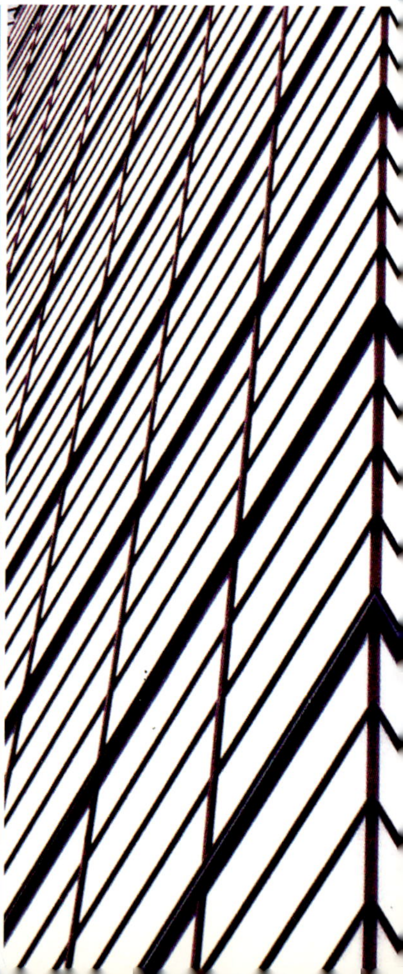

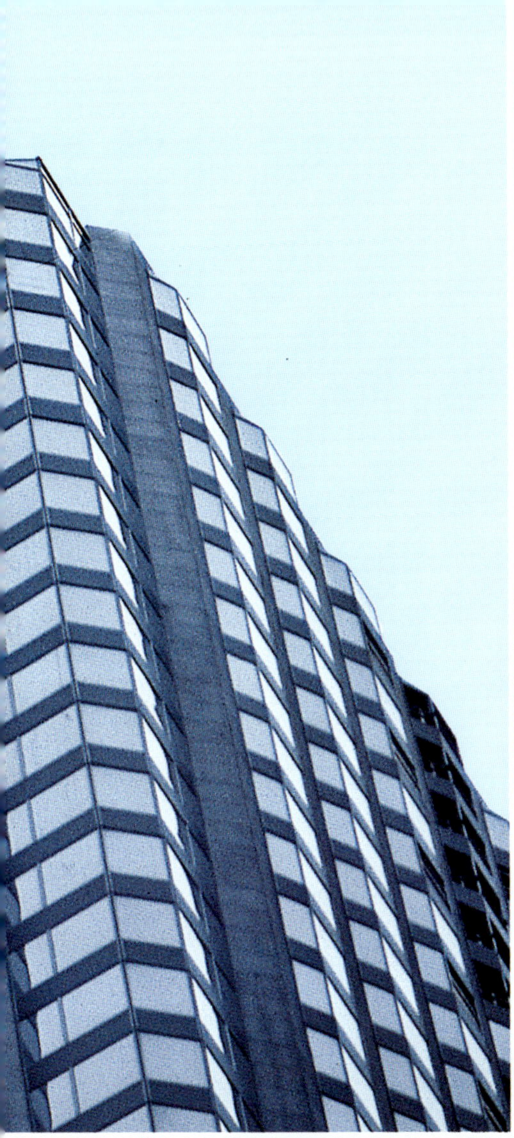

The façades of modern skyscrapers provide good examples of simple two-dimensional patterns. Most of our examples are from the United States, besides Great Britain, Israel, and France.

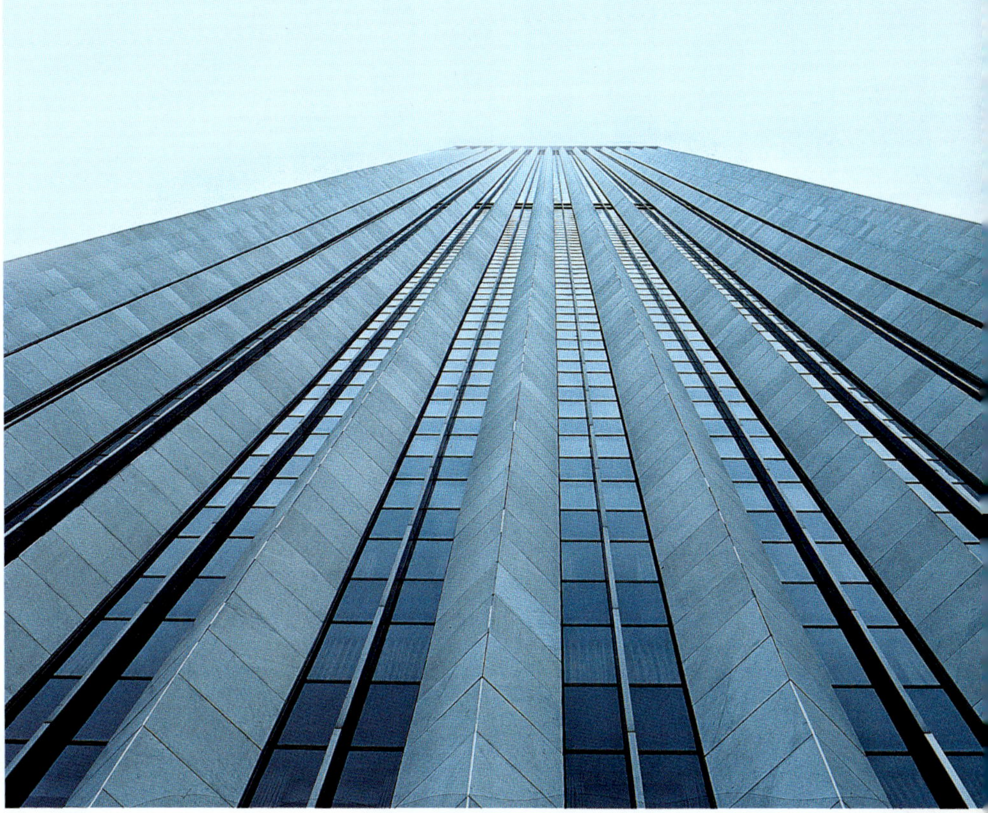

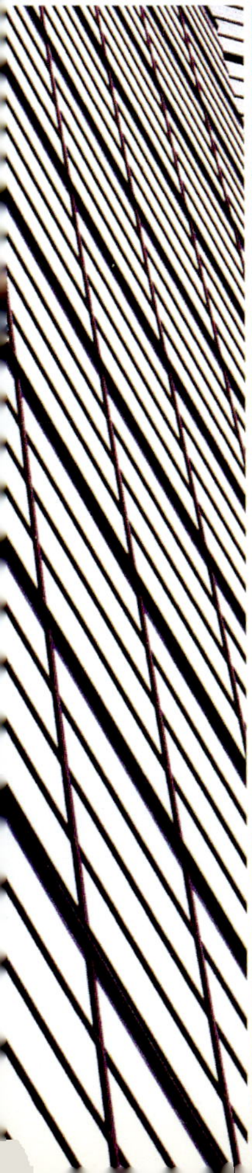

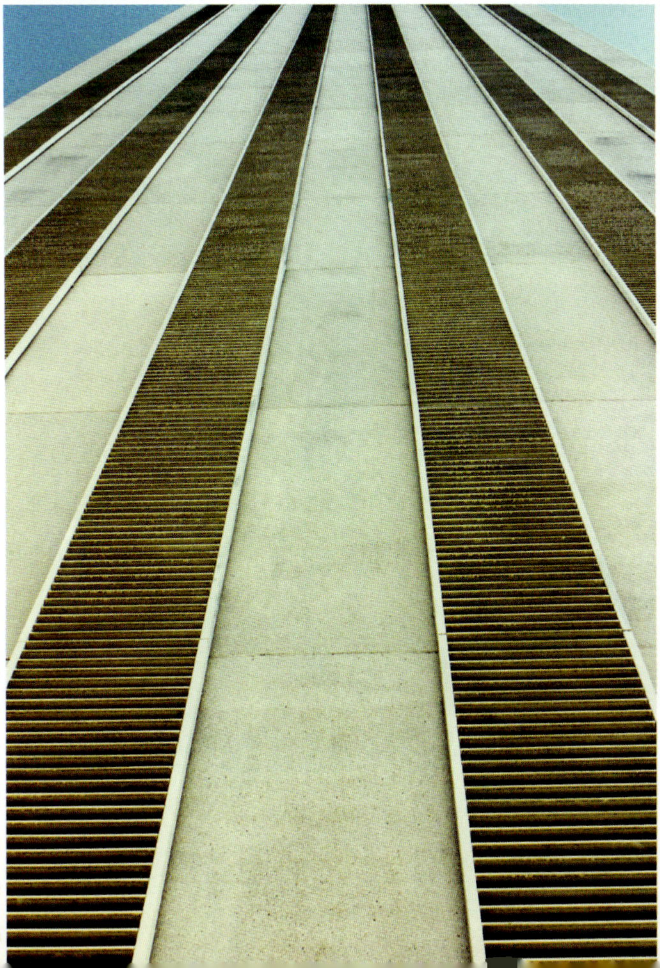

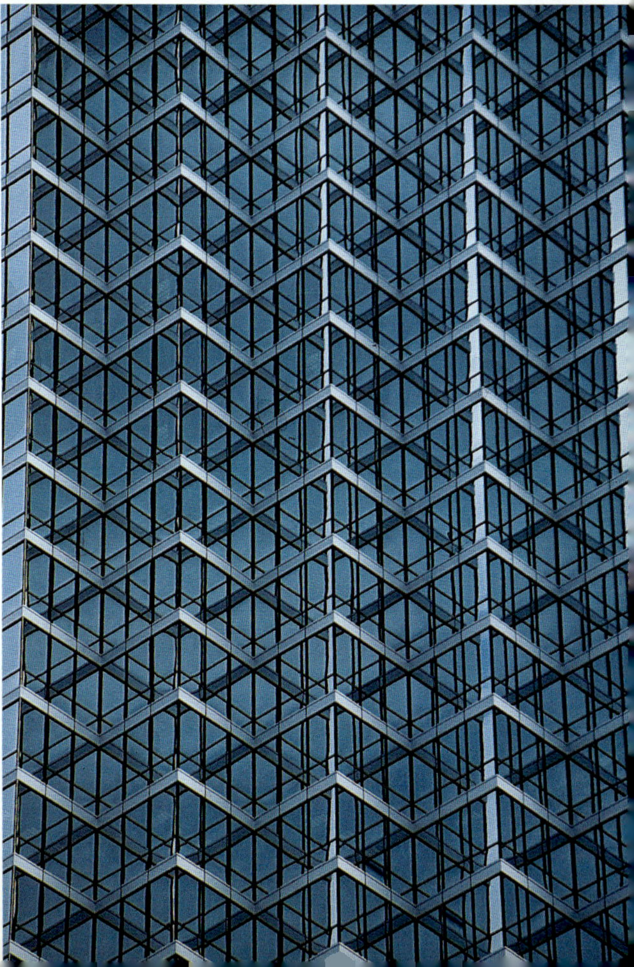

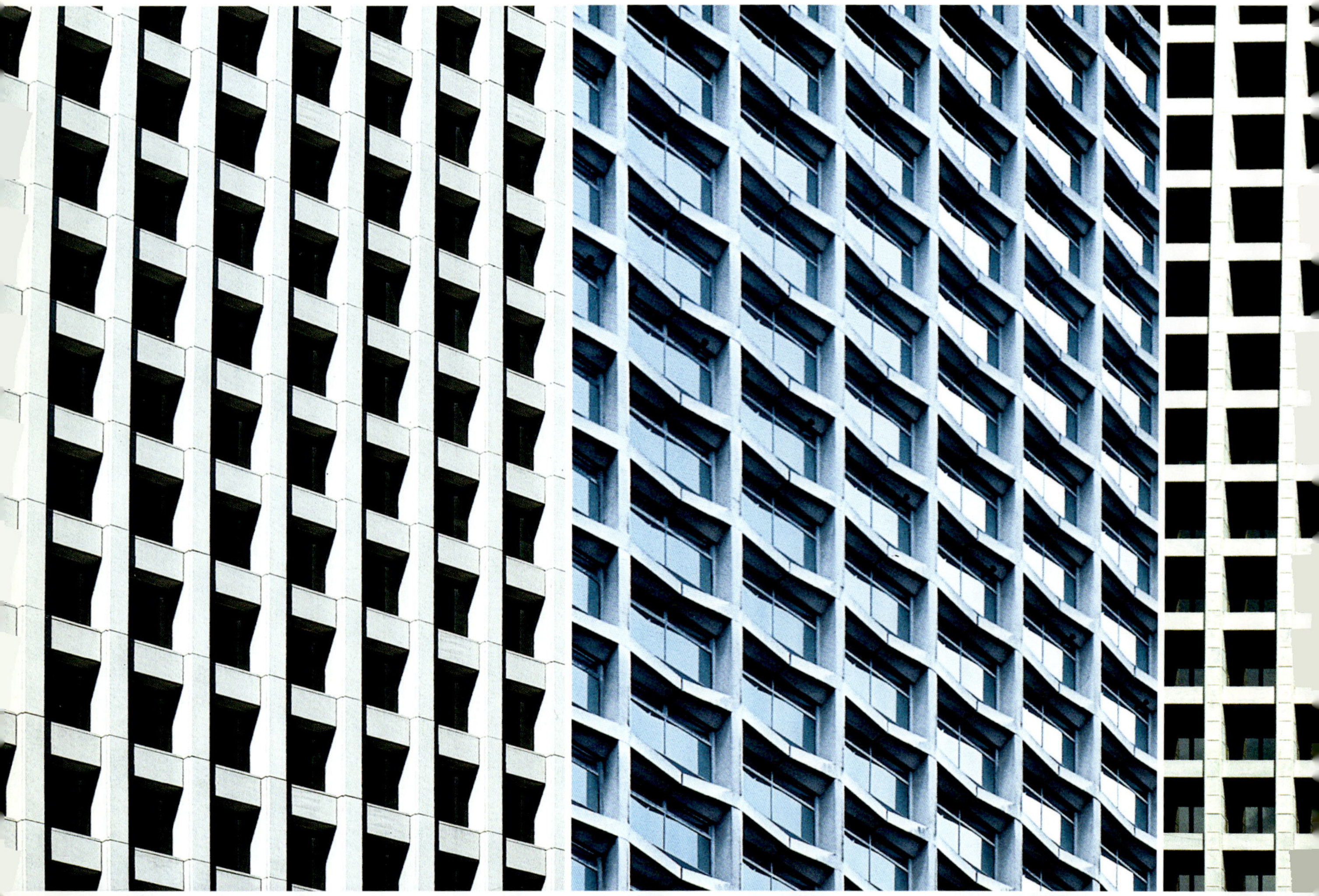

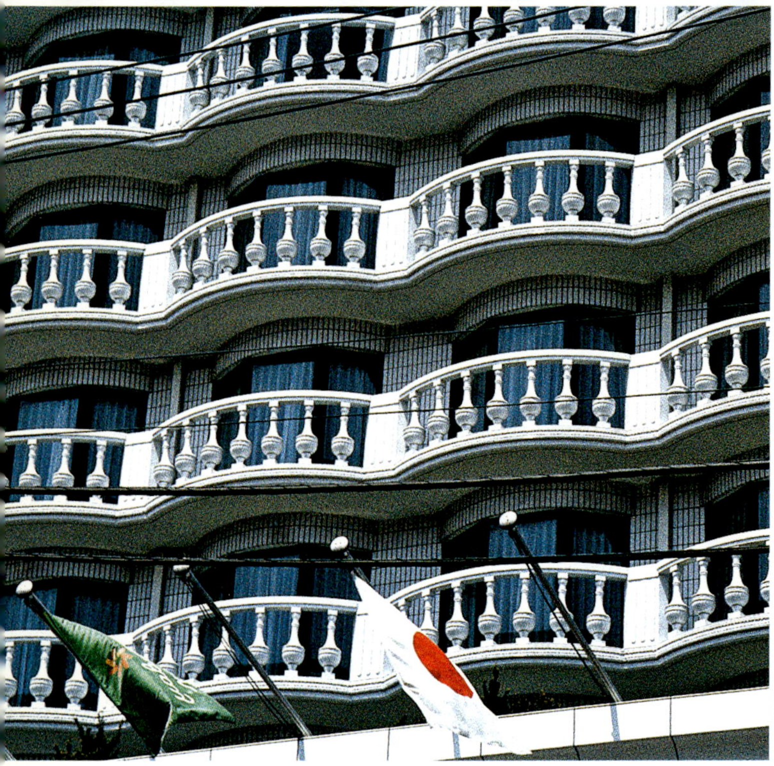

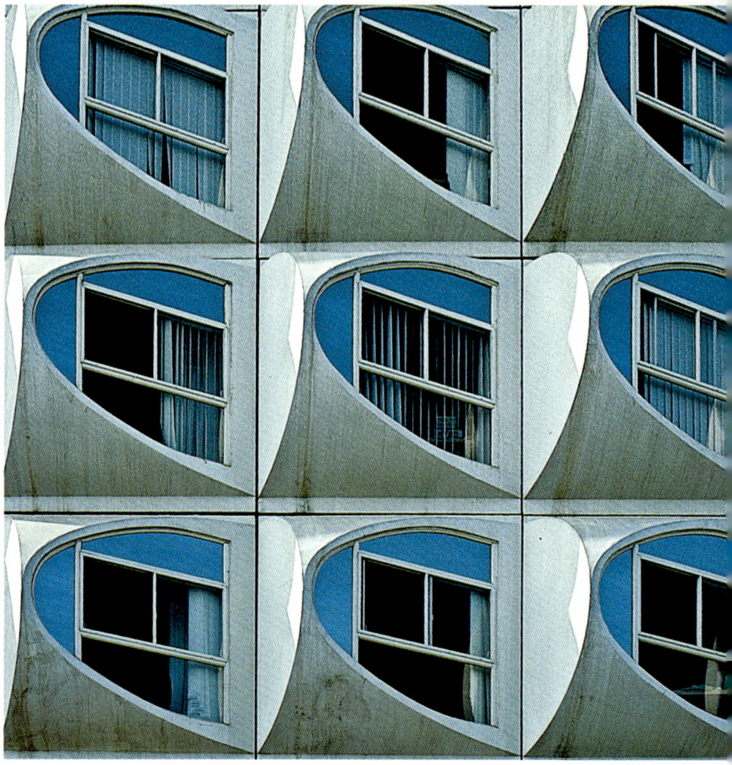

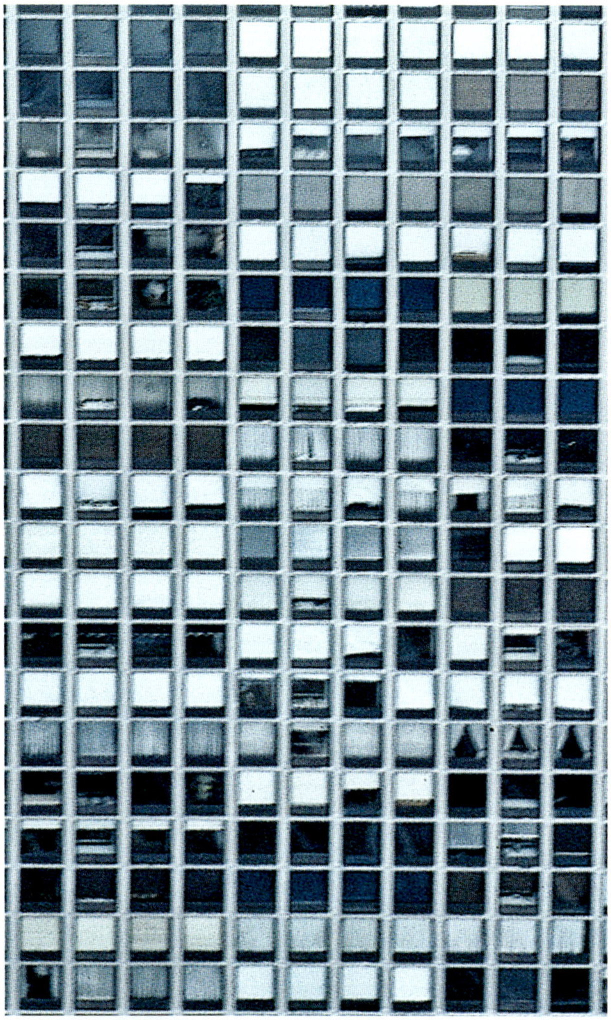

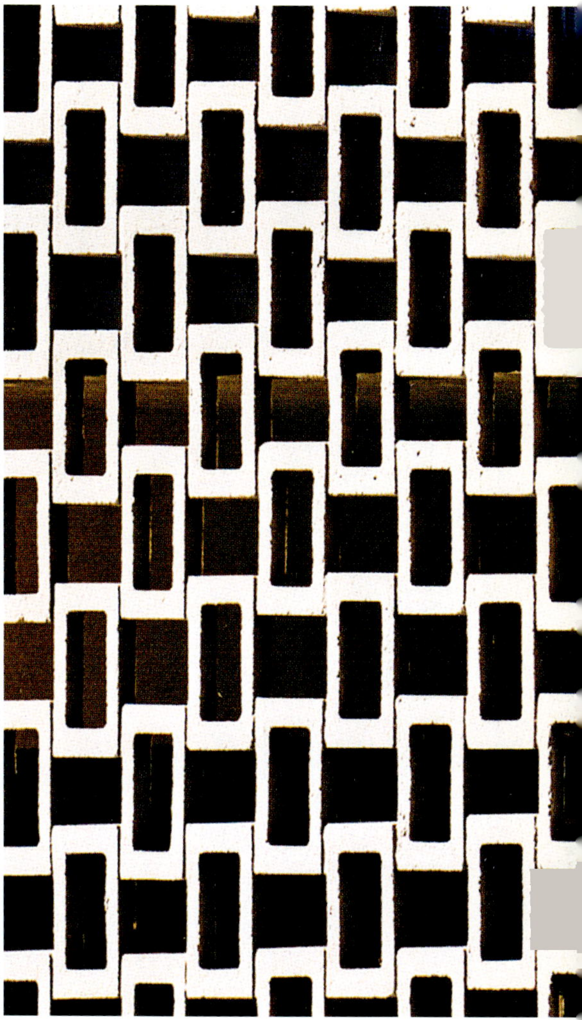

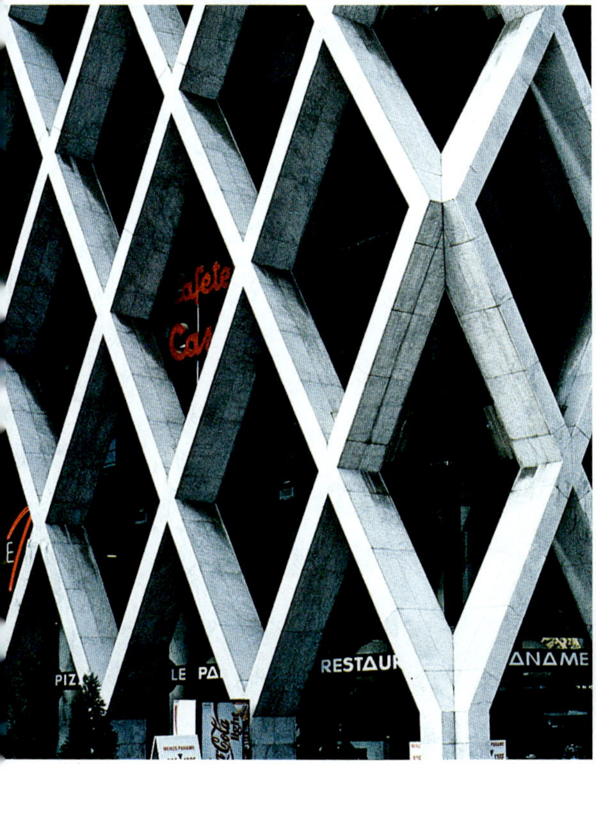

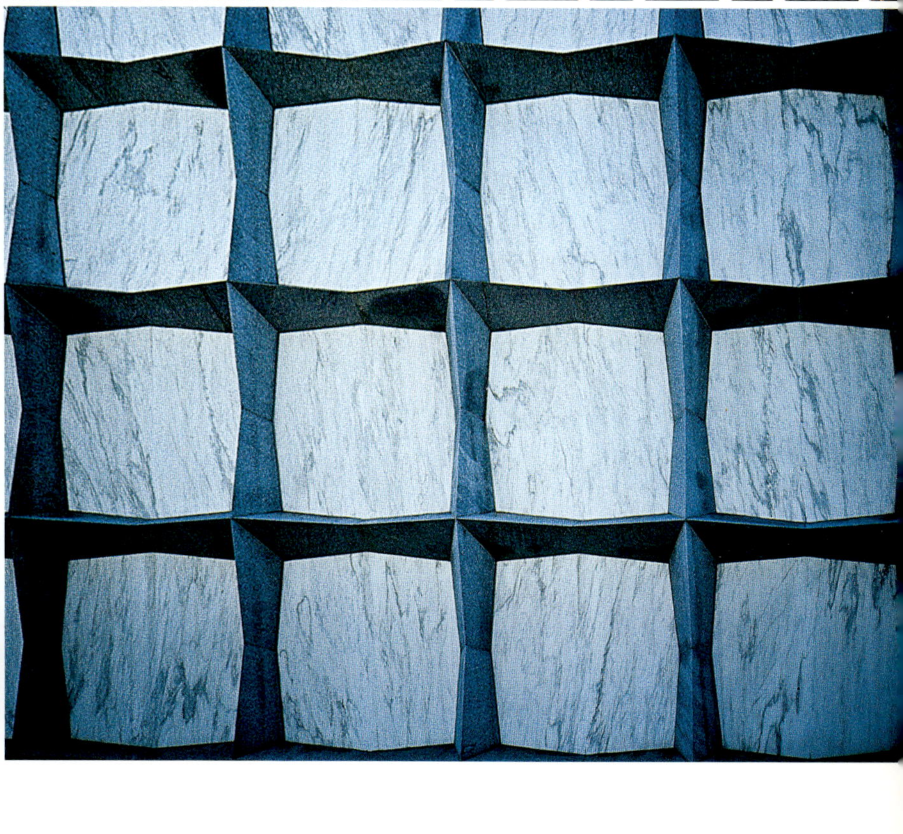

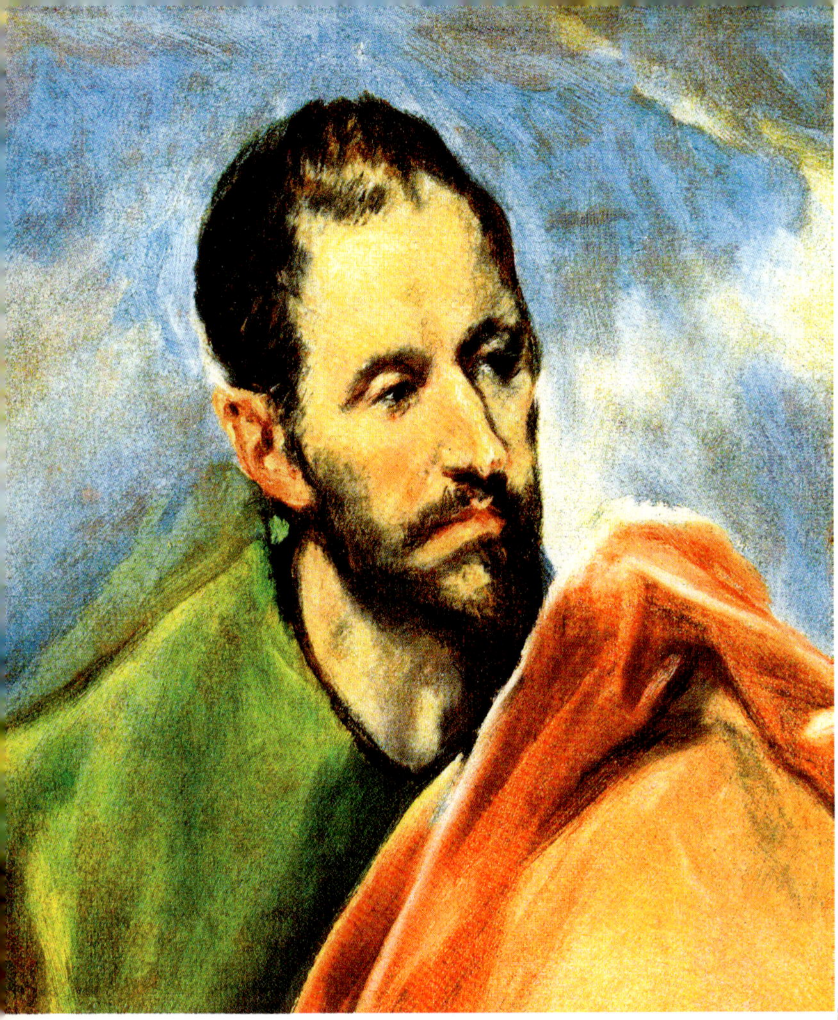

1 El Greco
2 Rembrandt
3 Waldemar Swierzy
4 Roy Lichtenstein
5 Ernst Ludwig Kirchner
6 Henri Matisse
7 Seymour Chuast
8 Vincent van Gogh
9 Christo
10 Francis Bacon
11 István Szőnyi
12 Salvador Dali
13 Amadeo Modigliani
14 József Nemes Lampért
15 Gustav Dore
16 Georges Braque
17 Waldemar Swierzy
18 Carlo Corra
19 Georges de la Tour
20 Joan Miro

El Greco, *Study Head*
(courtesy of the Museum of Fine Arts, Budapest)

The artist István Orosz provided an intriguing kind of repetition in a painting titled *Hommage à El Greco* with 20 variations of El Greco's famous *Study Head*. We offer this image on page 183 as a puzzle to guess who are the artists whose styles are imitated by Orosz. The answers are given at the top right of this page.

The façades with highly symmetrical patterns convey an atmosphere of dull uniformity against which the famous architect Hundertwasser made a statement with his buildings with conspicuously diverse façades.

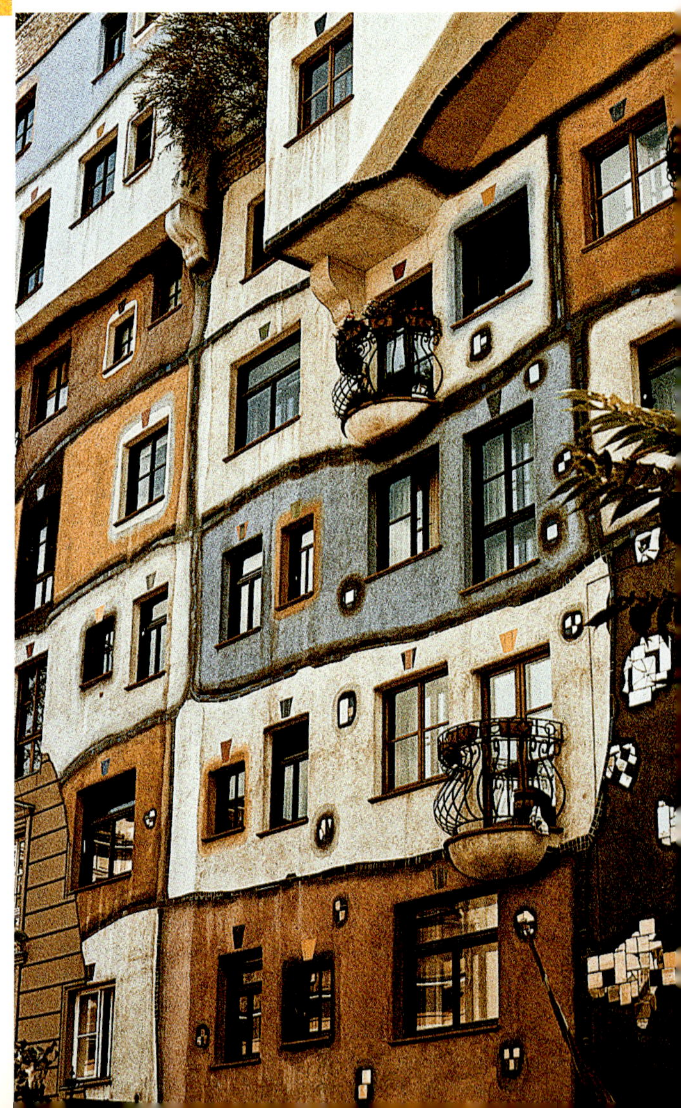

Hundertwasser house in Vienna

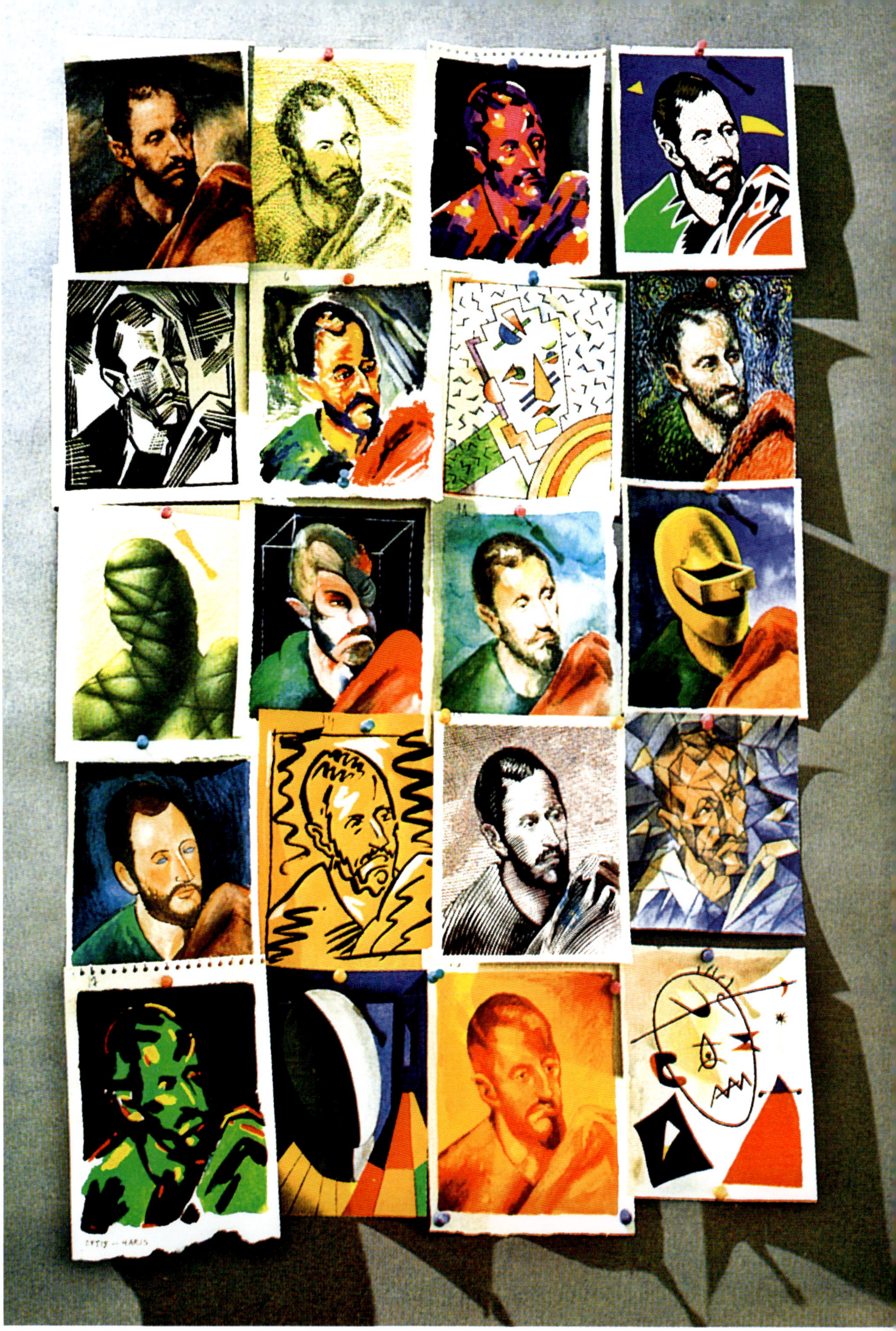

István Orosz, *Hommage à El Greco*

(courtesy of the artist)

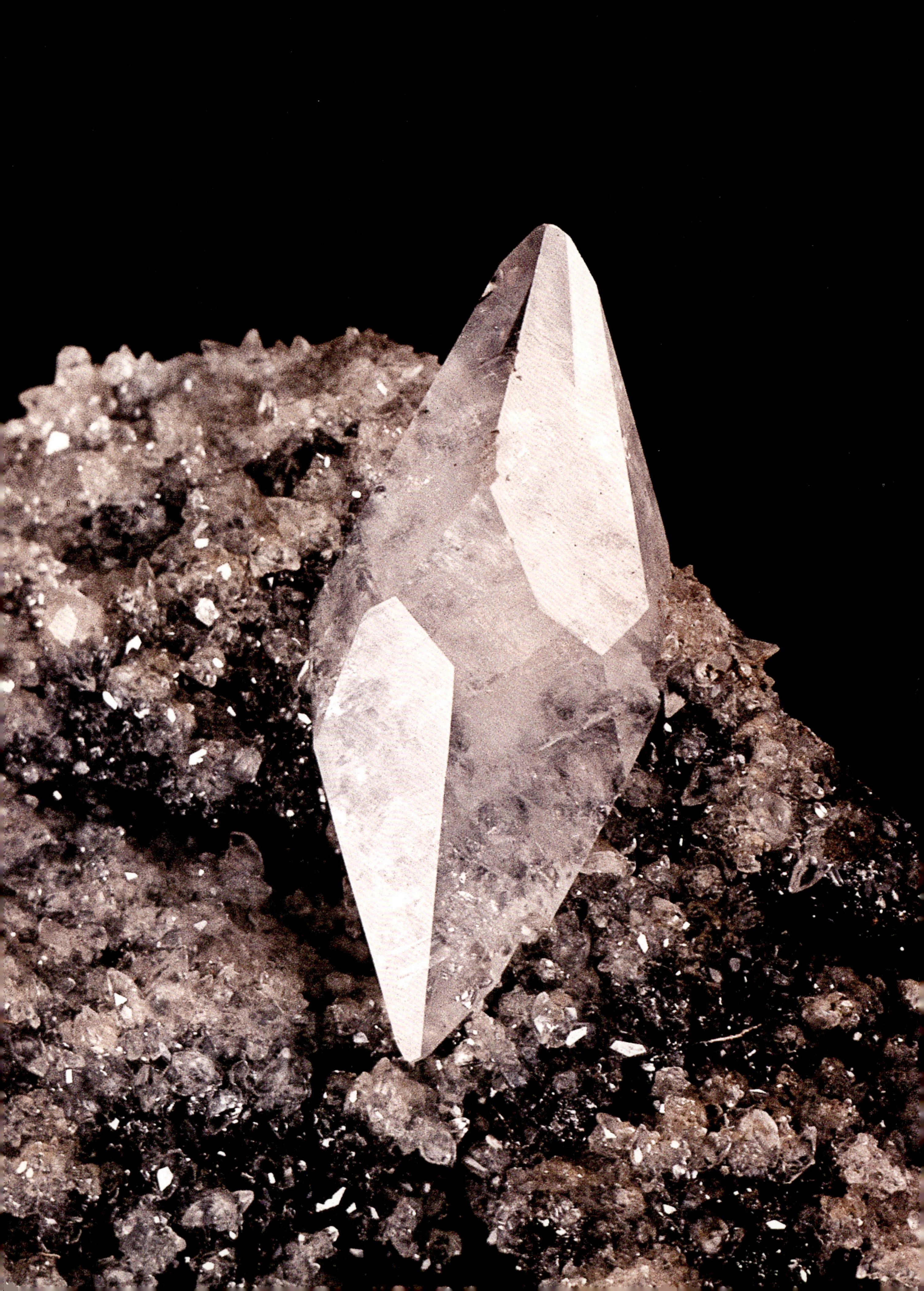

Crystals

Crystals have always fascinated people. Karel Ĉapek, the Czech writer, wrote the following after his visit to the mineral collection at the British Museum in London (K. Ĉapek, *Anglické Listy*, Československý Spisovatel, Prague, 1970):

"There are crystals as huge as the colonnade of a cathedral, soft as mould, prickly as thorns; pure, azure, green, like nothing else in the world, fiery, black; mathematically exact, ...There are crystal grottos, ...architecture and engineering art ... Egypt crystallizes in pyramids and obelisks, Greece in columns; the Middle Ages in vials; London in grimy cubes ...To equal nature it is necessary to be mathematically and geometrically exact."

The snowflake is a water crystal. The word crystal comes from the Greek *krystallos*, meaning "clear ice." The name originated from the mistaken belief that the beautiful transparent quartz stones found in the Alps were formed from water at extremely low temperatures. Crystals generally have beautiful symmetrical shapes.

We will return to this question below, but already here we note that it is the internal structure that is decisive for a crystal and its external appearance may occur in a variety of shapes.

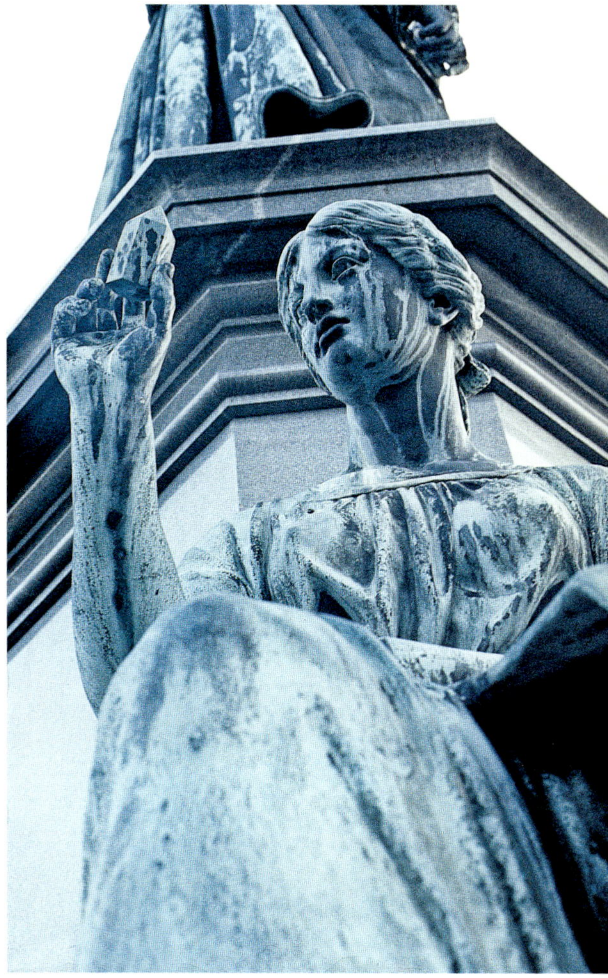

Woman holding a crystal as part of the Linneaus sculpture composition in Stockholm

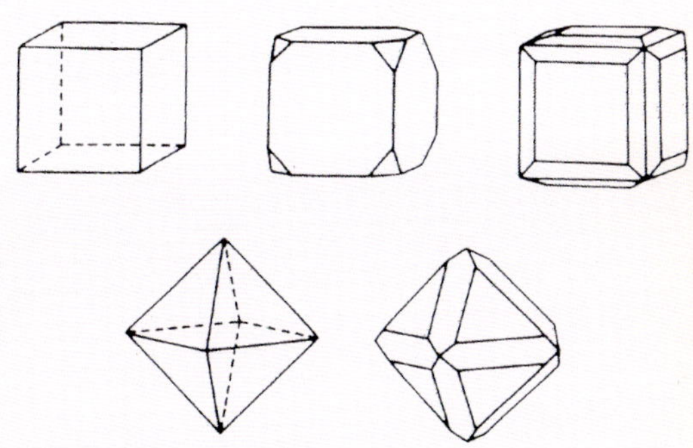

Shapes of sodium chloride (NaCl) crystals

Calcite crystal from Gyöngyösoroszi, Hungary
(from the mineral collection of Eötvös Loránd University in Budapest, courtesy of István Gatter)

Crystals are grouped into 32 classes according to their external appearance. Each is represented by one example (after M. Hargittai, I. Hargittai, *Symmetry through the Eyes of a Chemist.* 3rd Edition. Springer, 2009).

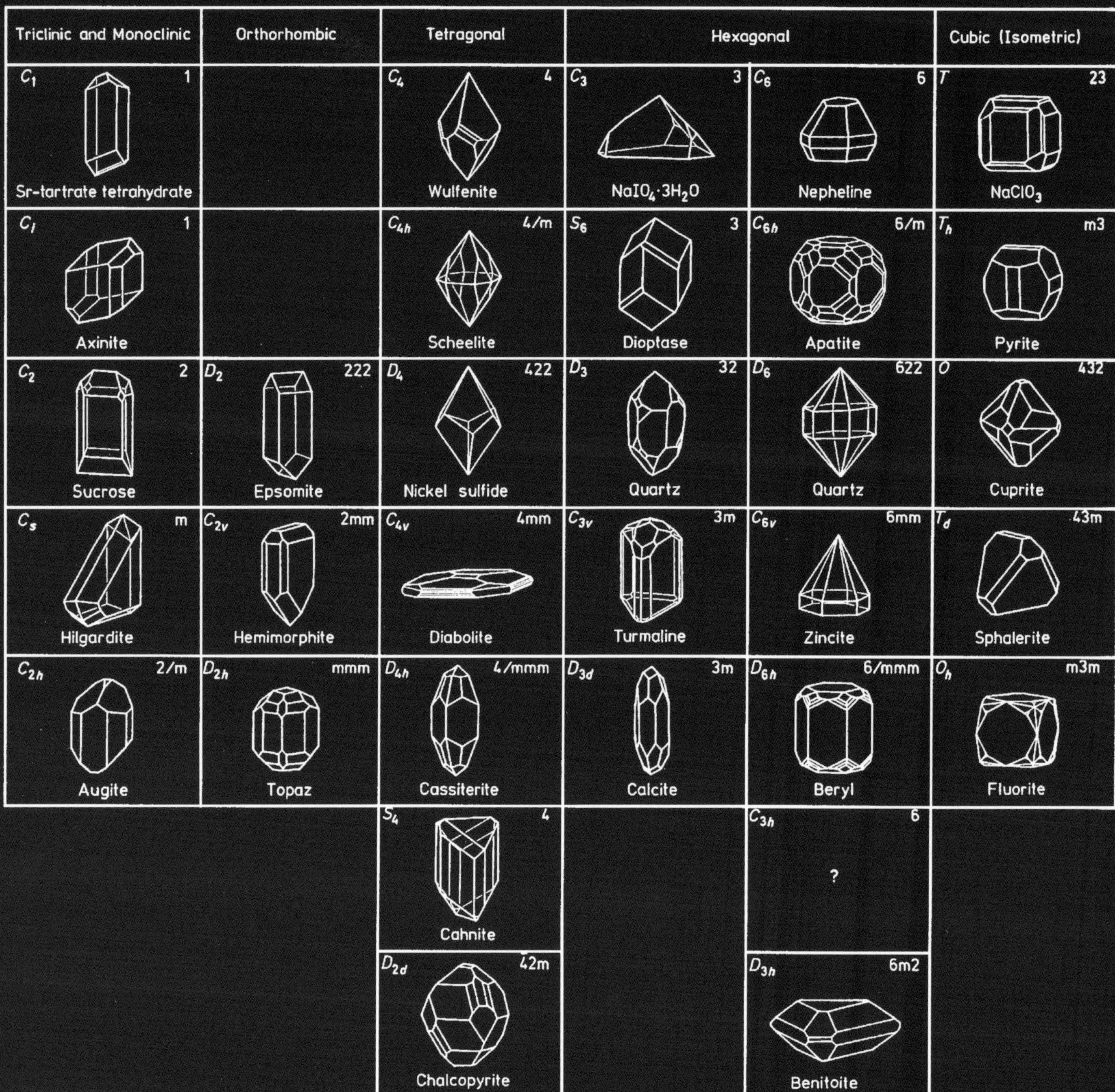

The next 6 pages display crystals from the mineral collection of Eötvös Loránd University in Budapest.
(courtesy of Tamás Váczi)

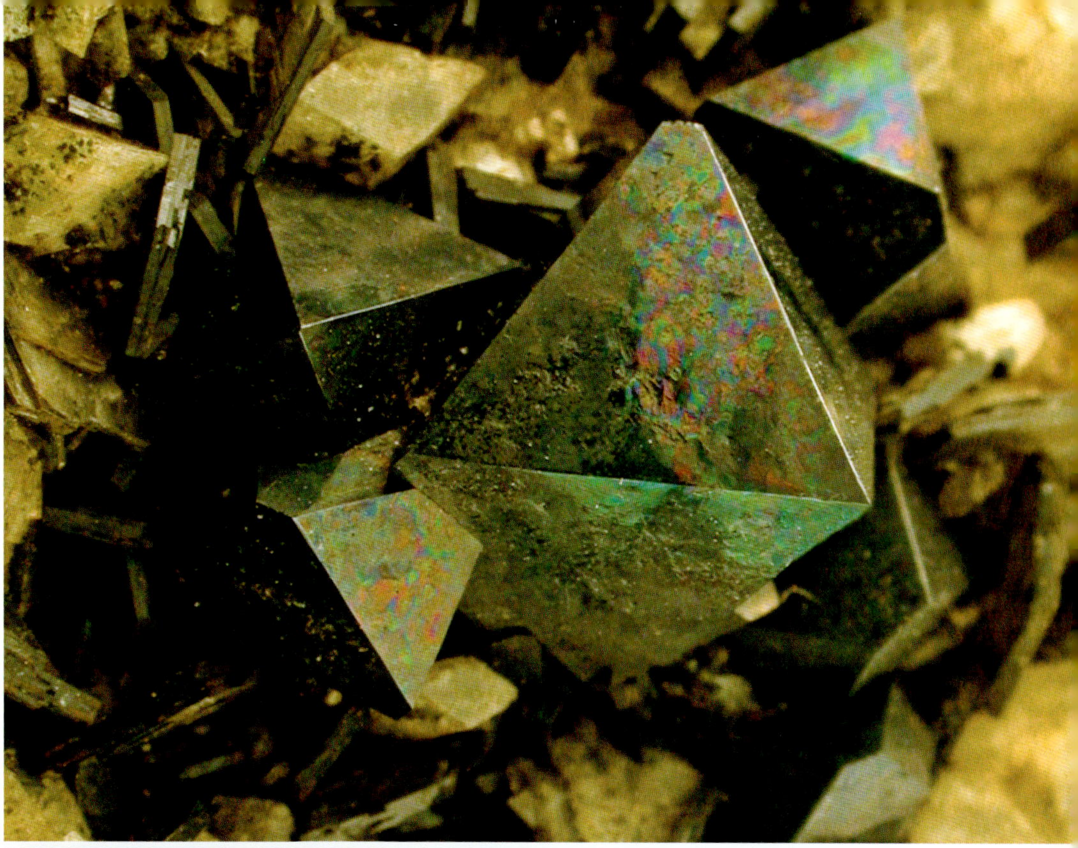

Magnetite from Switzerland

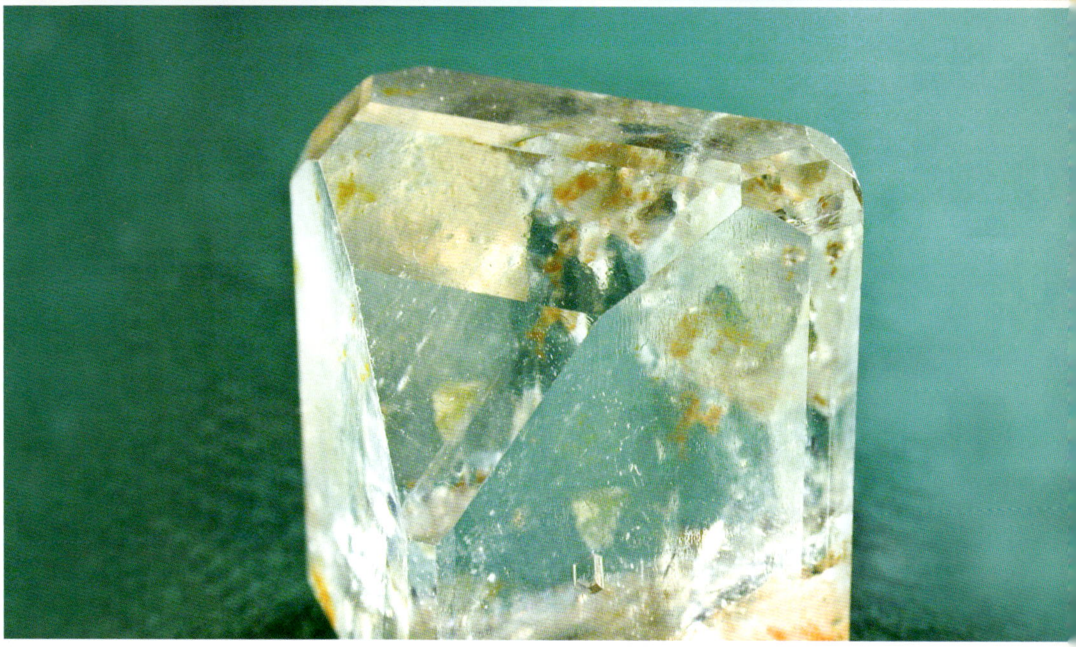

Topaz from Nigeria

Crystals have two kinds of symmetry, external and internal. The external symmetry is due to the internal structure, but there is no simple relationship between the shape of a crystal and the internal arrangement of its building elements. The symmetry of the internal arrangement is characterized by periodicity in three directions. What 7 was for repetition in one direction and 17 for repetition in two directions, it is 230 for repetition in three directions. This means that there are 230 different possibilities to build up a three-dimensional structure that is periodic in all three directions of space.

The internal arrangement of the building elements determines the properties of a crystal rather than its external shape. A piece of glass may be cut into a beautiful shape just as a piece of diamond. However, regardless of the shape it is made into, the atoms in glass are not in a regular arrangement whereas in diamond they are. Glass breaks easily whereas diamond is exceptionally hard. The difference in strength comes from the difference in internal structure.

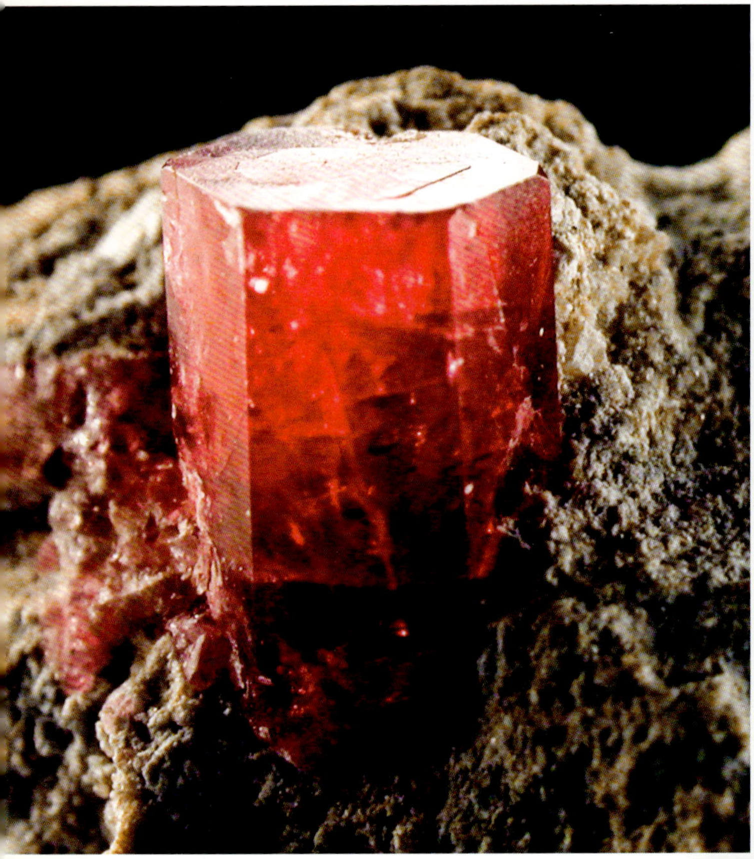

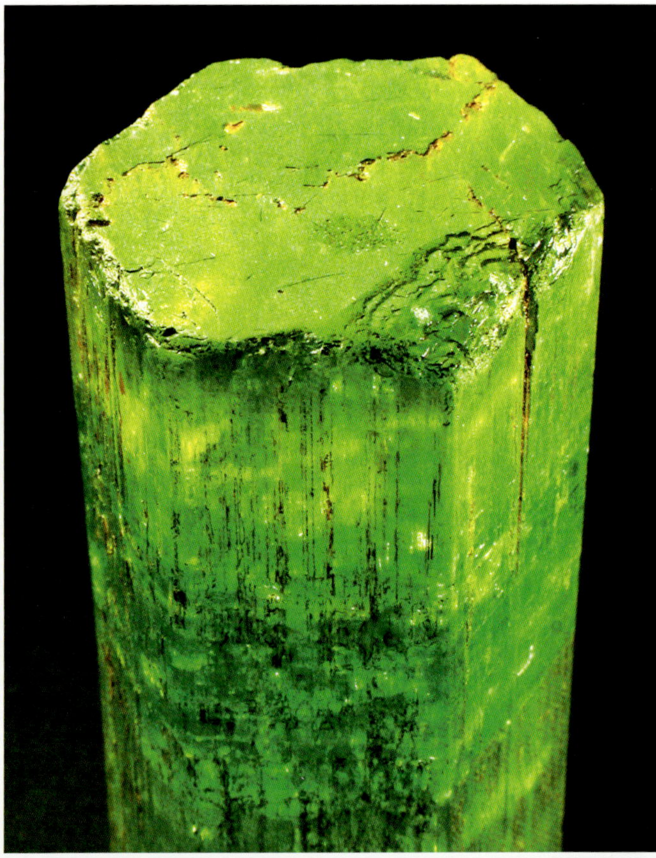

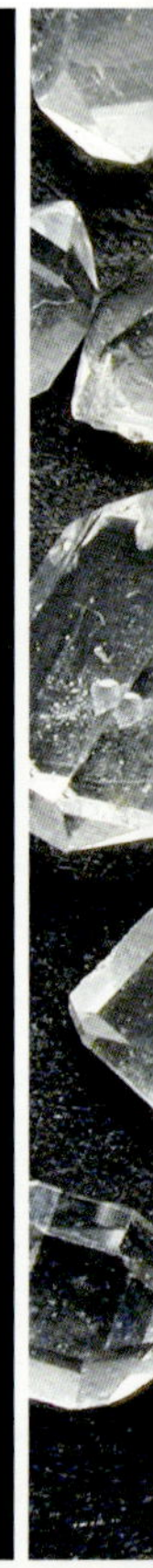

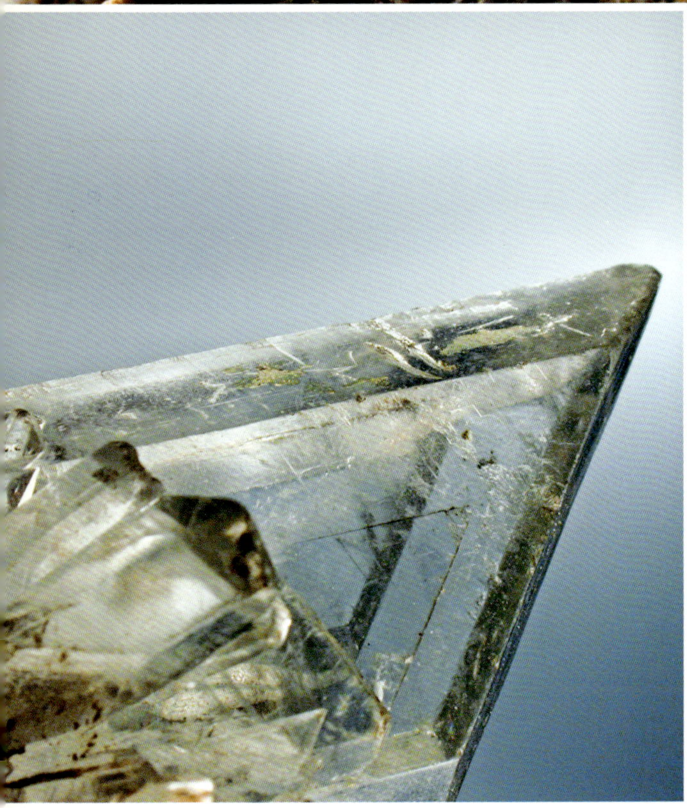

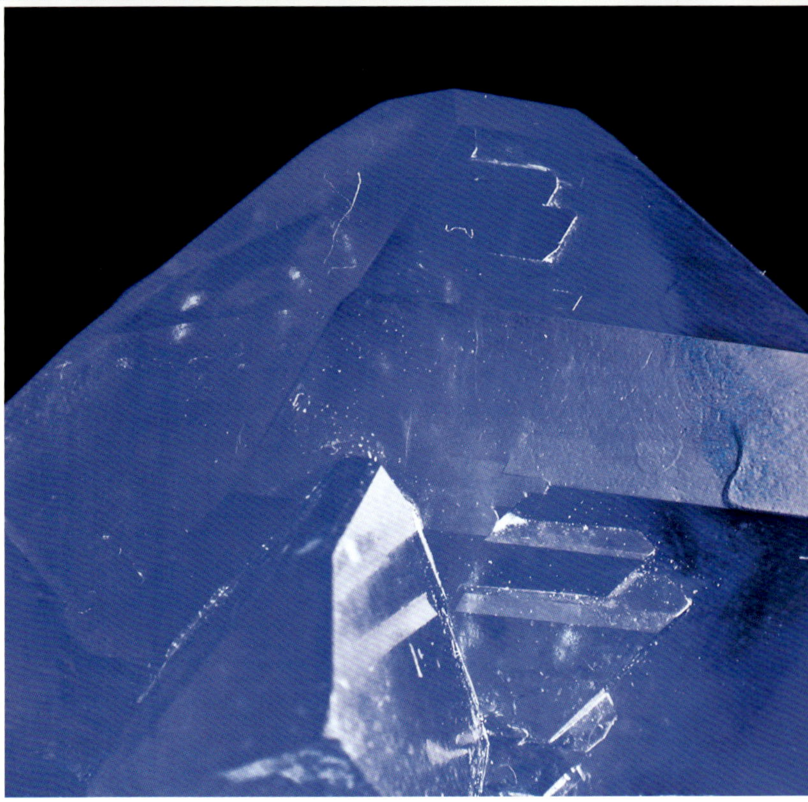

Top left: Beryl from Utah

Top right: Beryl from Siberia, Russia

Bottom left: Gypsum from Hungary

Bottom right: Calcantite (artificially produced)

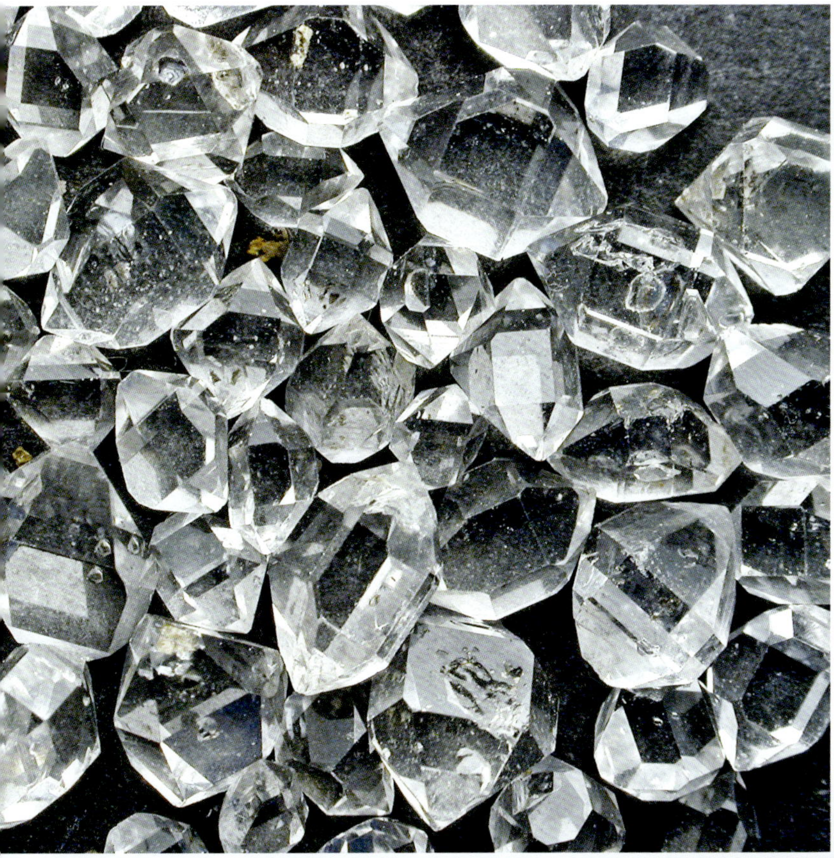

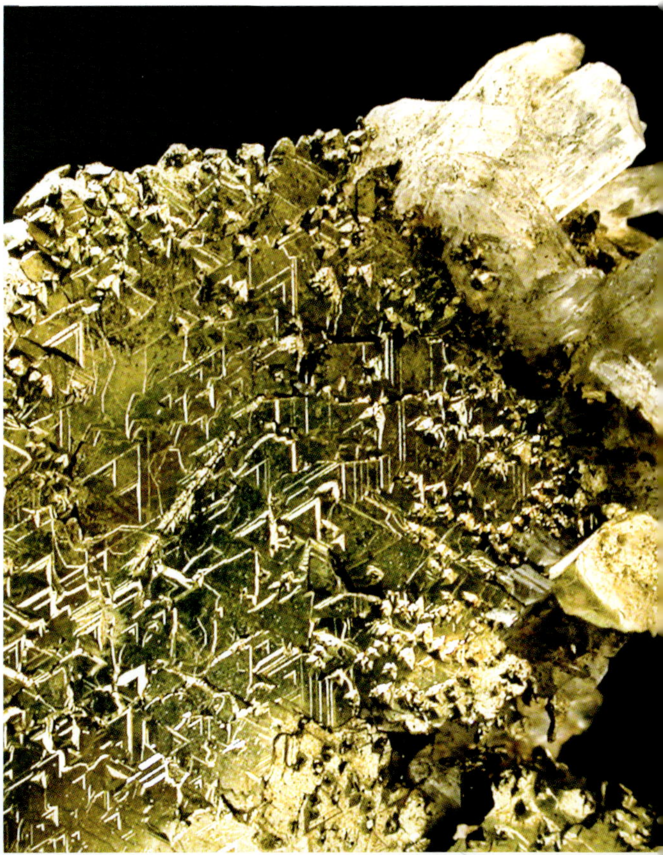

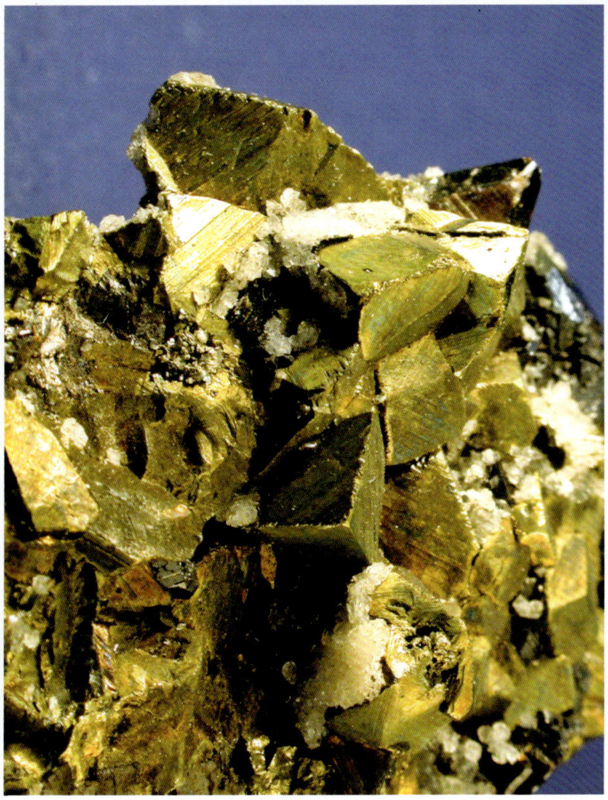

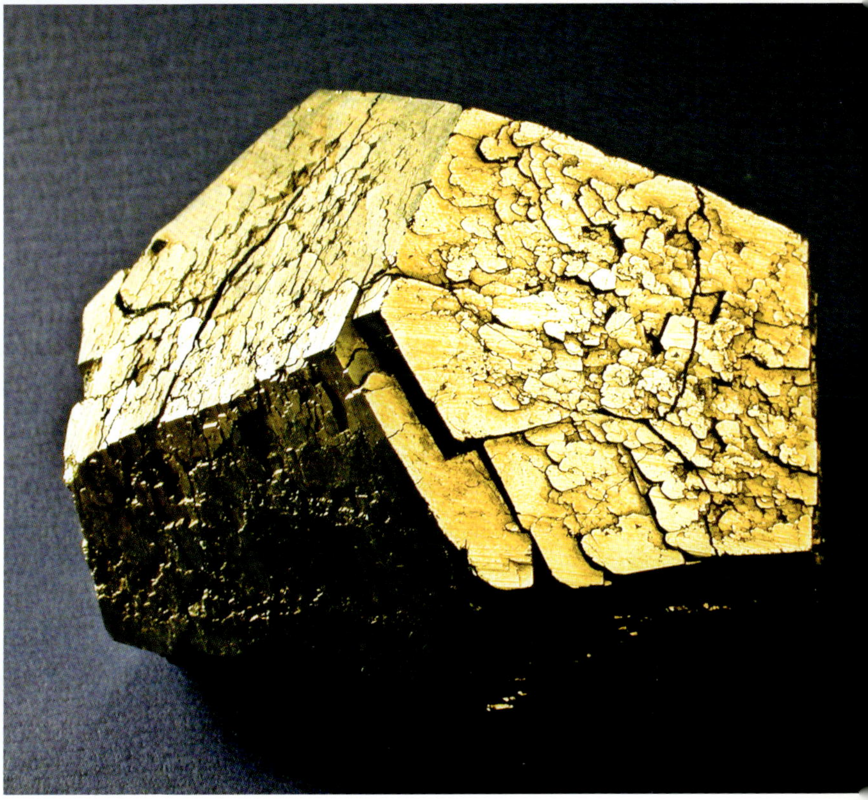

Top left: Quartz from the Ukraine

Top right: Gold from Romania

Bottom left: Calcopyrite from Bulgaria

Bottom right: Pyrite from the Island of Elba, Italy

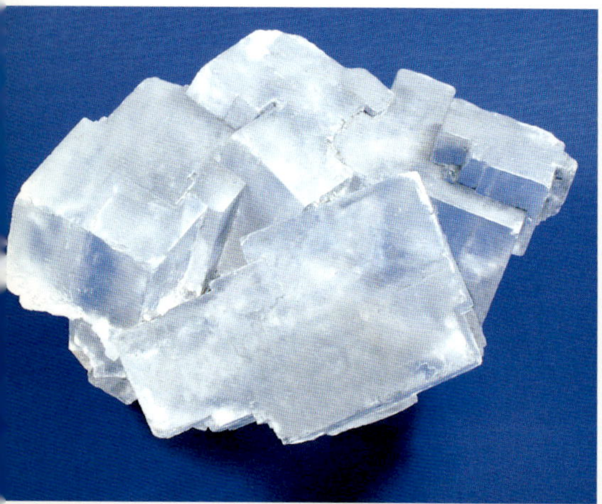

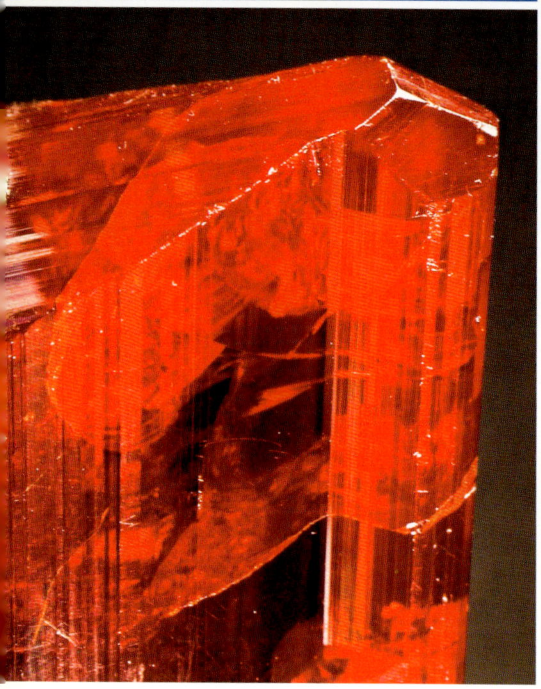

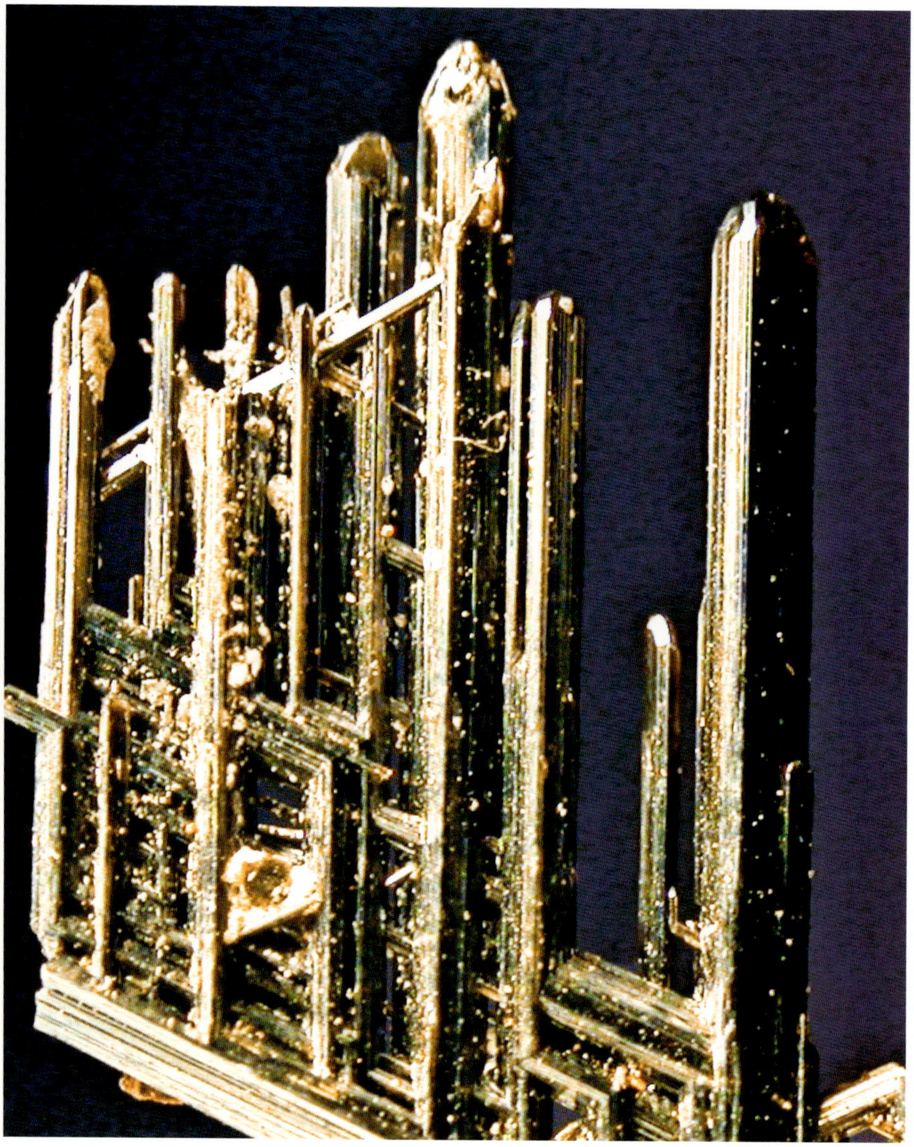

Top left: Halite from Romania

Bottom left: Realgar from China

Right: Rutile from unknown source

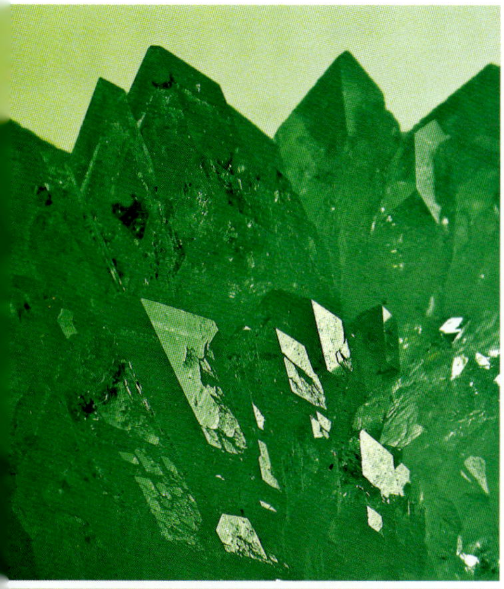

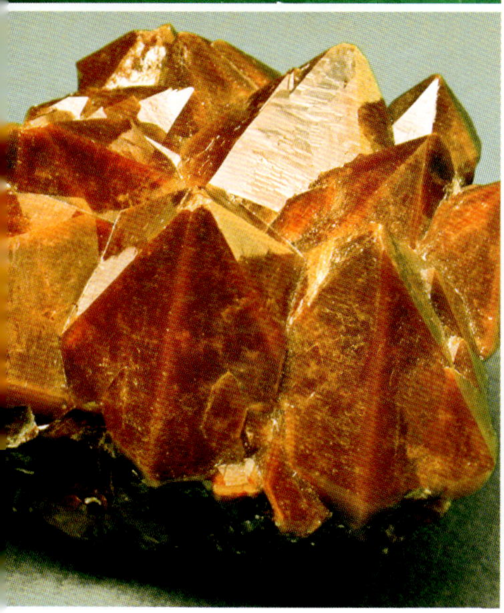

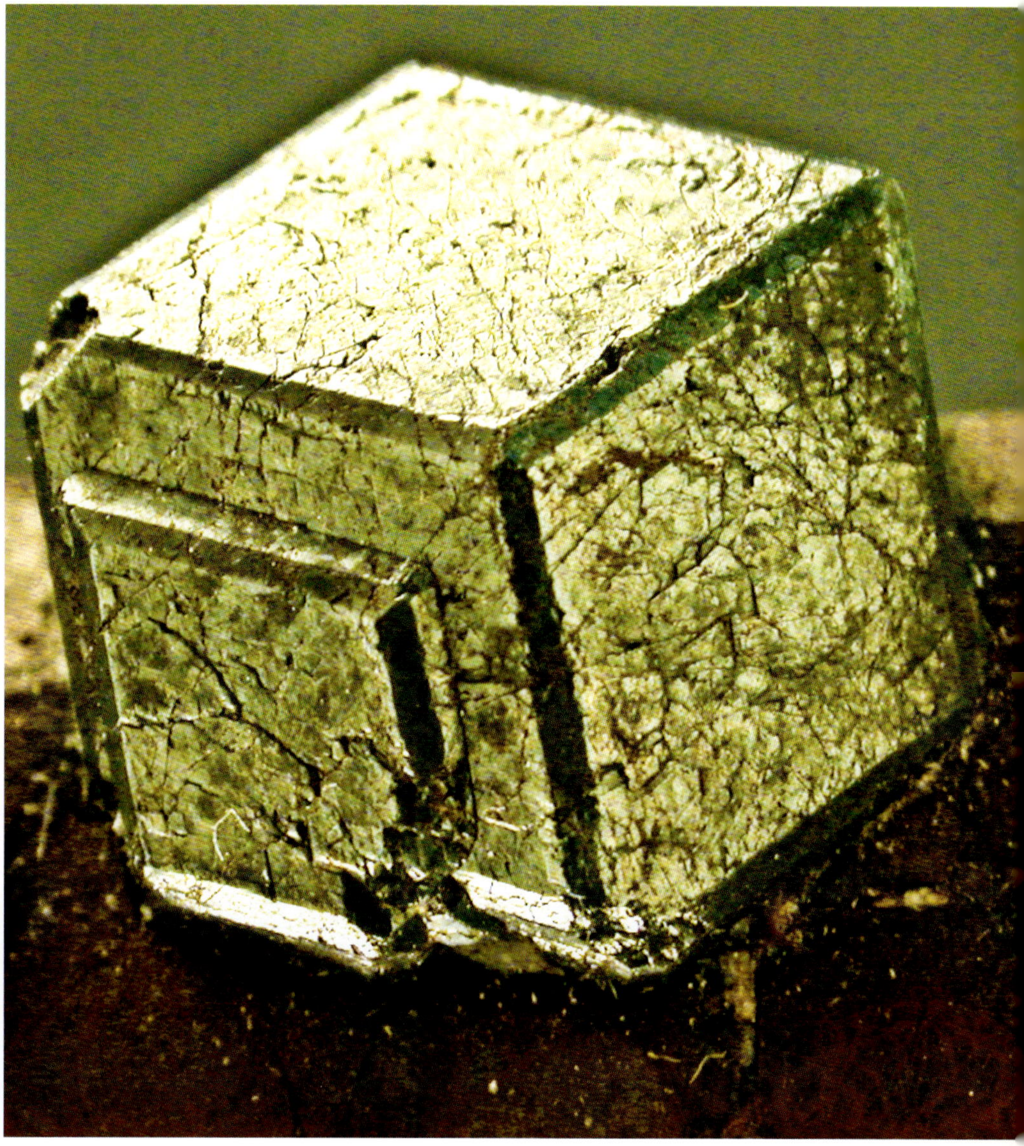

Top left: Apophillite from India

Bottom left: Quartz from Kazakhstan

Right: Magnetite from Slovakia

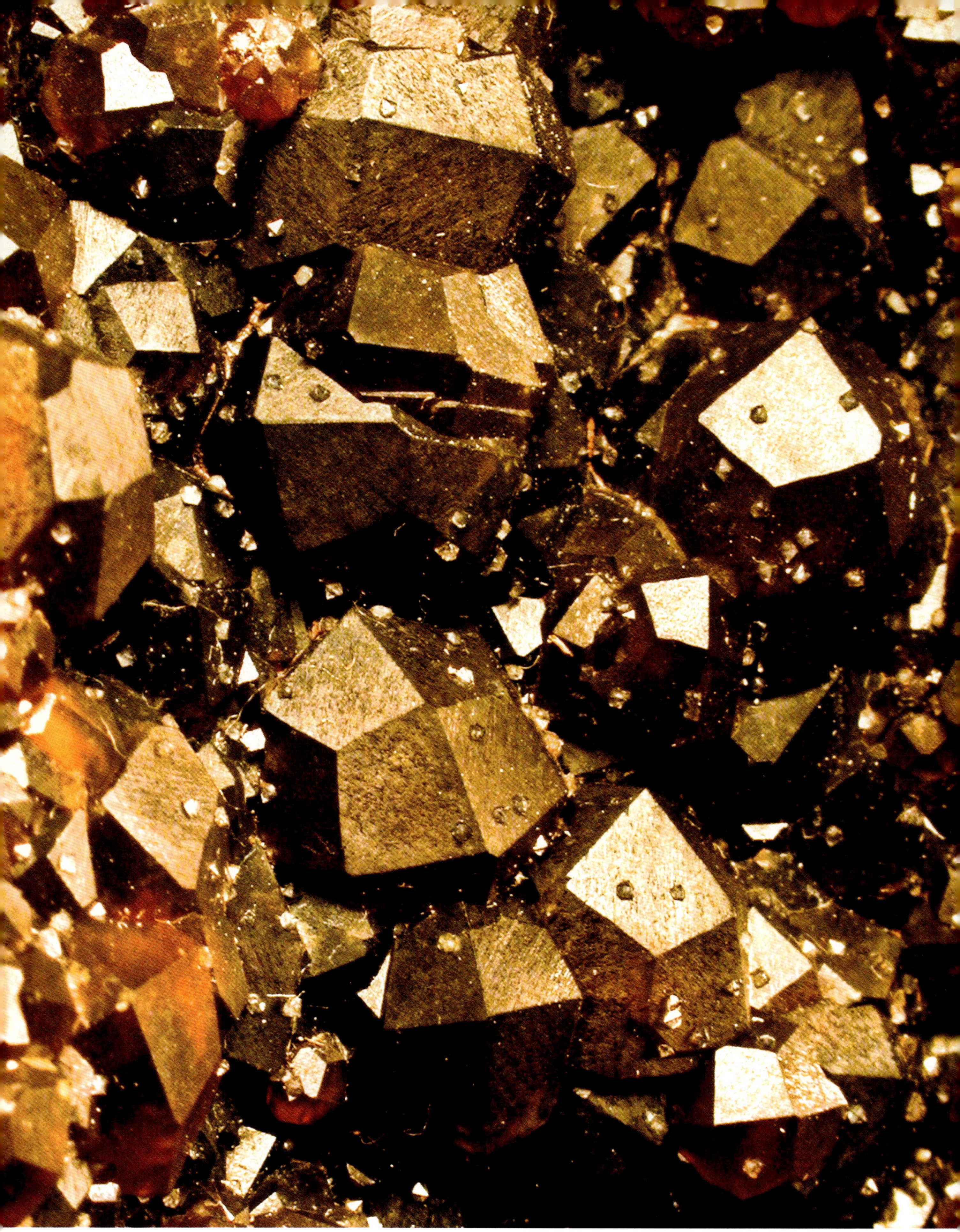

Andradite from Romania

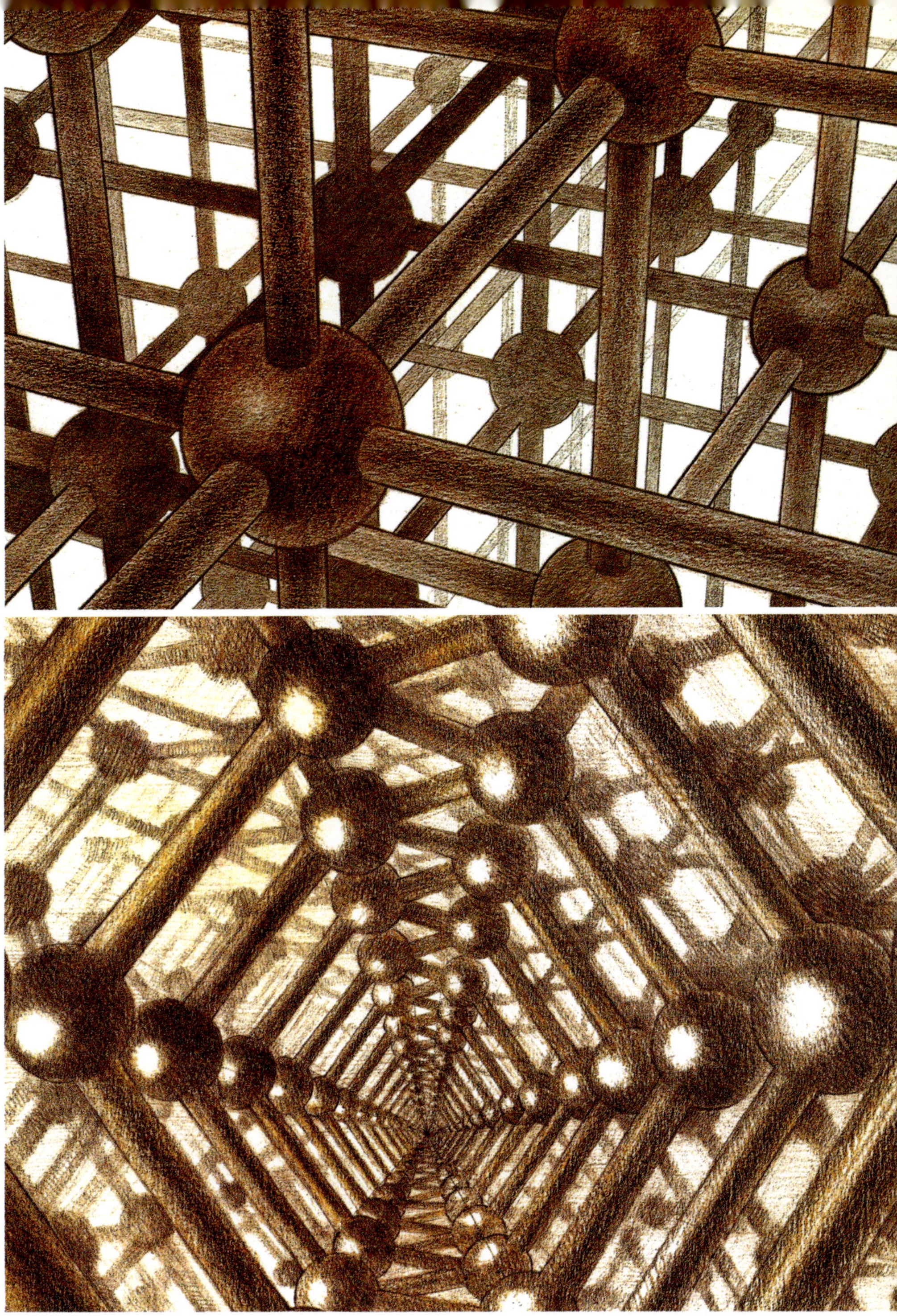

Illustrations of the internal arrangements of atoms in crystals by Ferenc Lantos (Pécs, Hungary)

While it was easy to illustrate the patterns with repetitions in one and two directions, it is hardly possible to do the same for most of the 230 possibilities for repetitions in three directions. Two simple examples are depicted here. In both, the lattice positions are occupied by spheres representing atoms. In the top drawing these atoms are arranged in the apexes of cubes. In the bottom drawing, the atoms form a three-dimensional hexagonal pattern.

Quasicrystals

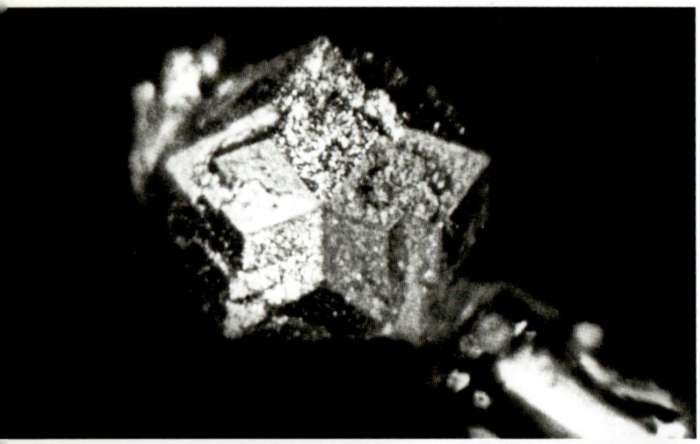

Quasicrystal in an aluminum-lithium-copper alloy

(courtesy of F. Dénoyer, Paris)

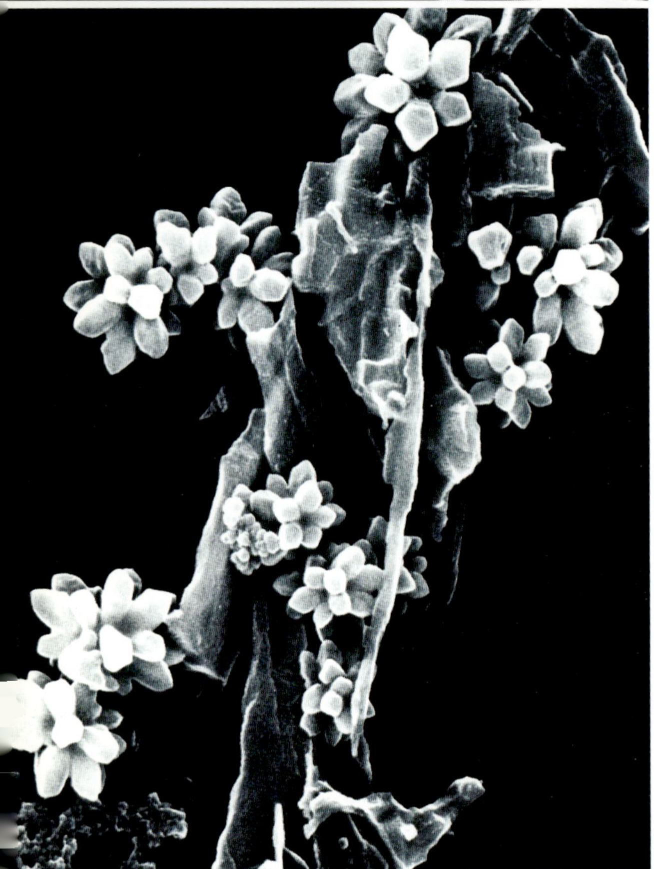

In a piece of diamond — the diamond being a crystal — the carbon atoms build up a structure which is regular and periodic. In a piece of glass, there is neither regularity, nor periodicity in the arrangement of the silicon and oxygen atoms. These are two extreme cases and for a long time, until the mid-1980s, it was assumed that all solids belong to one of the two cases. A quarter of a century ago, however, new structures were discovered that showed regularity but not periodicity. These were new structures for us, but they have been around in nature all the time. Today they are called *quasicrystals* and a few examples are depicted here. They are conspicuous in having fivefold symmetry that used to be considered impossible in the internal structure of crystals before the discovery of quasicrystals.

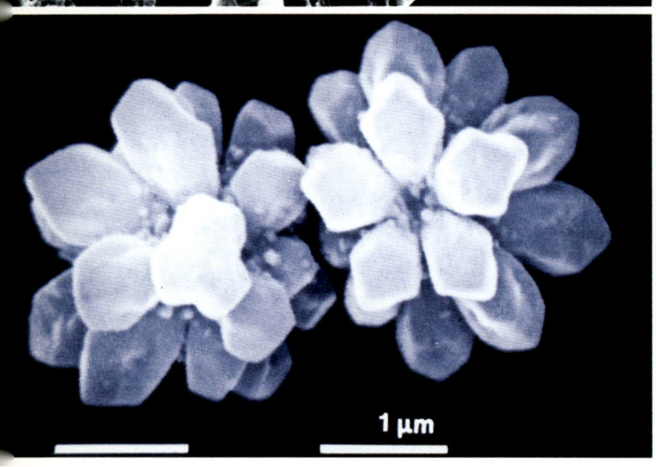

Quasicrystals in aluminum-manganese alloys

(courtesy of Ágnes Csanády, Budapest)

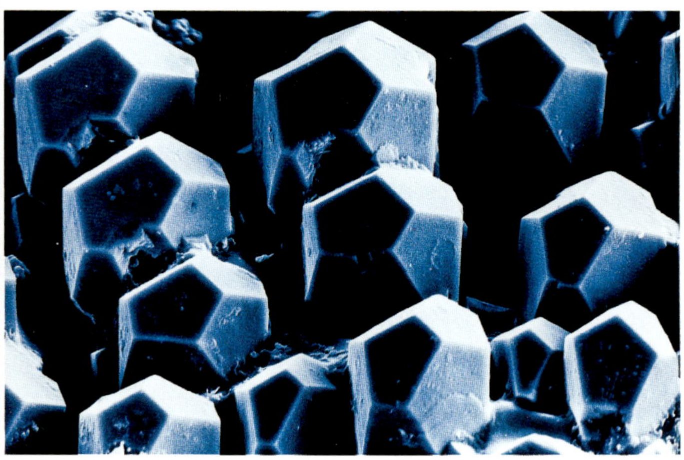

Quasicrystal in aluminum-copper-rubidium alloy

(courtesy of Hans-Ude Nissen, Zurich)

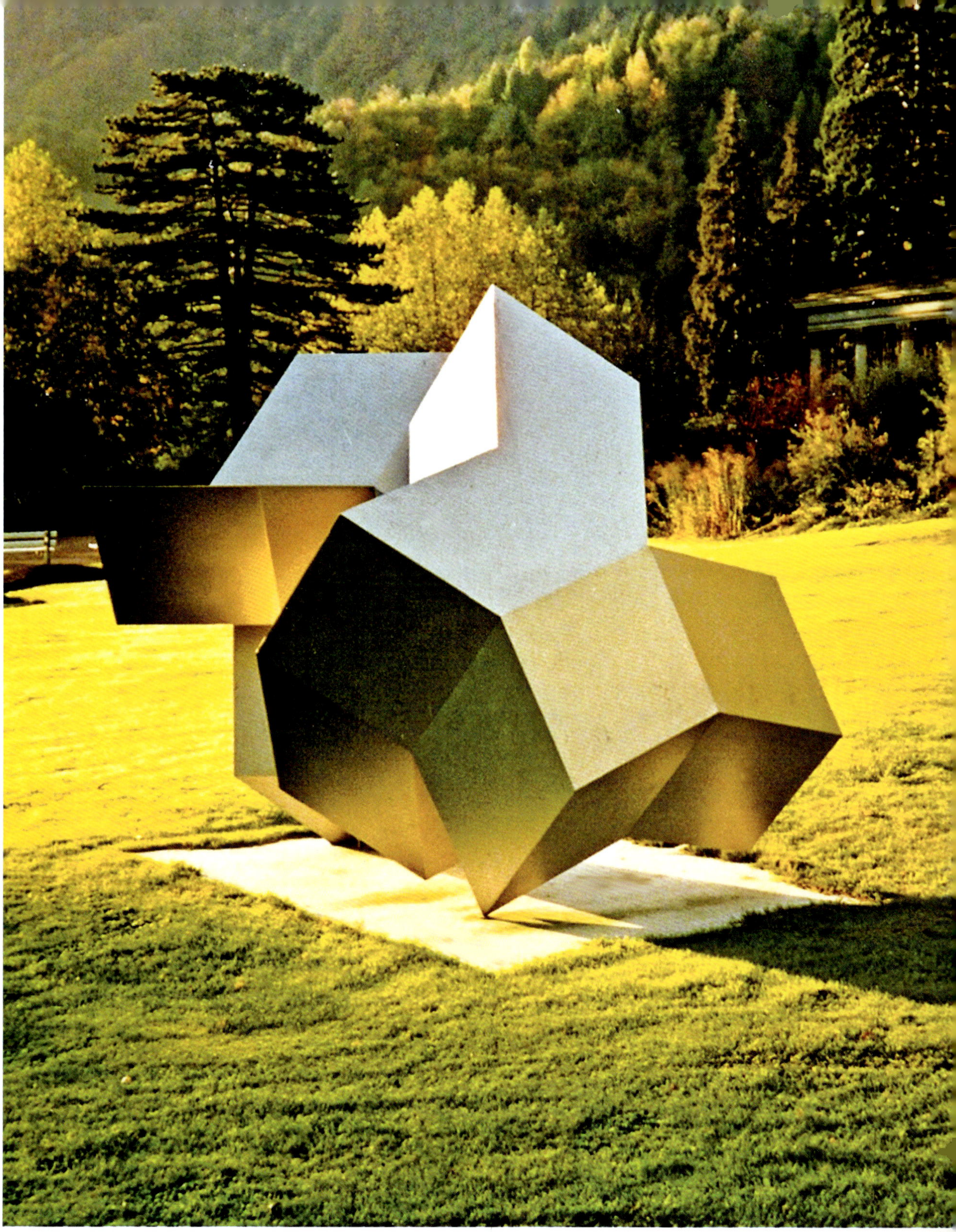

Sculpture resembling quasicrystals by Peter Hächler, Switzerland

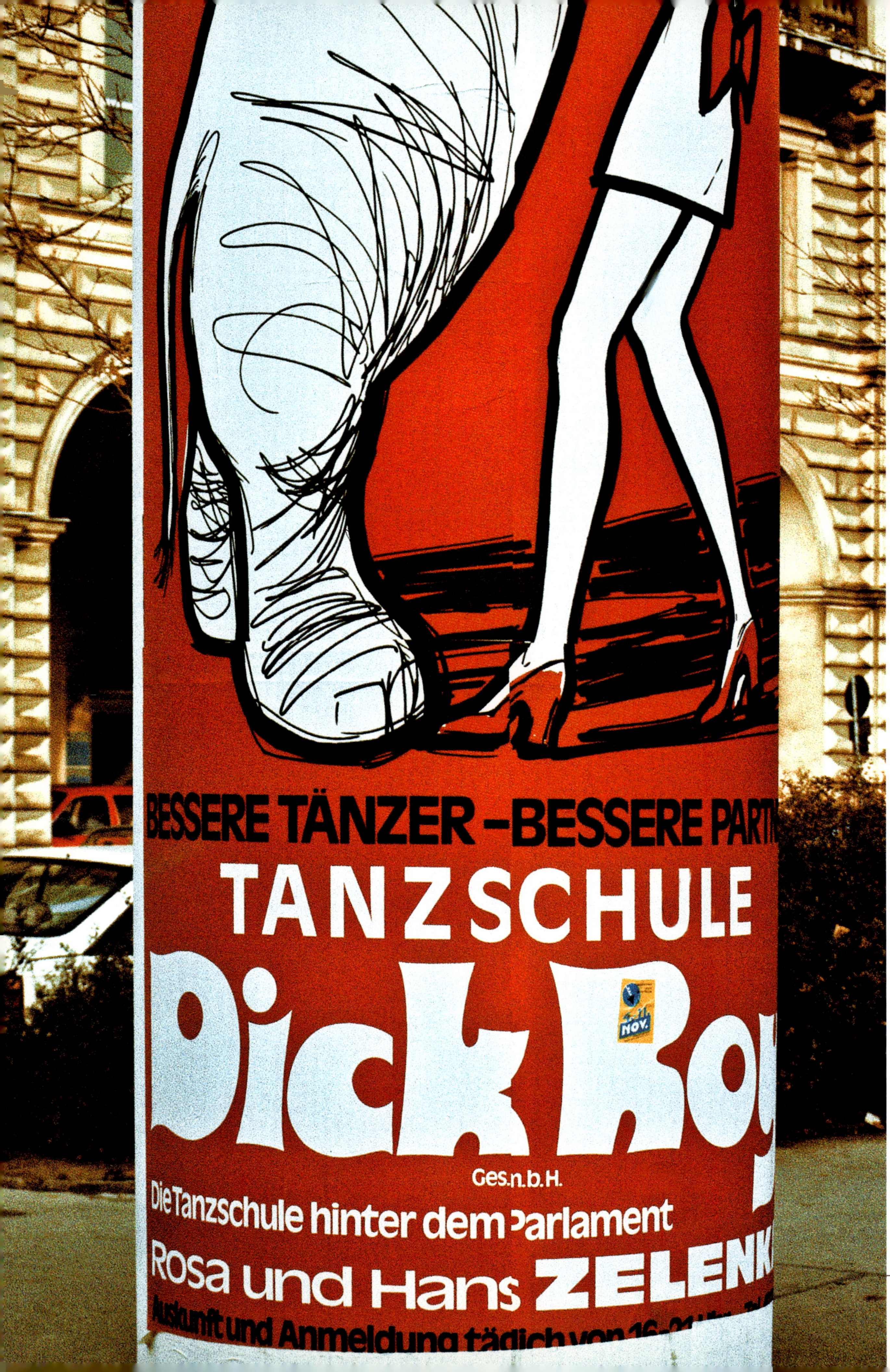
BESSERE TÄNZER – BESSERE PART
TANZSCHULE
Dick Ro
Ges.n.b.H.
Die Tanzschule hinter dem Parlament
Rosa und Hans ZELENK
Auskunft und Anmeldung

Antisymmetry

Consider these deer, numbered 1 to 4 from left to right. Between 1 & 2, there is mirror symmetry and so is between 3 & 4. But how about 2 & 3? If we take their shapes only, there is mirror symmetry between them. However, in this case, the reflection does not only recreate the deer, it also turns its coloring into its opposite. Such kind of symmetry is called antisymmetry. In other words, antisymmetry is when a symmetry relationship is accompanied by one of the properties turning into its opposite. The property may be color, shape, any trait, and there is no limit to what we may consider to be a property and its opposite. In the poster depicted on page 196, the clumsiness of the elephant turns into dancing skills of the lady in the process of reflection. This demonstration of antisymmetry is meant to attract customers for the dancing school. Antisymmetry has proved to be an efficient approach in advertising and political campaigns.

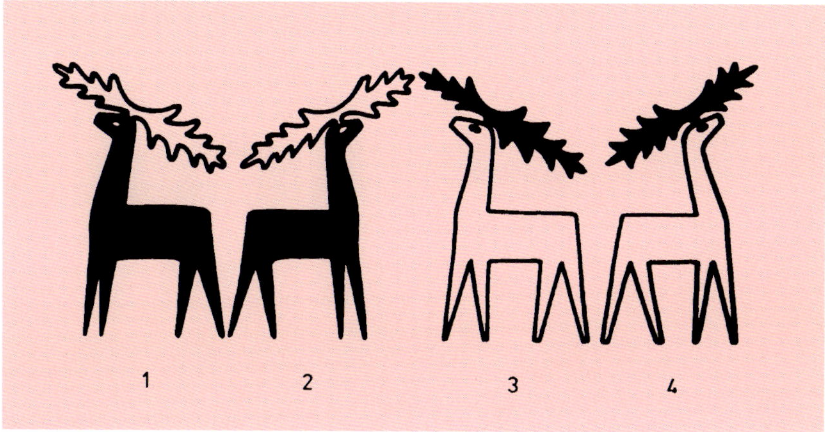

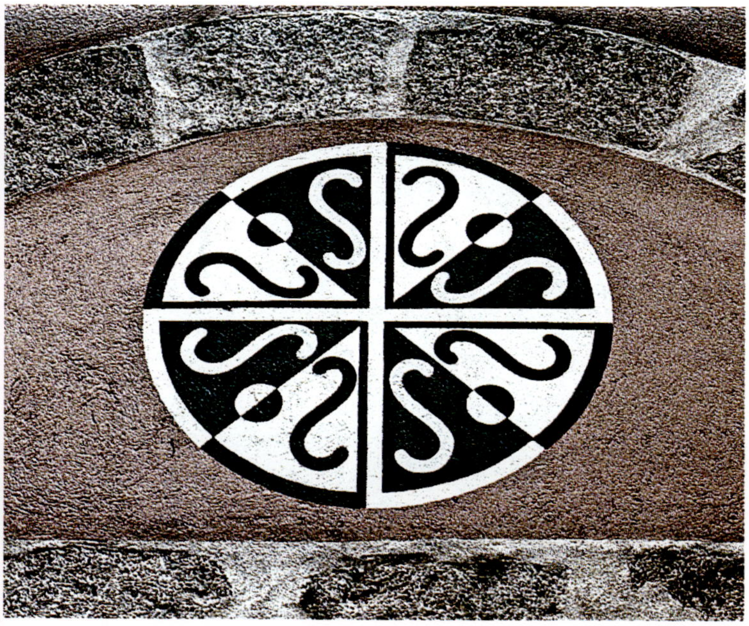

Decorations on a Spanish house

Color changes accompany reflections as well as rotations.

Poster advertising a dancing school in Vienna

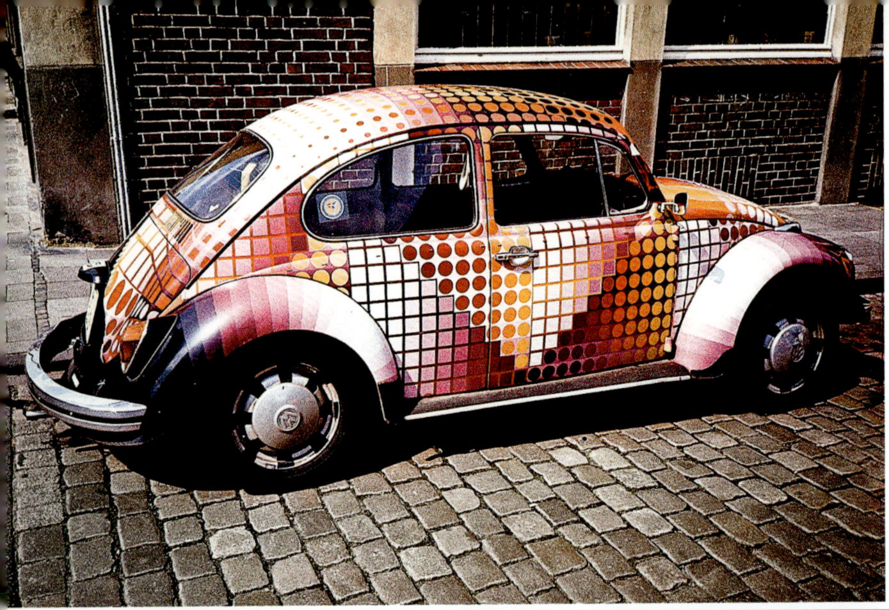

Op-art style pattern on a car; in addition to the color changes, there is a transformation between squares and circles.

A curious example of antisymmetry is part of the tower of a gatehouse at Park Güell, the famous park in Barcelona built by Antoni Gaudi. If we look only at the detail of the tower depicted below to the left, it has a symmetry that is glide reflection, combined with antisymmetry and similarity symmetry.

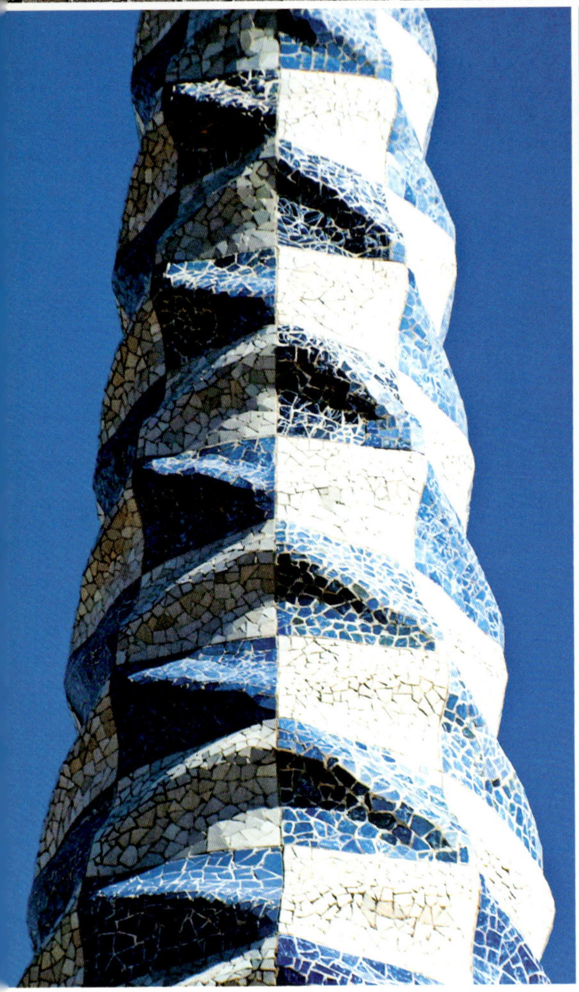

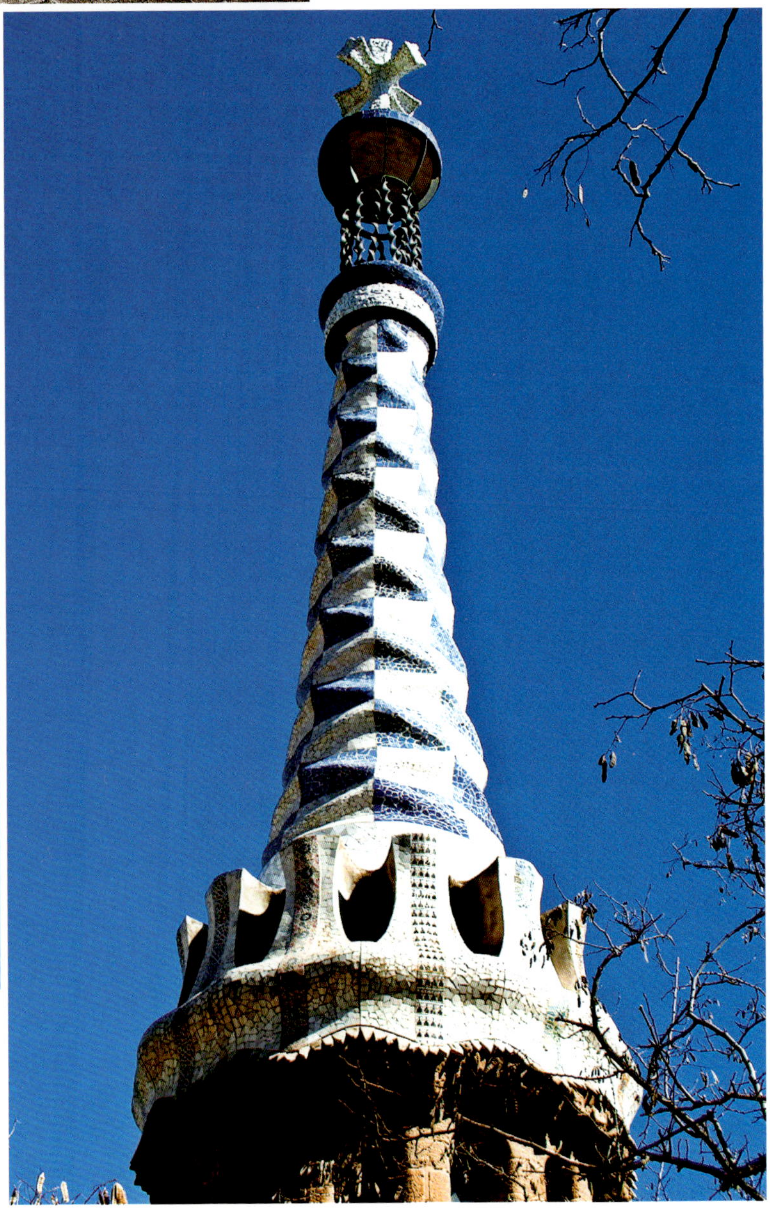

Details of a tower in Park Güell in Barcelona by Antoni Gaudi

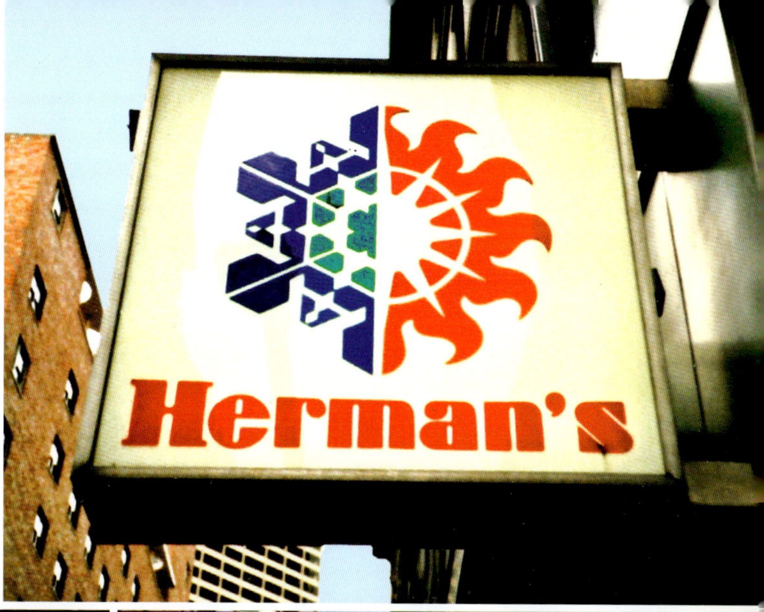

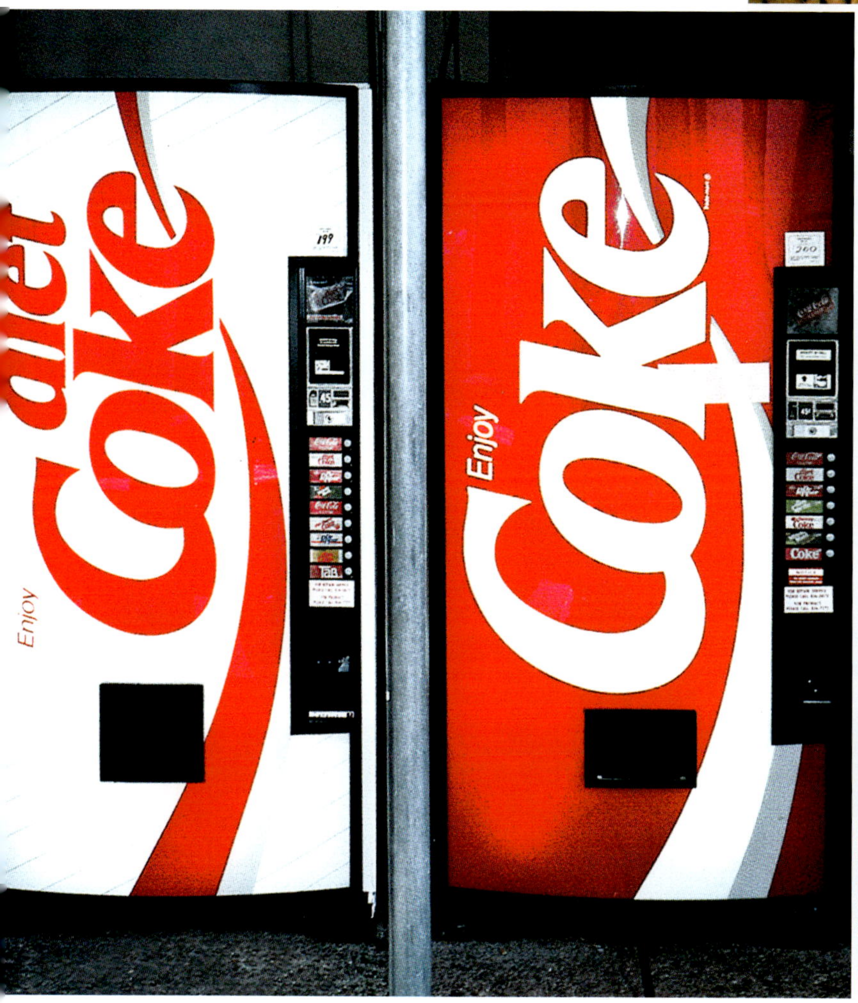

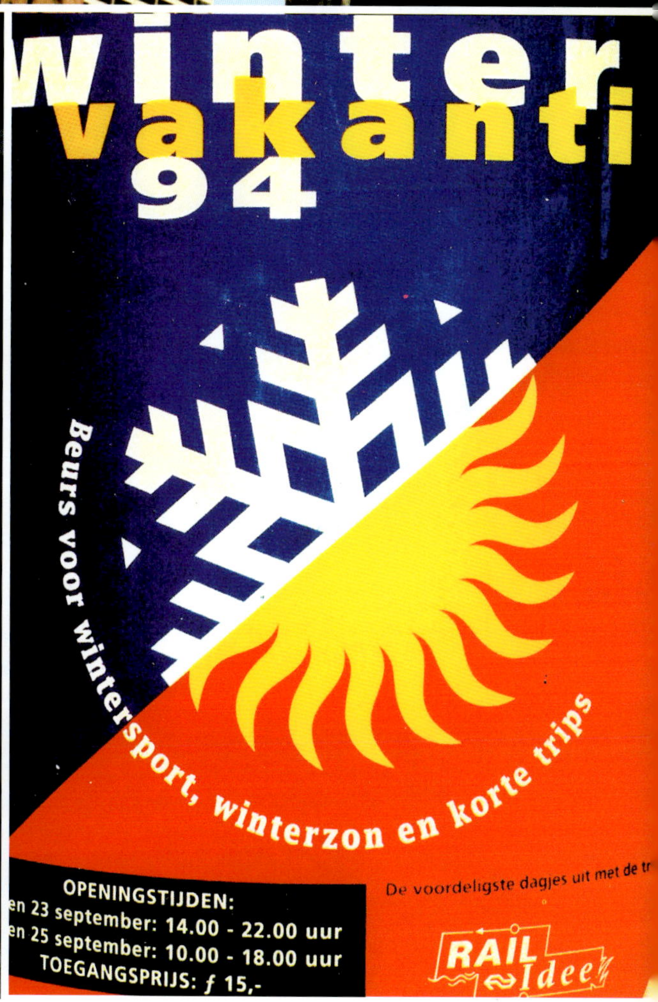

The logo of the store for sporting goods in Boston (top) and the Dutch poster advertising winter holidays (bottom right) both utilize the antisymmetric relationship between half a snowflake and half of the Sun. There is an antimirror symmetry between them.

The two Coke machines are not connected by symmetry, but express a relationship in which a peculiar property — the sugar content — is turning into its opposite. The reversal of colors also indicates that they are each other's opposites.

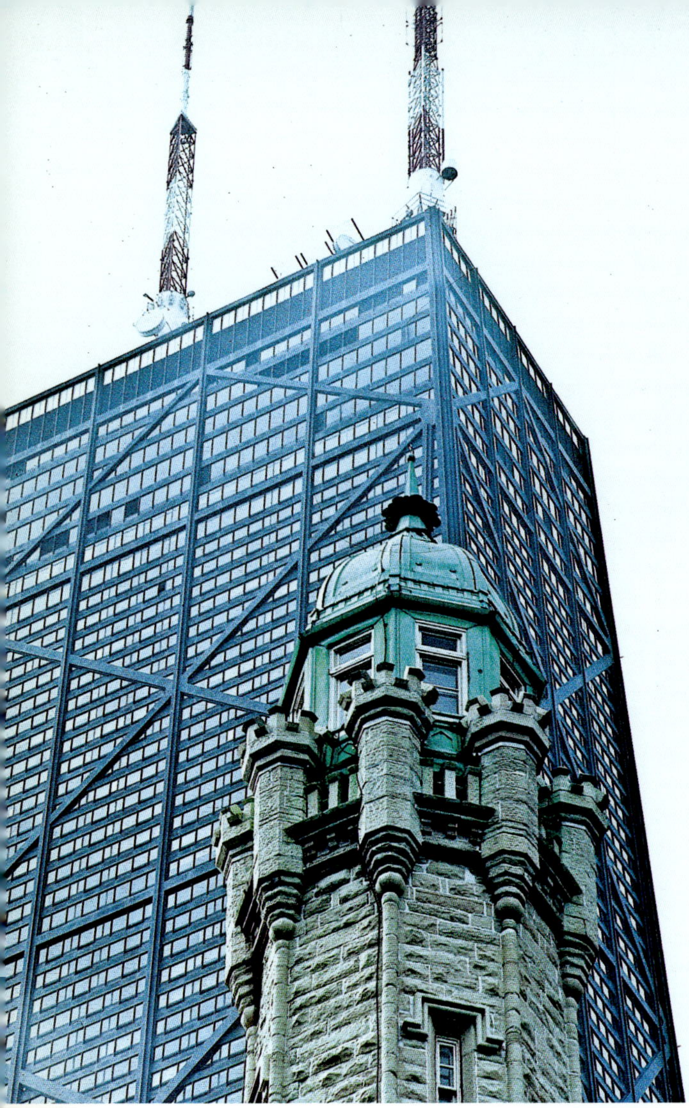

Provided we are willing to go to an increasing degree of abstraction, an appealing anti-relationship can be recognized between old and new; modern and ancient; and so on.

Old and new buildings stand side by side in our examples from three American cities.

Boston

New York City

Dallas, Texas

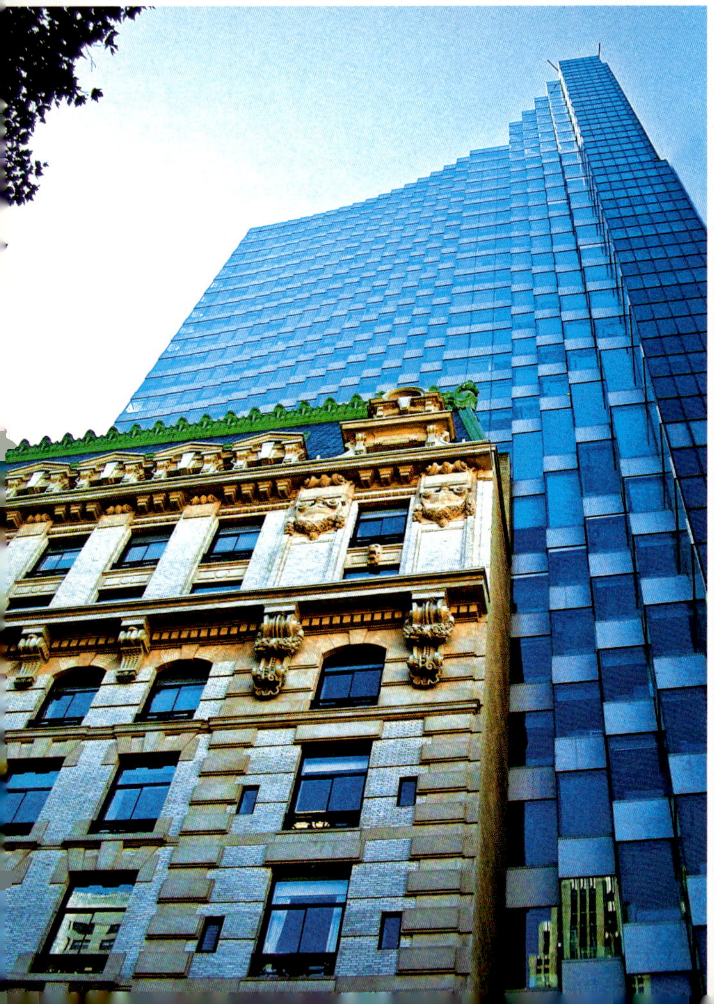

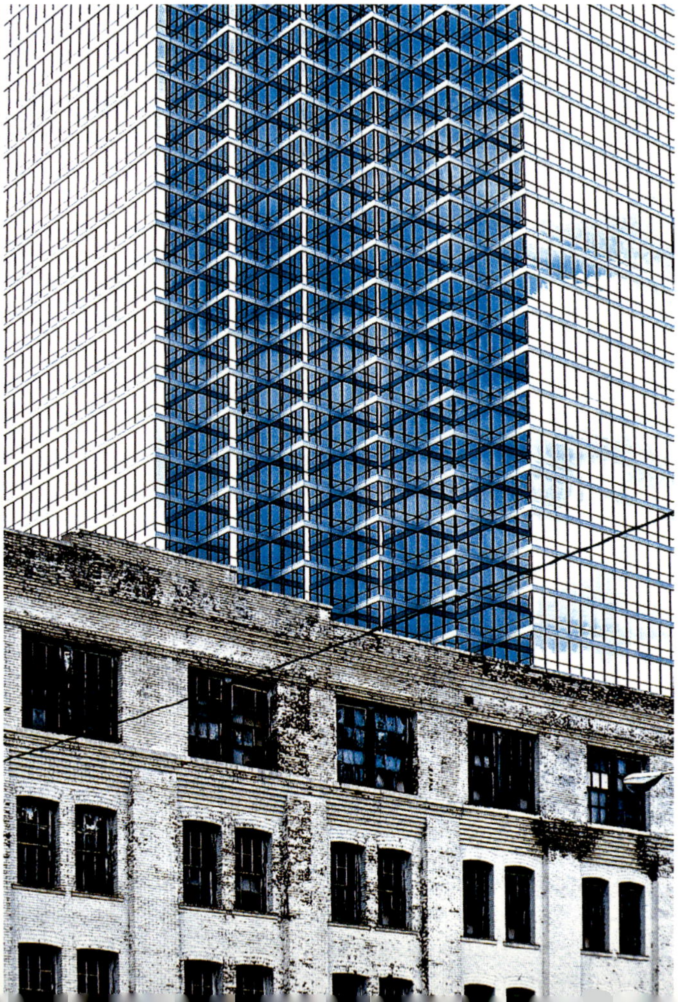

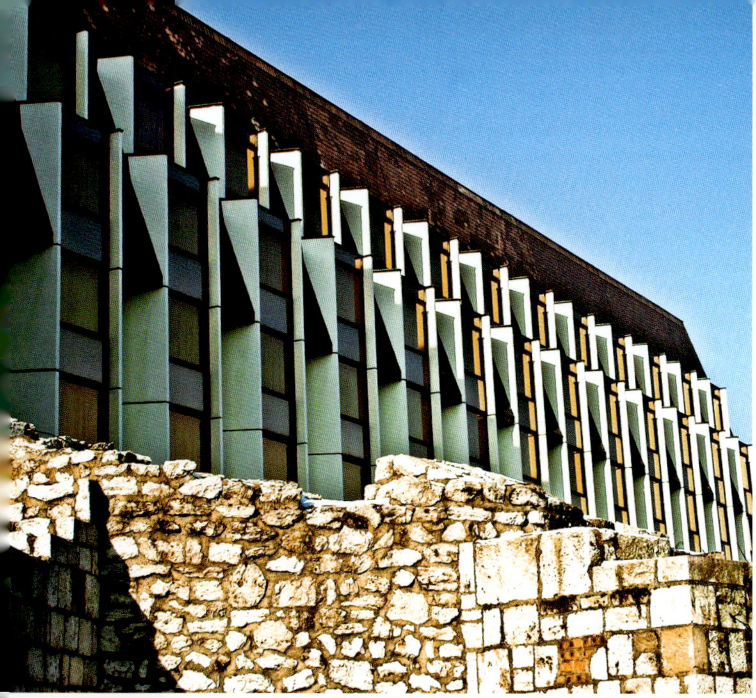

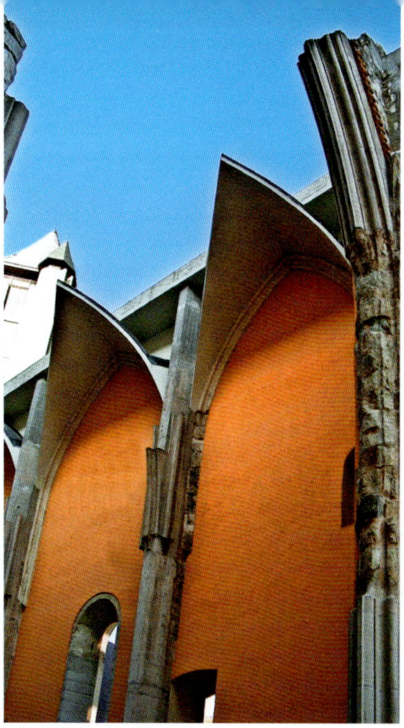

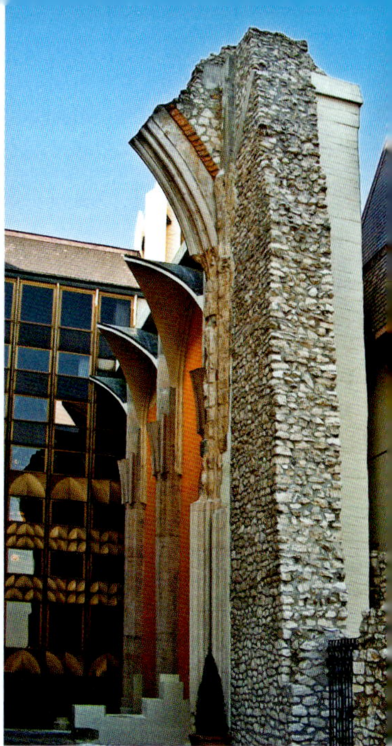

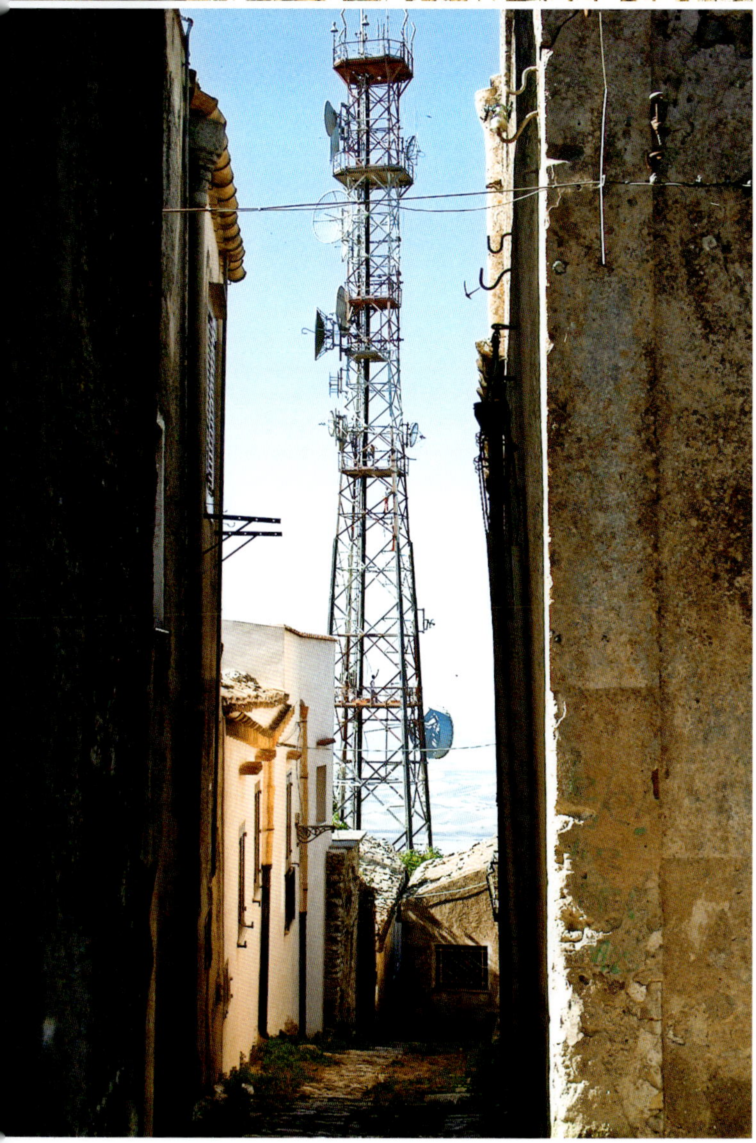

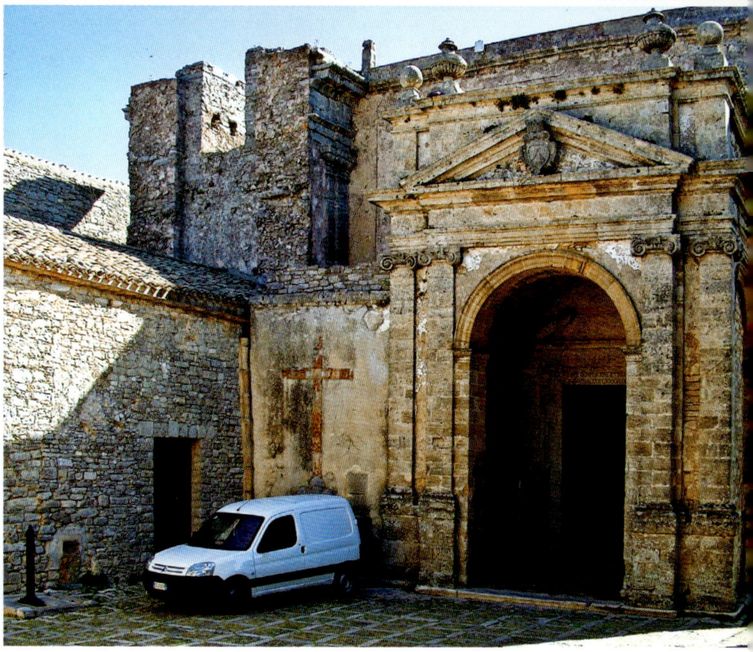

Top: Three images of Hotel Hilton of Budapest behind ancient ruins

Bottom left: NATO locator in medieval Erice, Sicily

Bottom right: Medieval building with modern car in Erice, Sicily

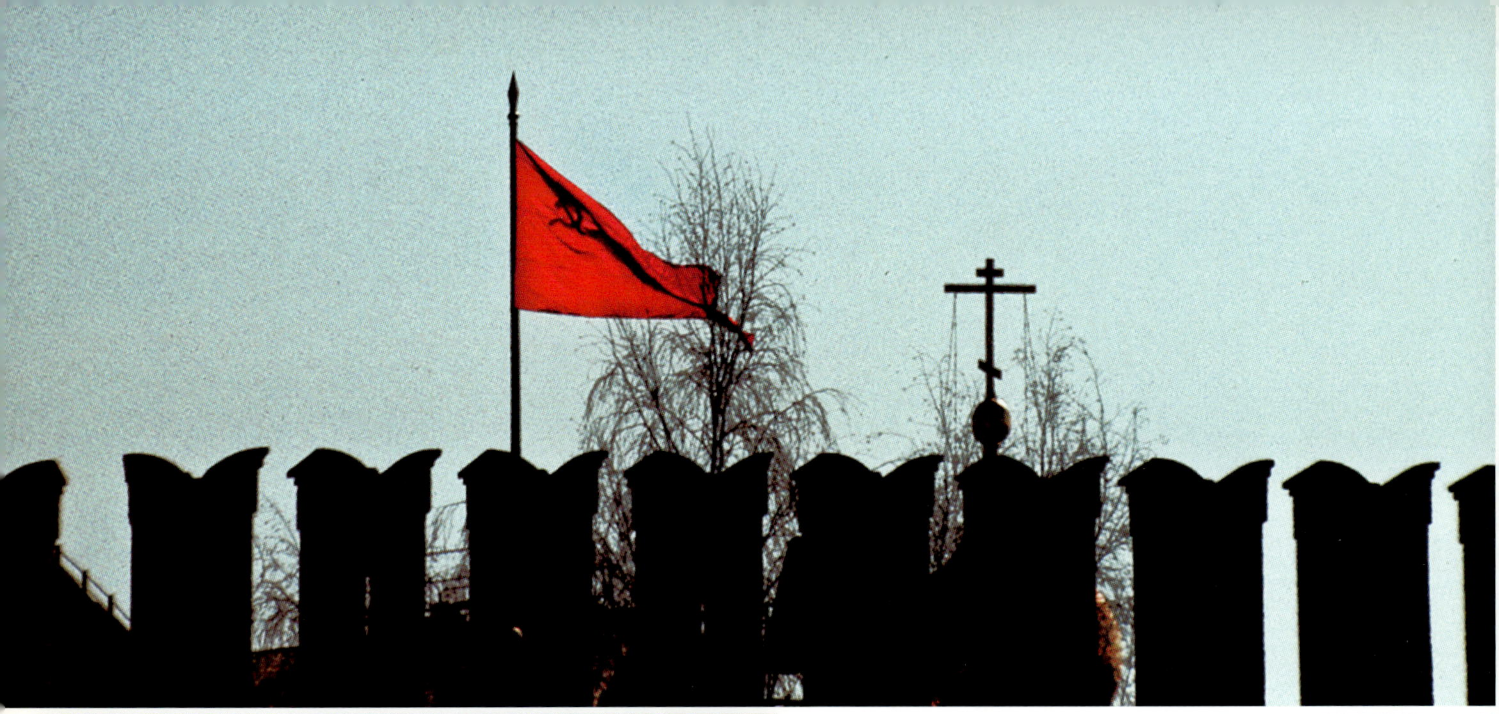

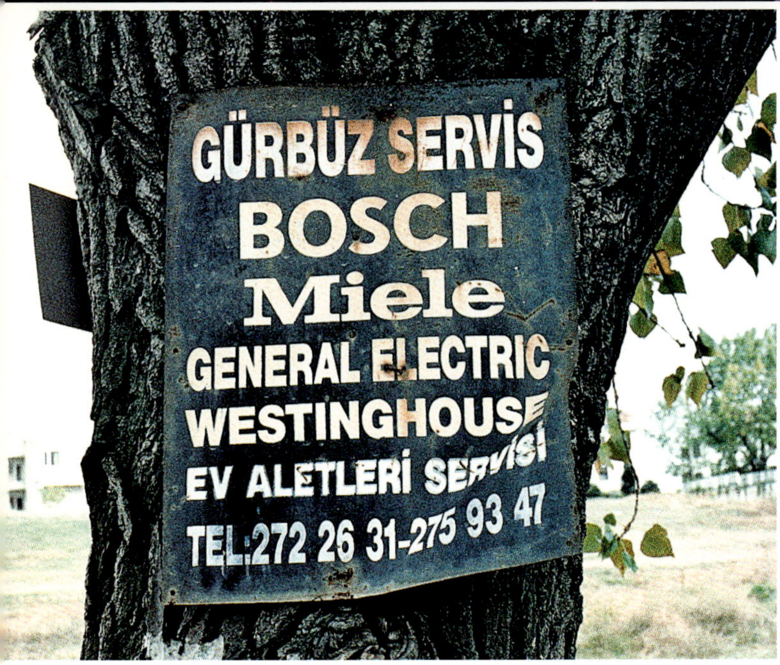

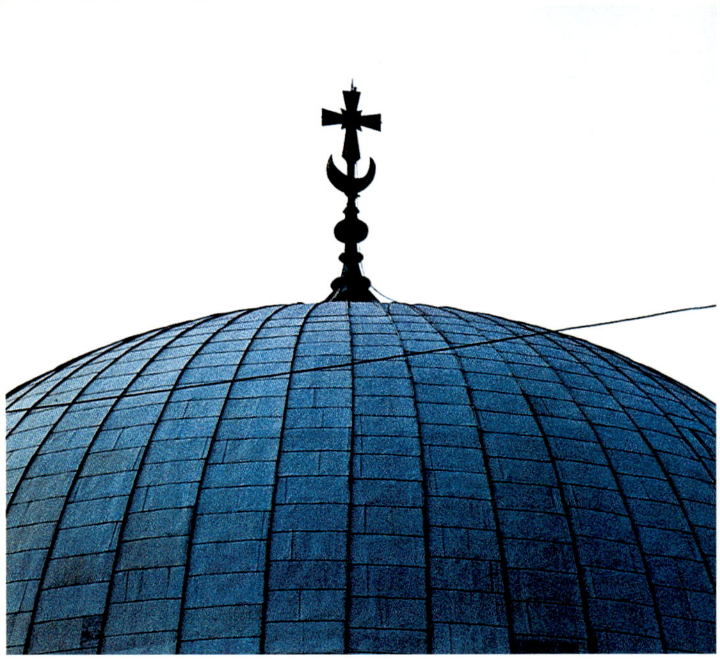

Top: The hammer-and-sickle-decorated red flag of the Soviet Union is in contrast with the eastern orthodox cross behind the Kremlin wall in Moscow, Russia (then, Soviet Union)

Bottom left: The rusted Turkish advertisement in Istanbul is in contrast with the internationally known companies it advertises

Bottom right: The Roman Catholic cross contrasts the crescent (from the times of Turkish occupation) beneath it at the top of the former mosque in Pécs, Hungary

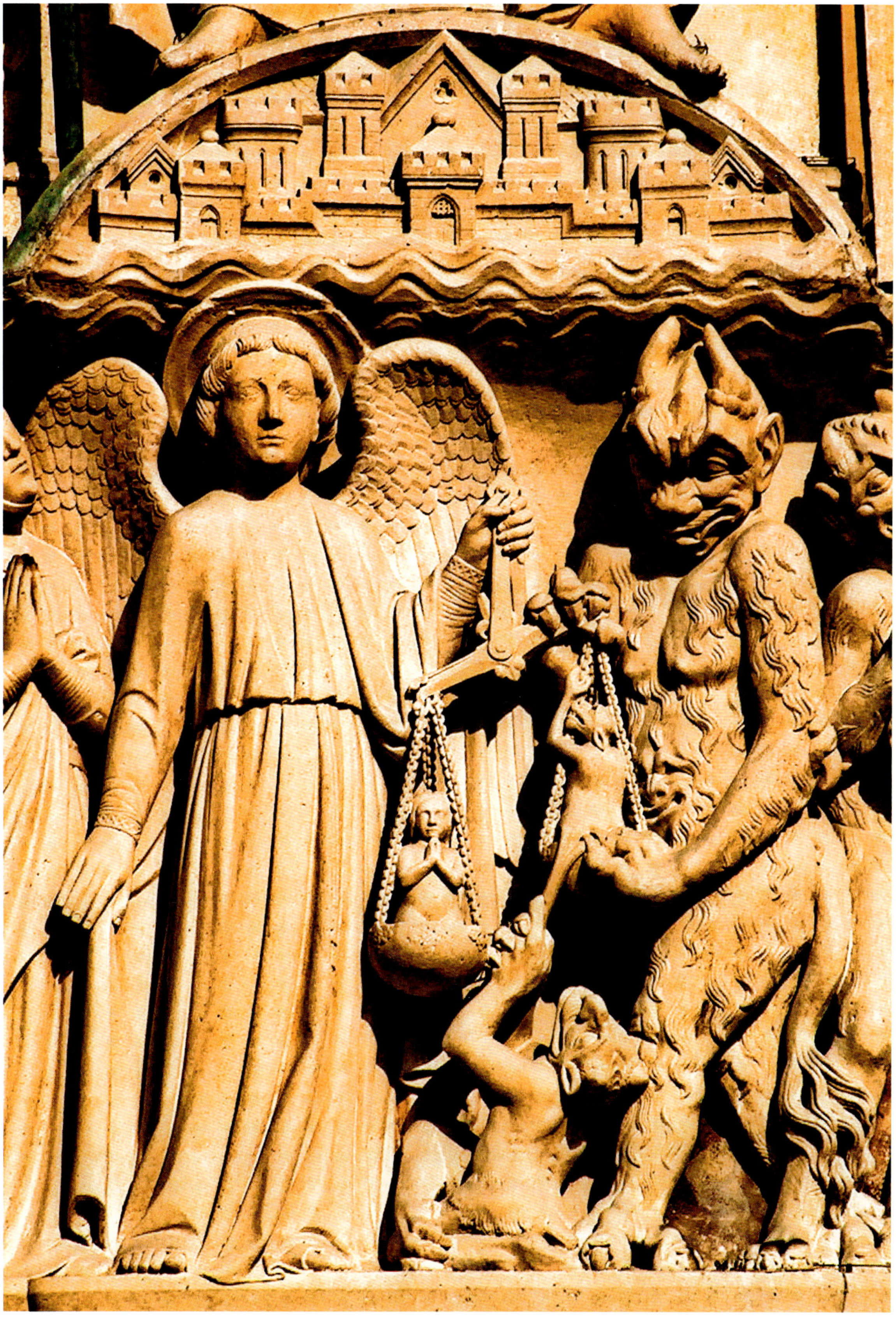

Detail of the relief decorating the front of the Notre Dame Cathedral in Paris

The ensemble of the angel and devil expresses the contrast between good and evil.

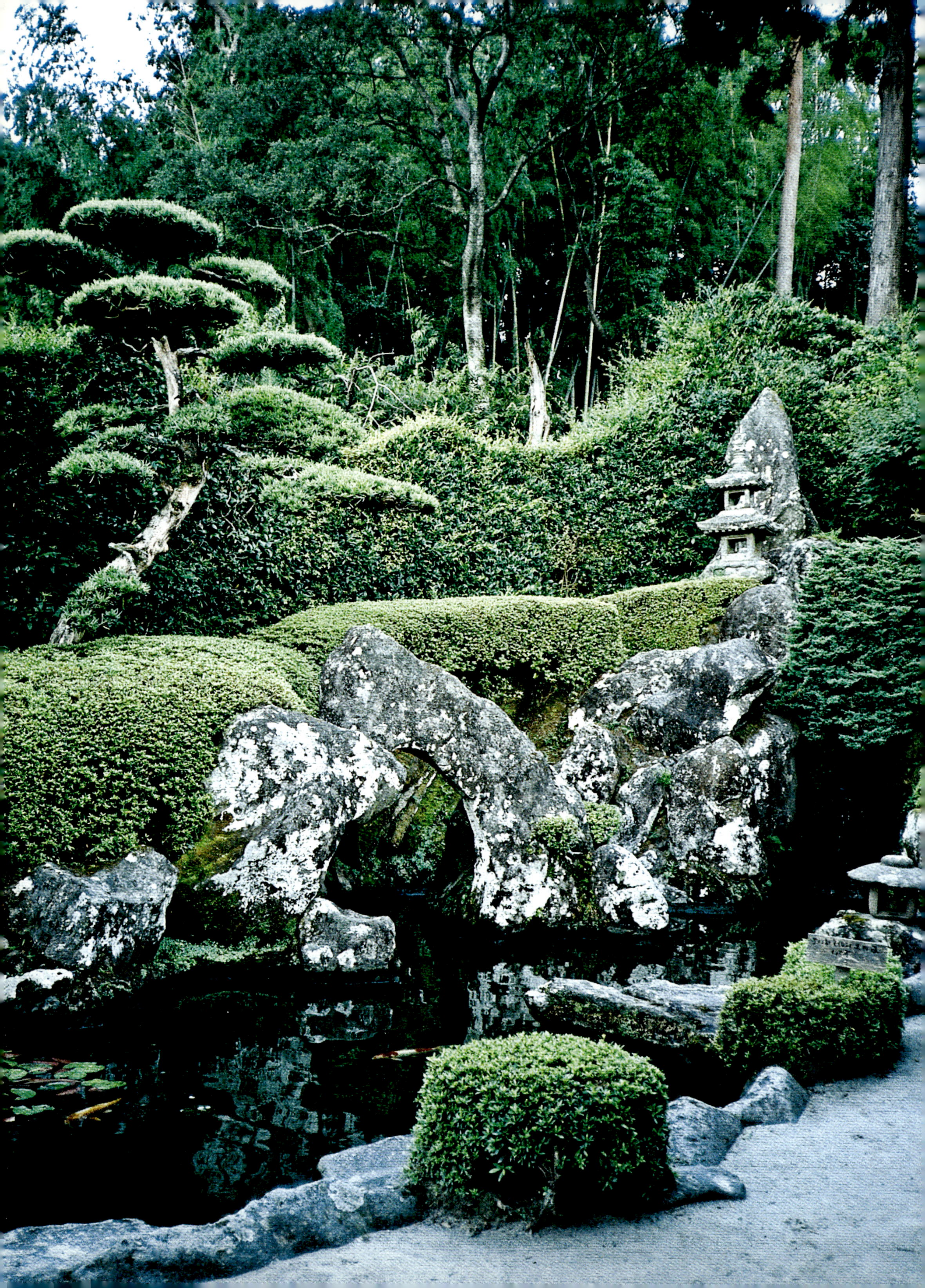

Epilogue

What we presented about symmetry in this book is only the tip of the iceberg. This is literally so, because we focused on the world around us that can be seen mostly with the naked eye. There are worlds beyond that; suffice it to mention the world of fundamental particles that constitutes the foundations from which everything else is built. Also, we have concentrated on symmetry whereas the absence of symmetry can be as important or even more important in our existence than symmetry. However, symmetry is a good reference point for noticing its absence and recognizing its meaning.

We are also aware of the dangers of overemphasizing symmetry. We can easily fall victim to *symmetritis*, the illness of seeing or at least looking for symmetry everywhere. Such overextended attention may become obsessive and at the same time it may be confining and irritating. This is why it is useful from time to time let ourselves relax and realize the beauty of imperfection. It has been noted that perfection may not be the most suitable characteristic for human habitat. This is why a medieval Japanese author issued this caveat, "In everything… uniformity is undesirable." This is why "builders of antiquity purposely and secretly introduced minute variation from absolute symmetry in their columnar structures" in the words of Thomas Mann with which he concludes his description of the beauty of snowflakes in *The Magic Mountain*.

Symmetry is an excellent tool in helping to understand our world, but is merely one of its components, so we should not overestimate its importance. However, it has boundless capacities to help us appreciate the wonders of what is around us both as offered by nature and created by humans.

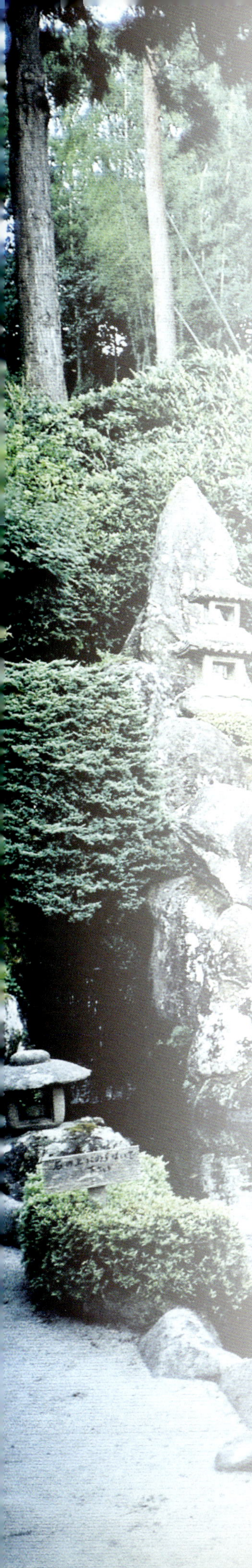

Japanese garden

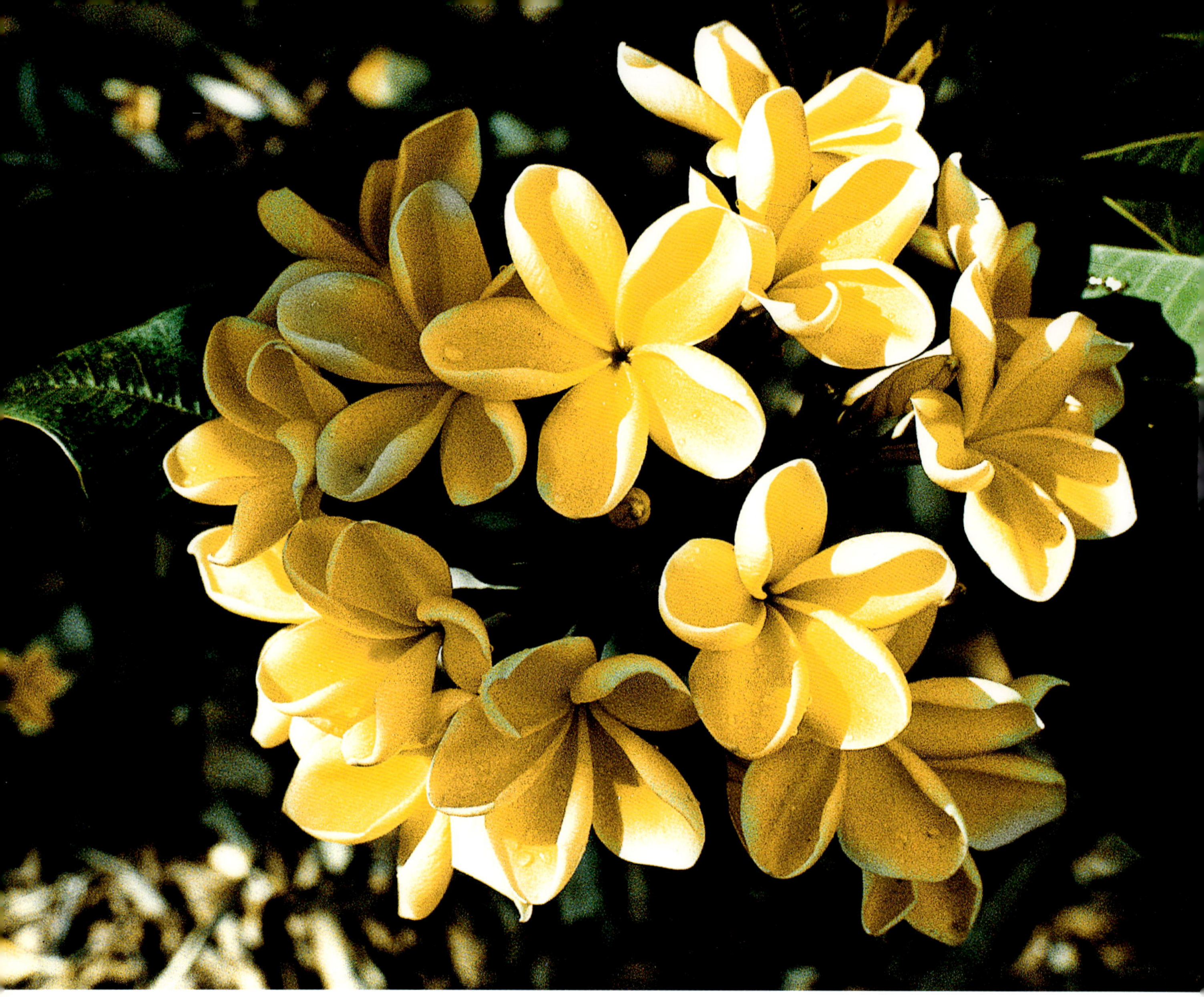

Acknowledgments

Our interest in symmetry has grown out of our research and teaching activities and we express our thanks to the Hungarian Academy of Sciences for its unfailing and generous support for over forty years. We also appreciate the backing of the Budapest University of Technology and Economics and the Hungarian National Scientific Research Funds.

Most of the images in this book are our own photographs. Where this is not the case we indicated the sources and in doing so we tried to be as accurate and meticulous as possible and regret any oversight that may have occurred. Here we list again the names of individuals and organizations that helped us with images and advice in completing the collection we are presenting in this book and express our appreciation to them:

Individuals:
Mária Apagyi (Pécs); Zoltán Bagosi (Budapest); Michael Bartell (Atlanta, Georgia); the late György Buday (London); Anna Rita Campanelli (Rome); Ágnes Csanády (Budapest); Gloria Del Bel (Buffalo, New York); Françoise Dénoyer (Paris); Aldo Domenicano (Rome); Jose Elguero (Madrid); the late Lima de Freitas (Portugal); István Gatter (Budapest); Sándor Görög (Budapest); Eszter Hargittai (Evanston, Illinois); Herbert Hauptman (Buffalo); Sándor Kabai (Hungary); János Kerényi (Budapest); R. Klockenkamper (Dortmund); Mária Kolonits (Budapest); Reiko Kuroda (Tokyo); Ferenc Lantos (Pécs); the late Györgyi Lengyel (Budapest); Alan L. Mackay (London); Benoit Mandelbrot (New Haven, Connecticut); J. Morral (Storrs, Connecticut); Hans-Ude Nissen (Zurich); István Orosz (Budakeszi, Hungary); Charles O. Perry (Norwalk, Connecticut); Tamás Váczi (Budapest); László Vámhidy (Pécs); Zoltán Varga (Budapest); and Ilya Yanov (Jackson, Mississippi).

Our special thanks are due to Mária Kolonits (Budapest) for her creative and dedicated assistance in the course of producing this version of *Visual Symmetry*.

Institutions and Organizations:
American Museum of Natural History, New York; Budapest Zoo; Museum of Fine Arts, Budapest; The Hermitage Museum, St. Petersburg; Rodin Museum, Paris; Caedmon of Whitby Publishers, Whitby, UK; Dover Publications; NASA; and the Mineral Collection of Eötvös Loránd University, Budapest.

The forerunners of this book were two volumes: Our *Pictorial Symmetry* book in Hungarian was published in 2005 by Galenus in Budapest and *Symmetry: A Unifying Concept* in English was published in 1994 by Shelter Publications in Bolinas, California. We appreciate the assistance and encouragement of the two publishers, Mr. István Fári and Mr. Lloyd Kahn, respectively. The present project relies much on these previous publications.

We are grateful to World Scientific and Dr. K. K. Phua personally for undertaking the publication of the present book. We have enjoyed a most fruitful cooperation of close to two decades with him and World Scientific in which we consider the present book an important milestone. We owe special thanks to Ms. Kim Tan and Mr. Ng Chin Choon for their dedicated work and cooperation during the production of this book.

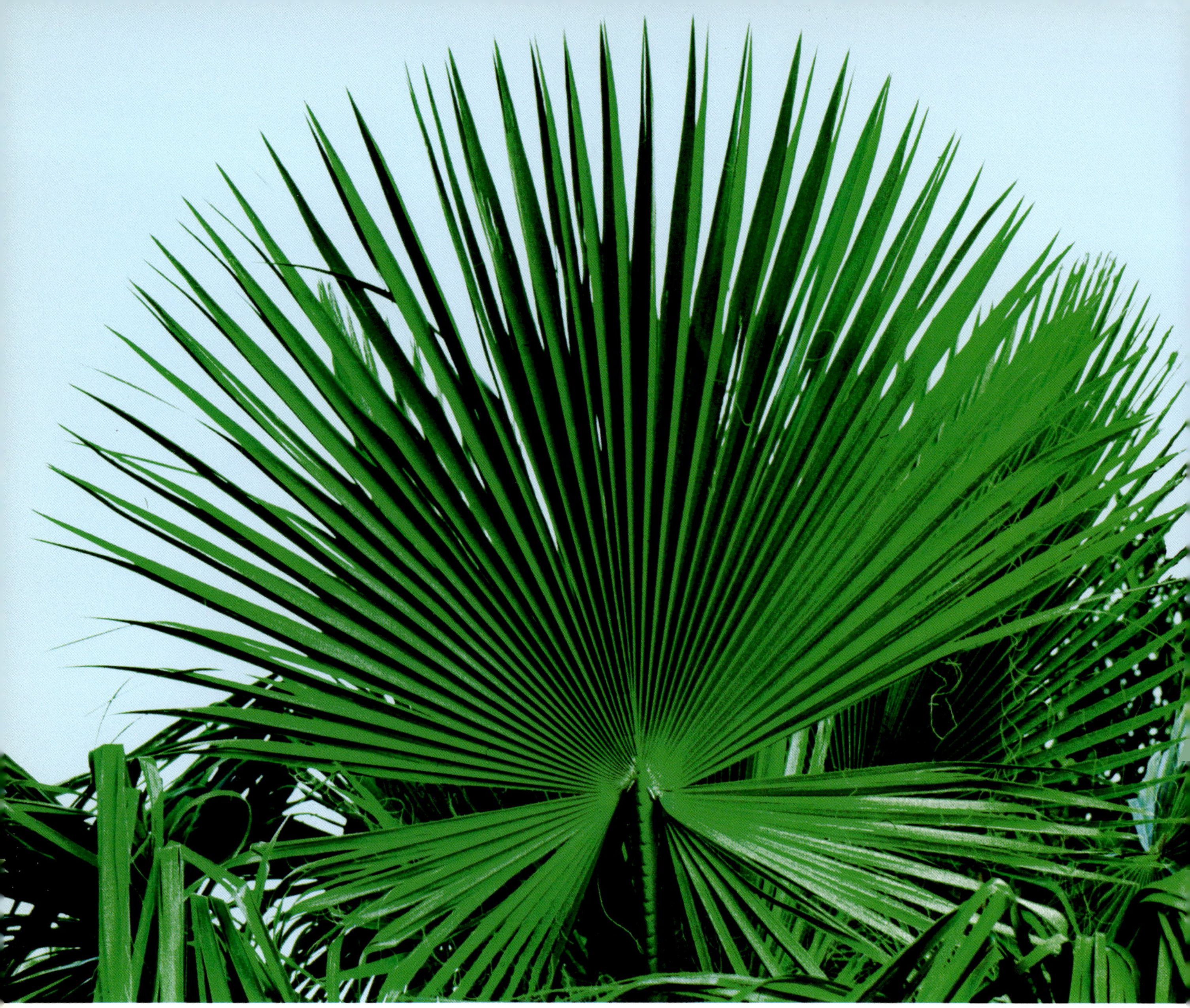

Also by the Authors

M. Hargittai, I. Hargittai, *Symmetry through the Eyes of a Chemist*, Third Edition. Springer, Berlin– Heidelberg–New York, 2009; Russian translation of the first edition: Mir, Moscow, 1989.

I. Hargittai, *The DNA Doctor: Candid Conversations with James D. Watson*. World Scientific, Singapore, 2007.

I. Hargittai, *The Martians of Science: Five Physicists Who Changed the Twentieth Century*. Oxford University Press, New York, 2006 (soft cover, 2008).

I. Hargittai, *Our Lives: Encounters of a Scientist*. Akadémiai Kiadó, Budapest, 2004; German translation, *Wege zur Wissenschaft*, Lj-Verlag, Merzhausen, Germany, 2006.

M. Hargittai, I. Hargittai, *Pictorial Symmetry* (in Hungarian: *Képes szimmetria*), Galenus, Budapest, 2005.

I. Hargittai, M. Hargittai, B. Hargittai, *Candid Science I–VI: Conversations with Famous Scientists*. Imperial College Press, London, 2000–2006.

I. Hargittai, *The Road to Stockholm: Nobel Prizes, Science, and Scientists*. Oxford University Press, Oxford, 2002 (soft cover, 2003); Japanese translation: Morikita, Tokyo, 2007; Chinese translation: Shanghai Scientific & Technological Education Publishing House, Shanghai, 2007.

M. Hargittai, *Cooking the Hungarian Way*, Second Edition. Lerner Publications, Minneapolis, 2002.

I. Hargittai, T. C. Laurent, eds., *Symmetry 2000* (in two volumes). Portland Press, London, 2002.

I. Hargittai, M. Hargittai, *In Our Own Image: Personal Symmetry in Discovery*. Kluwer/Plenum, New York, 2000.

M. Hargittai, I. Hargittai, eds., *Advances in Molecular Structure Research*, Vols. 1–6. JAI Press, 1995–2000.

I. Hargittai, M. Hargittai, *Symmetry: A Unifying Concept*. Shelter, Bolinas, California, 1994. Abridged German version: *Symmetrie: Eine neue Art, die Welt zu sehen*. Rowohlt Taschenbuch Verlag, Reinbek, 1998.

R. J. Gillespie, I. Hargittai, *The VSEPR Model of Molecular Geometry*. Allyn & Bacon, Boston, 1991; Russian translation: MIR, Moscow, 1992; Italian translation: Zanichelli, Bologna, 1994.

I. Hargittai, ed., *Fivefold Symmetry*. World Scientific, Singapore, 1992.

I. Hargittai, C. A. Pickover, eds., *Spiral Symmetry*. World Scientific, Singapore, 1992.

I. Hargittai, ed., *Quasicrystals, Networks, and Molecules of Fivefold Symmetry*. VCH, New York, 1990.

I. Hargittai, ed., *Symmetry 2: Unifying Human Understanding*. Pergamon Press, Oxford, 1989.

M. Hargittai, I. Hargittai, *Discover Symmetry!* (in Hungarian: *Fedezzük föl a szimmetriát*), Tankönyvkiadó, Budapest, 1989; *Upptäck symmetri!* (in Swedish), Natur och Kultur, Stockholm, 1998.

I. Hargittai, M. Hargittai, eds., *Stereochemical Applications of Gas-Phase Electron Diffraction* (in two volumes). VCH Publishers, New York, 1988.

I. Hargittai, B. K. Vainshtein, eds., *Crystal Symmetries, Shubnikov Centennial Papers*. Pergamon Press, Oxford, 1988.

I. Hargittai, ed., *Symmetry: Unifying Human Understanding*. Pergamon Press, Oxford, 1986.